2nd International Conference on Sustainable Environment and Disaster Management (2nd SUSTAIN 2025)

E3S Web of Conferences Volume 652 (2025)

Malang, East Java, Indonesia
28 August 2025

ISBN: 979-8-3313-2722-4

Printed from e-media with permission by:

Curran Associates, Inc.
57 Morehouse Lane
Red Hook, NY 12571

CURRAN ASSOCIATES INC.
proceedings
.com

Some format issues inherent in the e-media version may also appear in this print version.

For additional information, please contact EDP Sciences – Web of Conferences
at the address below.

EDP Sciences – Web of Conferences
17, Avenue du Hoggar
Parc d'Activité de Courtabœuf
BP 112
F-91944 Les Ulis Cedex A
France

Phone: +33 (0) 1 69 18 75 75
Fax: +33 (0) 1 69 28 84 91

contact-edps@webofconferences.org

Additional copies of this publication are available from:

Curran Associates, Inc.
57 Morehouse Lane
Red Hook, NY 12571 USA
Phone: 845-758-0400
Fax: 845-758-2633
Email: curran@proceedings.com
Web: www.proceedings.com

TABLE OF CONTENTS

2nd International Conference on Sustainable Environment and Disaster Management

Center for Environment and Disaster Mitigation Research Institute and Community Engagement
Universitas Negeri Malang

Jalan Semarang No 5. Malang 65145 Phone.+62-341-551312 Email: sustain@um.ac.id
Email:sustain@um.ac.id
Email:sustain@um.ac.id

SUSTAIN

POSTER

Statement of Peer review

In submitting conference proceedings to E3S Web of Conferences, I certify to the Publisher that I adhere to the **Policy on Publishing Integrity** of the journal in order to safeguard good scientific practice in publishing.

1. All articles have been subjected to peer review administered by the proceedings editors.
2. Reviews have been conducted by expert referees, who have been requested to provide unbiased and constructive comments aimed, whenever possible, at improving the work.
3. Proceedings editors have taken all reasonable steps to ensure the quality of the materials they publish and their decision to accept or reject a paper for publication has been based only on the merits of the work and the relevance to the journal.

Title, date and place of the conference

2nd SUSTAIN, 28th August 2025
Universitas Negeri Malang
Malang, East Java, Indonesia

Proceedings editor(s):

Melati Julia Rahma, SP., M.Ling.

Dr. Siti Zulaikah S.Pd., M.Si.

Dr. Heni Masruroh , S.Pd, M.Sc.

Date and editor's signature

Thursday, 9 October 2025, Malang, East Java, Indonesia

Melati Julia Rahma, SP., M.Ling **Dr. Siti Zulaikah S.Pd., M.Si.** **Dr. Heni Masruroh , S.Pd, M.Sc**

Sodong Cave as a Geomythological Space: Integrating Geological Values and Sacred Narratives in The Ijen Geopark

Asyifa Zahra Nur Anggraeni [1], *Melati Julia* Rahma[2*], and *Firman Sauqi Nur* Sabila[3]

[1,2]*Geography , Universitas Negeri Malang, JL. Semarang, Malang Indonesia*
[3]*Geology Geopark Banyuwangi, Jl. Gajah Mada, Banyuwangi, Indonesia*

Abstract. The conservation of geological sites and geodiversity in the Ijen Geopark, particularly Sodong Cave, is essential for preserving geological heritage that is vulnerable to degradation. Sodong Cave is a lava cave formed by the differential cooling of ancient Ijen volcanic activity. It is identified as part of the Kalibaru Formation, which formed 30 million years ago, and requires protection to ensure the sustainability of its internal ecosystem. This study explores the phenomenon of geomythology by integrating geological values with local sacred narratives, offering an innovative approach to conservation. It is the first study to focus on documenting the geomythology of Sodong Cave as part of the UNESCO Global Geopark Ijen natural heritage. Using a mixed-methods approach, the study aims to identify the geological characteristics of Sodong Cave, understand community perceptions of the sacred narratives associated with it, and develop the potential for incorporating these values into conservation strategies. With seven springs considered sacred, Sodong Cave is not only a geological site but also a spiritual space that influences community behavior in maintaining cleanliness and promoting environmental preservation. The study's results indicate that the concept of geomythology, which combines aspects of geology and local mythology, indirectly contributes to conservation efforts within the community, particularly in the Sodong Cave area. These conservation efforts are based on unwritten rules that have evolved into written regulations, serving as a form of environmentally friendly conservation. This is manifested in the community's social practices, such as ritual footstep tracing. This is supported by the presence of two sacred graves of elderly figures, validated through measurable calculations of the distance between objects using Euclidean distance, resulting in a classification of class one with a radius range of 0–335 meters. In addition, research using this approach is expected to contribute as a reference that supports sustainable geotourism development policies by utilizing local wisdom as a means of protection from the threat of damage and attracting tourists to learn more about conservation values that are integrated with the culture of the community.

*Corresponding author: melatijuliarahma.fis@um.ac.id

1 Introduction

Geological heritage is a landscape with high scientific value that is constructed to illustrate past phenomena [5]. Essentially, geological heritage necessitates conservation efforts aimed at understanding, preserving, and utilizing vulnerable geological sites. Conservation efforts play a crucial role in maintaining the sustainability of geological sites, particularly in the Ijen Geopark area, which boasts geological diversity from volcanic activity and also preserves sacred narratives as a living space developed generation after generation by local communities [10]. Conservation approaches typically focus on the physical aspects of geology, overlooking the cultural elements that shape how communities perceive and interact with their surroundings. Therefore, a geomythological approach is necessary in managing geological sites, especially in the context of geotourism, by integrating geological and cultural aspects as part of sustainable conservation efforts [3].

Geomythology can be defined as an approach that combines geology with local legends or myths that have developed within a community, and it can be used to uncover information related to the past that aligns with geological conditions [6]. The global relevance of geomythology on an international scale can be found in Las Cuevas Cave, Belize, which reveals the ritual practices of the ancient Maya tribe with the integration of the physical landscape and mythological space in rain-making rituals performed throughout the year in each season using jute snails as offerings to the rain god, who is believed to be the spirit of water in controlling the water cycle and fertility. This differs from the geomythological approach in the Ijen Geopark area, particularly in Sodong Cave, which is a type of lava cave resulting from the cooling of basaltic lava flows from the ancient Ijen Volcano. These stem from the local community's beliefs, which include unique cultural values that indirectly originate from the community's own beliefs and have fostered an attitude of environmental preservation. A distinctive feature of Sodong Cave is that it is surrounded by seven springs. The local community believes that this place is sacred because it is said to have been a place of meditation for Mbah Kecling or Kyai Hasan. He is one of the respected elders in the area, still revered to this day, which has led to the creation of a sacred narrative to maintain the cleanliness of the location. This narrative includes warnings against littering and prohibitions against women who are menstruating from entering the cave.

However, there has not been much research on this topic in Indonesia. This is based on previous findings that sacred narratives are often neglected in conservation planning, as they are often included in minor priority criteria, making this research approach necessary. This study aims to identify the geological characteristics of Sodong Cave, understand the community's perception of the sacred narratives surrounding it, and describe the potential for integrating geological and mythological values as a means of environmental conservation. This is expected to promote local wisdom as a means of protection from the threat of damage and attract tourists to learn more about conservation values integrated with community culture.

2 Methodology

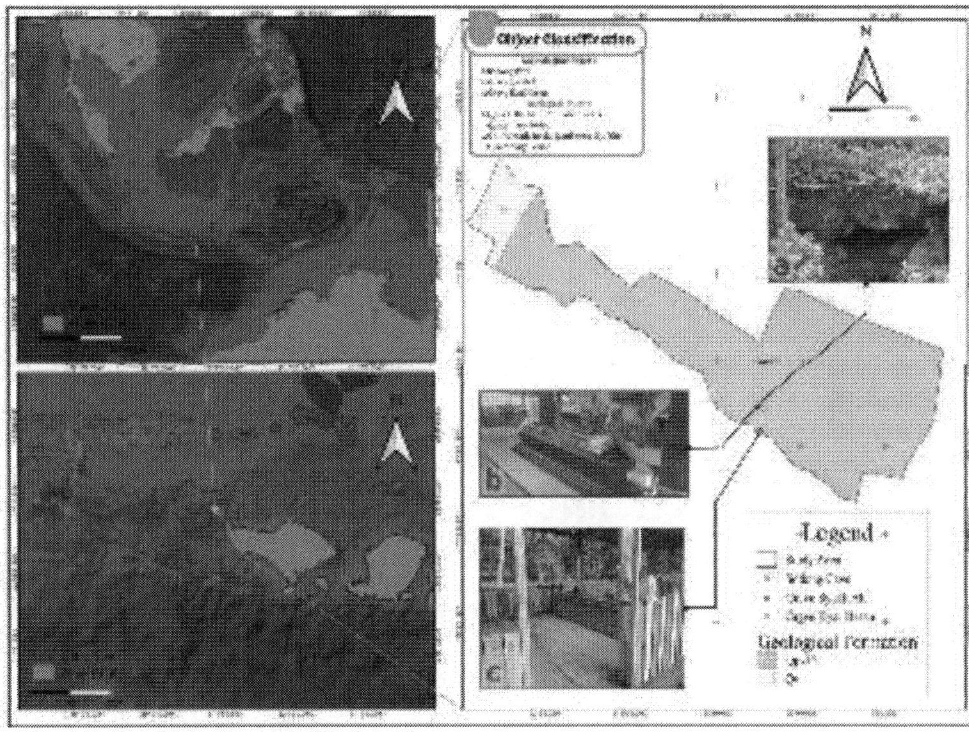

Fig. 1. Research location map of Sodong Cave

Sodong Cave is a lava site that has been managed as a tourist attraction since 2018 by the local tourism group in Pelinggihan Hamlet, Grogol Village, Giri Subdistrict, Banyuwangi Regency, East Java, with coordinates at 8° 11' 13.17'' S 114° 19' 27.64 E. This area has a unique feature in the form of a lava cave with seven springs surrounding the cave entrance. Additionally, there are two graves of elderly figures that add a spatial uniqueness in establishing a spiritual dimensional pattern. This study employs a mixed-method approach. By definition, a mixed-method approach combines qualitative and quantitative methods to strengthen the research findings, making them more comprehensive by exploring validated qualitative variables using quantitative approaches [8]. The qualitative approach is used to describe phenomena, meanings, and perceptions of research subjects within a specific context [2]. In this case, in-depth interviews with four key informants were selected purposively based on their specific expertise and strategic positions in the geomythology knowledge ecosystem in the Sodong Cave area, including: local elders (≥65 years old) with knowledge of history and spiritual traditions, Ijen Geopark managers for a conservation perspective, geology academics specializing in geomorphology-volcanology, and members of the Sodong Cave management tourism awareness group.

The results of the description of these phenomena were then reinforced with a quantitative approach based on Euclidean distance calculations, measured using ArcGIS software, which facilitated spatial analysis of the proximity between research objects. The use of Euclidean distance as the primary metric to validate the sacred spatial patterns in the Sodong Cave area was based on the mathematical capability of the formula $d(x,y) = \sqrt{\sum(x-y)^2}$, which is used to calculate the direct distance between coordinate points with high accuracy. Qualitative data understanding is validated quantitatively based on the spatial pattern of proximity between objects that form sacred narratives. The integration of mixed methods was chosen based on the urgency to understand the geological conditions and sacred narratives that have formed in Sodong Cave, located in the Ijen Geopark area of Banyuwangi.

This research was conducted in Pelinggihan Hamlet, Grogol Village, Giri District, with data collection carried out over three months from March to May 2025. Data collection focused on a temporal structure designed to maximize efficiency through a systematic stage that integrated observation and in-depth interviews. The observation stage in March included a comprehensive field survey for the initial mapping of sacred sites and documentation of the physical conditions of the Sodong Cave area. The interview phase, conducted in April and May 2025, consisted of in-depth interviews lasting 180-240 minutes per informant, divided into two separate sessions with a one-week interval to allow for reflection and elaboration of answers. The interviews focused on community elders to obtain an authentic perspective of indigenous knowledge. Meanwhile, May is allocated for interviews with geopark managers and geology academics to obtain technical data that complements the physical conditions and sacred narratives of Sodong Cave. The analysis unit focuses on Sodong Cave as both a physical geological object and a sacred cultural space, as well as on local narratives from the surrounding village community, including the tombs of Kyai Hasan and Syekh Ali Magrobi.

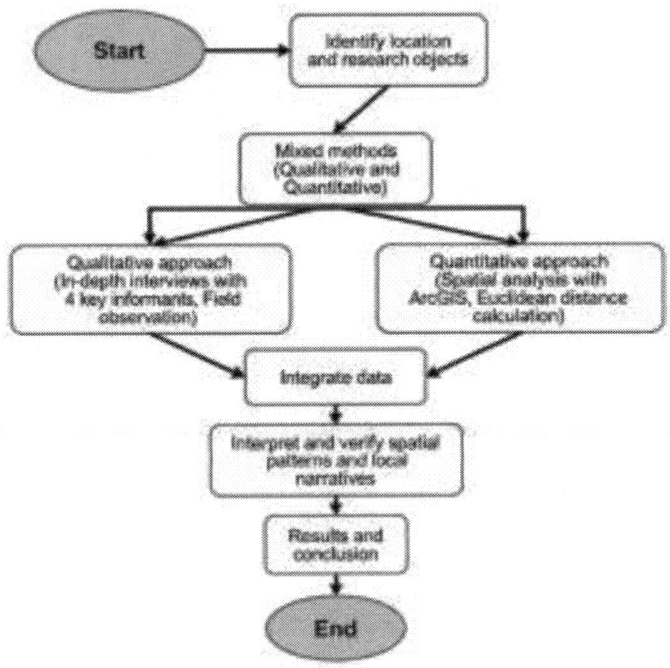

Fig. 2. Research Flowchart

3 Results

3.1 Geomorphology and Geological Characteristics of Sodong Cave

Sodong Cave is one of the lava caves that, from a geological perspective, is classified as part of the Kalibaru Formation. The Kalibaru Formation is the result of volcanic eruptions that occurred during the Quaternary period, formed from the ancient Ijen Volcano's eruptions. The rock structures in this formation are classified as Pleistocene-aged rocks [11]. Essentially, the Kalibaru Formation was formed after the Miocene epoch ended, with the resumption of volcanic activity in the form of eruptions that brought breccia, conglomerate, sandstone, and tuff, dominated by plagioclase, pyroxene, and olivine minerals. Its formation process was influenced by lava flows and pyroclastic deposits that underwent cooling and deposition with relatively deep aquifer potential due to fractures in the rock structure within the formation.

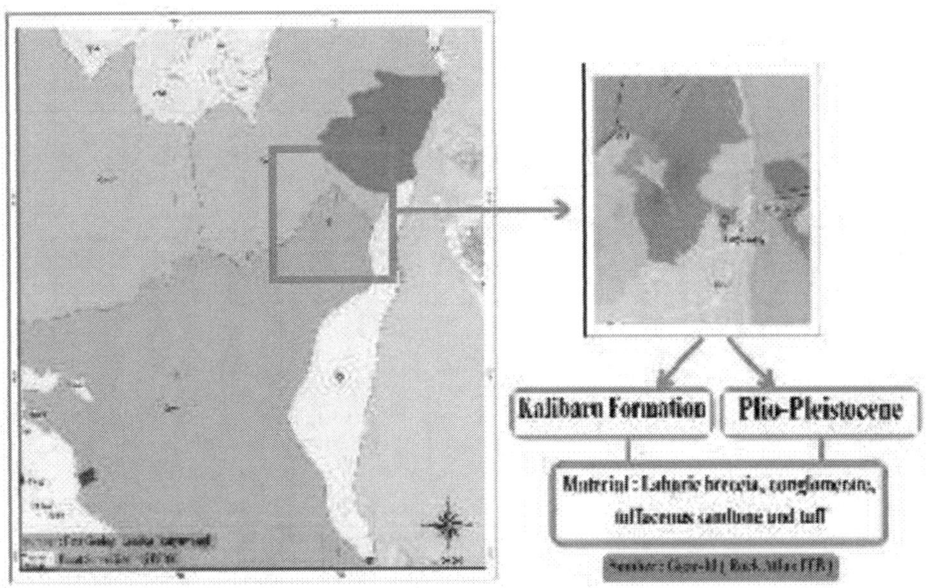

Fig. 3. Geology of Sodong Cave

In addition to being part of the Kalibaru formation, Sodong Cave is also identified as a lava cave formed by the process of basaltic lava flow undergoing differential cooling. This process occurs when the upper part of the lava flow hardens first due to friction with the air, while the lower part, which is still liquid, continues to flow and eventually forms a cave-like cavity. Caves formed by cooling lava flows usually have relatively low viscosity with a minimal and homogeneous mineral composition. Based on the morphological conditions of Sodong Cave, there is a cave mouth structure that tends to be broad and relatively short, which is consistent with one of the characteristics of lava caves, namely a flat and broad structure. Based on its location, Sodong Cave is identified as being in the middle zone of the lava system, so there are no autobreccia structures, which are usually found near the eruption source or what is known as the proximal zone. In addition, lava cave structures generally have a branching pattern resembling human veins with overlapping networks connected both

https://doi.org/10.1051/e3sconf/202565201001

laterally and vertically, supporting the assumption that there are other cave branches below the ground surface [4]

Fig. 4. Morphology of Sodong Cave
Source: Research Documentation, 2025

Based on the results of observations, several unique configurations were found inside the lava cave, identified by their structure and constituent materials. This can be seen in (Figure 5.a.1), which shows the presence of hydrothermal alteration processes that have caused physical changes, with the rocks appearing to have white spots. Physical changes can occur due to mineral precipitation processes when the lava flow cools, carrying mineral ions that react with the host rock, forming spots that typically produce bright colors. There are also groundwater fractures in the area (Figure 5.a.2), indicating the potential of an aquifer that the local community has managed through the construction of water pipes to connect the water source flowing to lower elevations for use as a daily necessity by the local community. The condition of these fractures indicates that Sodong Cave is still hydrogeologically active, as there is a large flow of water from the rock fissures located at the bottom of the cave. The presence of water also suggests that the aquifer system is not closed and is still connected to the local groundwater network beneath the ground surface.

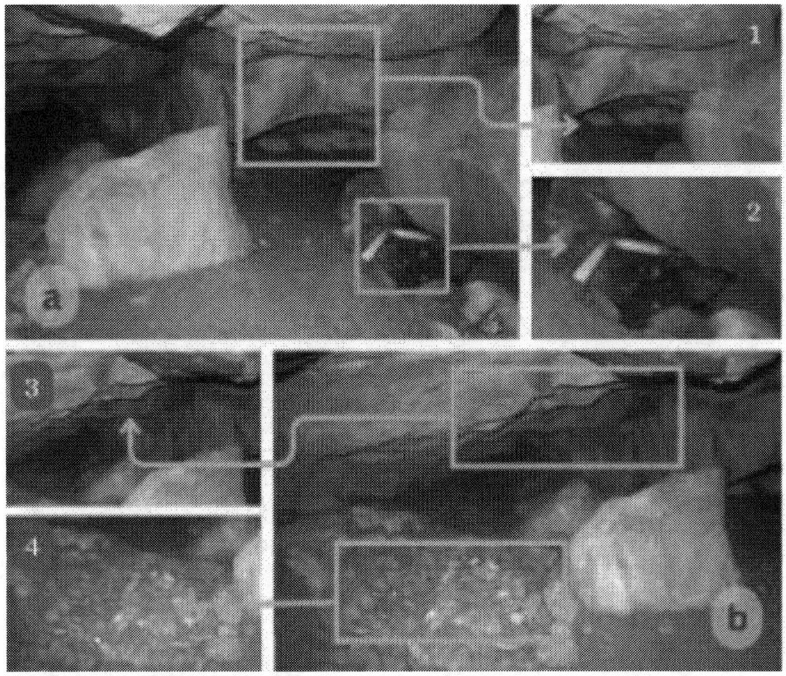

Fig. 5. Lava Cave Configuration
a.1) alteration; a.2) groundwater fractures,
b.3) cave ceiling structure; b.4) cave floor debris
Source: Research Documentation, 2025

Then, when viewed from the structure of the cave ceiling (Figure 4.b.3), it can be seen that Sodong Cave has a massive rock structure with thin layers and a wavy texture, indicating that the lava flowed rapidly and irregularly during cooling, forming a relatively rough surface texture. Several parts of the cave ceiling exhibit a greenish color due to biological activity, specifically moss, influenced by humidity, temperature, and low light intensity, which allows moss species to thrive between the cave ceilings. In section (Figure 4.b.4), debris resulting from physical weathering was found, caused by the detachment of rock fragments and boulders from the cave ceiling, a process known as breakdown [15]. The presence of detached material on the cave floor is also caused by erosion from water flowing into cracks or crevices in the cave ceiling, leading to accelerated weathering that causes material to detach from the cave ceiling and fall toward the cave floor. The processes of weathering, collapse, and accumulation of detached material indicate that the geological conditions of Goa Sodong are still naturally preserved with cycles that continue to occur over time.

https://doi.org/10.1051/e3sconf/202565201001

3.2 Sacred Narrative According to the Community

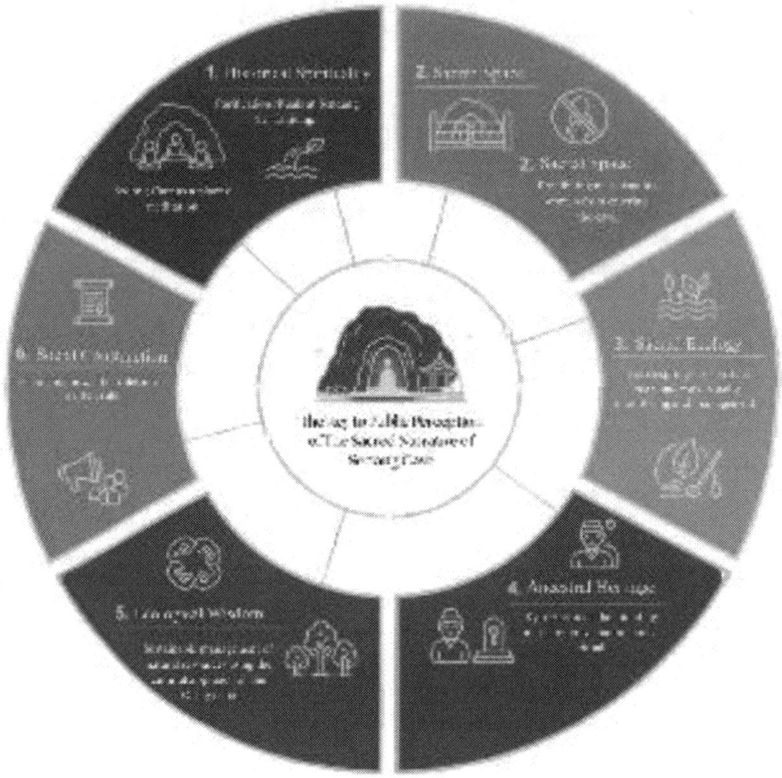

Fig. 6. Diagram of key dimensions of public perception of the sacred narrative of Sodong Cave
Source: *Results of Research Data Processing, 2025*

Culturally, Sodong Cave has high historical and spiritual value for the local community. This value is inextricably linked to the customs of the indigenous Banyuwangi people, known as the Osing tribe or descendants of the kings of the Blambangan Kingdom. The traditional community's belief in interpreting the legacy of their ancestors through natural features, such as caves, is rooted in local wisdom that is respected and preserved as best as possible through customary law. Spatially, the influence of the Osing tribe is reinforced by its distribution pattern, which is divided into several areas, including the districts of Giri, Glagah, Kabat, Rogojampi, Songgon, Singojuruh, and Genteng. The existence of the Osing tribe, which persists today, is found in the village of Grogol in the Giri district, where its distribution pattern is driven by the geographical conditions of the area, which is located within the vicinity of an active volcano with relatively fertile soil resulting from volcanic activity, utilized for agricultural needs in the farming sector. In this context, local wisdom has been developed to preserve the ecological condition of water sources, an essential element in meeting daily needs and irrigating rice fields. The local wisdom that has been created is inseparable from the belief of the Osing tribe that the utilization of nature is not only focused

https://doi.org/10.1051/e3sconf/202565201001

on exploitation but also requires exploration of sustainability, which is upheld in local principles for long-term sustainability [14].

Based on in-depth interviews, it was found that the name "Sodong" originates from the word "sidang." This term was used to describe the conditions in ancient times when Goa Sodong was used as a meditation site and believed by the local community to be a place where religious leaders gathered to discuss and spread Islamic teachings. The site was chosen due to its location in a highland area with dense vegetation and proximity to a water source. Additionally, its location in the middle of the forest further solidified its use as a meditation site due to its seclusion and natural beauty.According to community information, there is an elder figure who once performed a pilgrimage and meditated in Goa Sodong, namely Kyai Hasan, who is better known to the community as Mbah Kecling. He was the son of Wong Agung Wilis, one of the warriors of the Blambangan Kingdom who made significant contributions during the Dutch colonial period. This information is supported by the presence of Kyai Hasan's grave and the grave of Syekh Ali Muhammad Al Magrobi, both of which are located near the Sodong Cave tourist site. The two of them are believed by the community to be the first people to settle in the area, which eventually became the place where the community resides to this day.

Fig. 7. Illustration of Sodong Cave Mythology
Source: *Results of Research Data Processing (2025)*

The existence of these two tombs forms a spatially sacred narrative pattern that describes the community's beliefs in terms of the spatial dimension, encouraging the formation of religious values and thereby sanctifying the Goa Sodong area. This is reinforced by the theory of sacred geography, where natural features such as caves are interpreted spiritually and mythologically, thereby creating sacred spaces that influence settlement patterns, natural resource management practices, and community perspectives regarding traditions or rituals [9]. Culturally, Goa Sodong is not only physically interpreted as a reconstruction of the lava flow process but also as a sacred space formed by social practices in the form of sacred narratives, leading to the emergence of local wisdom in the form of prohibitions or rituals that must not be violated.

Fig. 8. The sacred narrative of Sendang Sri Tanjung and the written rules for adapting sacred narratives
Source: *Research Documentation*, 2025

The local wisdom formed from this sacred narrative is built through local mythology that links Goa Sodong with the activities of saints and scholars in spreading Islamic teachings, as well as the existence of the tombs of holy figures such as Mbah Kecling and Syekh Ali Muhammad Al Magrobi, which reinforce the religious dimension of the area. According to the theory of social construction, the sacred meaning and value of a place are formed through social interaction processes that establish unwritten rules [12]. The unwritten rules that are still upheld today to prevent them from fading away include the prohibition for menstruating women to enter the cave and the rule to maintain cleanliness by not littering around the cave and the seven springs as part of the ritualization that preserves the sanctity of the sacred space. This is aimed at protecting the clean water source named Sendang Sri Tanjung, which is believed to be a space for purification before beginning the pilgrimage ritual to the two graves of the elders believed to be the ancestors of the Grogol Village area. As the cave's function evolved from a meditation site to a tourist destination, the community collaborated with students in a community service program to formalize the unwritten rules into written guidelines for visitors. These guidelines aim to preserve the sacred narrative and ecological conditions of the cave area and the surrounding springs.

3.3 Integration of Sacred Narratives According to the Community

Sodong Cave is a geological phenomenon formed from the cooling of lava flows, creating a network of lava caves with significant ecological and hydrological value. The presence of springs inside the cave has given it an important role in the local ecosystem, contributing to the sustainability of water resources for the surrounding community. Integrating the physical conditions, such as geological value, with local wisdom in the form of sacred narratives developed within the community can serve as a strategic approach to its management, aimed at preserving both the quantity and quality of the lava cave and the spring. Integration efforts encourage the formation of conservation based on local wisdom, resulting in the emergence of habits of maintaining the cleanliness and preservation of the area for both the local community and visiting tourists. Basically, the local wisdom conservation approach is relatively easy for the community to understand and follow because it has been formed and is ingrained in the daily lives of the community. Local wisdom in the form of sacred narratives indirectly shapes social norms conceptualized within geomythology, thereby supporting the sustainable conservation of Goa Sodong as a lava site within the Ijen Geopark. The community enforces prohibitions against littering and restricts access for menstruating women as a form of respect for the religious and ecological values of this site. From a geological perspective, these rules indirectly contribute to the protection of water sources and cave ecosystems from contamination, thereby preserving the structure and configuration of the lava cave. Indirectly, the conservation of Goa Sodong is not only based on a scientific approach but also through social mechanisms rooted in belief and tradition, thereby forming geomythology as a representation of sustainable, environmentally friendly conservation [13].

The conservation strategy for Sodong Cave, which integrates geology and sacred narratives, can serve as a model for managing the Ijen Geopark area in maintaining the sustainability of lava sites. Understanding the relationship between geology and local beliefs can help maintain a balance between ecological and cultural aspects. Additionally, the presence of geomythology represents a significant, unique potential developed by the community in promoting Sodong Cave tourism to visitors. Indirectly, the development of tourism also involves geomythology as a form of environmentally friendly and sustainable conservation that can maintain geological stability and preserve the sacred narratives that have existed since ancient times. This aligns with the management of Goa Sodong tourism, which began to develop in 2018 with the contribution of the local tourism community. Water source management is divided into two categories: water supply and distribution for tourist attractions such as adult and children's swimming pools, and water supply and distribution for household activities and rice field irrigation. Both management systems are integrated into a network of PVC pipes that transport water from the central water source to downstream areas, taking into account slope gradients to ensure optimal water flow volume distribution. In addition, there is management of tourist facilities to support attractions and visitor accessibility, which undergoes annual evaluation. The evaluation activities are intended as routine management and maintenance efforts carried out by the manager to maintain the sustainability of Sodong Cave as a geomythology-based tourist attraction. These efforts are carried out so that the relationship between the condition of the facilities and accessibility to tourist attractions remains stable and tends to experience an increase in the number of visitors each year. Development is focused on wooden architecture supported by woven ornaments, as shown in (Figures 9.d, e, f), to reinforce the cultural impression as one of the advantages of tourism wrapped in the concept of geomythology.

 https://doi.org/10.1051/e3sconf/202565201001

Fig. 9. Management of Sodong Cave Tourism
Source: Research Documentation, 2025

The use of the geomythology concept was validated using calculations of the distance between Goa Sodong and other mythological objects in the form of two ancestral tombs. This was used to determine the spatial dimension patterns associated with local mythology based on the direct distance radius to the reference objects generated from data processing using the Euclidean distance tool in ArcGIS Pro software. The distances obtained from this analysis are based on the Pythagorean theorem principle, classified according to the nearest and farthest distances [7]. Based on calculations classified into 15 distance radius classes, the farthest distance is within the range of 3,905–4,724 meters, while the closest distance is 0–335 meters. The Goa Sodong site and other mythological sites are classified in the first class with a range of 0–335 meters, visualized by a dark green symbol in Figure 8. This indicates that the Goa Sodong area, with its two graves as mythological objects, has a strong connection to geomythology, making the conservation strategy based on local wisdom a significant potential that can be developed to ensure the sustainability of the lava site.On the other hand, other areas classified as the furthest distance still have the same potential, but require ongoing collaboration with Goa Sodong, which has become the main object of origin for geomythology.

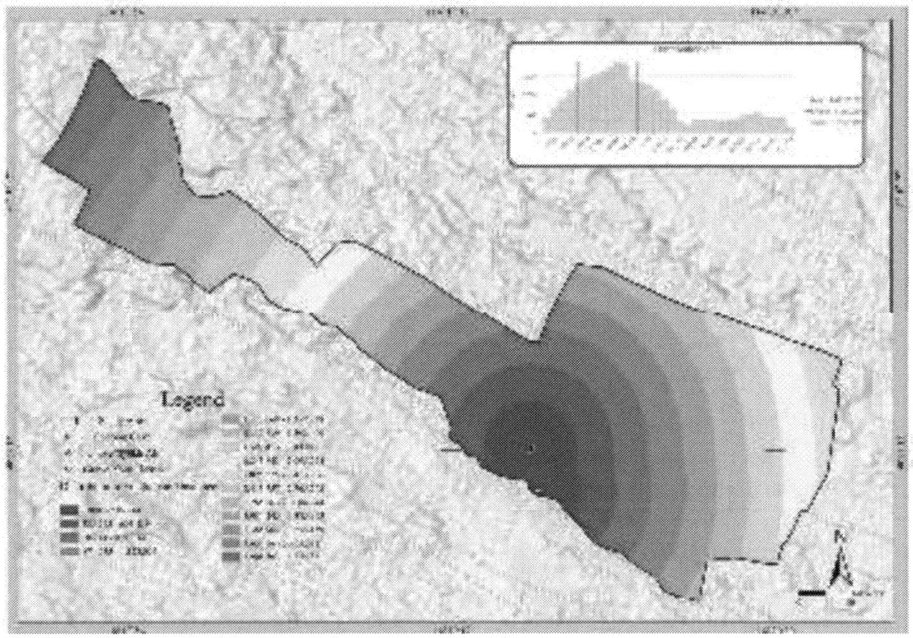

Fig. 10. Euclidean Distance Geomythology Spatial Pattern of Sodong Cave
Source *: Results of Research Data Processing (2025)*

In the Euclidean distance analysis process, a histogram graph was obtained containing information on the distribution of distance values with a positively skewed distribution tendency. The mean or average distance between objects in meters was found to be 1,610.5, with a median value of 1,381.6. These results are interpreted to mean that most objects are relatively close to the main object point. In addition, the level of variation in distance between points is reflected in the standard deviation value of 1.016.3. This value explains the spatial diversity in the distribution of essential objects, such as Goa Sodong and the tombs of elders.

Table 1. Euclidean Distance Analysis

Parameter	Value	Description
Mean	1,610.57	Average distance between objects in meters.
Median	1,380.6	Half of the data has values below 1,380, so the data is in the extreme range.
Standart Deviation	1,016.3	Shows fairly high distance variation with distribution in line with the average.
Skewness	0.85	Data skewed to the right indicates a fairly significant distance.

Parameter	Value	Description
Kurtosis	2.93	The distribution is somewhat peaked, approaching a normal distribution.
Count	26,038	Number of valid values
Nulls	64,962	Data with no values
Max / Min	4,273.4 / -1	Negative data indicates data that has not been filtered

Source : *Results of Research Data Processing (2025)*

The results of calculations based on classification in maps and histograms provide an overview of the zones of accessibility for physical and mythological objects integrated into the concept of geomythology. These zones are quantitatively measurable and can serve as a reference for the future development of Goa Sodong tourism, ensuring the preservation of culture-based conservation with geomythology as a means of safeguarding the lava site for the sustainability of the ecosystem within it.The zoning can also serve as evidence validating the integration of spatial and mythological narrative approaches in geotourism development, which involves two leading indicators: geological objects and mythological objects [1].

4 Discussion

Research findings highlight several key aspects of establishing local wisdom-based conservation in the Sodong Cave area, which is within the Ijen Geopark. The unique geological structure and diverse landscape are among the factors that support the lava site's existence as tangible heritage, protected by a sacred narrative that ensures its long-term sustainability, and informed by a socio-cultural approach developed by the surrounding community. The integration of physical geology and mythology also encourages the development of environmentally friendly and long-term sustainable culture-based tourism management. This is inseparable from the role of community groups that continue to develop tourism systems, from facility management to the use of natural architectural concepts. Collaboration with this geomythological integration approach can ensure effective natural resource management and contribute to long-term tourism sustainability by considering the environmental aspects, including the geological and socio-cultural conditions that shape Sodong Cave tourism as a tourist attraction, and providing education regarding the importance of conserving lava sites as representations of past events that have significant potential for long-term development and preservation.

When compared to other international geoparks, the Sodong Cave model shares similarities with management practices in places such as the Surtshellir lava cave in Iceland and Mount Etna in Italy. In Iceland, ancient mythological stories about journeys to the underworld led the community to naturally restrict activities in the cave since Viking times, and this informal protection later became the basis for the modern conservation system in Vatnajökull National Park. Similarly, on Mount Etna, Greek-Roman legends about the fire god Vulcan created zones that were considered off-limits by the Sicilian people, and these zones later became core protection areas in the management of the UNESCO Geopark. These three places show a similar pattern, namely that the spiritual belief of the community in the sanctity of a place has proven to be effective in protecting the environment long before formal conservation regulations existed.

The findings of this study are important for improving the UNESCO Global Geoparks guidelines, particularly section 3.4 on Community Involvement and Local Economic Development. The way Sodong Cave transformed unwritten rules into official conservation protocols through joint discussions between elders, managers, geologists, and tourism awareness groups can serve as an example for other Geoparks in managing sites as tourist attractions that involve indigenous knowledge. The concept of "sacred buffer zones," which originated from the community's perception of spiritual space and was subsequently validated by Euclidean distance analysis using ArcGIS software, also demonstrates the integration of traditional knowledge with modern science, addressing the major challenges faced by Ijen Geopark in managing its geological heritage.

5 Conclusion

Sodong Cave is a tangible representation of the integration between geology and sacred narratives that can form the basis of a culture-based conservation strategy using a geomythological approach. Geologically, this cave was formed from basaltic lava flows resulting from the ancient Ijen volcano's volcanic activity, and it features seven springs that play a crucial role in the local ecosystem. Meanwhile, sacred narratives developed within the community, such as the cave as a place for meditation and purification at the Sri Tanjung spring and the presence of Mbah Kecling's grave, encourage the emergence of sacred geography patterns. These patterns are validated using Euclidean distance spatial analysis, which shows accessibility results within the closest first-class distance range. This accessibility fosters a connection between geological and mythological objects, creating a spatial pattern with relatively high sacred value that helps maintain ecological balance within the area.

The geomythological approach in interpreting Goa Sodong reveals that local myths and beliefs significantly contribute to the preservation of the lava site, which is one of the natural heritage sites within the Ijen Geopark area. Therefore, the conservation of Sodong Cave is necessary, not only from a scientific perspective, but also through community involvement in maintaining a balance between ecological and spiritual aspects, thereby promoting the long-term sustainability of the lava site for both conservation and tourism development. This geomythological approach can be applied to other locations in the Ijen Geopark that also have spiritual stories, such as Ijen Crater with the legend of Dewi Ganga, Kalibendo Waterfall, which is believed to be a place of asceticism, and other unmapped lava caves. Thus, the geomythology approach not only enriches scientific knowledge about the relationship between humans and nature but also provides a practical way to manage geoparks that respect both scientific knowledge and local community wisdom.

References

1. Abdurahman, O., & Sukamto Kadarisman, D. (2022). Geomythology as a Geotourism Attraction, Case Study: The Sangkuriang Legend in The Bandung Highland and Its' Surrounding Areas Based on Geological to Hermeneutics Interpretation. *International Journal of Geotourism Science and Development*, *2*(1), 34–40. https://doi.org/10.58856/ijgsd.v2i1.17
2. Bahiyah, U., & Gumiandari, S. (2024). Metode Penelitian. In *General and Specific Research* (Vol. 4, Issue 2). https://adisampublisher.org/index.php/edu/article/view/744/784

3. Berliandaldo, M., & Fasa, A. W. H. (2022). Pengelolaan Geowisata Berkelanjutan Dalam Mendukung Pelestarian Warisan Geologi: Perspektif Collaborative Governance. *Inovasi*, *19*(1), 79–97. https://doi.org/10.33626/inovasi.v19i1.529
4. Calvari, S., Giudice, G., Maugeri, R., Messina, D., Morgavi, D., Miraglia, L., La Spina, A., & Spampinato, L. (2024). Complex lava tube networks developed within the 1792–93 lava flow field on Mount Etna (Italy): insights for hazard assessment. *Frontiers in Earth Science*, *12*(September), 1–26. https://doi.org/10.3389/feart.2024.1448187
5. dos Santos Costa, S. S., do Nascimento, M. A. L., & da Silva, M. L. N. (2024). Geological and mining heritages in the Seridó UNESCO Global Geopark: Ediacaran to Cambrian mineral deposits revealed by historical mines in Northeast Brazil. *International Journal of Geoheritage and Parks*, *12*(2), 311–332. https://doi.org/10.1016/j.ijgeop.2024.04.004
6. Fotiou, E. (2023). Transreligiosity and Religious Revitalization in Modern Greece: Bridging Religion and Science through Geomythology. *Religions*, *14*(6). https://doi.org/10.3390/rel14060754
7. Kusuma, A. P., & Oktavianto, A. D. (2022). Analisis Metode Euclidean Distance dalam Menentukan Koordinat Peta pada Alamat Rumah. *Jurnal Teknologi Dan Manajemen Informatika*, *8*(2), 108–115. https://doi.org/10.26905/jtmi.v8i2.8871
8. McCusker, K., & Gunaydin, S. (2015). Research using qualitative, quantitative or mixed methods and choice based on the research. *Perfusion (United Kingdom)*, *30*(7), 537–542. https://doi.org/10.1177/0267659114559116
9. Moyes, H. (2012). *Sacred Darkness A Global Perspective on the Ritual Use of Caves*.
10. Permanadewi, S., Samodra, H., Irzon, R., Prabowo, A., & Ansori, C. (2024). Quantitative assessment for geosites of the Ijen Geopark in Banyuwangi Regency, East Java, Indonesia. *International Journal of Geoheritage and Parks*, *12*(1), 63–74. https://doi.org/10.1016/j.ijgeop.2024.01.003
11. Prabowo, A., & Permanadewi dan Hanang Samodra, S. (2024). Analisis Batuan Gunungapi di Daerah Banyuwangi dan Sekitarnya, Provinsi Jawa Timur Berdasarkan Aspek Geokimia Volcanic Rocks Analysis in Banyuwangi and Surrounding Areas, East Java Province Based on Geochemistry Aspects. *Jurnal Geologi Dan Sumberdaya Mineral*, *21*(4), 197–207. https://doi.org/10.33332/jgsm.geologi.v22.4.197-207
12. Pramono, M. S., Tampake, T., Lattu, I. Y. ., & Molewe, A. C. (2024). Peter L. Berger and Thomas Luckmann's social construction of the institutionalization of tolerance values in the Nyadran Perdamaian tradition. *AL MA'ARIEF : Jurnal Pendidikan Sosial Dan Budaya*, *6*(1), 49–61. https://doi.org/10.35905/almaarief.v6i1.9759
13. Setiawan, E., & Triyanto, J. (2021). Integrasi Kearifan Lokal Dan Konservasi Masyarakat Sekitar Desa Penyangga Taman Nasional Alas Purwo. *Jurnal Analisa Sosiologi*, *10*(2), 452–470. https://doi.org/10.20961/jas.v10i2.51888
14. Sumarmi, S. (2015). Local Wisdom of Osing People in Conserving Water Resources. *Komunitas*, *7*(1), 43–51. https://doi.org/10.15294/komunitas.v7i1.3429
15. Valerio, A., Tallarico, A., & Dragoni, M. (2008). Mechanisms of formation of lava tubes. *Journal of Geophysical Research: Solid Earth*, *113*(8), 1–14. https://doi.org/10.1029/2007JB005435

https://doi.org/10.1051/e3sconf/202565201002

Exploring the Potential and Development Strategies of Waterfall Tourism in Ngawi Regency

Clara Anastasya[1*] and *Choirul* Amin[2]

[1,2] Faculty of Geography, Universitas Muhammadiyah Surakarta, Indonesia

Abstract. Ngawi Regency in East Java, Indonesia, possesses abundant natural tourism potential, particularly in the form of waterfall attractions. However, the development of these sites has not reached optimal levels due to issues such as inadequate infrastructure, limited accessibility, and minimal community involvement. This study aims to (1) analyze the characteristics of waterfall tourism in Ngawi Regency, (2) assess the internal and external potentials of these destinations, and (3) formulate development strategies using SWOT analysis. A qualitative descriptive method was employed, supported by field observations, interviews, questionnaires, and documentation. The findings reveal that Srambang, Suwono, and Pengantin waterfalls hold high development potential based on their combined internal and external scores, while Teleng and Ringin Jenggot waterfalls are categorized as having moderate potential. The SWOT analysis indicates that Ngawi's waterfall tourism is positioned in Quadrant I, suggesting a strategy focused on growth by leveraging strengths to seize external opportunities. Key recommendations include improving infrastructure, enhancing digital promotion, empowering local communities, and integrating sustainability into development planning. With a comprehensive strategy, Ngawi's waterfall tourism can contribute significantly to regional development and ecological preservation.

1 Introduction

Ngawi Regency has various tourist attractions, one of which is natural tourism in the form of waterfalls. Tourist attractions in Ngawi Regency have their own charm. Waterfalls are one of the natural tourist attractions that are often the destination of tourists. Ngawi Regency, which is located in East Java Province, has various unique and interesting waterfall tourist destinations [1, 2]. Geography and tourism are two different fields of science, but on the other hand, these two fields of science are interrelated [3, 4]. Tourism geography involves the study of geographical aspects or factors that influence the development and promotion of tourism in an area, such as geographical conditions, natural and cultural diversity, infrastructure, accessibility, and climate. Tourism is an industry related to people's travel and

* Corresponding author: ca122@ums.ac.id

https://doi.org/10.1051/e3sconf/202565201002

accommodation in a place for the purpose of vacationing, doing business, or recreation [5, 6]. Several waterfalls in this area, such as Suwono Waterfall, Srambang Waterfall, Pengantin Waterfall, Teleng Waterfall, and Ringin Jenggot Waterfall, have unique natural beauty which is the main attraction. In addition to offering stunning views, this location is often a place for activities such as photography, trekking, and enjoying a calming natural atmosphere. Suwono Waterfall is a tourist destination located in the Ngawi hills. The surrounding environment is filled with green vegetation. This location is one of the main choices for tourists who want to enjoy the beauty of nature and seek tranquility. then Srambang Waterfall is located on the slopes of Mount Lawu, this waterfall is famous for its cool air and beautiful panorama. This destination also offers a light trekking route for tourists who want to enjoy the beauty of the surrounding forest. The name Pengantin Waterfall comes from the flow of water that forms two parallel paths, resembling a bridal couple. Then Teleng Waterfall is famous for its unique natural rock formations. This location is relatively hidden, but it is one of the favorite tours for tourists who do adventure, and the last one, especially the Ringin jenggot waterfall, has a natural atmosphere with large banyan trees growing around the location. The name "Ringin Jenggot" itself comes from this banyan, which is the icon of the location. The rapid flow of water adds to the charm of this place as a natural tourist destination. Ngawi Regency as part of the East Java Province has a wealth of natural resources that have great potential to be developed into a tourist destination, especially nature tourism.

Effective management of tourism potential, both internally and externally, is essential to create sustainable and competitive attractions [7, 8]. Internal potential includes all aspects that come from within the tourist attraction itself, such as the uniqueness of nature, cultural richness, infrastructure, and supporting services. Some internal variables are the main attractions/attractions of tourist attractions, the strength of the attractions of tourist attraction components, tourist activities at the location, the diversity of supporting attractions, the physical condition of the tourist attraction directly and the cleanliness of the tourist environment. Meanwhile, external potential includes factors that are outside the tourist attraction, such as accessibility, promotion, government support, to the interest of global tourists. A deep understanding of these two aspects is the main key to developing tourist destinations that are able to attract visitors, improve the tourist experience, and support the empowerment of local communities. Some external variables are the relationship between places, assistance in tour packages, development and marketing of tourist attractions, travel time to the district capital, availability of public transportation to get to the location of the waterfall object, infrastructure to reach the tourist attraction, facilities to fulfill physical/basic needs at the location of the tourist attraction, facilities to organize the social needs of tourists at the location of the object, several equipment facilities containing parking areas, toilets, information services, and souvenirs.

The development of waterfall tourism that is operated wisely can encourage the creation of a sustainable local economy, create new jobs, and strengthen the identity and local wisdom of the surrounding community. 2025 is an important momentum for Ngawi Regency to start optimizing the tourism sector as one of the driving forces of the regional economy. With support from various parties, including the central government through policies to accelerate the development of tourist villages and digital promotion, Ngawi Regency needs to develop a mature strategy based on local potential. This will not only increase the number of tourist visits, but also have a positive impact on environmental conservation, strengthening local identity, and improving the welfare of the community around the tourist location.

The strategy for developing waterfall tourism objects in Ngawi Regency needs to be designed comprehensively, taking into account natural potential, local characteristics, and challenges in the field [9, 10]. One of the main focuses is improving infrastructure and accessibility to waterfall locations such as Pengantin Waterfall, Teleng and Ringin Jenggot. Many of these destinations are still hidden and difficult to reach by public or private vehicles due to damaged or inadequate road conditions. The local government together with related agencies need to repair roads, build small bridges, and provide more basic facilities such as parking lots, toilets, and neatly arranged stalls so that tourists feel comfortable and safe when visiting.

This is an important basis before implementing promotional strategies or other tourism activities. Furthermore, development strategies must prioritize the active participation of local communities. Residents around tourist locations can be empowered as the main actors in managing tourist areas, starting from guide service providers, homestay managers, to small business actors such as culinary and souvenirs. The government can hold regular training on tourism services, cleanliness, security, and simple financial management so that people are able to run their businesses independently and professionally. Strengthening institutions through the formation and development of Tourism Awareness Groups (Pokdarwis) is also very important so that there is an organizational structure that manages tourism collectively and responsibly. Thus, the economic benefits of the tourism sector can be directly felt by the surrounding community, while creating a sense of ownership and concern for the sustainability of the waterfall environment. To support successful development, the promotional aspect must also be carried out creatively and consistently. In the digital era like today, the use of social media, online tourism platforms, and collaboration with content creators can be effective means to introduce the beauty of waterfalls in Ngawi to a wider market. Aggressive strategies can fail if they are not accompanied by tourism HR training, strengthening community capacity, and sustainability of local government programs. Therefore, aggressive strategies in Ngawi also need to be accompanied by capacity building programs, so that SWOT results can truly be realized in the field [11, 12].

In developing strategies and analyzing a problem, it is often necessary to identify various related factors which are then mapped systematically. This process will be more effective if using the appropriate approach. One of the most frequently used and proven effective methods is the SWOT (Strengths, Weaknesses, Opportunities, Threats) analysis, which involves structured stages to understand the internal and external factors that influence a situation [13, 14].

2 Method

The data collection method of this research includes primary and secondary data. Primary data can be obtained by direct observation in the field of tourist attractions, while secondary data is obtained from parties and tourism offices related to Ngawi Regency, the method used is a qualitative approach with field observation, interviews, questionnaires and documentation.

The data analysis method uses SWOT analysis to design the development of tourist attractions. In the data collection process, direct observations were made to several waterfall locations that were the objects of research to record physical conditions, infrastructure, accessibility, and ongoing tourist activities. This observation was supplemented with documentation in the form of photos and field notes. In addition, semi-structured interviews were conducted with informants who were considered to know the conditions of tourist attractions, such as managers, visitors, community leaders, and the local Tourism Office. To obtain supporting data, questionnaires were also distributed to visitors to determine their perceptions of service quality, attractiveness, and feasibility of tourist locations. The results of the questionnaire were then analyzed thematically and used as comparative material with the results of observations and interviews.

All collected data is then analyzed using the SWOT (Strengths, Weaknesses, Opportunities, Threats) analysis approach to determine the strategic position of the tourist attraction. Identification of internal and external factors is carried out systematically, then formulated into the IFAS and EFAS matrices. The findings from this analysis will be the basis for compiling a tourism development plan that is in accordance with the potential and challenges faced by each tourist attraction.

To strengthen the analysis results, the EFAS (External Factor Analysis Summary) and IFAS (Internal Factor Analysis Summary) matrices are used, which are designed to assess the weight and assessment of each strategic factor. The IFAS matrix functions to examine factors within the organization, including advantages and disadvantages, while the EFAS matrix functions to assess external factors, such as opportunities and risks. Descriptive qualitative studies with information obtained through observation, interviews, and documentation are assessed using a descriptive approach.

3 Result and Discussion

3.1 Characteristics of tourism

A. Suwono Waterfall

Suwono Waterfall is one of the natural tourist attractions that is the main attraction in Janti Village, Ngrambe District. This tourist attraction is known for its clear water flow, flowing from a height of about 15 meters and surrounded by cliffs and green trees. Suwono waterfall tourism has quite a lot of complete facilities ranging from changing rooms, quite a lot of public toilets, a fairly large parking area that is very safe for groups of tourists using tourist buses and other supporting facilities such as prayer rooms, canteens and gazebos for resting. Suwono Waterfall provides various activities that can be done by tourists, such as playing in the water in the natural pool formed under the waterfall, trekking along the forest path, to camping in the open area available in the surrounding area. There are also educational tour packages that raise the theme of preserving the environment and local ecosystems, suitable for students and nature lover communities. Access to Suwono Waterfall is quite good. The road to the location is mostly paved, although there are some points that need to be passed on foot because of the hilly land contour. Two-wheeled and four-wheeled vehicles can still reach the main parking area, and there are signs that help tourists get to the location.

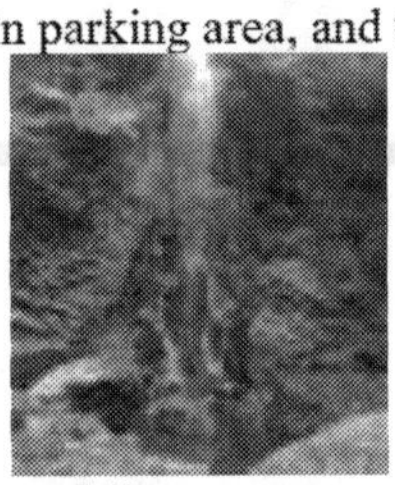

Fig.1. Suwono Waterfall

B. Srambang Waterfall

Srambang Waterfall is one of the natural tourist attractions located on the slopes of Mount Lawu. This waterfall is about 25 meters high and has clear water and a very natural environment, making it a favorite tourist destination for both local and out-of-town tourists. This tourist attraction began to be developed and officially opened in 2017. Its main attraction lies in the beauty of the waterfall combined with the green natural landscape, thematic parks, and contemporary photo spots scattered along the path to the waterfall. The facilities in this tourism area are quite complete and neatly arranged. Visitors can enjoy a large parking area,

 https://doi.org/10.1051/e3sconf/202565201002

toilets, prayer rooms, first aid gazebos, food stalls, gazebos, flower gardens, and the iconic bamboo bridge. In addition, there is a pedestrian path with concrete stairs and safety fences that facilitate access to the waterfall. In terms of accessibility, Srambang Waterfall is very easy to reach. The road to the location is paved and can be passed by two-wheeled and four-wheeled vehicles. There are also clear directions from the center of Ngawi to the entrance to the tourist area. Srambang Waterfall not only presents natural beauty, but also shows how nature tourism can be managed sustainably. The combination of natural beauty, well-organized facilities, and educational nuances make Srambang one of the leading tourist destinations on the slopes of Mount Lawu.

Fig.2. Srambang Waterfall

C. Pengantin Waterfall

Pengantin Waterfall is one of the unique natural tourist destinations and has a distinctive story value in the Ngawi Regency area. Located on the slopes of Mount Lawu, this waterfall is named "Pengantin" because it consists of two parallel and side-by-side water flows, resembling a pair of brides standing side by side. This tourist attraction began to be developed in early 2010 as part of the natural tourism potential development program in Ngawi. Its management is carried out collaboratively between the village government and the local community, while maintaining the sustainability of the surrounding nature. With a height of about 20 meters. The facilities available in this tourist area include a parking area, toilets, prayer rooms, food stalls, and several gazebos for relaxing. In addition, there is a trekking path that has been arranged so that visitors can reach the waterfall comfortably and safely. Several photo spots are also provided to add to the interesting and memorable travel experience. Activities that can be done at Pengantin Waterfall include playing in the water in the natural pool, taking selfies near the twin streams of the waterfall, camping. In terms of accessibility, the location of Pengantin Waterfall can be reached by two-wheeled or four-wheeled vehicles.

Fig.3. Pengantin Waterfall

https://doi.org/10.1051/e3sconf/202565201002

D. Teleng Waterfall

Teleng Waterfall is located in Ngrayudan Village, Jogorogo District, Ngawi Regency. This destination is located in a protected area with a natural atmosphere that is still very protected. It has a height of about 15 meters. Teleng Waterfall has not long been developed as a tourist destination and is still classified as a semi-natural natural tourist attraction. Its management is mostly carried out by the surrounding community independently, with support from the village government for the development of basic infrastructure. The facilities at the location are not as complete as other large tourist attractions, but some facilities such as a simple parking area, natural stone seats, and a path to the waterfall are already available. In terms of accessibility, the road to Teleng Waterfall is mostly village roads and footpaths. Visitors need to walk for about 10 to 15 minutes from the parking area to reach the main location.

Fig.4. Teleng Waterfall

E. Ringin Jenggot Waterfall

Ringin Jenggot Waterfall is one of the natural tourist destinations located in Ngrayudan Village, Sine District, Ngawi Regency, East Java. Located in a hilly area surrounded by tropical forest, this waterfall is about 10 to 12 meters high, with clear water flow. Tourism development in this area is still semi-natural and managed by the local community independently. Although it does not yet have complete facilities like commercial tourist attractions, visitors can still find footpaths, simple resting places, and directions to the location. This location is also often used as a place for environmental education activities, flora-fauna observation, and light camping. Access to Ringin Jenggot Waterfall requires a trip by private vehicle to a certain point, followed by a 10-20 minute walk through forest paths and residents' plantations. The facilities at this Ringin Jenggot waterfall are inadequate because it is still completely natural, some inadequate facilities such as public toilets, parking, gazebos and canteens are not available.

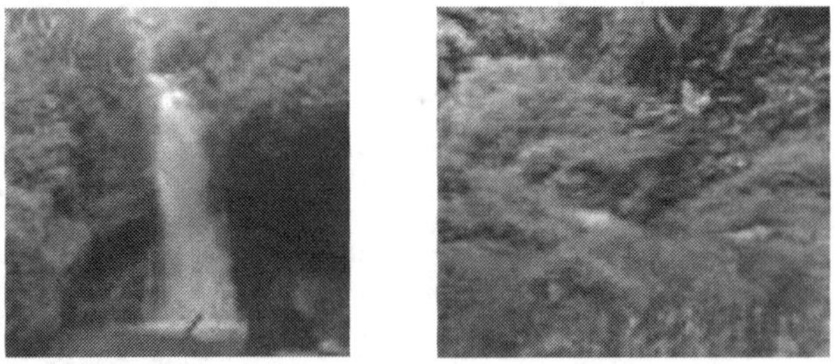

Fig.5. Ringin Jenggot Waterfall

 https://doi.org/10.1051/e3sconf/202565201002

3.2 Internal and External Potential

Referring to the results of observations, questionnaires and interviews with waterfall tourism managers, it is known that the total score and classification of internal potential (from within) and external potential (outside the object) of waterfall tourism in Ngawi Regency, which consists of 5 natural waterfall tourism objects in Ngawi Regency.

Table 1. Classification of Potential Waterfall Tourism Objects in Ngawi Regency

Number	Name Of tourist attraction	Average Internal Score	Average External Score	Total Score Average (Internal+Eksternal)	Potential Classification
1	Suwono Waterfall	2,05	1,8	3,85	Tall
2	Srambang Waterfall	2	1,98	3,98	Tall
3	Pengantin Waterfall	1,8	1,65	3,45	Tall
4	Teleng Waterfall	1,3	1,51	2,81	currently
5	Ringin Jenggot Waterfall	1,16	1,11	2,27	currently

 https://doi.org/10.1051/e3sconf/202565201002

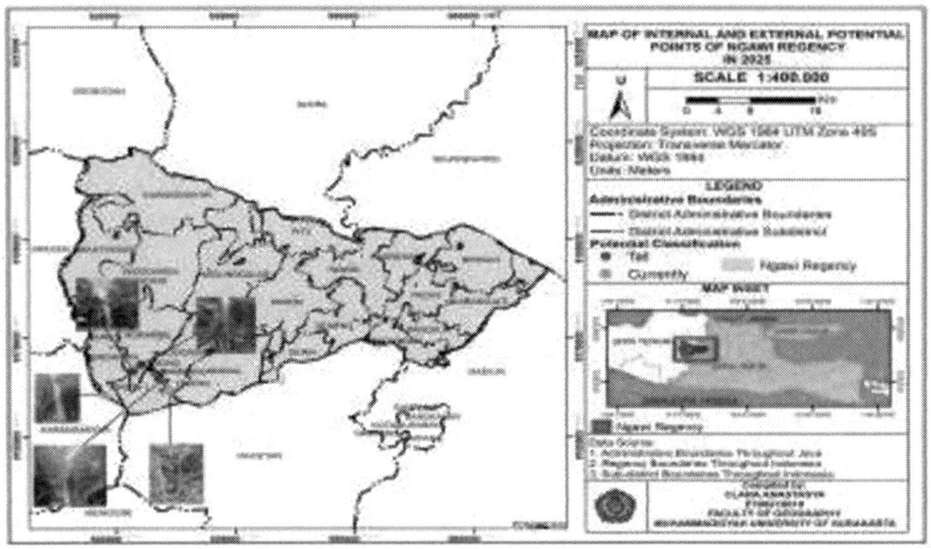

Fig. 6 Map of Internal and External Potential Points of Waterfall in Ngawi Regency in 2025

The 2025 Ngawi Regency Internal and External Potential Point Map classifies areas based on potential levels with two main colors: red for high potential and pink for medium potential. The red dots are Suwono, Srambang and Pengantin waterfalls which have superior natural tourism potential such as waterfalls and relatively good accessibility. Meanwhile, the pink dots are Teleng and Ringin Jenggot waterfalls whose potential is still developing, requiring additional management and infrastructure so that they can be further developed as superior destinations.

3.3 Waterfall Tourism Development Strategy

The strategy for developing waterfall tourism is based on the SWOT analysis method which includes IFAS and EFAS analysis and the SWOT matrix. Internal and external factors are used to evaluate the strengths, weaknesses, opportunities, and threats in the waterfall tourism sector in Ngawi Regency. This section will formulate a strategy for developing waterfall tourism in Ngawi Regency.

Table 2. SWOT Analysis of Waterfall Tourism Objects in Ngawi Regency

Internal Factors		**External Factors**	
Strength	***Weakness***	***Opportunity***	***Threats***
The natural elegance of the waterfall is a tourist attraction	Promotion and marketing are not yet optimal	Establishing cooperation with private parties, such as hotels and travel agents	The decline in the number of tourists in certain seasons, causes a decline in the number of tourists
Complete supporting facilities for tourist destinations	tourist accommodation such as lodging is still very limited	potential to collaborate with online tourism platforms for wider promotion	environmental damage caused by tourists
The clarity of the water and the calm atmosphere are suitable for tourists	Seasonal dependence affects the number of tourist visits	Increasing public awareness of the importance of preserving nature and the environment	Competition with other tourist destinations
There are various tourist attraction activities	Accessibility of public transportation and transportation routes is lacking	Increasing tourist interest in seeking more natural and cool tourist destinations	Pandemic or outbreak of infectious disease
Good transportation accessibility and easy to reach	Directions to the waterfall are still lacking	Support from the government in community-based tourism development programs	Changing traveler preferences
The average entrance ticket price is affordable and free.	Access roads to some of the waterfalls are still narrow	Increasing tourist interest in seeking unique natural tourism	
waste in the tourist area is managed well	Lack of development in managing waterfalls		

Internal Factors		External Factors	
Strength	*Weakness*	*Opportunity*	*Threats*
There is an award ceremony for champions in the nature tourism category			
The natural beauty of the environment around the waterfall is still natural and far from city pollution.			

Based on Table 2, it can be seen that waterfall tourism in Ngawi Regency has strengths, weaknesses, opportunities, and threats. The strengths in the table can be used to attract tourists, especially in Ngawi Regency itself. Weaknesses can be used as evaluation material for existing tourist attractions so that they can be improved in the future. Opportunities can be used to maximize existing internal strengths and also to take advantage of external momentum. Threats can be used as considerations in formulating adaptive policies.

Table 3. Internal Factor Analysis Summary (IFAS)

Strength	**Weight**	**Rate**	**Score**	*Weakness*	**Weight**	**Rate**	**Score**
The natural beauty of the waterfall is a tourist attraction	0,071	3	0,214	Promotion and marketing are not optimal	0,071	3	0,214
Complete supporting facilities for tourist destinations	0,071	3	0,214	Tourist accommodation such as lodging is still very limited	0,047	2	0,095
The clarity of the water and the calm atmosphere are suitable for tourists.	0,071	3	0,214	Seasonal dependence affects the number of tourist visits	0,047	2	0,095
There are various tourist attraction activities	0,047	2	0,095	Accessibility of public transportation and transportation	0,071	3	0,214

Strength	**Weight**	**Rate**	**Score**	*Weakness*	**Weight**	**Rate**	**Score**
				routes is lacking			
Good transportation accessibility and easy to reach	0,047	2	0,095	Directions to the waterfall are still lacking	0,047	2	0,095
The average entrance ticket price is affordable and free.	0,071	3	0,214	Access roads to some of the waterfalls are still narrow	0,071	3	0,214
Waste in the tourist area is managed well	0,047	2	0,095	Lack of development in managing waterfalls	0,071	3	0,214
There is an award ceremony for champions in the nature tourism category	0,071	3	0,214				
The natural beauty of the environment around the waterfall is still natural and far from city pollution.	0,071	3	0,214				
Total	**0,571**	**24**	**1,571**		**0,428**	**18**	**1,142**

Based on the results of the Internal Factor Analysis Summary (IFAS) in Table 3 above, it shows that the total score for strengths and weaknesses is 1.571 and 1.142, where the figure shows that the strength score is higher than the weakness score, so this can be utilized to become a more optimal and sustainable tourism potential.

Table 4. External Factor Analysis Summary (EFAS)

Opportunity	**Weight**	**Rate**	**Score**	*Threats*	**Weight**	**Rate**	**Score**
Establish cooperation	0,068	2	0,137	The decrease in the number	0,103	3	0,310

Opportunity	Weight	Rate	Score	*Threats*	Weight	Rate	Score
with private parties, such as hotels and travel agents				of tourists in certain seasons causes a decrease in the number of tourists			
Potential to collaborate with online tourism platforms for wider promotion	0,103	3	0,310	Environmental damage caused by tourists	0,103	3	0,310
Increasing tourist interest in seeking more natural and cool tourist destinations	0,103	3	0,310	Pandemic or outbreak of infectious disease	0,068	2	0,137
Support from the government in community-based tourism development programs	0,068	2	0,137	Changing traveler preferences	0,103	3	0,310
Increasing tourist interest in seeking unique natural tourism	0,103	3	0,310				
TOTAL	**0,551**	**16**	**1,517**		**0,448**	**13**	**1,206**

External Factor Analysis Summary (EFAS) analysis based on Table 4 shows that there are scores for opportunities and threats, namely 1.517 and 1.206, where the opportunity score is superior to the threat score, so this can be utilized to make a waterfall tourist attraction in Ngawi Regency better in the future. There are many ways that can be used to optimize, one of which is by maximizing digital promotion, especially through social media and collaboration with travel influencers.

 https://doi.org/10.1051/e3sconf/202565201002

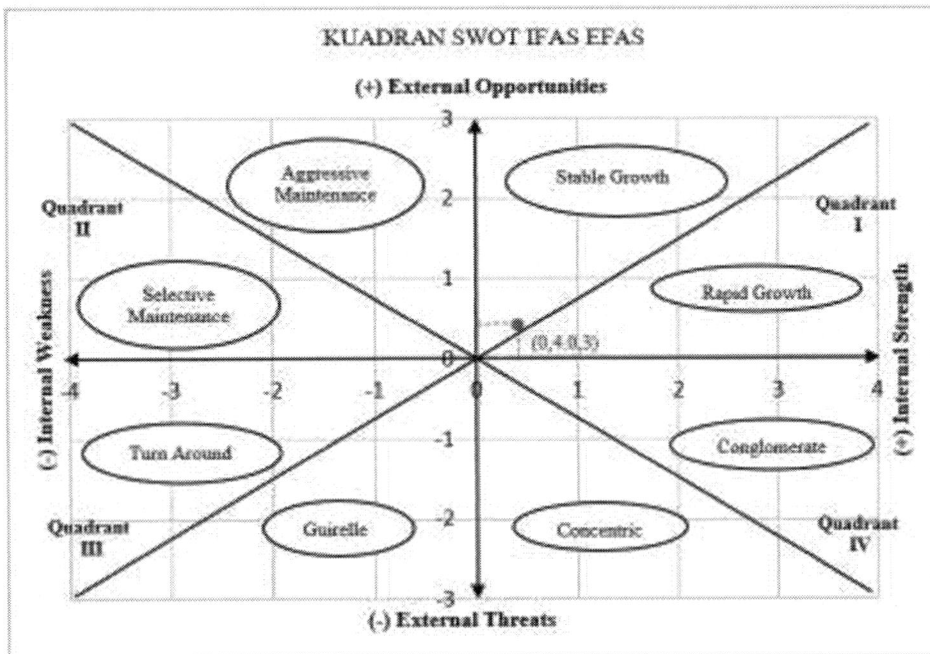

Fig.7. IFAS and EFAS Quadrant Matrix of Waterfall Tourism in Ngawi Regency

Based on the SWOT IFAS and EFAS matrix analysis for waterfall tourism in Ngawi Regency, the coordinate point is at position (0.4; 0.3), this position is obtained from the calculation process of factors X and Y. The value of factor X which is the IFAS value that forms a horizontal line and the total strength score is obtained minus the total weakness score. The value of factor Y is the EFAS value that forms a vertical line and is obtained from the total opportunity score minus the total threat score. As in the following equation:

X = Strength − Weakness
= 1,571 − 1, 142
= 0,429
Y = opportunity − Threat
= 1,571 − 1, 206
= 0,365

Points X and Y are located in Quadrant I. This quadrant shows that internal conditions are relatively strong (more strengths than weaknesses) and the external environment provides greater opportunities than threats. Thus, the most appropriate strategy is an approach that focuses on growth, namely optimizing potential to seize opportunities. This strategy can be in the form of developing tourism facilities, increasing accessibility, digital promotion through social media, to collaboration with travel influencers to reach a wider range of tourists. Several strength factors that influence this position include the natural beauty of the waterfall which is still natural and has high visual appeal, the diversity of waterfall tourist attractions in Ngawi, and the potential for developing nature-based and adventure tourism. In addition, support from the surrounding community and attention from the local government to the tourism sector are also important strengths. By utilizing these strengths, the development of waterfall tourism in Ngawi Regency has a great opportunity to grow sustainably and become a leading destination in East Java.

The results of the SWOT analysis indicate that waterfall tourism in Ngawi Regency holds considerable potential for development, primarily due to its rich natural beauty, relatively good accessibility for several locations, and growing public interest in nature-based tourism. The positioning of this tourism in Quadrant I of the SWOT matrix—where strengths outweigh weaknesses and opportunities outweigh threats—supports the adoption of an aggressive (S-O) strategy that leverages internal strengths to seize external opportunities [13].

Waterfalls such as Srambang and Suwono exemplify successful integration of natural attractions with relatively well-developed infrastructure, including paved roads, rest facilities, and clear signage. These findings align with [15], who argue that complete and well-maintained infrastructure is a key driver for increasing tourist satisfaction and revisitation rates in waterfall destinations. Srambang's thematic parks and photo spots also demonstrate the growing trend of "Instagrammable tourism," which enhances destination appeal, particularly among younger demographics.

However, the disparity in development across sites is evident. While Srambang and Suwono scored above 3.85 in combined internal and external evaluations, Teleng and Ringin Jenggot scored below 3.0, indicating the need for targeted improvements. As noted by [16] uneven development in ecotourism areas often stems from inadequate institutional support, low community participation, and limited budget allocation for remote destinations. In Ngawi's case, the latter two factors were particularly prominent in the management of semi-natural sites such as Ringin Jenggot, where facilities like toilets, signage, and parking areas remain insufficient.

Community involvement is another crucial element. The study identifies early community engagement, especially in sites like Pengantin Waterfall, where village-based management models are emerging. This supports earlier findings by [17], who emphasized that community-led ecotourism initiatives can simultaneously foster environmental stewardship and local economic empowerment. The formation and strengthening of Pokdarwis (Tourism Awareness Groups) is vital for sustaining local participation and embedding a sense of ownership among residents [18].

Promotion remains a major weakness across all waterfall destinations. Despite the growing role of digital platforms in tourism marketing, several sites still rely on conventional methods or lack active promotional campaigns altogether. This is consistent with [19] who reported that tourism development strategies often fail when aggressive marketing is not supported by strong human resource capacity and digital literacy. In response, collaborative efforts with travel influencers and online travel platforms are recommended to expand outreach and attract non-local tourists, particularly from urban areas of East Java and Central Java.

Environmental sustainability is a double-edged issue in Ngawi's waterfall tourism. On one hand, the natural environment remains largely unspoiled, offering clean air, biodiversity, and a tranquil setting, all of which are increasingly sought after in post-pandemic tourism trends [20]. On the other hand, increased tourist traffic without adequate visitor management plans risks damaging fragile ecosystems through waste, deforestation, and erosion—issues highlighted by [21] in their study on river-based ecotourism. Hence, Ngawi's tourism strategy must integrate environmental safeguards, such as visitor caps, eco-friendly infrastructure, and strict waste management systems.

Lastly, seasonal dependency remains a notable threat. Many waterfall destinations experience fluctuating tourist numbers depending on rainfall patterns and school holiday periods. This cyclical visitation pattern mirrors the findings of [22], who observed that nature-based tourism in Papua is highly sensitive to weather and seasonality. Ngawi's strategy could benefit from diversifying attractions—such as cultural festivals, agro-tourism, or wellness retreats—to reduce this dependency and encourage year-round visitation.

In summary, while Ngawi Regency holds a promising position for waterfall tourism development, a multi-pronged strategy is essential. This includes upgrading infrastructure, empowering communities, launching digital marketing campaigns, and ensuring environmental sustainability. Only through such an integrated approach can Ngawi transform its natural assets into sustainable economic and cultural capital

4 Conclusion

There are 5 waterfall tourist attractions in Ngawi Regency, it is known that the majority of waterfalls that are tourist destinations have advantages in terms of natural authenticity, beautiful landscapes, and an atmosphere that is still pristine and unpolluted. Waterfalls such as Pengantin Waterfall, Ringin Jenggot, Suwono, Srambang, and Teleng are part of the local natural wealth that has not been fully optimized as a leading destination. Entrance tickets to tourism are quite affordable, waterfall tourism has several rides such as photo spots, flower gardens, trekking paths, children's play areas, gazebos and natural pools. Based on the results of the study, the highest total score with an average score of internal and external potential of 3.85 is Srambang Waterfall and 3.98 Suwono Waterfall, the next position is Pengantin Waterfall with an average score of 3.45, the average score with a moderate classification of Teleng Waterfall is 2.81 and Ringgin Jenggot Waterfall has a score of 2.27. The score results show that Srambang, Suwono and Pengantin Waterfalls have the highest potential, which reflects readiness in terms of natural attractions, accessibility, and facility support, while Ringin Jenggot Waterfall still requires improvement and strengthening in terms of management and supporting facilities.

The position of the SWOT matrix shows that the waterfall in Ngawi Regency is in quadrant 1, namely in the Stable Growth section, namely a defensive development strategy and this quadrant shows that internal conditions are relatively strong, more strengths than weaknesses. The strategies used are maximizing strengths to take advantage of opportunities (S-O), maximizing strengths to overcome threats (S-T), overcoming weaknesses to take advantage of opportunities (W-O), and maximizing strengths to overcome threats (S-T) with further strategy elaboration.

References

1. Pujiasmanto, B., I. Aliyah, and N. Miladan. "Development of organic agro tourism for optimizing agricultural resources in Ngawi, East Java, Indonesia." In *IOP conference series: earth and environmental science*, vol. 637, no. 1, p. 012061. IOP Publishing, 2021.
2. Yustie, Renta, Ricky Angga Ariska, and Fadilla Purwitasari. "TOURISM DEVELOPMENT TOWARDS INTERNATIONAL COMPETITIVE TOURISM IN PROMOTING THE ECONOMY OF EAST JAVA." *Journal of Developing Economies (JDE)* 7, no. 2 (2022).
3. Terkenli, Theano S. "Tourism and landscape." *The Wiley Blackwell Companion to Tourism* (2024): 166-180.
4. Somantri, Lili. "Spatial Modelling Analysis for Potential Expansion of Cipanas City Candidate Area as Autonomous City." In *Forum Geografi*, vol. 38, no. 1, pp. 57-73. 2024.
5. Cretu, Carmen-Mihaela, Anca-Gabriela Turtureanu, Carmen-Gabriela Sirbu, Florentina Chitu, Emanuel Ştefan Marinescu, Laurentiu-Gabriel Talaghir, and Daniela Monica

Robu. "Tourists' perceptions regarding traveling for recreational or leisure purposes in times of health crisis." *Sustainability* 13, no. 15 (2021): 8405.

6. Amin, Choirul, Wahyu Tyas Pramono, Jumadi Jumadi, Dewi Novita Sari, and Maxim GM Samson. "Navigating urban poverty: The role of livelihood capital in the livelihood strategies of urban beggars in Indonesia." *Social Sciences & Humanities Open* 11 (2025): 101298.
7. Shpak, Nestor, Oksana Muzychenko-Kozlovska, Maryana Gvozd, and Włodzimierz Sroka. "Simulation of the influence of external factors on the level of use of the regional tourism potential: A practical aspect." *Administrative Sciences* 11, no. 3 (2021): 85.
8. Otcheskiy, Ivan, Lyailya Mutaliyeva, Elena Yudina, Diana Stepanova, Aleksandr Shelygov, and Elvira Seminskaya. "Developing tourist destination potential under influence of internal and external factors." *Anais Brasileiros de Estudos Turísticos: ABET* 13, no. 1 (2023): 12.
9. Pujiasmanto, B., I. Aliyah, and N. Miladan. "Development of organic agro tourism for optimizing agricultural resources in Ngawi, East Java, Indonesia." In *IOP conference series: earth and environmental science*, vol. 637, no. 1, p. 012061. IOP Publishing, 2021.
10. Zhang, Can, and Massoomeh Hedayati Marzbali. "Mapping Traditional Village Cultural Landscape Research: A Systematic Bibliometric Analysis." In *Forum Geografi*, vol. 38, no. 3, pp. 379-394. 2024.
11. Suyadi, Aman, Erna Handayani, Naelati Tubastuvi, Ira Hapsari, and Meilana Eka Trinowo. "Capacity building model for rural lower catagories: SWOT analysis." *International Journal of Research in Business and Social Science* 11, no. 6 (2022): 609-617.
12. Abdullah, Fadli Daud, Budi Sulistiyo, and Syahid Mujahid Wiwaha. "SWOT Strategy in Strengthening the Role of Village Community Empowerment Institutions: An Islamic Law Perspective." *Jurnal Al Maesarah* 3, no. 2 (2024): 64-77.
13. Pereira, Leandro, Miguel Pinto, Renato Lopes da Costa, Álvaro Dias, and Rui Gonçalves. "The new SWOT for a sustainable world." *Journal of Open Innovation: Technology, Market, and Complexity* 7, no. 1 (2021): 18.
14. Efe, Burak, Ömer Faruk Efe, and Alessio Ishizaka. "A model proposal to examine the effects of ships to marine pollution in terms of internal and external factors." *Soft Computing* 26, no. 5 (2022): 2121-2134.
15. Xingjun, Zhang. "An Empirical Study on the Tourist Satisfaction." *Journal of Digitainability, Realism & Mastery (DREAM)* 1, no. 07 (2022): 47-56.

Valuing Ecosystem Services in a UNESCO Global Geopark: Spatial Assessment of Tourists' Willingness to Pay for Biodiversity Conservation in Ciletuh, Indonesia

Purna Hindayani[1], Fitri Rahmafitria[2], Moh. Dede[2,] Riko Arrasyid[3], Armandha Redo Pratama[2], Zuzy Anna[4]*

[1]Catering Industry Management, FPIPS, Universitas Pendidikan Indonesia, Bandung, Indonesia
[2]Resort and Leisure Management, FPIPS, Universitas Pendidikan Indonesia, Bandung, Indonesia
[3]Geography Education, FPIPS, Universitas Pendidikan Indonesia, Bandung, Indonesia
[4]SDGs Center, Universitas Padjadjaran, Bandung, Indonesia

Abstract. Biodiversity ecosystem services are shared resources whose value can be measured through willingness to pay (WTP). This study aims to assess the willingness to pay of tourists for the conservation of biological resources in the UNESCO Global Geopark Ciletuh which is one of the leading tourist destinations in Indonesia. Furthermore, knowing how much visitors are willing to pay (WTP) for natural ecosystems can help policy makers manage biological resources and the environment which are the main foundation of sustainable nature-based tourism. The survey was conducted on 376 visiting tourists. The method used is a spatial assessment of the environmental ecosystem service approach, willingness to pay (WTP) for the uniqueness of biodiversity and factors that influence the WTP value using logit regression. The results of the mapping of ecosystem services supporting biodiversity show a very high-high category of 42,17% of the total area of the Ciletuh Geopark. The results also show that more than half of the respondents 68.8% are willing to pay for the conservation of biological resources in the Ciletuh Geopark. The average willingness to contribute to the general conservation fund is around $ 1.27 (Rp. 20,896). It was found that WTP for conservation was influenced by factors closely Introduction

1 Backgrounds

Indonesia stands as one of the world's biodiversity hotspots, with extensive geological diversity and unique cultural heritage [1]. Moreover, Indonesia is one of the most highly populated nations globally and is classified as The Third-largest tropical mega-biodeversity country [2] . Geopark Ciletuh-Palabuharatu reconized as a UNESCO Global Geopark (UGGp) since 2018 which exemplifies the intricate interplay between geodivercity, biodiversity, and sustainable development. Geopark is a protection concept that emphasizes geological heritage, geological diversity, and natural landscapes, all of which are integrated into the concept of conservation, education, and sustainable development through a bottom-up participatory approach [3].

* Corresponding author: purnahindayani@upi.edu

The Geopark Ciletuh-Palabuharatu is not only important for ecological and cultural reasons but also as a driver of economic development through tourism. Tourism in the Geopark Ciletuh-Palabuharatu is characterized by its biodiversity, including its geology, and is being developed as a tourist destination with comparative and competitive advantages and diverse natural resources [4]. Tourism in the Geopark Ciletuh-Palabuahrantu has increased every year and even increased after the COVID-19 pandemic ended. However, while tourism in the Geopark Ciletuh-Palabuharatu provides economic benefits to local communities and regional development, it can also negatively impact conservation efforts if not managed wisely and sustainably.

In addition, there are tourist destinations that only focus on mass tourism, so that this type of tourism management model has a negative impact on tourist destinations, because some economic benefits are at the expense of environmental, social and cultural balance [5] including tourists' negligence towards the environment due to accessibility [6], which ultimately causes pressure on the environment such as habitat destruction [7], pollution [7] and disturbance of biodiversity [7], [8] which is an important element in geoparks. Aziza et al., (2023) [8] stated that tourist activities in Geopark Ciletuh-Palabuhanratu cause various environmental impacts, including physical alterations in water quality, soil compaction due to trampling, accumulation of waste, an increase in the average air temperature, and a decrease in humidity levels, thus affecting the composition of biodiversity.

Furthermore, unplanned tourism infrastructure development and environmentally unfriendly tourism activities risk violating conservation principles. Therefore, it is crucial to maintain environmental conservation in the Geopark Ciletuh-Palabuharatu area as a sustainable tourism destination that preserves the environment, unique geology, biodiversity, and educational value of the area. This encourages tourists to participate in preserving biodiversity in the geopark destination by paying more to support the protection and sustainable management of the area [9], [10].In this case, tourists are directly involved in their commitment to biodiversity conservation and the development of environmentally conscious geopark tourism destinations.

Numerous studies have discussed tourists' willingness to pay (WTP) for biodiversity conservation and environmental protection [9], [10], most of which focus on socioeconomic status and WTP. This study focuses on spatial analysis of biodiversity ecosystem service providers in the Geopark Ciletuh Palabuharatu as a database of willingness to pay for conservation purposes, along with socio-demographic factors of tourists' WTP. Therefore, this study aims to determine tourists' WTP for levies used in policies to improve destination sustainability and maintain conservation. It should be acknowledged that using the contingent valuation method to measure WTP presents methodological challenges, such as hypothetical bias, where respondents may provide answers that do not fully reflect their actual behavior[9], [11], [12]. Nevertheless, this study is presented as an application that attempts to provide robust and valid estimates while addressing the limitations of the method. Theoretically, WTP is a measure of economic value that reflects changes in social welfare, so this understanding provides deeper context for the importance of this study's results in supporting policymaking oriented toward conservation and sustainable tourism development. Based on tourists' willingness to pay (WTP) for biodiversity conservation and environmental protection, estimated revenue is calculated to evaluate the impact of the proposed entrance fee and recommend restructuring methods for the current entrance fee system.

 https://doi.org/10.1051/e3sconf/202565201003

2 Methodology

This research was conducted at the Ciletuh-Palabuhanratu Geopark in West Java Province, Indonesia. Respondents using a convenience sample were 376 tourists (both domestic and international) visiting the Geopark Ciletuh-Palabuhanratu, while secondary data were obtained from relevant agencies, including land cover maps, ecoregion maps, and administrative maps of the Ciletuh-Palabuhanratu Geopark, along with other data.

2.1 Spatial Assessment of biodiversity Ecosystem services the Geopark Ciletuh-Palabuhanratu

This study uses a quantitative method with a spatial approach to obtain numerical data and process it into a map depicting the biodiversity ecosystem services map in the Geopark Ciletuh-Palabuharatu. The data used are secondary data, namely ecoregion data and land cover maps that are processed through a process. The data analysis stage begins with the identification of ecosystem services from both maps, namely land cover maps and ecoregions, assessment and weighting of ecosystem scores based on expert judgment of ecoregion and land cover data, spatial analysis and calculation of the ecosystem services index providing biodiversity, and spatial visualization of biodiversity ecosystem services. Then, the assessment and weighting of biodiversity ecosystem services by experts for land coner and ecoregions is conducted using the pairwise comparison method through the Analytical Hierarchy Process (AHP). Mapping ecosystem services using a land-use-based proxy approach, which utilizes expert judgment from multiple disciplines, can produce comprehensive assessments [13]. The experts involved in this assessment process consist of four people, namely regional planning experts, environmental geography experts, biologists and environmental management experts. Furthermore, after being assessed, it is overlaned and then divided into 5 classifications ranging from very low, low, medium, high and very high.

2.2 Willingness to pay for biodiversity conservation in the Geopark Ciletuh-Palabuhanratu

This research approach uses the concept of tourist willingness to pay, namely the monetary value a person gives to goods and services produced by land and environmental resources, especially the quality and protection of ecosystem services. In this study using a binary logit model, in making WTP decisions, socio-demographic factors such as gender, age, education, income, and number of dependents play a role as dependent variables that influence WTP [9], [10]. In addition, tourism attributes, such as total costs incurred during the tour also influence WTP [10], [12]. This study uses a closed-ended dichotomous choice format with a double dichotomous choice where each respondent is given an initial bid. Follow-up questions will depend on the respondent's response and are used to conduct the bid elicitation process. If the first bid is answered yes, the bid is higher than the initial bid. Meanwhile, if the first bid is answered no, the bid is lower than the initial bid. The bid determination aims to provide convenience for visitors who are respondents in estimating their willingness to pay and a more accurate bid value in determining the willingness to pay value. The range of bid values offered starts from IDR 3,000, IDR 5,000, IDR 6,000, IDR 8,000, IDR 10,000, IDR 15,000, IDR 20,000, up to a maximum of IDR 25,000, the resulting bid value is based on the results of interviews and field observations. Analysis of factors influencing the tourist's willingness using binary logit regression using SPSS software is presented as follows.

$$WTP=\beta_0 + \beta_1 Bid + \beta_2 X_1 + \beta_3 X_2 + \beta_4 X_3 + \beta_5 X_4 + \beta_6 X_5 + \beta_7 X_6 + \beta_8 X_7 \quad (1)$$

WTP = Tourist's Willingness to Pay
β_0 = Constant
β_1 *Bid= Bidding Price (Rp.5.000-Rp.25.000)*
β_2 =Age
β_3 =Gender
β_4 =Income
β_5 =Education
β_6 =Marriage
β_7 =Tourist Origin
β_8 =Total Cost during trip
e = Galat error

In terms of methodological limitations of the study, two of the most important are the limited data on the tourist population of the Ciletuh-Palabuharatu Geopark and the inherent bias associated with stated preference studies, particularly contingent valuation. However, with a sample size of 367, this survey sample is considered to provide a fairly representative picture of the visitor profile at the site. The contingent valuation method itself has several known weaknesses, including the hypothetical nature of payments and the presence of a starting point bias, where respondents' willingness-to-pay estimates can be influenced by the reference price provided by the interviewer. Nevertheless, the use of a double-dichotomous choice in questions related to willingness-to-pay is believed to be closer to actual purchasing behavior than open-ended questions, especially when the entrance fee is used for conservation efforts at the Ciletuh-Palabuharatu Geopark.

3 Results and Discussion

3.1 Assessment spatial for environmental ecosystem services maps for biodiversity conservation

The biodiversity ecosystem services mapping data in Geopark Ciletuh shows the distribution of ecosystem service categories based on administrative sub-areas (districts) and their levels, ranging from "very low" to "very high." Ciemas district has the largest area with a variety of ecosystem service categories, followed by Simpenan, Ciracap, and Cisolok. The total area assessed in this study covers approximately 121,735 hectares.

.

Table 1. Extent of Ecosystem Services for Biodiversity Conservation.

No	Geopark Ciletuh-Palabuharatu Subdistrict	Categories of Ecosystem Services					Total (Ha)
		Very Low	Low	Moderate	High	Very High	
1	CIEMAS	12928,81	692,55	8141,97	6948,33	537,55	29249,21
2	CIKAKAK	2106,05	6,63	863,34	5189,95	3307,55	11473,52
3	CIRACAP	5961,57	339,47	4127,66	4725,31	57,09	15211,09
4	CISOLOK	1280,34		1910,50	10830,17	3160,03	17181,05
5	PELABUHAN RATU	2729,75	14,99	2986,66	3464,53	76,74	9272,67

 https://doi.org/10.1051/e3sconf/202565201003

No	Geopark Ciletuh-Palabuharatu Subdistrict	Categories of Ecosystem Services					Total (Ha)
		Very Low	Low	Moderate	High	Very High	
6	SIMPENAN	4292,62	75,29	5244,07	7449,27	213,77	17275,01
7	SURADE	3549,48	5,72	3243,54	5160,97	18,86	11978,59
8	WALURAN	8790,89	34,13	1076,53	127,00	64,85	10093,39
Total (Ha)		**41639,50**	**1168,78**	**27594,26**	**43895,54**	**7436,45**	**121734,53**

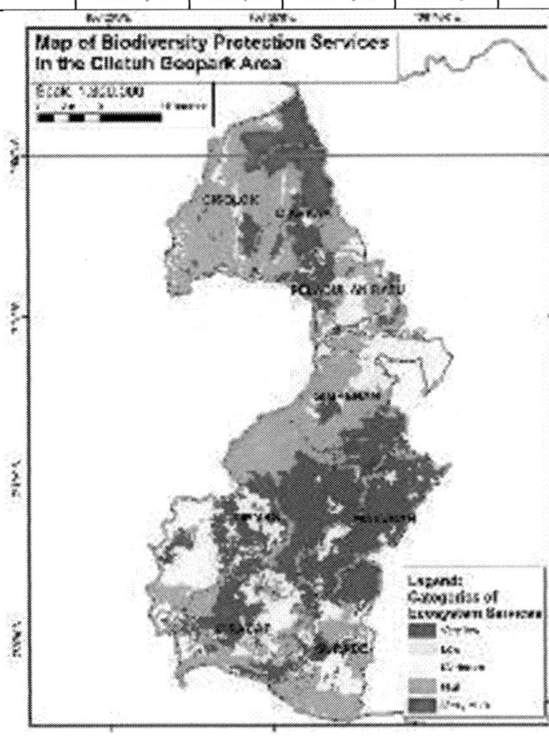

Fig. 1. Ecosystem Services Map for Biodiversity Conservation Geopark Ciletuh-Palabuharatu

The results of the mapping of ecosystem services supporting biodiversity show a very high-high category of 42,17% of the total area of the Geopark Ciletuh-Palabuharatu. Moreover, The mapping results illustrate that the distribution of biodiversity ecosystem services is uneven—for example, the "very high" category is mostly found in Ciemas and Cisolok, while other sub-districts such as Waluran and Pelabuhan Ratu have smaller proportions of the highest category. A higher class (High-Very High) on the biodiversity ecosystem services map indicates greater species diversity and the ability to foster the development of flora and fauna. Furthermore, a higher class indicates a greater regional capacity to support biodiversity conservation activities, and vice versa[14].This type is a crucial basis for analyzing conservation priorities and area management to ensure key ecosystem services are sustainably maintained. The spatial integration in this study is carried out by mapping the biodiversity ecosystem services within the Geopark Ciletuh-Palabuharatu area, which serves as a critical foundation for informing respondents that the geopark tourist locations belong to zones with high to very high biodiversity levels. This information helps respondents better understand the ecological conditions of the areas they visit, enabling them to determine their willingness to pay (WTP) values more accurately, particularly distinguishing between tourists visiting "Very High" zones (such as Ciemas and Cisolok) and those in "Moderate" zones. Thus, this spatial approach connects the specific ecosystem conditions with tourists' economic responses in supporting sustainable conservation funding.

3.2 Positioning Willingness to Pay (WTP) of tourists in Environmental Conservation in the Ciletuh-Palabuhanratu Geopark and the socio-demographic factors that influence it

The findings reveal that a majority of tourists (259 people or 68.8%) depicts a willingness to pay to support the conservation in Geopark Ciletuh-Palabuharatu. The high level of willingness underscores a pronounced environmental awareness among visitors regarding the critical importance of preserving natural ecosystems and promoting sustainable tourism. Based on logit regression analysis, the Fit model, meanwhile The Nagelkerke R^2 value shows 0.786, which means that the ability of the independent variable to explain the dependent variable is 78.6%, while the remaining 21.4% is due to other factors outside this research.

Table 2. Variables in the Equation

		B	S.E.	Wald	df	Sig.	Exp(B)
Step 1[a]	Bid	.001	.000	85.242	1	.000	1.001
	Age	.647	.274	5.590	1	.018	1.909
	Gender	.655	.429	2.332	1	.127	1.925
	Income	.000	.000	.716	1	.397	1.000
	Education	.100	.190	.278	1	.598	1.105
	Marriage	-.505	.505	.997	1	.318	.604
	Tourist	11.573	48.410	.057	1	.811	106140.068
	Total Cost	.000	.000	1.326	1	.250	1.000
	Constant	-16.790	48.529	.120	1	.729	.000

a. Variable(s) entered on step 1: Bid, Age, Gender, Income, Education, Marriage, Tourist, Total cost.

$$g(x)=-16.79+0.001+0.647+0.655+0.000+0.100-0{,}505+11{,}57+0{,}000+e$$

Based on the results of the Omnibus Test of Model Coefficients, with a Chi-square value of 307.487 and a p-value of 0.000, the regression model using eight independent variables (bidding price, age, gender, income, education, marital status, tourist origin, and total trip cost) collectively had a significant influence on the dependent variable, WTP. The Wald test determined which independent variables contributed significantly to the model. The results showed that the bid variable, with a Wald value of 85.242 and a p-value of 0.000, was highly significant, indicating a strong influence. The Age variable, with a Wald value of 5.590 and a p-value of 0.018, was also significant. Meanwhile, the remaining variables did not have a significant influence. Furthermore, the willingness to pay (WTP) for conservation in the Ciletuh Geopark was found to be influenced by the number of bidding prices offered to become the value of the willingness to pay for conservation in the Ciletuh-Palabuharatu Geopark. On average, tourist respondents are willing to contribute approximate $1.27 (equivalent Rp. 20,896) to the general conservation fund in form of entrance fee. In addition, significant variables in Age indicate variations in conservation valuation across different age cohorts. Understanding these determinants is vital for the design and implementation of equitable and efficient funding mechanisms tailored to enhance biodiversity conservation and sustainable tourism in the Geopark Ciletuh-Palabuharatu. Moreover, The bid and age variables are used to predict a person's willingness to pay. In this case, the bid variable with an odds value of Exp(B) = 1.001 means that every 1 unit increase in the bid value can increase the respondent's willingness to pay by 0.1%. Meanwhile, the age variable with an odds value of Exp(B) = 1.909 shows that every 1-year increase in age almost doubles the respondent's willingness to pay. Thus, the bid and age variables are significant factors influencing the WTP value, where BID makes a small but consistent contribution, while age has a significant influence on respondents' willingness to pay decisions.

The results of the WTP value research are expected to form the basis for implementing fees, whether for entrance tickets or donations, that will be used by tourism management parties to create sustainable tourism plans. Moreover, Rahmafitria et al. (2024) [15] stated careful planning is needed to balance market preferences and conservation efforts. In this case, tourism planning is needed that takes conservation aspects into account when making decisions and policies. The government has a significant role in formulating policies that align with the area's protection function while responding to tourist market demand. Furthermore, inappropriate management and accessibility planning can accelerate the destruction of protected areas which is a conservation area[15].

4 Conclusion

Geopark Ciletuh-Palabuharatu, recognized as a UNESCO Global Geopark since 2018, integrates geological conservation, biodiversity, and community-based sustainable development. The number of tourists, both domestic and international, has increased significantly, resulting in both positive and negative impacts. However, the negative impacts of tourism activities and unwise tourism management can lead to habitat destruction, pollution, and biodiversity disruption. This study, conducted through mapping ecosystem services for biodiversity conservation in the Ciletuh-Palabuharatu Geopark, showed that almost 42% were categorized as high and very high in supporting biodiversity in the location. Furthermore, in maintaining high and very high biodiversity conservation areas, survey results showed that tourists agreed to pay around $1.27 for conservation funds, which can be in the form of tickets and fees. This important finding is useful for policymakers in financing inclusive and effective conservation schemes to support responsible and sustainable tourism development.

References

1. N. Firdaus, Supriatna, and J. Supriatna, "Ecosystem services research trends in Indonesia: a bibliometric analysis," *Biodiversitas*, vol. 23, no. 2, pp. 1105–1117, 2022, doi: 10.13057/biodiv/d230255.
2. I. Sanka *et al.*, "Synthetic biology in Indonesia: Potential and projection in a country with mega biodiversity," *Biotechnol. Notes*, vol. 4, no. December 2022, pp. 41–48, 2023, doi: 10.1016/j.biotno.2023.02.002.
3. UNESCO, "Statutes of the International Geoscience and Geoparks Programme," *UNESCO Program. Meet. Doc.*, p. IGGP/2015/ST, 2015, [Online]. Available: https://unesdoc.unesco.org/ark:/48223/pf0000260675.page=1
4. Y. Yanuar, Z. Anna, P. Hindayani, M. F. Rosana, Z. Zulfiadi, and A. Sudradjat, "Prospective Analysis of Sustainable Development Strategy of Geopark Tourism of Ciletuh-Palabuhanratu West Java Indonesia," *Int. J. Curr. Innov. Res.*, vol. 4, no. 2(a), pp. 1033–1039, 2018, doi: 10.24327/IJCIR.
5. J. I. Pulido-Fernández and Y. López-Sánchez, "Are tourists really willing to pay more for sustainable destinations?," *Sustain.*, vol. 8, no. 12, 2016, doi: 10.3390/su8121240.
6. F. Rahmafitria, V. Sukmayadi, and H. Purboyo, "The Real and Actual Tourism

Accessibility in Protected Areas," *IOP Conf. Ser. Earth Environ. Sci.*, vol. 501, no. 1, 2020, doi: 10.1088/1755-1315/501/1/012047.

7. M. Stefanica, C. B. Sandu, G. I. Butnaru, and A. P. Haller, "The nexus between tourism activities and environmental degradation: Romanian tourists' opinions," *Sustain.*, vol. 13, no. 16, pp. 1–19, 2021, doi: 10.3390/su13169210.
8. B. Aziza, E. K. S. Harini Muntasib, and R. Hermawan, "Impact of Tourism Activity in Ciletuh-Palabuhanratu Geopark (Cpg), Sukabumi, West Java," *Media Konserv.*, vol. 28, no. 1, pp. 43–50, 2023, doi: 10.29244/medkon.28.1.43-50.
9. C. Jurado-Rivas and M. Sánchez-Rivero, "Investigating Change in the Willingness to Pay for a More Sustainable Tourist Destination in a World Heritage City," *Land*, vol. 11, no. 3, 2022, doi: 10.3390/land11030439.
10. P. Hindayani, A. S. Bratanegara, and A. R. Pratama, "Willingness To Pay (WTP) Wisatawan dalam Pelestarian Lingkungan Paska Perubahan Cagar Alam Kamojang ke Taman Wisata Alam Kamojang," *Ilmu Lingkung.*, vol. 22, no. 1, pp. 43–49, 2024, doi: 10.14710/jil.22.1.43-49.
11. B. Witt, "Tourists'willingness to pay increased entrance fees at Mexican protected areas: A multi-site contingent valuation study," *Sustain.*, vol. 11, no. 11, 2019, doi: 10.3390/su11113041.
12. D. Boto-García, P. Mariel, J. B. Pino, and A. Alvarez, "Tourists' willingness to pay for holiday trip characteristics: A Discrete Choice Experiment," *Tour. Econ.*, vol. 28, no. 2, pp. 349–370, 2022, doi: 10.1177/1354816620959901.
13. S. Maynard, D. James, and A. Davidson, "The development of an ecosystem services framework for South East Queensland," *Environ. Manage.*, vol. 45, no. 5, pp. 881–895, 2010, doi: 10.1007/s00267-010-9428-z.
14. N. Ambarwati, D. Hastuti, and L. R. Wijayanti Faida, "Carrying Capacity Based on Ecosystem Services for Biodiversity Conservation in the Petungkriyono Essential Ecosystem Area Candidate," *J. Ilmu Kehutan.*, vol. 17, no. 1, pp. 96–105, 2023, doi: 10.22146/jik.v17i1.5762.
15. F. Rahmafitria, H. P. H. Putro, A. Rosyidie, P. Dirgahayani, H. P. D. Setiyorini, and G. R. Nurazizah, "Navigating Conservation Access: Unraveling Ignorant Tourist Behavior and Typologies in Komodo National Park," *J. Manaj. Hutan Trop.*, vol. 30, no. 2, pp. 237–245, 2024, doi: 10.7226/jtfm.30.2.237.

https://doi.org/10.1051/e3sconf/202565201004

Harnessing Internal and External Potentials for Sustainable Tourist Village Development: A Case Study from Selo District, Central Java

Salma Adilla Widyawati[1*] and *Choirul* Amin[2]

[1,2] Faculty of Geography, Universitas Muhammadiyah Surakarta, Indonesia

Abstract. Tourist villages are essential in promoting sustainable rural tourism by utilizing local natural and cultural resources. Selo District in Boyolali Regency, Central Java, Indonesia, is strategically located between Mount Merapi and Mount Merbabu and holds strong potential for tourist village development. However, limited infrastructure, low promotional efforts, minimal community involvement, and disaster risks have hindered its growth. This study aims to analyze the internal and external potentials of six tourist villages in the district and formulate appropriate development strategies. A mixed-method approach was used, employing SWOT analysis with IFAS and EFAS matrices to evaluate strengths, weaknesses, opportunities, and threats. The findings reveal that all villages have high internal and external potential. The analysis places the district in a Stable Growth position (Quadrant I), indicating the need for gradual, well-planned development. Recommended strategies include enhancing promotional efforts, improving tourism facilities, strengthening community capacity, and fostering stakeholder collaboration. This study underscores the importance of leveraging local potential and external support to develop sustainable tourism in rural and disaster-prone regions.

1 Introduction

Rural tourism has gained increasing global recognition as a sustainable approach to regional development, especially in areas with rich natural and cultural resources. Tourist villages offer a platform for preserving local traditions, empowering communities, and stimulating rural economies through tourism-based activities. Unlike mass tourism, tourist villages emphasize small-scale, community-driven initiatives that are environmentally friendly and culturally sensitive [1, 2]. This form of tourism is particularly suitable for countries like Indonesia, where rural communities often reside in areas of exceptional ecological and cultural value.

The Indonesian government has acknowledged the potential of tourist villages by integrating their development into national tourism and economic policies. Following the setbacks of the COVID-19 pandemic, the Ministry of Tourism and Creative Economy

* Corresponding author: ca122@ums.ac.id

launched several initiatives to promote community-based tourism as a means of recovery and long-term resilience [3, 4]. Tourist villages are now viewed as tourism destinations and strategic tools to accelerate rural development, reduce urban-rural disparities, and promote sustainable livelihoods [5, 6].

Selo District, located in Boyolali Regency, Central Java, is a compelling example of a region with strong potential for tourist village development. Its unique geographical position between Mount Merapi and Mount Merbabu offers breathtaking landscapes, a cool climate, and adventure tourism opportunities such as hiking and paragliding. Additionally, the district boasts rich cultural traditions, agricultural activities, and local culinary products that can attract diverse segments of domestic and international tourists. Several villages in Selo have initiated tourism activities, yet overall development remains fragmented and underutilized due to infrastructural challenges, insufficient promotion, low managerial capacity, and disaster risk exposure [7].

Strategic planning is essential to maximize the existing strengths and address the barriers to sustainable tourism in the area. One practical approach is the SWOT analysis, which examines internal strengths and weaknesses alongside external opportunities and threats. This framework allows for more precise strategy formulation and priority setting when enhanced through quantitative tools such as the IFAS (Internal Factor Analysis Summary) and EFAS (External Factor Analysis Summary). Applying this methodology in the context of Selo can help identify the most appropriate growth model and ensure that tourism contributes to long-term resilience and community well-being [8, 9].

This study aims to identify and analyze the internal and external potentials of six tourist villages in Selo District and formulate development strategies using a SWOT-based IFAS-EFAS approach. The research seeks to provide actionable recommendations for stakeholders—local governments, tourism managers, and communities—to support sustainable rural tourism in disaster-prone mountain regions. By leveraging local assets and aligning them with broader tourism trends and policy support, Selo District can position itself as a model for integrated, community-based tourist village development in Indonesia.

2 Method

This study employed a mixed-methods approach, combining qualitative and quantitative techniques to comprehensively understand the internal and external potentials of tourist villages in Selo District, Central Java. This approach enables the integration of numerical data with contextual insights, essential for formulating strategic development recommendations in tourism planning [10].

The research focused on six designated tourist villages in Selo District: Samiran, Bhumi Damalung, Lencoh, Bumi Klakah, Tomsu, and Arga Kencana. These villages were selected using purposive sampling, targeting key informants such as village tourism managers, local leaders, and community members with in-depth knowledge of the respective tourism areas.

Primary data were collected through direct observation, semi-structured interviews, and documentation. Observations assessed physical facilities and attractions, while interviews provided qualitative insights into community participation, management practices, and challenges. Secondary data were obtained from government reports, tourism office publications, and academic literature relevant to rural tourism development.

The collected data were analyzed using the SWOT analysis framework, supported by the IFAS (Internal Factor Analysis Summary) and EFAS (External Factor Analysis Summary) matrices. Each village was evaluated based on key variables:

- Internal factors: tourist attractions, institutional structure, and supporting facilities.
- External factors: accessibility, promotion, and management environment.

Scoring was applied to assess the weight and rating of each factor, producing composite scores for internal strengths and weaknesses, as well as external opportunities and threats. These values were then plotted in a SWOT quadrant matrix to determine the strategic position of the tourist villages. The resulting coordinates (X = strength minus weakness, Y = opportunity minus threat) informed the recommendation of an appropriate growth strategy (e.g., aggressive, conservative, diversification, or stable growth) based on Rangkuti's [11] model of strategic planning.

This methodological framework was chosen for its ability to combine structured decision-making with participatory insights, making it particularly suitable for complex development contexts such as community-based tourism in rural, disaster-prone areas.

3 Result and Discussion

3.1 Internal Potential and External Potential of Tourist Villages in Selo District, Boyolali Regency

The classification of Tourism Village Potential is divided into three categories: low, medium, and high. A low classification is assigned if the average score is 0.0-1.0, a medium classification if the average score is 1.1-2.0, and a high classification if the average score is 2.1-3.0. Based on this potential classification, it can be observed that the potential of each tourist village falls into one of three categories: low, medium, or high. The following are the overall results of the internal and external potential assessments of tourist villages in Selo District, as presented in the table below.

Table 1. Overall Results of Internal and External Potential Assessment of Tourism Villages in Selo District

Number	Name of Tourist Village	Total Internal and External Potential Scores	Average Score	Classification
1.	Samiran Tourism Village	40	2,67	High
2.	Bhumi Damalung Tourism Village	35	2,33	High
3.	Lencoh Tourism Village	38,3	2,56	High
4.	Bumi Klakah Tourism Village	37,3	2,49	High
5.	Tomsu Tourism Village	32	2,13	High
6.	Arga Kencana Tourism Village	39,3	2,62	High

Based on the average score of each tourist village in Table 1 above, it can be seen that each tourist village has a high potential classification, with an average score ranging from 2.67 to 2.13. Samiran Tourism Village has an average score of 2.67, Bhumi Damalung Tourism Village has an average score of 2.33, Lencoh Tourism Village has an average score of 2.56, Bumi Klakah Tourism Village has an average score of 2.49, Tomsu Tourism Village has an average score of 2.13, and Arga Kencana Tourism Village has an average score of 2.62. Based on the potential classification, a mapping of the internal and external potential points of tourist villages in Selo District, Boyolali Regency, in 2025 can be made, as follows below.

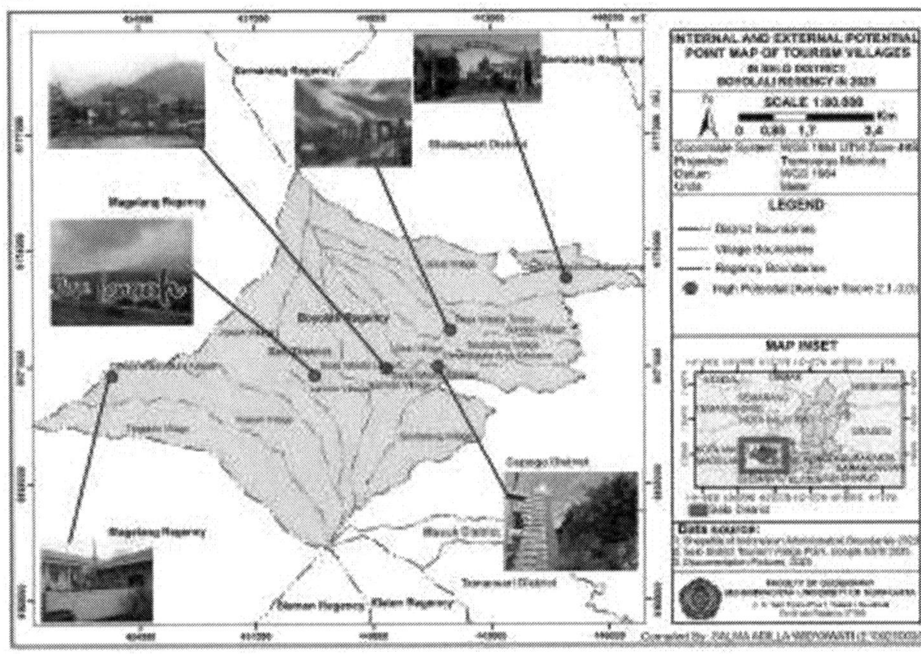

Fig. 1 Map of Internal and External Potential of Tourism Villages in Selo District, Boyolali Regency in 2025.

Tourism Villages in the Selo District have the potential to develop into leading tourism destinations. Based on the results of the observations conducted, the assessment yielded internal and external potential scoring values for the Tourism Villages in Selo District. Internal Potential is assessed through the variables of Tourism Attraction, Institutions, and Supporting Facilities, while External Potential is evaluated based on the variables of Location and Accessibility, Promotion, and Management.

Based on the assessment of tourist villages in Selo District, it can be concluded that all tourist villages evaluated fell into the "High" category for both Internal Potential and External Potential. Although there were variations in the number and average scores, all tourist villages achieved relatively high scores, with an average ranging from 2.13 to 2.67. This reflects that, overall, tourist villages in Selo District possess strong and significant potential for development and are worthy of continued efforts as leading destinations.

The internal potential of tourism villages in the Selo District includes three main aspects: tourism attractions, institutions, and supporting facilities. Based on internal potential, the tourist attractions in these villages are exciting and have great potential for development. This attraction is the main strength, as the tourist villages in the Selo District possess extraordinary natural beauty, being located between Mount Merapi and Mount Merbabu. All tourism villages offer natural tourism that utilizes this potential. In addition to its natural beauty, several types of attractions are very engaging, ranging from natural tourism (Merapi Garden Flower Park, Sanjaya Hill, Nganten Mountain Tourism, Cilenguk Hill, New Selo, Sumber Songo Waterfall, and Gancik Hill Top) to artificial tourism (cow milking, vegetable picking, village tours to UMKM, Bhumi Damalung Camping Ground, Bukit Cilenguk Camping Ground, Polkadot Bridge, Klangi Waterpark, and Embung Manager), as well as special interest tourism (trekking and paragliding), among others. These various types of attractions differ significantly from those in other tourist villages due to their natural wealth and appeal. Beyond natural attractions, the diversity of local culture, arts, and products also enhances their internal potential. In terms of institutions, each tourist village in the Selo District has a Tourism Awareness Group (Pokdarwis), which serves as the main driver and regulates tourism activities in the Selo District tourist villages. Meanwhile, supporting facilities in most of them have begun to develop and are quite complete, such as homestays, places to eat, and other supporting tourism facilities. Samiran Tourist Village, Lencoh Tourist Village, and Arga Kencana Tourist Village have the most comprehensive supporting facilities. In contrast, Bhumi Damalung Tourist Village, Bumi Klakah Tourist Village, and Tomsu Tourist Village still need improvements, particularly for homestay facilities and dining options around the tourist area.

The external potential includes variables of location and accessibility, promotion, and management. Based on external potential, the location and accessibility of tourism villages in Selo District are not far from the capital of Boyolali Regency. The distance from Selo District to the capital of the regency is approximately 21 km and takes around 30-60 minutes. The accessibility itself is generally quite good, but the quality of the roads and directions could probably be improved. In terms of promotion, several tourism villages have utilized social media for marketing; however, they have not been managed optimally and consistently.

Overall, the internal and external potentials of the Tourism Village in Selo District greatly support the development of nature- and culture-based tourism destinations. However, for more optimal development, improvements are needed in the aspects of promotion, management, and enhancement or addition of supporting tourism facilities.

3.2 Development Strategy for Tourism Villages in Selo District, Boyolali Regency

To determine the tourism village development strategy needed in Selo District, we utilized the interview data from the Interview Guide. This tourism village development strategy employs the SWOT analysis method alongside IFAS and EFAS analysis and the SWOT matrix. These internal and external factors are used to analyze the strengths, weaknesses, opportunities, and threats of the Tourism Village in Selo District. The following strategy outlines the development of the Tourism Village in Selo District.

Table 2. SWOT Analysis of Tourism Villages in Selo District, Boyolali

Internal Factors		External Factors	
Strength	**Weakness**	**Opportunity**	**Threat**
• Natural beauty due to its location between Mount Merapi and Merbabu, making nature the main attraction. • Features cultural and religious	• Supporting facilities for tourist villages, such as lodging and homestays, are still very lacking and have not been widely distributed in	• The potential for developing new tourism, such as camping grounds, agrotourism, jeep adventures, revitalization	• Competition from other tourist destinations/tourist villages around Selo District and Boyolali Regency. • Natural disasters such as volcanic eruptions and landslides can

Internal Factors		External Factors	
Strength	**Weakness**	**Opportunity**	**Threat**
tourism attractions. • Offers special interest tourism attractions, including paragliding and trekking. • Contains unique and diverse artificial attractions. • Several tourist villages are well-equipped with complete supporting facilities for tourist destinations. • Availability of local culinary products, special foods, crafts, and souvenirs. • The local community strongly supports the development of tourist villages and destinations. • There is cooperation between the government, private sector, and tourist village managers. • Some residents create crafts, souvenirs, or special culinary delights by utilizing available resources. • The Pokdarwis Organization (Tourism	each tourist village. • Promotion is not optimal and not sustainable. • Community skills in managing tourist villages remain low. • Dependence on the season and weather affects the level of tourist visits and tourism activities. • Limited funds hinder the development of new tourist facilities and attractions in tourist villages. • Tourism management is still not optimal or maximized. • There are no iconic tourist destinations in Selo District to increase tourist visits.	of graves, and educational tourism that utilizes natural resources. • Support from the government and the private sector in facilitating development. • The potential for developing local products and special foods into unique souvenirs and culinary delights. • The potential to create integrated, attractive, and unique tourism packages in each tourist village. • Increasing tourists for nature tourism. • Increasing tourists for unique and authentic rural tourism (tourist villages).	affect the sustainability of tourism. • The risk of visitors engaging in independent tourism without tour packages offered by tourist villages. • Dependence on investors, as some villages are still awaiting agreements with investors, causing delays in tourism activities. • Extreme weather and the rainy season greatly affect tourism activities such as paragliding and influence tourist visits.

Internal Factors		External Factors	
Strength	**Weakness**	**Opportunity**	**Threat**
Awareness Group) plays a role in managing tourist villages.			

Based on Table 2 above, it can be seen that the Tourism Village in Selo District has strengths, weaknesses, opportunities, and threats. The following is a SWOT Analysis with Internal Factor Analysis Summary (IFAS) analysis.

Table 3. SWOT Analysis Summary of Internal Factor Analysis (IFAS)

Strenght	Weight (b)	*Rate* (r)	Score (b×r)	Weakness	Weight (b)	*Rate* (r)	Score (b×r)
Natural beauty due to its location between Mount Merapi and Merbabu, making nature the main attraction.	0,06818	3	0,20455	Supporting facilities for tourist villages, such as lodging and homestays, are still very lacking and have not been widely distributed in each tourist village.	0,06818	3	0,20455
Features cultural and religious tourism attractions.	0,06818	3	0,20455	Promotion is not optimal and not sustainable	0,04545	2	0,09091
Offers special interest tourism attractions, including paragliding and trekking.	0,06818	3	0,20455	Community skills in managing tourist villages remain low.	0,04545	2	0,09091
Contains unique and diverse artificial attractions.	0,06818	3	0,20455	Dependence on the season and weather affects the level of tourist visits and tourism activities.	0,04545	2	0,09091
Several tourist villages are well-equipped with complete supporting facilities for tourist destinations.	0,06818	3	0,20455	Limited funds hinder the development of new tourist facilities and attractions in tourist villages.	0,06818	3	0,20455

https://doi.org/10.1051/e3sconf/202565201004

Strenght	Weight (b)	*Rate* (r)	Score (b×r)	Weakness	Weight (b)	*Rate* (r)	Score (b×r)
Availability of local culinary products, special foods, crafts, and souvenirs	0,04545	2	0,09091	Tourism management is still not optimal or maximized.	0,06818	3	0,20455
The local community strongly supports the development of tourist villages and destinations.	0,04545	2	0,09091	There are no iconic tourist destinations in Selo District to increase tourist visits.	0,06818	3	0,20455
There is cooperation between the government, private sector, and tourist village managers.	0,06818	3	0,20455				
Some residents create crafts, souvenirs, or special culinary delights by utilizing available resources.	0,04545	2	0,09091				
The Pokdarwis Organization (Tourism Awareness Group) plays a role in managing tourist villages.	0,04545	2	0,09091				
Total	**0,59091**	**26**	**1,59091**	**Total**	**0,40909**	**18**	**1,09091**

The IFAS analysis in Table 3 shows that the total scores for strengths and weaknesses are 1.590 and 1.090, respectively, indicating that the total score for strengths is higher than that for weaknesses. Based on the total scores of strengths and weaknesses above, it can be inferred that the total IFAS score is 0.5. This total score is obtained by subtracting the total score of weaknesses from the total score of strengths.

Table 4. SWOT Analysis External Factor Analysis Summary (EFAS)

Opportunity	Weight (b)	*Rate (r)*	Score (b×r)	*Threat*	Weight (b)	*Rate (r)*	Score (b×r)
The potential for developing new tourism, such as camping grounds, agrotourism, jeep adventures, revitalization of graves, and educational tourism that utilizes natural resources.	0,10714	3	0,32143	Competition from other tourist destinations/tourist villages around Selo District and Boyolali Regency.	0,07143	2	0,1428
Support from the government and the private sector in facilitating development.	0,07143	2	0,14286	Natural disasters such as volcanic eruptions and landslides can affect the sustainability of tourism.	0,10714	3	0,32143
The potential for developing local products and special foods into unique souvenirs and culinary delights.	0,07143	2	0,14286	The risk of visitors engaging in independent tourism without tour packages offered by tourist villages.	0,07143	2	0,14286
The potential to create integrated, attractive, and unique tourism packages in each tourist village.	0,10714	3	0,32143	Dependence on investors, as some villages are still awaiting agreements with investors, causing delays in tourism activities.	0,07143	2	0,14286
Increasing tourists for nature tourism.	0,10714	3	0,32143	Extreme weather and the rainy season greatly affect tourism activities such as paragliding and influence tourist visits.	0,10714	3	0,32143
Increasing tourists for unique and authentic rural tourism (tourist villages).	0,10714	3	0,32143				
Total	**0,57143**	**16**	**1,57143**	**Total**	**0,42857**	**12**	**1,07143**

The EFAS analysis in Table 4 shows that the total score of opportunities and threats is 1.571 and 1.071, respectively. This indicates that the total score of opportunities is greater than the total score of threats. Based on the total scores of opportunities and threats above, it can be seen that the total score of EFAS is 0.5. This total score is obtained by subtracting the total score of threats from the total score of opportunities.

Based on the results of the IFAS-EFAS analysis, a quadrant matrix is created, as shown in Figure 3.2. As can be seen in the results of the image below, it indicates that the IFAS and EFAS analyses are at points 0.5 and 0.5. These two points are derived from the calculation results of factors X and Y. The value of factor X represents the factor that forms a horizontal line, and the value of factor X is obtained from the total strength score minus the total weakness score. Meanwhile, the value of factor Y represents the factor that forms a vertical line, and the value of factor Y is obtained from the total opportunity score minus the total threat score, as shown in the following equation:

$$X = Strength - Weakness$$
$$= 1{,}59091 - 1{,}09091$$
$$= 0{,}5$$
$$Y = Opportunity - Threat$$
$$= 1{,}57143 - 1{,}07143$$
$$= 0{,}5$$

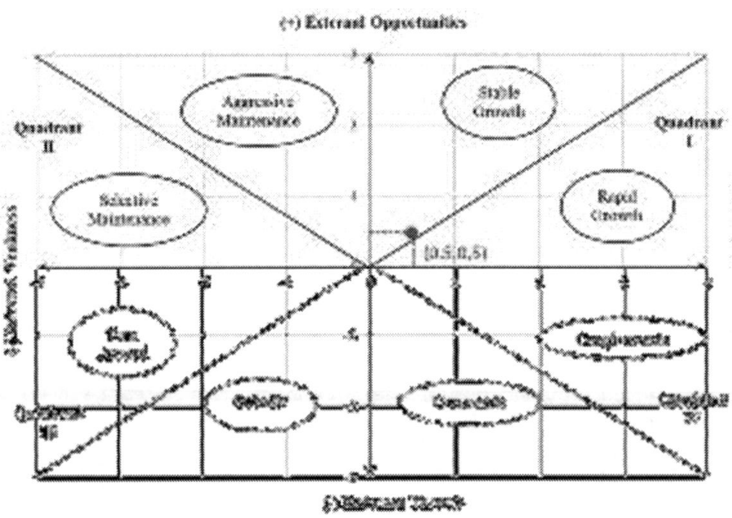

Fig. 2 IFAS EFAS SWOT Analysis Quadrant Matrix for Selo District Tourism Village.

Based on Figure 2 above, points X and Y are located at (0.5, 0.5), indicating that the score for X is 0.5 and for Y is 0.5. These results show that the Selo District Tourism Village is in Quadrant I, in the (Stable Growth) section. According to Rangkuti (2001), Quadrant I describes a very profitable situation due to significant opportunities and strengths. In conditions like this, the right strategy is to implement an aggressive growth policy to maximize existing potential. In Quadrant I, there are two growth strategies: Rapid Growth and Stable Growth. Based on the results of this study, it is situated in the Stable Growth section. According to Mutmainnah Ali et al. (2022), the Stable Growth Strategy is an approach used to maintain and sustain the growth rate that has been achieved previously [12]. According to Murniati (2024), the Stable Growth Strategy is a development approach that is carried out in stages, with target setting adjusted to the existing situation and conditions [13].

These results also support several previous studies that used the SWOT approach and the IFAS-EFAS matrix in formulating tourism development strategies. In comparison, research conducted by Andri Krisna Dianto et al. (2024) on Mondoluku Tourism Village, Gresik, also showed that the tourism village was in Quadrant I with X and Y coordinate scores (0.75; 0.95). The strategy suggested in this study is to collaborate with the tourism office and the surrounding community, utilize information technology, and develop tourist attractions based on local potential. This strategy aims to support village independence by maximizing its strengths and opportunities [14]. Furthermore, another study conducted by Putri Yolanda et al. (2024) on the Harau Valley Nature Tourism Park (TWA) in Lima Puluh Kota Regency also found that the area is in Quadrant I with an IFAS-EFAS score (0.34; 1.24). The recommended strategy is an aggressive one (growth-oriented strategy) by optimizing internal strengths, such as its status as a conservation area and the attraction of waterfalls, as well as utilizing investment opportunities and strategic locations. This study also emphasizes the importance of improving facilities and infrastructure, digitalizing tourism information, and involving the community to realize sustainable tourism management [15].

Based on the overall research above, it can be concluded that the use of a strategic approach through SWOT, IFAS, and EFAS analysis provides a strong foundation for formulating and compiling policies for developing tourist villages. The position of the Tourist Village in Selo District, which is in Quadrant I, shows that this tourist village is in a favorable condition for sustainable development. The appropriate strategy is a stable growth strategy, as it encourages development that is carried out in stages and is planned. This approach also helps maintain a balance between efforts to increase tourism potential and environmental preservation, thus avoiding the risk of damage due to management that is too rapid and unfocused. This strategy is very relevant for promoting the progress of tourist villages without overlooking the sustainability of the resources they possess. The following is a strategy formulation based on the Stable Growth strategy for Tourist Villages in Selo District:

a. Strategy that combines Strength and Opportunity (SO)

Utilizing the potential of the natural beauty and scenery of Selo District as a tourist attraction, developing integrated and attractive tour packages that combine natural beauty, culture, and all types of tourism, including artificial tourism, education, and special interests, to attract visitors. Promoting and developing new tourist attractions such as campgrounds, agrotourism, jeep adventures, glamping, and educational tourism through government and private support. Utilizing the support of local communities and tourism awareness groups to develop agricultural-based educational tourism and local products such as special foods, crafts, and souvenirs as additional attractions.

b. Strategy that combines Strength and Threat (ST)

Creating a community-based tourism disaster mitigation program by leveraging local community support and cooperation with the private sector and government. Establishing iconic tourist destinations or events to differentiate from other tourist villages while strengthening safety and comfort standards to mitigate the impact of uncontrolled independent tourism.

c. Strategy that combines Weakness and Opportunity (WO)

Optimizing digital promotion by recruiting or creating a dedicated team for tourism village promotion as official social media administrators. Utilizing government and private support opportunities for training and funding. Improving community skills through local potential-based tourism training, such as agrotourism and local product education. Developing local products into typical souvenirs to enhance the economy of tourist villages and improve tourism management. Submitting capital assistance programs from the government and private sector to establish tourism support facilities such as homestays, lodgings, and rest areas in tourism villages that lack complete supporting facilities.

d. Strategy that combines Weakness and Threat (WT)

Increasing coordination among tourism villages to build collective strengths in facing competition with other tourism villages and the risks of independent tourism. Diversifying tourism attractions during the rainy season to ensure that tourism activities continue during that period.

4 Conclusion

Based on the assessment of tourist villages in the Selo District, it can be concluded that all tourist villages assessed were classified in the "High" category for both Internal Potential and External Potential. Although there were variations in the number and average scores, all tourist villages received relatively high scores, with an average score ranging from 2.13 to 2.67. Samiran Tourism Village received an average score of 2.67, Bhumi Damalung Tourism Village had an average score of 2.33, Lencoh Tourism Village scored an average of 2.56, Bumi Klakah Tourism Village received an average score of 2.49, Tomsu Tourism Village had an average score of 2.13, and Arga Kencana Tourism Village received an average score of 2.62. This reflects that, overall, tourist villages in the Selo District have strong and even potential for tourism and are worthy of continuous development as leading destinations. The Internal and External potential of tourism villages in the Selo District strongly supports the development of nature- and culture-based tourist destinations. However, for more optimal growth, improvements are needed in aspects of promotion, management, and enhancement or addition of supporting tourism facilities.

Based on the results of the IFAS-EFAS SWOT analysis for points X and Y, namely (0.5, 0.5), these results indicate that the Tourism Villages of Selo District are located in Quadrant I in the Stable Growth section. This means that the Tourism Villages in Selo District possess greater internal strengths than weaknesses and have more external opportunities than the threats faced. Thus, the recommended strategy is a stable growth strategy with an emphasis on gradual but planned development. By implementing strategic measures, it is hoped that the Tourism Villages in Selo District can grow sustainably, remain highly competitive, and contribute significantly to the local economy and the preservation of regional culture.

References

1. Chinawat, Suriya. *"Community Development Initiatives and Sustainable Tourism in Rural Thailand: A Case Study of Chiang Mai." Journal of World Economy* 3, no. 1 (2024): 19-24.
2. Somantri, Lili. *"Spatial Modelling Analysis for Potential Expansion of Cipanas City Candidate Area as Autonomous City."* In *Forum Geografi*, vol. 38, no. 1, pp. 57-73. 2024.
3. Yuda, Radikal, Muhamad Ansori, and Umar Shodiq Al Faruqi. *"Community Resilience through Creative Economy in Indonesia." Indonesia Post-Pandemic Outlook: Rethinking Health and Economics Post-COVID-19* (2022): 185.
4. Amin, Choirul, Priyono Priyono, Umrotun Umrotun, Maulida Fatkhiyah, and Suliadi Firdaus Sufahani. *"Exploring the prevalence of protective measure adoption in mosques during the COVID-19 pandemic in Indonesia." Sustainability* 13, no. 24 (2021): 13927.
5. Chen, Xiuling, and Jie Li. *"Facilitating knowledge-driven economic and social development: The significance of demographic transformation in tourism villages in China." Journal of the Knowledge Economy* 15, no. 3 (2024): 13890-13918.

6. Raihan, A. N., U. Umrotun, and M. Musiyam. *"Analysis of Strategies of Geo-tourism Development in Boyolali, Indonesia."* In *IOP Conference Series: Earth and Environmental Science*, vol. 1357, no. 1, p. 012003. IOP Publishing, 2024.
7. Ait Zamzami, Hamza, Mohammed Elaanzouli, Ayoub Zahrani, Jamila Saidi, Taieb Boumeaza, and Muhammad Musiyam. *"Tourism Investment and Land Use Change in the Northern Piedmont of the Western High Atlas (1989-2022)."* In *Forum Geografi*, vol. 38, no. 1, pp. 83-93. 2024.
8. Jamil, Ainul Huda, Mazzlida Mat Deli, Ummu Ajirah Abdul Rauf, Maryam Jamilah, Suguna Sinniah, and Siti Intan Nurdiana Wong Abdullah. *"Resilience and sustainability in tourism-forestry systems through risk management framework: review and concept." Journal of Law and Sustainable Development* 12, no. 1 (2024).
9. Nurfitri, Ina, Sugeng Utaya, and Dwiyono Hari Utomo. *"Sustainable Tourism Management Strategy at Punaga Beach: A SWOT and AHP Approach for Environmental, Socio-Economic, and Cultural Sustainability." JAMBURA GEO EDUCATION JOURNAL* 6, no. 1 (2025): 37-47.
10. Hu, Haisheng, and Keshuai Xu. *"Visualizing the development of research on tourism resilience with mixed methods." Sage Open* 12, no. 3 (2022): 21582440221126684.
11. Rangkuti, Freddy. *SWOT–Balanced Scorecard*. Gramedia Pustaka Utama, 2013.
12. Ali, Mutmainnah, Nasiah Badwi, Sukri Nyompa, Ramli Umar, and Ichsan Invanni. *"Strategi Pengelolaan Agrowisata Pango-Pango Kabupaten Tana Toraja di Masa Pandemi Covid-19." LaGeografia* 21, no. 1 (2022): 68-79.
13. Murniati, Murniati. *"Model Strategi Peningkatkan Kunjungan Wisatawan dalam industri Pariwisata; Analisis SWOT dan Implementasi Praktis."* YUME: Journal of Management 7, no. 2 (2024): 1089-1100.
14. Dianto, Andri Krisna, Dwiyana Anela Kurniasari, and Wilfridus Romian Marang. *"Perumusan Strategi Pengembangan Desa Wisata Mondoluku Kabupaten Gresik Menuju Kemandirian Desa."* INOVASI: Jurnal Ekonomi, Keuangan, dan Manajemen 20, no. 1 (2024): 179-186.
15. Yolanda, Putri, Rinekso Soekmadi, Nandang Prihadi, and Rizky Pratama Adhi. *"Analisis Keberlanjutan Taman Wisata Alam Lembah Harau di Kabupaten Lima Puluh Kota."* Jurnal Teknologi Lingkungan 25, no. 1 (2024): 048-058.

Ecological Balance Analysis of Batang Toru Forest in North Sumatra After Hydropower Development in 2017 and 2024

Sylvia Nurmalasari[1] *and Rohman Hakim*[2*]

[1]Faculty of Geography, Muhammadiyah University Surakarta, Surakarta 57162, Central Java, Indonesia
[2]Faculty of Geography, Muhammadiyah University Surakarta, Surakarta 57162, Central Java, Indonesia

Abstract. The Batang Toru forest conservation area in North Sumatra, where a 510-Megawatt hydroelectric power plant is being built, is controversial because it is located in a tropical rainforest and is home to the endangered Tapanuli orangutan. The hydropower development, which is considered environmentally friendly and has the potential to reduce energy dependence that produces carbon emissions, has also triggered other concerns, especially regarding the impact of development on forest environments that are vulnerable to environmental damage and potential disasters. The results showed that ecological damage due to development did not affect the health of the river, but showed the impact of a decrease in the population of Tapanuli orangutan species from the original 767 species to 400-600 species and a reduction in forest land cover area of around 76950,63 Ha (31.15%). Thus, the aim of this research is to study the post-construction effect of the Batang Toru hydropower plant on the Batang Toru forest ecosystem, including river health as well as the population of protected orangutan *(Pongo tapanuliensis)* species within it, and to provide management strategies through SWOT analysis to mitigate the impacts of hydropower development.

1 Introduction

The Batang Toru forest conservation zone is a tropical rainforest ecosystem in North Sumatra that possesses a remarkably high degree of biodiversity and ecological roles. This region is home to diverse flora and fauna, including 310 species of birds, 80 species of reptiles, 64 species of amphibians, and hundreds of tree species, with an average height reaching 30 meters. The Tapanuli orangutan species *(Pongo tapanuliensis)* holds significant ecological and conservation importance. This orangutan species, recognized only in 2017[13], is regarded as critically endangered due to its extremely restricted habitat solely in Batang Toru. Consequently, this forest serves as both an ecosystem for biodiversity and a crucial conservation area for the survival of *Pongo tapanuliensis*.

* Corresponding author: rh208@ums.ac.id

The Batang Toru forest is administratively located in three districts: North Tapanuli, Central Tapanuli, and South Tapanuli. Ecologically, it forms part of a tropical rainforest ecosystem characterized by significant rainfall, averaging 2000 mm annually. According to landscape ecology theory, ecosystems of tropical rainforests with intricate vegetation structures are vital for sustaining ecological balance and ensuring habitat connectivity. Should forest fragmentation take place, it will generate edge effects that diminish habitat quality and threaten the survival of endemic species like orangutans.

A potential cause of forest fragmentation in Batang Toru is the establishment of a 510 MW Hydroelectric Power Plant (PLTA) managed by PT North Sumatra Hydro Energy since 2017. This initiative is viewed as a type of renewable energy advancement that has the potential to supply 15% of North Sumatra's electricity requirements[6], and in the framework of ecological modernization, it is frequently regarded as a collaboration between economic development and environmental conservation[9]. Conversely, in reality, extensive hydropower projects in protected regions frequently lead to deforestation, forest degradation, and the loss of biodiversity[2]. In Batang Toru, this initiative is seen as contentious due to its construction requiring the clearing of forested areas and river redirection, which could lead to habitat fragmentation and a decrease in environmental quality.

As a result of the project, various environmental issues have arisen, including diminished forest cover, fragmentation of orangutan habitats, hunting, and the movement of orangutans to agricultural areas and human settlements. Such issues can affect the availability of food and genetic variation for orangutans, thereby elevating their risk of extinction[12]. Moreover, diverting the river for hydropower may disrupt the ecological balance of the Batang Toru River. According to river ecology theory, alterations in river hydrology impact water quality factors like temperature, dissolved oxygen, and substrate conditions, which, in turn, affect the survival of aquatic species[1]. If not managed correctly, this condition may cause lasting ecological disruption

Numerous research works have examined the condition of the Batang Toru ecosystem and the risks posed by hydropower projects[5][10][8]. Nonetheless, the majority of research continues to concentrate on the initial phases of development, whereas extensive studies investigating the effects following construction on forest cover alterations, river health, and the Tapanuli orangutan population remain scarce. Indeed, these studies hold great significance in establishing a scientific foundation for developing suitable management approaches moving forward.

Thus, this study aims to examine the aftereffects of the Batang Toru hydropower plant on the forest ecosystem, focusing on alterations in forest land cover, the state of the Batang Toru River ecosystem, and the population decline of Tapanuli orangutans *(Pongo tapanuliensis)*. Moreover, this research offers management strategies via SWOT analysis to alleviate the adverse effects of hydropower development.

2 Research Method

2.1 Research Area

The research site covers the Batang Toru forest area in North Sumatra Province. Administratively, this area is included in three districts at once, namely North Tapanuli, Central Tapanuli and South Tapanuli. The forest area reaches 120.00 ha and has an astronomical position between 98° 50'-99° 18' East Longitude and 1° 26'-10° 56' North Latitude.

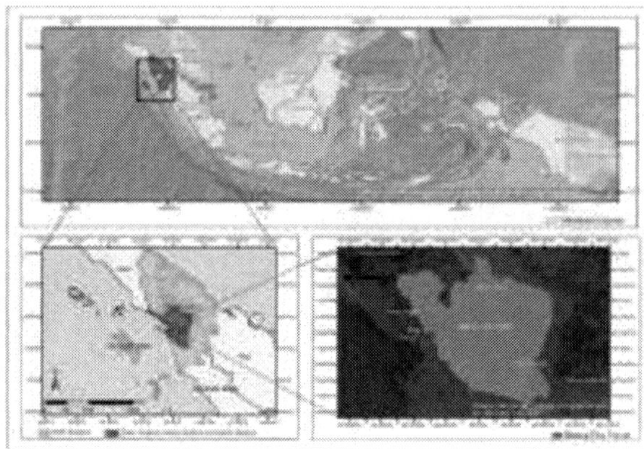

Fig. 1. Administrative map of the Batang Toru forest area.

Source : Author (2024)

2.2 Data Processing and Data Analyse Procedures

All information utilized in this research, encompassing details on alterations in forest cover for 2017 and 2024, river health status, the Tapanuli orangutan population, and SWOT matrix evaluation, will undergo both quantitative and qualitative analysis. Land cover change analysis was conducted using Landsat 8 satellite imagery from the USGS EarthExplorer platform for the years 2017 and 2024. The procedure included image pre-processing through radiometric and atmospheric adjustments to enhance image quality. Land cover classification is carried out using the supervised classification technique with the Maximum Likelihood Classifier (MLC) algorithm, selected for its capacity to account for spectral variability among classes. The subsequent flowchart illustrates the processing of the land cover map for the Batang Toru forest region.

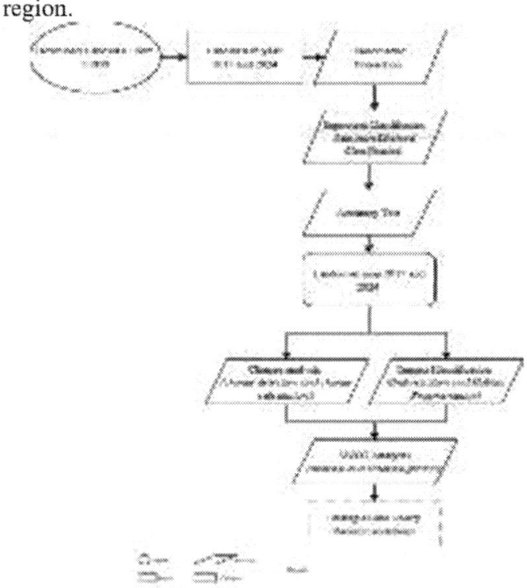

Fig. 2. Research flowchart
Source : Author (2024)

The classification results mentioned need to be validated by conducting an accuracy assessment with a confusion matrix and calculating overall accuracy and the Kappa coefficient. The validation data was derived from the visual analysis of high-resolution images and supplementary field information from official sources. The minimum classification accuracy was established at a threshold of 85% to guarantee the dependability of the findings[3].

The Batang Toru River's water quality was assessed through the Singscore index, an integrative approach for evaluating river ecological conditions based on physical-chemical and biotic factors[10]. This approach was selected because it offers a straightforward and effective quantitative evaluation to characterize the health condition of water bodies. Classification using the Singscore method can be seen below.

$$\textbf{SingScore} = \frac{\sum_{i}^{s}=1\ ai}{s}\ x\ 20 \qquad (1)$$

S = total number of taxa in the sample and ai is the tolerance value for the i-th taxon

Singscore categories are as follows: 0–79 Poor; 80–99 Fair; 100–119 Good; and 120+Very Good

However, the Singscore approach has drawbacks, such as its sensitivity to seasonal changes, constraints of the biological parameters employed, and a tendency to simplify ecosystems into a single score[11]. Consequently, the Singscore results from this study were merged with descriptions of physical-chemical parameters (temperature, pH, dissolved oxygen, substrate) to create a more detailed overview.

A SWOT analysis was performed to pinpoint internal elements (strengths and weaknesses) and external elements (opportunities and threats) that influence the management of the Batang Toru ecosystem following the establishment of the hydroelectric power facility. Factors were identified by integrating literature reviews and analyzing policy documents. Consequently, the SWOT analysis in this research was influenced not only by the researcher's insights but also encompassed various literature reviews pertinent to area management. This procedure sought to enhance the validity of the findings and reduce personal bias. A SWOT matrix was subsequently created to devise mitigation and conservation strategies that were suitable and pertinent to the local setting.

3 Result and Discussion

3.1 Effect of Development on River Health

In addition to affecting changes in forest ecology, including the biology of flora and fauna, the construction of the Batang Toru hydropower plant also affects the quality of water sources in the area. As with land change, human behavior factors greatly influence changes in the watershed environment, including water balance[4].

Due to the development project, many local communities are concerned about fluctuations that could impact water quality along the Batang Toru river. In the Batang Toru forest area, there is a main river that is the largest river with a length of 69.32 km and abundant water availability[5]. This main river has its headwaters in Tarutung, Humbang Hasundutan Regency and flows into the Indonesian Ocean. Based on the pattern of the Batang Toru river, it has a parallel pattern, which is utilized daily by the population for agricultural needs, plantations, and industrial activities.

Based on research conducted by Muhtadi & Leidonald (2020) discussing the status of water quality in Batang Toru, the water quality of the Batang Toru River is good, with parameters that are favorable for aquatic organisms based on water quality descriptions *(in situ measurements).* The table of river water quality status in Batang Toru is calculated through the singscore values displayed in table and bar graph bellow.

Table 1. Status of waters in the Batangtoru river based on singscore values

Location	Singscore (average)	Category
Down stream	84	Fair
Middle stream	92	Fair
Upstream	104	Good
Tributaries	126	Excellent
Sub tributaries	124	Excellent

Source : Muhtadi & Leidonald (2020)

Graphic 1. River water quality values in Batang Toru

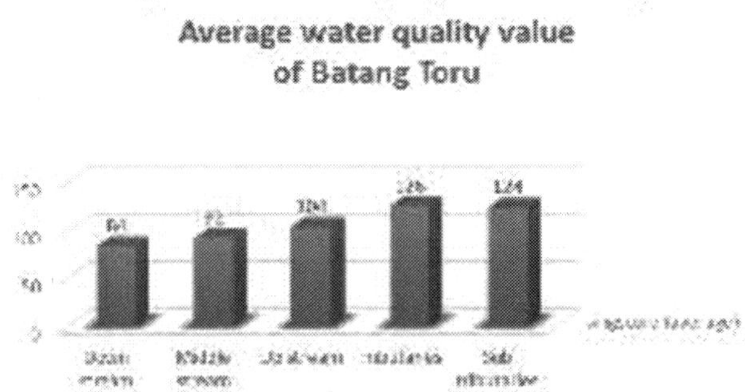

Source : Muhtadi & Leidonald (2020)

Based on singscore calculation above, the condition of the Batang Toru river water has reached its highest score by being obtained in the *"very good"* category. For the category of singscore values in water quality, it is divided into: 0-79 (Poor), 80-99 (Fair), 100-119 (Good), and 120+ (Excellent). Another thing that can strengthen the evidence of these calculations is shown in the results of temperature measurements which range from 20.5°C - 29.0°C. The highest temperature was found in the upstream area with an average temperature of almost 29.0°C. This is due to the influence of surrounding conditions such as land cover and altitude. As for the low temperature range, it has more low tree cover than other areas.

While other measurement results such as pH obtained values ranging from 7.1 - 8.7. This value shows an alkaline concentration due to low organic matter, especially in the upstream area, causing a low level of decomposition of organic matter. Then in the concentration of dissolved oxygen in the Batang Toru river flow, the oxygen concentration ranged from 2.8 - 8.7 mg/L. This value also shows that the oxygen concentration in the Batang Toru river area is quite high in all areas except in the estuary area. This is a result of the strong currents at this location and the low levels of turbidity and flow. As water movement allows oxygen diffusion from air to water, The dissolved oxygen content of flowing water is higher than that of stagnant water.

In addition to these physical-chemical characteristics, other influences are found in the surrounding habitat. Such as substrate conditions and current speed, where the substrate in the Batang Toru river tends to vary from sandy, rocky and gravelly. The upstream area of the river tends to have a rocky substrate with swift water flow conditions, the middle area with a sand and gravel substrate, while the downstream area has a muddy sand substrate. In this case, the condition of the substrate greatly affects the brightness and color of the water.

Observations by Muhtadi & Leidonald (2020) show that waters tend to be murky due to the influence of sandy or muddy substrates.

Based on the description above, the condition of the Batang Toru river water quality shows good quality, so that along the river flow there are also still found river water organisms such as the Macroinvertebrate family types ranging from Parathelphusidae, Heptageniidae, and Perlidae. However, due to the continued development of hydropower plants, there is the potential for changes in forest land use to built-up land. It is inevitable that one day it can pose a risk of a decrease in groundwater quality, this can be influenced by continuous deforestation which can have an impact on soil erosion. So it is necessary to have well-implemented environmental conservation efforts to maintain the river ecosystem as well as its water quality.

However, when examining comparable cases in different areas, hydropower development frequently exhibits similar ecological impact patterns. For instance, building dams on the Mekong River has led to significant alterations in water flow and seasonal flooding patterns, impacting fish migration and decreasing the productivity of fisheries that provide livelihoods for millions[15]. In the Amazon, the building of the Belo Monte dam has led to a decline in water quality and a reduction in aquatic biodiversity in the Xingu River[7]. And then in the Congo Basin, proposals for constructing the Inga III dam have sparked comparable worries regarding the decline of aquatic biodiversity and the displacement of indigenous communities. In contrast, the Batang Toru exhibits a comparable pattern, though on a regional level, where altering the river flow could harm the ecosystem's quality and disrupt the aquatic food chain.

3.2 Forest Landscape Change and Impacts on Tapanuli Orangutans

The design of the Batang Toru hydropower plant has been in the public spotlight, mainly due to issues related to its impact on forest ecosystems that are vulnerable to threats of environmental damage and potential disasters.In addition to affecting the balance of forest ecosystems and their surroundings, other problems also arise, namely related to deforestation and forest fragmentation which can trigger a decline in the orangutan population. In 2017, researchers discovered a new species of tapanuli orangutan *(Pongo tapanuliensis)* which is genetically different from its relatives, namely the Sumatran orangutan and Bornean orangutan[12].

The existence of this new species could be interpreted as an advantage, but it could also be a disadvantage if no conservation efforts are made for this rare species. Recorded, Tapanuli orangutans lived from the highlands of West Lake Toba in North Sumatra to the Bukittinggi region in West Sumatra. However, today they only live in the Batang Toru ecosystem in North Sumatra, mainly in three forest blocks separated by rivers and roads. Most of the Tapanuli orangutan population lives in unprotected forest areas, although they also live in the Lubuk Raya Wildlife Reserve in the south[12]. Below is a map of the Batang Toru forest area, which is home to the Tapanuli orangutan species.

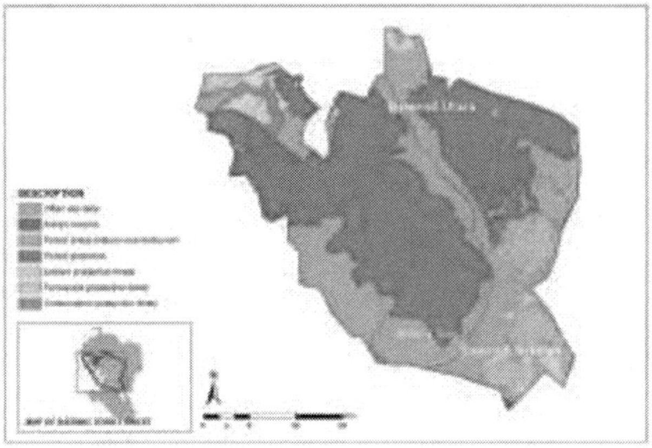

Fig.3. Batang Toru forest zone map
Source : Author (2024)

Land clearing and project development activities have also removed approximately 618 hectares of vegetation cover, consisting of various types of vegetation cover (forests, plantations, and agricultural land)[8]. The effect of reduced land cover due to project development will certainly affect the balance of the forest ecosystem. One of them is the reduction of habitat for the orangutan species Pongo tapanuliensis. This condition can also be seen through the following figure which displays a visualization of the land cover map in the Batang Toru forest area.

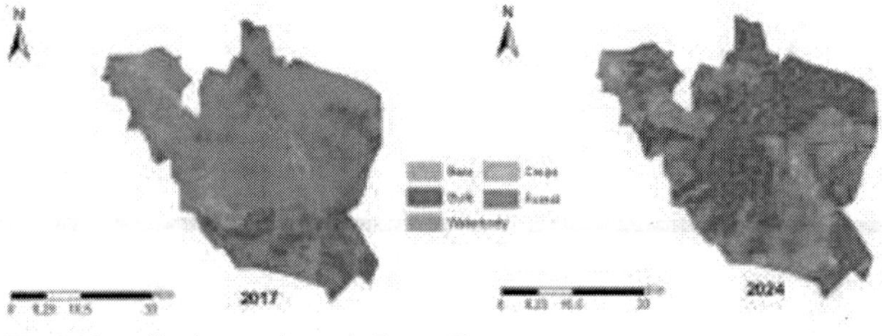

Fig. 4. Map of land cover change in Batang Toru
Source : Author (2024)

Based on the display of the land cover map in the Batang Toru forest area above, it is displayed with different color symbology as information from each land cover, including settlements (red), vacant land (orange), rice fields (light green), waters (blue), and forest (solid green). Through the map, it can be seen that there has been a change in land cover, especially in forest vegetation. The solid green (forest) color in 2017 tends to be more widespread compared to the appearance in 2024. This can also be seen through the following table of land cover area and land cover change in 2017 and 2024.

https://doi.org/10.1051/e3sconf/202565201005

Tabel 2. Land cover area and land cover change in Batang Toru area in 2017 and 2024

Land cover	2017		2024	
	Area (Ha)	Percentage (%)	Area (Ha)	Percentage (%)
Built	26261,5	10,65%	90988,8	36,97%
Crops	13519,19	5,48%	26218,1	10,65%
Forest	197934,43	80,30%	120983,8	49,15%
Waterbody	3914,1	1,58%	703,2	0,28%
Bare	4850,26	1,96%	7210,9	2,93%

Source : Author (2024)

Based on the calculation of the percentage of land cover change in the Batang Toru area above, it shows that there was an 31.15% decrease in land cover area in the forest area. Previously in 2017 the total area of forest reached 197934.43 Ha and decreased seven years later (2024) with a total area of 120983.8 Ha. This change occurred due to a lot of new land clearing, especially the opening of built-up land, including settlements, industries, and project areas such as the Batang Toru hydropower development project. As a result, an area of 6,116 hectares was affected by the development of this project, which is included in the conservation forest area. This shows that project development has a major impact on forest land cover change. In addition, the balance of the local ecosystem can be affected by these changes in land cover, especially with the reduction of orangutan habitat which can threaten the behavioral life of this endemic animal.

In 2017, when this species was declared a rare endemic fauna, research by Wich (2017) stated that the Tapanuli orangutan population was still in the range of 767 individuals whose habitat was spread over 3 blocks including the western block which was home to 581 individuals, the eastern block with 162 individuals, and the Sibual-buali Nature Reserve with 24 individuals. From the distribution of the Tapanuli orangutan population, around 85% are under protection status, but around 15% are in unprotected areas or on land with other designations. However, by 2024, there will only be around 400-600 Tapanuli orangutans in West Batang Toru (including Lubuk Raya Wildlife Reserve and Sibual-buali Wildlife Reserve, which are separated from the West Block and estimated to have a population of less than 50 individuals), and around 150-160 Tapanuli orangutans in East Batang Toru.

The loss of forest cover caused by the construction of a hydroelectric power plant in Batang Toru highlights the worldwide challenge of balancing renewable energy development with biodiversity protection. In the realm of ecological modernization, renewable energy is frequently viewed as a remedy for carbon emissions. Nonetheless, recent research indicates that large dam projects can be considered *"green energy that is not green"* as they lead to the destruction of primary forests, carbon emissions from land degradation, and the extinction of native species[14]. This is clear in Batang Toru, where assertions of clean energy conflict with the reality that the Tapanuli orangutan population has decreased by 40% over the past seven years.

In comparison to the Amazon situation, the habitat destruction for distinct freshwater species caused by hydropower dams generates controversies akin to the habitat loss of orangutans in Batang Toru. In the Mekong, the primary issue concerns not just biodiversity but also the socio-economic reliance of local communities on the river. Simultaneously, in the Congo Basin, political factors and international energy interests contribute to the intricacy of the conflict. Batang Toru exemplifies this worldwide trend, where renewable energy confronts a critical challenge fulfilling energy demands or protecting invaluable tropical biodiversity.

3.3 Development Impact Management Strategy

In an effort to avoid forest destruction and population decline of the Tapanuli orangutan *(Pongo tapanuliensis)*, a management strategy is needed to address the impacts of the Batang Toru hydropower development on the forest ecosystem. The Batang Toru ecosystem is among the most pristine forests in Sumatra, which is also habitat to the critically endangered species Tapanuli orangutan *(Pongo tapanuliensis)* and establishment of a hydropower project that puts it in significant danger. Therefore, in order to maintain the balance of the forest ecosystem and the survival of the Tapanuli orangutan population, a sustainable management strategy is needed. The strategy is assisted by using SWOT matrix analysis.

The SWOT analysis consisting of the development worksheet described above is divided into four parts, namely Strengths, Weaknesses, Opportunities, and Threats. By dividing again based on internal and external factors, to make it easier to take the right strategy. The determination of categories in SWOT analysis such as strengths and weaknesses based on a focus on internal factors, while opportunities and threats are based on the influence of external factors. SWOT analysis serves to illustrate the compatibility between the resources owned (strengths and weaknesses) with environmental conditions (opportunities and threats). This suitability helps in maximizing the utilization of strengths and opportunities, while minimizing weaknesses and threats. The analysis will be described as follows:

Tabel 3. Internal strategic factors

Strengths	Weaknesses
The water quality of the Batang Toru River is relatively good based on Singscore values, especially in tributaries and upstream areas.	Significant deforestation (31.15% decline in forest cover between 2017–2024).
Renewable energy potential that can reduce dependence on fossil-based energy.	Fragmentation of the Tapanuli orangutan habitat and a 40% decline in population in seven years.
The existence of endemic species with high conservation value (Tapanuli orangutans).	Initial management analysis is still predominantly top-down and does not sufficiently involve local communities and indigenous peoples.

Source : Author (2024)

3.3.1 Strengths

The relatively good condition of the river water is obtained through the results of the singscore which shows an *"excellent"* category in sub-tributaries and tributaries and is also supported by physical-chemical conditions that support a healthy aquatic ecosystem. This abundant natural resource potential can help reduce dependence on fossil-based energy use. There is a rare species of Tapanuli orangutan that makes the Batang toru forest area its home. The type of Tapanuli orangutan species is relatively small considering that the fauna is a newly discovered endemic species.

3.3.2 Weaknesses

Deforestation has disrupted the balance of the Batang Toru forest ecosystem, including the reduction of 6,116 hectares (31,15%) of forest vegetation cover that is home to the Tapanuli orangutan. Another impact that can occur is the threat of extinction of the Tapanuli orangutan population. Strategic management of mitigation issues is still dominated by a top-down approach and does not sufficiently involve local communities and indigenous peoples. As a result, existing problems will not be resolved at their root.

In addition to the internal strategic factors above, there are external strategic factors in sustainable management that can be carried out in these problems as can be seen in tabel 4 below.

Tabel 4. External strategic factors

Opportunities	Threats
Conservation development through habitat restoration, wildlife corridors, and forest rehabilitation.	The potential extinction of the Tapanuli orangutan (Pongo tapanuliensis) in the medium term.
Collaboration with research institutions and international NGOs that highlight similar cases.	Social conflicts between local communities, indigenous peoples, the government, and corporations.
Utilization of the area as a center for environmental education and biodiversity-based ecotourism.	The long-term decline in ecosystem functions (water, carbon, and food supply) due to deforestation.

Source : Author (2024)

3.3.3 Opportunities

Conservation development can potentially help in better forest management, especially in habitat restoration and determining special corridors for Tapanuli orangutans, as well as enacting regulations for protected animal protection and ecosystem rehabilitation. Partnerships with research institutions and international NGOs can be made to design effective evidence-based approaches, such as habitat restoration, invasive species management, and ecological engineering. In addition, this partnership can also be a bridge in designing policies that support ecosystem protection based on scientific evidence. A place to develop educational tours related to nature conservation, such as about the importance of protecting rare species and unique ecosystems in the Batang Toru forest.

3.3.4 Threats

The extinction of the endemic species of Tapanuli orangutan due to the reduction of forest area as a habitat for this endemic species has caused a decline in the population of *Pongo tapanuliensis*. The occurrence of conflicts against infrastructure development that occurs between the community and the project manager regarding the need for infrastructure development with nature conservation areas. The consequences of large-scale deforestation can lead to a long-term decline in the ecosystem's function as a provider of water, carbon, and food.

Based on the results of the Strengths, Weaknesses, Opportunities, and Threats (SWOT) analysis, the SWOT matrix analysis is then described which can be used as a reference for determining strategies. The SWOT matrix analysis table can be seen as follows:

Tabel 5. Matriks SWOT

	Strengths (S)	Weaknesses (W)
	The water quality of the Batang Toru River is relatively good based on Singscore values, especially in tributaries and upstream areas. Renewable energy potential that can reduce dependence on fossil-based energy. The existence of endemic species with high conservation value (Tapanuli orangutans).	Significant deforestation (31.15% decline in forest cover between 2017–2024). Fragmentation of the Tapanuli orangutan habitat and a 40% decline in population in seven years. Initial management analysis is still predominantly top-down and does not sufficiently involve local communities and indigenous peoples.

Opportunity (O) Conservation development through habitat restoration, wildlife corridors, and forest rehabilitation. Collaboration with research institutions and international NGOs that highlight similar cases. Utilization of the area as a center for environmental education and biodiversity-based ecotourism.	Combining renewable energy capabilities with creative conservation strategies, like green hydropower designs that take habitat connectivity into account (S1,O1) Leveraging the status of national strategic initiatives to draw international funding for research and conservation efforts (S2,O2) Creating ecotourism centered around Tapanuli orangutans that enhances both conservation efforts and the local economy (S3,O3)	Minimizing habitat fragmentation via scientifically supported ecological restoration (landscape connectivity) (W1,O1) Enhancing the involvement of local communities and indigenous populations in conservation strategies (W2,O2) Involving global organizations to enhance technical skills in biodiversity assessment (W3,O3)
Threats (T) The potential extinction of the Tapanuli orangutan (Pongo tapanuliensis) in the medium term. Social conflicts between local communities, indigenous peoples, the government, and corporations. The long-term decline in ecosystem functions (water, carbon, and food supply) due to deforestation.	Utilizing the conservation significance of the Tapanuli orangutan as a key species to ease possible conflicts with development initiatives (S1,T1) Leveraging the relatively decent water quality of rivers as a key metric in promoting sustainable management (S2,T2) Maximizing renewable energy potential by aligning electricity demands with biodiversity commitments (S3,T3)	Carrying out forest rehabilitation and community-based reforestation initiatives to diminish ecological threats (W1,T1) Creating a multi-party platform (government, community, NGOs, academics) to reduce socio-political tensions (W2,T2) Creating long-term strategies for mitigation that guarantee energy advancement does not compromise tropical biodiversity (W3,T3)

Source : Author (2024)

Based on the results of the SWOT matrix above, there are various strategies that can be carried out based on the opportunities and weaknesses of each analysis result. The description is done by comparing the strengths with opportunities and likewise with weaknesses with opportunities. Furthermore, also on the advantages with threats and weaknesses with threats. So that a good management strategy is produced in monitoring, mitigation, and conservation that can be implemented in a sustainable manner.

4 Conclusion

This research indicates that the development of the Batang Toru hydropower facility has greatly affected the forest and river ecosystems. Examination of images from 2017 to 2024 shows a 31.15% reduction in forest cover, resulting in habitat fragmentation and a 40% decrease in the Tapanuli orangutan population. The state of the Batang Toru River remains fairly good, although water quality has deteriorated in the vicinity of the construction site.

This pattern aligns with the situations of hydropower facilities in the Mekong, Amazon, and Congo, which demonstrate that a significant trade-off exists between supplying renewable energy and biodiversity loss.

Along with its ecological effects, the building of the Batang Toru hydropower facility presents socio-political issues. Local communities, environmental groups, and indigenous populations have risen against national energy policy, often sidelining their perspectives. A SWOT analysis indicates that opportunities like global collaboration, ecotourism, and the incorporation of local knowledge can serve as mitigation strategies. Consequently, the Batang Toru situation illustrates both regional concerns and worldwide difficulties in achieving harmony between energy transition and the preservation of tropical biodiversity.

References

1. Allan, J. D., & Castillo, M. M. (2007). Stream Ecology: Structure and Function of Running Waters. Springer. https://doi.org/10.1007/978-1-4020-5583-6
2. Bologna, M., & Aquino, G. (2020). Deforestation and world population sustainability: a quantitative analysis. Scientific Reports, **Vol10(1)**, 7631. https://doi.org/10.1038/s41598-020-63657-6
3. Congalton, R. G., & Green, K. (2019). *Assessing the accuracy of remotely sensed data: principles and practices*. CRC press.
4. Hakim, R. (2024). Analysis of The Influence of Environmental Behavior on the Water Balance of Kaligarang Watershed, Central Java. *International Journal for Disaster and Development Interface*,**Vol4(1)**, 69–81. https://doi.org/10.53824/ijddi.v4i1.62
5. Hernaningsih, T. (2020). Analisis Kualitas Air Di Ruas Sungai Batang Toru Dengan Metode Storet Dan Indeks Pencemaran. *Jurnal Rekayasa Lingkungan*, **Vol*13*(2)**, 138-151.
6. KPPIP. (2023). Groundbreaking PLTA Mentarang Induk oleh Presiden, Wujud Komitmen Pemerintah untuk Menghadirkan Energi Hijau. https://kppip.go.id
7. Lees, A. C., Peres, C. A., Fearnside, P. M., Schneider, M., & Zuanon, J. A. S. (2016). Hydropower and the future of Amazonian biodiversity. Biodiversity and Conservation, **Vol25(3)**, 451–466. https://doi.org/10.1007/s10531-016-1072-3
8. Meijaard, E., Ni'matullah, S., Dennis, R., Sherman, J., Onrizal, & Wich, S. A. (2021). The Historical Range and Drivers of Decline of the Tapanuli Orangutan. PLoS One, **Vol16(1)**. https://doi.org/10.1371/journal.pone.0238087
9. Mol, A. P. J., & Sonnenfeld, D. A. (2000). Ecological Modernisation Around the World: Perspectives and Critical Debates. Routledge.
10. Muhtadi, A., & Leidonald, R. (2020). Habitat characteristics and water quality status in the Batangtoru Watershed, North Sumatra Province, Indonesia. IOP Conf. Series: Earth and Environmental Science, **Vol454(1)**, 012092. https://doi.org/10.1088/1755-1315/454/1/012092
11. Novita, S., Yuliana, N., & Fadilah, R. (2022). Evaluasi indeks kualitas air berbasis biotik di ekosistem sungai tropis. *Jurnal Biologi Tropis*, **Vol22(1)**, 33–42. https://doi.org/10.29303/jbt.v22i1.2001
12. Prasetyo, D., Hadiprakarsa, Y., Kuswanda, W., & Sugardjito, J. (2021). Population status of Tapanuli Orangutan (Pongo tapanuliensis) within the renewable energy development and its management implications. Forest and Society,**Vol5(2)**, 478–493. https://doi.org/10.24259/fs.v5i2.13529.

13. Wich, S. A., Fredriksson, G., Usher, G., Kühl, H. S., & Nowak, M. G. (2019). The Tapanuli orangutan: Status, threats, and steps for improved conservation. Conservation Science and Practice, **Vol1(6)**, e33. https://doi.org/10.1111/csp2.33
14. Winemiller, K. O., McIntyre, P. B., Castello, L., Fluet-Chouinard, E., Giarrizzo, T., Nam, S., ... & Sáenz, L. (2016). Balancing hydropower and biodiversity in the Amazon, Congo, and Mekong. Science,**Vol351(6269)**, 128–129. https://doi.org/10.1126/science.aac7082
15. Ziv, G., Baran, E., Nam, S., Rodríguez-Iturbe, I., & Levin, S. A. (2012). Trading-off fish biodiversity, food security, and hydropower in the Mekong River Basin. Proceedings of the National Academy of Sciences, **Vol109(15)**, 5609–5614. https://doi.org/10.1073/pnas.1201423109

Economic Circular through Waste Management Programs: A study in Giwangan Village, Yogyakarta

Bernadus Tresno Sumbodo[1], *Siti Rochmah* Ika[2*], *Sardi* Sardi[3], *Fitri* Nuryanti[4], *Ignatius Suprih* Sudrajat[5], and *Ari Kuncara* Widagdo[6]

[1,2,3,4]Janabadra University, Yogyakarta, Indonesia
[5]Sarjanawiyata Tamansiswa University, Yogyakarta, Indonesia
[6]Sebelas Maret University, Surakarta, Indonesia

Abstract. This study documents a village-level application of the triple helix model for urban waste management after the permanent closure of the Piyungan Landfill in Yogyakarta, Indonesia. It is among the first studies to explore how waste management programs at the village scale, such as biopore composting, longan tree planting, maggot cultivation, and urban fish farming, can contribute to the circular economy. Using a participatory action research approach with 60 participants from four neighborhood associations, the results show that nearly 100% of organic waste was successfully managed within the community. The programs produced 250 kg of compost, 50 kg of catfish, and 80 kg of tilapia in the first harvest. Furthermore, each tree yielded 85 kg of longans, benefiting the local community. These findings underscore the actual feasibility of community-based circular economy models and provide answers for other urban villages encountering similar waste management issues. The study indicates that combining biological waste solutions with community-driven efforts may serve as a sustainable approach to urban trash management, decreasing reliance on landfills and yielding economic advantages. This model could be scaled and adapted to other urban areas across Indonesia and globally to foster more sustainable, waste-conscious communities.

1 Introduction

Urban waste management poses a growing concern in Indonesian cities, especially in Yogyakarta, where high population density and business activity exacerbate the environmental and health consequences of inadequate trash disposal. The permanent closure of the Piyungan Landfill in April 2024 signified a crucial turning point for the region's waste management system, necessitating a shift from centralized disposal to decentralized, community-oriented strategies [1, 2]. Despite municipal regulations prioritizing sorting, source processing, and autonomous district management, there is a paucity of research

* Corresponding author: ika@janabadra.ac.id

investigating how village-level efforts might implement circular economy concepts in post-landfill urban environments [3, 4].

Giwangan Village, situated in the southern region of Yogyakarta City and featuring essential infrastructures like the City Main Market and Bus Terminal, presents a noteworthy case study for analyzing such projects. The village consists of seven administrative divisions, 13 citizen associations (RW), and 44 neighborhood units (RT), housing roughly 7,677 persons as of 2017 [5]. The density and complexity of socio-economic circumstances in Giwangan underscore the necessity for adaptable, community-oriented waste management strategies that incorporate environmental, social, and economic factors.

This research utilizes the triple helix concept, highlighting the relationship between government, community organizations, and academic institutions [6]. The village government in Giwangan supplied policy direction and infrastructure assistance, the community contributed labor and indigenous knowledge for waste management, and universities provided technical expertise, training, and program assessment. Early introduction of the triple helix framework is essential, as it illustrates the potential for coordinated multi-sectoral efforts to realize sustainable and scalable circular economy practices at the village level [7].

The activities in Giwangan encompassed biopore composting, longan tree cultivation, maggot farming, and urban aquaculture. These operations were intended to convert organic waste into economically profitable and environmentally friendly products. The initiatives associated waste reduction with income-producing activities, thereby decreasing household organic waste, improving food security, and generating supplementary livelihoods for local residents [4,8].

Data, both quantitative and qualitative, gathered from 60 participants across four neighborhood organizations using participatory action research (PAR) approach. The results indicated that the efforts effectively lowered nearly 100% of household organic waste. The projects produced measurable economic outputs, comprising 250 kg of compost, 50 kg of catfish, and 80 kg of tilapia in the initial harvest, in addition to 85 kg of longans per tree allocated to locals. The results demonstrate that village-level, triple helix-supported initiatives can yield both environmental and economic benefits, highlighting the possibility of decentralized waste management in urban settings.

This study offers a reproducible blueprint for other urban villages facing post-landfill waste concerns by documenting the execution and results of these activities. The Giwangan framework provides actionable insights for policymakers and practitioners, indicating that the integration of triple helix collaboration with community-based circular economy initiatives can diminish landfill reliance, augment economic advantages, and foster environmental sustainability in comparable urban settings in Indonesia and beyond [9].

2 Material and methods

2.1 Method

This study utilized a PAR methodology, which prioritizes the collaboration between researchers and community members to discern issues, execute interventions, and assess results. PAR was chosen because the study seeks to evaluate both the results of organic waste management initiatives and the methodologies by which community members engage as co-researchers. The approach is in line with the triple helix model, which combines input from government, academia, and community members to make interventions that are both long-lasting and able to grow [10]. The Giwangan Village Government gave policy advice and built infrastructure. Academic institutions gave technical help and training. Community groups and residents provided human resource, local knowledge, and ongoing feedback.

2.2 Participants

The study included 60 residents from four neighborhood associations (RW 6, 8, 10, and 11) in Giwangan Village, Yogyakarta, Indonesia. Participants were chosen based on their current participation or prospective involvement in the organic waste management initiatives. Janabadra University's Institutional Review Board gave its ethical approval, and all participants gave their informed consent before the study began. Participants were apprised of the research objectives, methodologies, potential risks, and benefits, with an emphasis on their voluntary participation.

2.3 Timeline and intervention

The PAR process took place over four stages over the course of two years:

Stage 1 (May 2023): First Survey – Quantitative surveys were conducted to evaluate baseline household waste management practices, awareness, and perceptions. This step showed where the analysis was lacking and helped plan the interventions.

Stage 2 (June 2023 – February 2024): Putting Biopore Composting and Longan Tree Planting into Action—Community members helped build biopore holes, learn how to compost, and plant longan trees. The first compost and longan harvests took place at the end of this stage.

Stage 3 (March 2024 to February 2025): Raising maggots and fish in cities People learned how to grow black soldier fly larvae (BSF) to feed catfish and tilapia in ponds. This stage included turning organic waste into aquaculture systems and keeping an eye on the first harvests.

Stage 4 (March 2025 – August 2025): Evaluation – Surveys and qualitative observations done after the interventions looked at how well they worked to cut down on organic waste, boost the economy, and get people more involved in their communities.

2.4 Collecting and analysing the data

Structured household surveys were used to collect quantitative data at the beginning (Stage 1) and after the program was put into action (Stage 4). The data comprised the quantity and classification of household waste, levels of participation, and perceptions regarding environmental and economic advantages. We used descriptive statistics (percentages, frequencies, and means) to sum up these numbers.

Qualitative data were collected through participant observation, focus group discussions, and open-ended survey questions to document residents' experiences and involvement with the interventions. Responses were coded thematically to find patterns in community participation, perceived benefits, and barriers to implementation.

3 Results and discussion

3.1 Initial survey

Table 1 presents the results of the initial survey of the 60 respondents regarding their household waste and how to handle their waste. The survey was held in the first two weeks

https://doi.org/10.1051/e3sconf/202565202001

of May 2023. The questionnaires were designed to portray the existing condition before the waste management activities were given to the Giwangan village society.

Table 1. Descriptive statistics of the initial survey

Questions	N	Answers	Percent (%)
On a household scale, what is the most common kind of waste?	60	Organic waste and food waste	55
		Everything derived from plastic, including plastic packaging and products	41
		Paper and paper products	4
Is the closure of the Piyugan landfill leaving you uncertain about the future of domestic trash management?	60	Yes	75
		No	25
How have you managed waste so far (before the last closure of Piyungan TPST)?	60	Transported by garbage personnels	53
		It has been sorted according to the type of waste	38
		Placed at the nearest waste depot	9
In your opinion, what is the problem of not managing household waste?	60	Do not know how to manage waste	48
		Do not have time to take care of the trash	24
		Lazy and very troublesome	28

Table 1 presents the results of the initial survey among residents in Giwangan Village. According to respondents, the highest waste generated by households was organic waste and food scraps (55%). When asked about the reasons for not managing their waste, resulting in increased household waste generation, the most common response was not knowing how to manage it (48%). As depicted in Table 1, residents' awareness of independent waste management remains quite low. An initial survey conducted among 60 residents in Giwangan Village showed that 75% of respondents were confused about how to manage their waste after the Piyungan Landfill closed, as 53% of respondents had simply dumped their trash and had it immediately picked up by garbage collectors.

Based on the initial survey results above, it can be concluded that even the majority of residents in villages that already have an environmentally friendly vision are still confused or unaware of how to manage their waste following the closure of the Piyungan Landfill. The initial survey results provide a general overview of the conditions in urban villages in Yogyakarta City. These baseline findings emphasized the significance of community education and engagement for successful implementation and guided the development of focused, participatory interventions, including urban aquaculture, biopore composting, and maggot cultivation.

3.2 Biopore, composting, and planting longan trees.

The Giwangan Village waste management project combines longan tree planting, home composting, and biopore construction as supplementary methods to manage organic waste and produce financial and environmental advantages.

To improve soil infiltration and hasten the breakdown of organic waste, households installed biopore holes. Through these biopores, organic waste from homes—mainly market leftovers and kitchen scraps—was turned into compost. On average, each household produced 250 kg of compost annually, which, if sold at local market prices (IDR 3,000 per kg), would equal IDR 750,000 per household. Approximately 30% of households sold compost, generating direct revenue, while the remaining 30% represents potential economic benefits. The majority of compost was used in home gardens or public green areas.

In order to produce fruits for domestic use, compostable leaves, and a greener urban landscape, longan trees were planted. Each tree produced an average of 85 kg of longan fruit per year by the third year, which was worth IDR 1,905,000 on the market. The total yearly economic benefits for households with three to five trees ranged from IDR 5,715,000 to IDR 9,525,000 per household. About 40% of the harvest was sold, representing realized income, and the remaining 60% was consumed at home (food security benefit).

Combining the yields of longans and compost, the estimated total annualized economic value per household is IDR 7,215,000–12,225,000, which is made up of IDR 5,715,000–9,525,000 from longan fruit (partially realized) and IDR 1,500,000–2,700,000 from compost (partially realized). These numbers demonstrate how organic waste programs at the village level have the potential to provide significant economic benefits at the household level in addition to environmental and social advantages.

In addition to their financial benefits, these interventions increase soil fertility, lessen the need for chemical fertilizers, and promote urban greening while encouraging environmental consciousness and community involvement. The Giwangan model shows how integrated biopore, composting, and urban tree cultivation can provide observable household-level circular economy benefits in urban environments by offering both actualized and potential economic outcomes. During the first harvest, the harvested longans were not sold but distributed evenly among residents of Giwangan Village. Figure 1 portrays a mass longan harvest attended by the Deputy Governor of DIY and the Mayor of Yogyakarta City and the longan trees near Village Giwangan Office.

Fig. 1. Mass harvest longan trees and the longan trees near Village Giwangan office.

3.3 Maggot cultivation and urban fish farming.

Geographically, Giwangan Village is home to the Bendung Lepen (river dam), located in Mrican Village. The Bendung Lepen was once the primary irrigation channel from the Gajah Wong River, irrigating agricultural land. It was in a dilapidated condition, littered with garbage, and smelled foul, even serving as a pigsty. Local residents and youth then took the initiative to clean it up and transform it into a beautiful tourist spot. The cleanup was carried out in stages, and since 2022, the Bandung Lepen has become a clean and attractive place, with ornamental fish swimming in the clear water.

Fig. 2. Bendung Lepen irrigation fisheries

In the second year, organic waste management activities included maggot cultivation and its integration into feed for urban fisheries. The community service program built two maggot houses, five catfish ponds in RW 8 and 10, and revitalized three tilapia ponds in the Bendung Lepen irrigation area. Maggot cultivation education and implementation began in the first year, but its integration with urban fisheries began in the second year. Fisheries in Bendung Lepen are based on the concept of raising fish in a tourist village.

The economic benefits of maggot cultivation and urban fisheries are calculated as follows. Maggots produce an average of 12 kilograms per day at a price of IDR 9,000 per kilogram. This quantity of maggots is used as catfish feed. Tilapia feed at Bendung Lepen is supplied by waste feed from the Environmental Agency (DLH) plus pellet feed from tourists. The first catfish harvest took place in April 2025, with yields from the first two ponds totaling 50 kilograms. At a price of IDR19,000 per kilogram, the potential income is IDR 950,000.

Meanwhile, the tilapia harvest at Bendung Lepen from the first two ponds reached 80 kilograms. If the price of tilapia is IDR 21,000 per kilogram, the potential income reaches Rp 1,680,000. Similar to longans, the first harvest is not yet sold but is distributed to residents who have actively participated in the organic waste management program. Figure 2 depicts the fisheries in Bendung Lepen.

3.4 Evaluation activities and post-survey.

Each year, organic waste management activities are evaluated by distributing questionnaires to participants. The results of the activity evaluation survey are shown in Table 2. As seen in the table, all respondents agreed to participate in the organic waste reduction program, including their family members. All respondents agreed that organic waste reduction infrastructure facilities such as maggot houses, biopore holes, fish ponds, and supplies of plant and fish seeds were provided on time and ready to use. There was an increase in the level of organic waste reduction from the first year, the majority at the level of >80%-90%, to nearly 100%. An increase in the perception of the program's usefulness also occurred in environmental benefits, from 85% to 95%, while the perception of economic benefits increased from 80% to 90%. In general, the organic waste reduction program was perceived as beneficial both economically and environmentally.

There are a number of reasons why the adoption rate is so high. First, the triple helix framework made it easy to work together. The government gave policy advice and built infrastructure, academic partners trained people and shared their technical knowledge, and community groups got people involved, which built trust and helped people learn from each other. Second, the economic incentives, such the chance to make money by selling compost, harvested longans, or fish, made families more likely to keep doing it. Third, the interventions used social capital by making it easier for people to participate through community networks, peer learning, and recognition at the neighborhood level for being engaged.

Despite these achievements, a number of drawbacks were noted. Particularly after initial funding, academic facilitation, and government support diminish or end, the long-term viability of these programs is still in doubt. To ensure that village-scale circular economy initiatives continue, future research should concentrate on strategies for maintaining participation, obtaining long-term funding, and creating independent community governance structures.

4 Conclusion

This study shows how community-based circular economy projects can successfully turn organic waste into useful products at the village level when they are carried out through a triple helix partnership between the local government, academic institutions, and community members. In addition to reducing household organic waste, the Giwangan Village case demonstrates how biopore composting, longan tree cultivation, maggot farming, and urban aquaculture also contribute to economic growth, environmental sustainability, and increased community involvement.

By describing how village-scale circular economy interventions can be operationalized in post-landfill urban settings, this study closes a significant scientific gap and offers empirical support for the triple helix framework's efficacy in developing nations. In practice, it provides a template for local and village leaders looking to decentralize waste management while encouraging sustainable livelihoods and community empowerment.

In the future, other urban villages in Indonesia and comparable international contexts can use the Giwangan model as a scalable and flexible framework. Other communities can adopt sustainable, participatory waste management programs that concurrently address social cohesion, economic development, and environmental challenges by adopting the integrated triple helix approach.

Table 2. Descriptive statistics of post-surveys.

Questions	**First year project**		**Second year project**	
	Answer	Percent	Answer	Percent
The presenters of the specific waste management training provide a clear explanation.	Yes	100	Yes	100
	No	0	No	0
Upon completion of the waste management course, I intend to actively engage in and put into practice the strategies taught.	Yes	100	Yes	100
	No	0	No	0
The infrastructure of waste management program is provided on time and ready to use	Yes	100	Yes	100
	No	0	No	0
What proportion of household organic waste has been diminished by this year's activities?	Nearly 100%	20	Nearly 100%	75
	>80%-90%	67	>80%-90%	20
	>70-80%	13	>70-80%	5
	>50-70%	0	>50-70%	0
My family engages in waste management by disposing of organic trash in the nearest biopore or maggot site.	Strongly agree	80	Strongly agree	90
	Agree	20	Agree	10
	Disagree		Disagree	
	Strongly disagree		Strongly disagree	
Waste management programs are environmentally beneficial	Strongly agree	85	Strongly agree	95
	Agree	15	Agree	5
	Disagree		Disagree	
	Strongly disagree		Strongly disagree	
Waste management programs are economically beneficial	Strongly agree	80	Strongly agree	90
	Agree	20	Agree	10
	Disagree		Disagree	
	Strongly disagree		Strongly disagree	

Acknowledgements

This study was funded by the Ministry of Research, Technology and Higher Education under the Regional-Based Community Service (PbW) Scheme year 2025.

References

1. Humas DIY, Pemda DIY Resmi Tutup TPA Piyungan, Berita Pemerintah Daerah Istimewa Yogyakarta. Accessed: Apr. 20 (2024)
2. D. Agus, DetikJogja (202 D. Agus, Alasan Pemda DIY Tutup Permanen TPA Piyungan, DetikJogja. Accessed: Apr. 20 (2024)
3. A. Oviedo, M. Mora, S. Ponsa, and J. Colón, Techno-economic evaluation of decentralized community composting as a management model to valorize organic matter in small and medium-sized towns, J. Mater. Cycles Waste Manag. (2025) https://doi.org/10.1007/s10163-025-02340-2
4. B. T. Sumbodo, A. N. Ma'rifah, S. Sardi, P. Setiawan, I. S. Sudrajat, A.K. Widagdo, A. N. Salsabila, and S. R. Ika, Waste Management. in an Urban Village through Biopore Organic Waste and Composting Education: A Lesson from Giwangan Village, Yogyakarta, in *Int. Conf. Manag. Muamalah*, Bangi, Malaysia, September 6, (2023)
5. Jogjakota.go.id, Profil Kelurahan Giwangan (2018)
6. H. Etzkowitz, *The Triple Helix: University–Industry–Government Innovation in Action* (Routledge, London, 2008)
7. M. Espuny, J. S. da M. Reis, E. C. B. Giupponi, A. B. T. Rocha, A. C. F. Costa, C. F. Poltronieri, and O. J. de Oliveira, Techno-economic evaluation of decentralized community composting as a management model to valorize organic matter in small and medium-sized towns, Recycling **10**, 1 (2025) 10.3390/recycling10020050
8. B. T. Sumbodo, Sardi, S. Raharjo, H. Prasetyanto, and S. R. Ika, Urban farmer communities empowerment through the climate village program in Sleman, Yogyakarta, IOP Conf. Ser. Earth Environ. Sci. **824**, 012116 (2021) 10.1088/1755-1315/824/1/012116
9. E. Mulyadi, D. Valeriani, and E. Fitriyanti, Implementation of the Triple Helix Model in Increasing Economic Opportunities for Waste Management in Pangkalpinang City, Equity J. Ekon. **12**, 39 (2024) 10.33019/equity.v12i1.258
10. H. Etzkowitz and L. Leydesdorff, The transformation of university-industry-government relations, Electron. J. Sociol. **5** (2001)

Green Pioneer Program: Promoting ESD Through Maggot-Based Waste Management and Responsible Consumption to Enhance Students' Sustainability Awareness at SMP Laboratorium UM in Indonesia

Heru Setiawan[1], *Fatchur* Rohman[2] *Frida Kunti* Setiowati[3], *Murni Sapta* Sari[4], *Bagus* Priambodo[5] *St* Muriati[6]

[1,6] Doctoral Student, Department of Biology, FMIPA, Universitas Negeri Malang, Indonesia
[2,3,4,5] Lecturer, Department of Biology, FMIPA, Universitas Negeri Malang, Indonesia

Abstract. SMP Laboratorium Universitas Negeri Malang, as a partner school of Universitas Negeri Malang, holds strong potential to serve as a model for sustainability education. Nevertheless, students and teachers still demonstrate limited understanding of key sustainability concepts, especially those related to SDG 12 (responsible consumption and production) and SDG 15 (life on land). To address this gap, this study implemented the Green Pioneer Program as an Education for Sustainable Development (ESD) initiative. The program was conducted from February to July 2025 and consisted of three main activities: an ESD seminar, a rried out from February to July 2025 and consisted of three main activities: an ESD seminar, Black Soldier Fly (BSF) maggot-based organic waste management training, and responsible consumption practices. Data were collected using surveys, observations, student reflections, and stakeholder feedback, and were analyzed thematically, supplemented by an independent sample t-test. The findings indicate that the Green Pioneer Program was successfully implemented, providing students with meaningful ecological knowledge and practical skills in sustainable practices. The program received positive responses from school leaders, teachers, and students. Although the t-test revealed no statistically significant difference in sustainability awareness between male and female students, participants demonstrated a relatively high frequency of sustainability practices following the program.

* Corresponding author: fatchur.rohman.fmipa@um.ac.id

1 Introduction

Environmental issues in Malang City, Indonesia, particularly related to domestic waste and pollution, have become increasingly urgent. The city generates approximately 620 tons of waste daily, with nearly 30% poorly managed [1]. Community engagement in 3R (Reduce, Reuse, Recycle) practices remains low [2], and widespread reliance on single-use plastics reflects limited awareness of environmental conservation and sustainable living [3]. This condition stems largely from unsustainable lifestyle patterns and low levels of environmental knowledge and sustainability awareness among citizens, including school-aged children [4][5][6].

SMP Laboratorium Universitas Negeri Malang (UM), a partner school of the university, presents a strategic opportunity to implement contextual sustainability education. The school has adopted the Pancasila Student Profile Project (P5) with the theme "Sustainable Lifestyle" under the Merdeka Curriculum. However, initial observations reveal limited understanding among students and teachers regarding sustainability concepts, particularly related to responsible consumption (SDG 12) and terrestrial conservation (SDG 15). Existing ESD- related programs at the school tend to focus narrowly on climate change, with limited project- based activities that yield tangible ecological impacts. In addition, organic waste—primarily leaf litter and food waste—makes up 60% of the school's daily waste, yet is largely unmanaged and disposed of without sorting, posing environmental and health concerns [7]. The survey was administered to 120 students from grades VII–IX at SMP Laboratorium Universitas Negeri Malang with the objective of assessing their understanding and practices related to sustainable lifestyles. The instrument consisted of 20 items developed from indicators of SDG 12 (responsible consumption and production) and SDG 15 (life on land), covering dimensions such as waste management, responsible consumption, and biodiversity conservation. Responses were collected using a five-point Likert scale and analyzed descriptively to identify students' baseline sustainability awareness and practices. The results indicate that students' overall understanding remains limited and their sustainable lifestyle practices are still inconsistent, thereby justifying the need for the Green Pioneer Program.

Addressing these challenges requires a contextual, action-oriented educational model. Education for Sustainable Development (ESD) equips learners with the knowledge, values, and skills needed to promote sustainable and equitable futures [8]. UNESCO has emphasized the role of ESD in achieving the Sustainable Development Goals since 1992 [9]. Integrating ESD into the P5 curriculum can support both environmental education and behavior change [10]. In response, the Green Pioneer Program was designed as a community engagement initiative that integrates Education for Sustainable Development (ESD) into school-based activities. The program aimed to strengthen students' ecological literacy, develop practical problem-solving skills, and promote sustainable behaviors aligned with SDG 12 (responsible consumption and production) and SDG 15 (life on land).

The program was implemented from February to July 2025 in three sequential stages. First, an ESD seminar introduced students and teachers to the concept of sustainability, responsible consumption, and ecological challenges faced by the local community. Second, students participated in hands-on training on Black Soldier Fly (BSF) maggot-based organic waste management, where they learned to cultivate BSF larvae, process food waste into animal feed, and reduce organic waste at school. Third, the program emphasized responsible consumption practices, including waste reduction campaigns, reflection sessions on daily consumption habits, and student-led initiatives to promote sustainable lifestyles within the school community. BSF larvae can efficiently decompose organic waste and generate high-value byproducts such as compost and animal feed [11][12]. The program empowers students to develop hands-on skills in waste management, build responsible consumption habits, and foster environmental empathy [13][14].

This study had two main objectives. First, it aimed to implement and evaluate the Green Pioneer Program as an Education for Sustainable Development (ESD) initiative that integrates experiential learning through maggot-based organic waste management and responsible consumption practices. Second, it sought to examine the extent to which the program enhances junior secondary school students' sustainability awareness and daily sustainable practices. More broadly, the study contributes to strengthening sustainability education at the junior secondary level and reinforces the university's role in fostering community-based environmental solutions.

2 Method

2.1 Research Design and Approach

This study employed a mixed-methods embedded design [15] integrated with a community-based participatory research (CBPR) framework. The approach was selected to capture both measurable outcomes and in-depth insights into the implementation and impact of the Green Pioneer Program, an Education for Sustainable Development (ESD) initiative. Quantitative and qualitative data were collected concurrently, with qualitative findings used to elaborate on the quantitative trends.

2.2 Time and Location

The study was conducted from February to July 2025 at SMP Laboratorium Universitas Negeri Malang, a private junior secondary school with institutional ties to Universitas Negeri Malang. The school was selected based on its potential as a model for contextual ESD implementation, supported by a strong academic infrastructure and its participation in the "Sustainable Lifestyle" theme of the Merdeka Curriculum.

2.3 Research Procedure

The study was carried out in three phases. Firstly, in the Preparation Phase, A situational analysis was conducted through surveys, interviews, and observations to assess students' baseline knowledge, waste conditions, and institutional readiness. A training module based on Kolb's Experiential Learning Theory was developed, combining theoretical and practical components of BSF maggot farming and responsible consumption. Secondly, in the Implementation Phase, the program began with an introductory ESD seminar, followed by pre-tests assessing students' sustainability awareness and behavioral frequency. Students participated in BSF cultivation training (media preparation, waste sorting, feeding, and composting) and were encouraged to adopt sustainable habits such as zero-waste lunches, using reusable items, and engaging in school gardening. Thirdly, in the Evaluation Phase, the impact of the program was assessed through post-tests, observations, interviews, and documentation.

2.4 Data Collection and Instruments

Data were collected using both quantitative and qualitative instruments. The quantitative data included pre- and post-test surveys measuring two constructs: (1) Sustainability Awareness (knowledge and attitudes), (2) Frequency of Sustainable

Practices (daily eco- behaviors). The instrument was adapted from previous validated scales (e.g., [26]) and contextualized for junior high school students. Descriptive statistics were used to analyze shifts in scores before and after the intervention. The qualitative data included: (1) Field observations using structured observation rubrics to assess student engagement and skill application during practical sessions. (2) Semi-structured interviews and focus group discussions with students, teachers, and the school principal to explore perceptions, experiences, and changes in environmental attitudes. (3) Documented visual data (photos and videos) and student reflections, which served both as learning evidence and supporting evaluation materials.

2.5 Data Analysis

Descriptive statistics were used to compare pre- and post-intervention scores and identify percentage increases in awareness and practices [15]. Thematic analysis was applied to qualitative data to capture patterns in student engagement, perception changes, and behavioral shifts. Triangulation of multiple data sources was used to enhance the validity and reliability of the findings.

3 Results and Discussion

3.1 Implementation of the Green Pioneer Program

The "Green Pioneer Program" is an Education for Sustainable Development (ESD) initiative conducted by the Biology Department team at Universitas Negeri Malang, located in Malang City, East Java Province, Indonesia. The program is funded by the Institute for Research and Community Services (LPPM) of Universitas Negeri Malang, and it was implemented from February to August 2025 through training in organic waste management and students' responsible consumption practices as a program for the implementation of SDG 12 (Responsible Consumption and Production) and SDG 15 (Life on Land) by empowering students to become agents of change for the environment. This program aims to enhance Students' Environmental Awareness, where students gain an understanding of the importance of protecting the planet through responsible consumption and production. It also aims to develop skills in organic waste management. Students engage in hands-on learning to convert organic waste into compost and maggot biomass [12]. It also cultivates sustainable living habits from an early age through real actions 3R, conserving land, water, and energy. Finally, it can empower students as agents of change (Green Pioneers) where Students not only learn but also inspire their peers and communities through environmental campaigns.

The Green Pioneer Program was designed as a community engagement initiative that integrates Education for Sustainable Development (ESD) into school-based activities. The program aimed to strengthen students' ecological literacy, develop practical problem-solving skills, and promote sustainable behaviors aligned with SDG 12 (responsible consumption and production) and SDG 15 (life on land).

The program was implemented from February to July 2025 in three sequential stages. First, an ESD seminar introduced students and teachers to the concept of sustainability, responsible consumption, and ecological challenges faced by the local community. Second, students participated in hands-on training on Black Soldier Fly (BSF) maggot-based organic waste management, where they learned to cultivate BSF larvae, process food waste into animal feed, and reduce organic waste at school. Third,

the program emphasized responsible consumption practices, including waste reduction campaigns, reflection sessions on daily consumption habits, and student-led initiatives to promote sustainable lifestyles within the school community.
Throughout the program, participants engaged in experiential learning, group discussions, and reflection journals. School leaders and teachers also played an active role by supervising activities and encouraging the integration of sustainability topics into classroom learning. By combining theoretical knowledge with practical application, the Green Pioneer Program not only addressed students' limited understanding of sustainability concepts but also fostered concrete behavioral change toward more sustainable lifestyles.

The module was complemented by educational media (posters, videos, infographics). Coordination with school leadership ensured logistical readiness and participant engagement. We hope that the module can be used as a guide to other schools as well to implement the same program. The example of a Module is shown in Figure 1.

Fig 1. Guidebook of Green Pioneer

The implementation phase consisted of a series of interactive activities, including an introductory seminar on ESD and sustainable waste management, practical sessions on BSF maggot farming, and behavior-building routines promoting sustainable habits at school and

at home. The result shows that the program is successfully implemented. Students actively participated in each stage of the program, demonstrating engagement and enthusiasm. Positive responses from school principals, teachers, and students indicate strong acceptance and relevance of the program. The school principal emphasized, *"This program supports our school's mission to promote sustainability and provides real solutions to waste issues at school."* One science teacher noted, *"The BSF maggot training and responsible consumption activities enriched our students' learning and helped connect science concepts to everyday life."* The overall implementation showed a high level of feasibility and acceptance, establishing the program as a replicable model for ESD in middle school settings.

https://doi.org/10.1051/e3sconf/202565202002

The Green Pioneer Program introduced maggot-based organic waste management using Black Soldier Fly (BSF) larvae as a core experiential activity to contextualize sustainability learning. By engaging directly in BSF maggot cultivation, students learned to process food waste efficiently while understanding ecological cycles and the concept of the circular economy [12]. This hands-on approach aligns with Kolb's experiential learning theory, where knowledge is constructed through concrete experiences, reflection, and active experimentation [12]. Students were involved in the full cycle of maggot cultivation— preparing media from bran and kitchen waste, feeding larvae with sorted organic waste, maintaining appropriate conditions (temperature and moisture), and harvesting maggots and compost. This process enabled them to visualize the transformation of waste into valuable products: organic fertilizer and animal feed. Research confirms that BSF larvae can reduce organic waste volume by up to 70% while producing high-quality compost and protein-rich biomass [11][13][14]. This method not only addresses school waste challenges but also introduces students to innovative, scalable environmental technologies. The documentation
of the ESD Seminar and the location for Maggot cultivation are shown in Figure 2.

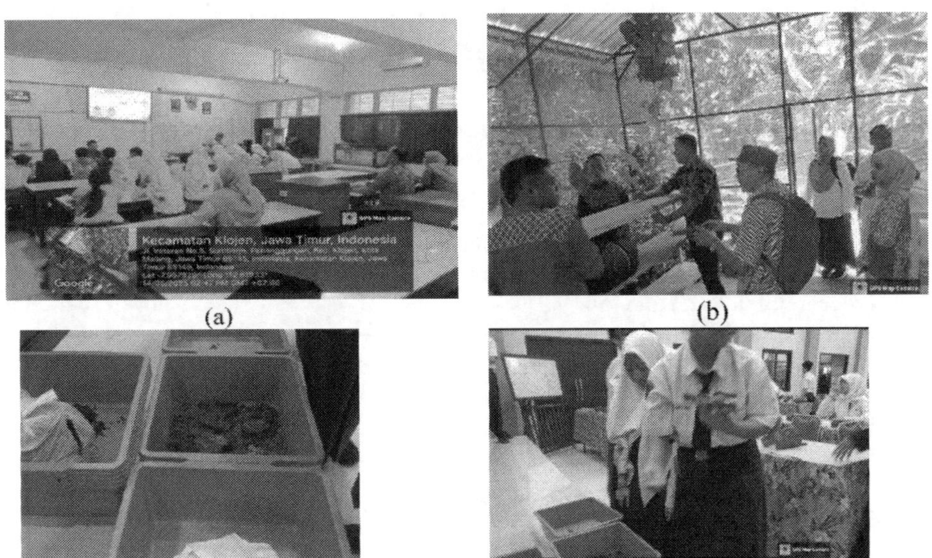

Fig 2. The implementation of the The Green Pioneer Program (a) ESD Seminar (b) location of maggot cultivation at the School (c) tools and equipment used (d) the training for students.

The learning experiences are essential in building eco-literacy, as students directly connect environmental issues with practical solutions [13]. Students also expressed enthusiasm; one participant shared, *"Now I understand how managing organic waste with maggots can help the environment—and it's fun too!"* These testimonies reflect that the Green Pioneer Program not only aligns well with the school's educational goals but also resonates with students' experiences and interests, fostering a sense of environmental responsibility and ownership in the learning process. Previous studies highlight that active involvement in environmental projects enhances students' environmental attitudes and behavioral intentions [14]. Moreover, this approach situates sustainability education in real-life contexts, increasing student engagement and internalization of pro-environmental values [4, 10]. The program effectively transforms waste from a problem into an educational resource, offering a meaningful model for ecological literacy at the school level. This integration of sustainability knowledge and practical application for some SDG targets illustrates the potential of contextualized, action-based learning in promoting responsible environmental behaviors among young learners [7].

3.2 Impact on Students' Sustainability Awareness

The Green Pioneer Program that has been implemented in SMP Laboratorium UM Malang also significantly enhanced students' sustainability awareness, which was measured through both quantitative and qualitative methods. Sustainability consciousness (SC) refers to awareness of sustainability issues across environmental, social, and economic dimensions, including beliefs, feelings, and actions. The Green Pioneer Program effectively improved students' sustainability awareness. Results showed a slight increase of 0.78 in students' SC scores after the Green Pioneer Program. Although the pretest scores were already relatively high (average above 3), the increase suggests that ESD ESD- based program positively influenced students' sustainability awareness. The measured aspects included sustainability knowledge, attitude, and behavior. Overall, the research results from the pre-test score to the post-test score of students' sustainability awareness showed an increase in N-Gain of 0.25, or in the low category. However, hypothesis testing showed no statistically significant difference ($p > 0.05$) between pre- and post-test scores, indicating that the improvement was not significant.

The independent sample t-test results revealed no statistically significant difference in sustainability awareness between male and female students, both before and after the program implementation ($p > 0.05$). This indicates that the Green Pioneer Program was equally effective in enhancing sustainability awareness across genders. Both male and female students showed similar levels of improvement in environmental knowledge, attitudes, and sustainable practices. These findings suggested that well-designed ESD interventions with contextual and hands-on approaches can engage diverse student groups effectively.

In addition to sustainability awareness, this study also measured students' Sustainability Practice Frequency during the ESD integration in the Green Pioneer program. Students rated their frequency of implementing sustainability practices during the ESD program with a 5- point Likert Scale. Based on the results of post-test measurements on SDG12 (Responsible Consumption and Production), it was found that students had a fairly high frequency of sustainability practice by saving energy and water, such as turning off lights or TVs (4.25) and turning off water taps when not in use (4.50). The habit of reusing items like bottles, lunch boxes, and shopping bags also showed high scores (4.12 and 4.10), which reflects awareness of sustainable consumption. However, low scores were seen in the indicators of reusing water for plants (3.16) and participating in recycling activities or waste banks (2.67), which indicates the need for further habituation through real practices at school in the program. Secondly, for the Sustainability practice of SDG15, students show positive behavior for not littering (4.78) and sharing environmental knowledge with others (4.18). However, scores remained low for active participation in school garden activities (3.34), planting local plants (3.46), and using compost (3.27), indicating that ecosystem conservation activities still need to be strengthened through integration into the curriculum and school programs. The frequency of students' sustainable lifestyle practices related to SDG12 and SDG15 after the program is shown in Table 1. Some examples of the documentation of students' sustainability practices are shown in Figure 3.

Table 1. Sustainability Practice Frequency of Students after the Green Pioneer Program

Statement for SDG12	f	Statement for SDG15	f
I bring food and drinks from home to reduce single-use packaging.	4.12	I care for plants at home and school regularly.	3.16
I choose local and eco-friendly products when shopping.	3.45	I help plant local trees or plants at school.	3.46
I only buy items I truly need, not just because they are trendy.	3.12	I use compost from organic waste for school gardening.	3.27
I reuse items like bottles, lunch boxes, and shopping bags.	4.10	I do not litter on the ground or around the school.	4.78
I turn off lights, fans, or the TV when not in use.	4.25	I protect beneficial insects and small animals in the ecosystem.	3.90
I turn off the water when brushing teeth or washing dishes.	4.50	I participate in school gardening as part of environmental action.	3.34
I reuse water (e.g., from rice washing) to water plants.	3.16	I share with friends or family about protecting soil and plants.	4.18
I separate organic and inorganic waste at school or at home.	3.87	I keep plants clean and avoid damaging them at school gardens/parks.	4.56
I join recycling, waste banks, or waste-reduction programs at school.	2.67		

 https://doi.org/10.1051/e3sconf/202565202002

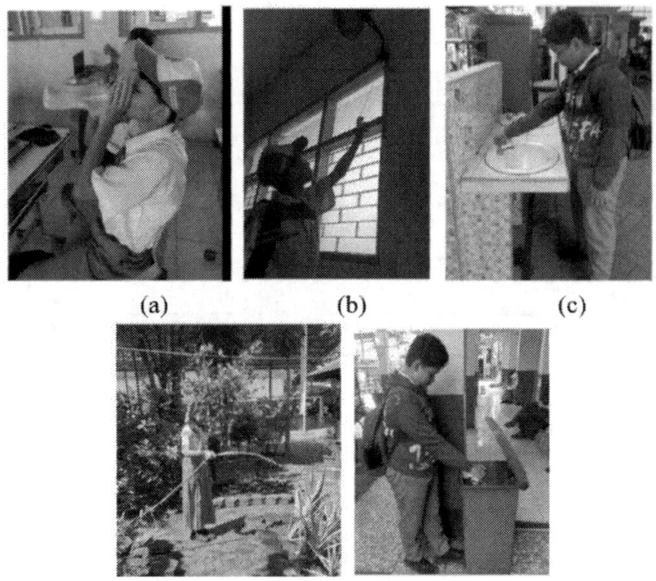

(a) (b) (c) (d) (e)

Fig 3. Some examples of Students' Sustainability Practice during the Green Pioneer Program: (a) using environmentally friendly product , (b) save energy/turning fan when not in use, (c) save water, (d) watering plant in garden, (e) throw rubbish in the dustbin.

Overall, students have demonstrated good awareness of responsible consumption, but require a more hands-on and collaborative approach to foster concrete actions in environmental and soil conservation in line with SDG12 and SDG15. Qualitative data further support these findings. During post-program interviews, students expressed greater environmental concern and a stronger sense of responsibility. One student shared:

"I used to throw away leftover food without thinking. After the Green Pioneer Program, now I understand it can be used to feed the maggots and turned into fertilizer. I even taught my family how to separate waste at home." (Student A, Grade 8).

Another student remarked:

"After learning about BSF and sustainability, I realized that my daily choices matter. I started planting vegetables using the compost we made at school. I also try to do some sustainable activities every day" (Student B, Grade 7).

These narratives reflect a shift not only in knowledge but in personal values and everyday behaviors. This aligns with [12], who found that emotional engagement and personal experience are crucial mediators in translating environmental knowledge into action. This is also supported by the result of observation data collected during practical sessions, which also showed behavioral changes. Most students actively participated in maggot farming and waste sorting. Teachers reported that some students independently initiated waste separation practices in other school areas, beyond the designated project. Furthermore, the practice-oriented nature of the program helped overcome abstract barriers often associated with ESD [14]. By providing tangible results—like seeing maggots consume food waste or using compost for planting—students were able to connect sustainability concepts with real-life impact [10]. This experiential reinforcement is consistent with findings by [4], who emphasize the role of perceived efficacy and meaningful engagement in building sustainable behaviors. The Green Pioneer Program not only delivered knowledge but fostered reflective, emotional, and behavioral transformation among students. By embedding sustainability into students' daily routines and enabling hands-on experiences, the program successfully cultivated a foundational sustainability mindset—one that has the potential to persist and influence broader social and environmental systems. This case illustrates the practical impact of ESD when designed with relevance, participation, and local ecological context in mind.

The Green Pioneer Program demonstrated tangible contributions to two Sustainable Development Goals: SDG 12 (Responsible Consumption and Production) and SDG 15 (Life on Land). Through daily sustainability practices, such as waste sorting, the use of reusable containers, students adopted more conscious consumption habits in both school and home settings. In relation to SDG 15, students participated in compost production and school gardening using the organic fertilizer generated from BSF maggot farming. This activity not only reduced the volume of school-based organic waste but also supported soil health and green space revitalization in the school environment [14, 15]. The use of compost directly contributed to ecological land management practices by promoting soil regeneration and minimizing chemical inputs. Moreover, students developed a personal connection with environmental stewardship through hands-on experience. These outcomes confirm that when ESD programs are contextual, participatory, and project-based, they can significantly contribute to multiple SDGs [12]. The program highlights how school-based interventions can serve as microcosms of sustainable development efforts at the community level.

The results of this study show that the Green Pioneer Program was successfully implemented through hands-on training of organic waste management and responsible consumption practices. Students gained meaningful ecological knowledge and practical skills. School leaders, teachers, and students responded positively. The findings of this study are consistent with previous research emphasizing the impact of experiential and contextual learning on enhancing students' environmental literacy and sustainability awareness. For example, [13] demonstrated that practical, project-based interventions—such as second-hand clothing markets—effectively improved students' responsible consumption behaviors, aligning with SDG 12. Similarly, [14] reported that ecological projects embedded in school activities contributed to students' understanding of the SDGs and fostered long-term behavioral change. These studies underscore that hands-on, real-world activities embedded within the school context are effective in translating knowledge into sustainable action.

The unique of the Green Pioneer Program from similar initiatives in other countries is the integrated use of maggot-based waste management as a core pedagogical tool, merging ecological education with waste reduction and resource regeneration. While maggot farming has been studied for its environmental and economic benefits [11, 13], its integration into an ESD framework at the junior secondary level remains limited in the literature. This program not only provided technical training but also daily sustainability habits through a structured series of behavioral interventions and ecosystem care practices. The program's success in engaging both genders equally also confirms its inclusive design. Thus, the study contributes a replicable, low-cost, and scalable model of ESD that promotes SDGs 12 and 15 while embedding ecological responsibility in everyday student behavior. This innovation highlights the importance of linking environmental education with practical, community-rooted action. The student action project based on Education for Sustainable Development (ESD) aims to foster sustainability awareness among junior high school students at UM Laboratory School. This program aligns well with Indonesia's curriculum of P5, where one of the themes is sustainable lifestyle, implemented in junior high schools over approximately four months. Schools play a key role in building students' sustainability awareness, particularly through subjects like biology, which teaches the interdependence between humans and their environment. According to the Theory of Planned Behavior [12], individuals form environmental attitudes based on valued objects—self, others, and the biosphere— interpreted through social-cognitive theory. This theory posits that attitudes are developed through experience. Research also supports that students with pre-existing environmental attitudes tend to show improvement after participating in environmental education programs [9, 10]. The KAP model (Knowledge, Attitude, Practice) by [14] emphasizes that knowledge influences attitudes, which in turn affect practices. In this study's context, ESD programs aim to enhance knowledge and attitudes that drive sustainable behaviors in alignment with the Sustainable Development Goals (SDGs).

Despite its successful implementation, the Green Pioneer Program had several difficulties. Students' lack of prior understanding and involvement in environmental sustainability was one major problem. The restricted supply of organic waste inputs, climate-related factors that impact maggot growth, and the requirement for ongoing supervision during the maintenance phase are examples of logistical restrictions that also occurred. Teachers observed that in the absence of consistent supervision, students frequently overlooked important composting or maggot care procedures. When considering the program critically, these difficulties highlight how crucial curricular integration, institutional support, and long-term monitoring are. ESD needs constant reinforcement; it cannot be a one-time action. Future initiatives should think about integrating sustainability into all topic areas and active participation of broader stakeholders to increase the impact and continuity after the program ends [12].

4 Conclusion

According to the purpose of this study, firstly, the Green Pioneer Program was successfully implemented as a contextual ESD model at SMP Laboratorium Universitas Negeri Malang. The program integrated experiential activities such as Black Soldier Fly (BSF) maggot-based organic waste management, responsible consumption practices, and sustainability campaigns, supported by active participation of students, teachers, and school leaders. This confirms that school-based community engagement can serve as an effective vehicle for embedding sustainability education in junior secondary contexts. Secondly, the program positively influenced students' sustainability awareness. Students demonstrated improvement in sustainability knowledge, attitudes, and practices, particularly in waste reduction, reuse of materials, and energy-saving habits. Although the statistical analysis revealed no significant gender differences, descriptive findings and qualitative reflections indicate meaningful behavioral changes aligned with SDG 12 and SDG 15. Future studies should investigate the long-term effects of such programs through longitudinal designs, compare different ESD approaches to identify the most effective models, and broaden stakeholder involvement by including parents and local communities. Such efforts will provide a more comprehensive understanding of how school-based ESD initiatives can contribute to cultivating sustainable lifestyles beyond the school environment.

Acknowledgments

This research was funded by the Institute of Research and Community Services (LPPM) Universitas Negeri Malang through the 2025 Internal Grant Scheme under contract number 24.2.1170/UN32.14.1/PM/2025. We extend our sincere gratitude to the Headmaster, teachers, and students of SMP Laboratorium UM for their active participation and collaboration. We also thank the UM Library for providing access to valuable academic resources and references that supported the development and implementation of this program.

References

1. Kurniawan, T. A., Meidiana, C., Othman, M. H. D., Goh, H. H., & Chew, K. W. Strengthening waste recycling industry in Malang (Indonesia): Lessons from waste management in the era of Industry 4.0. *J. Clean. Prod.* **382**, 135296 (2023).
2. Purba, H. D., Meidiana, C., & Adrianto, D. W. Waste management scenario through community based waste bank: A case study of Kepanjen district, Malang regency, Indonesia. *Int. J. of Environmental Science and Development*, **5**(2), 212. (2014)
3. García-González, E., Albareda-Tiana, S., Solís-Espallargas, C., & Jiménez-Fontana, R. Transformative education for sustainable consumption. *Trends High. Educ.* **1**(1), 1–15 (2022).
4. Heeren, A. J., Singh, A. S., Zwickle, A., Koontz, T. M., Slagle, K. M., & McCreery, A.
5. C. Is sustainability knowledge half the battle? An examination of sustainability knowledge, attitudes, norms, and efficacy to understand sustainable behaviours. *Int. J. Sustain. High. Educ.* **17**(5), 613–632 (2016).
6. Polas, M. R. H., Tabash, M. I., Bhattacharjee, A., & Dávila, G. A. Knowledge management practices and green innovation in SMEs: The role of environmental awareness towards environmental sustainability. *Int. J. Organ. Anal.* **31**(5), 1601–1622 (2023).
7. Sekito, T., Prayogo, T. B., Meidiana, C., Shimamoto, H., & Dote, Y. Estimating the flow of recyclable items and potential revenue at a waste bank: the case in Malang City, Indonesia. *Environment, Development and Sustainability*, **21**(6), 2979-2995. (2019).
8. Domingo, J. L., & Nadal, M. Domestic waste composting facilities: A review of human health risks. *Environ. Int.* **35**(2), 382–389 (2009).
9. Fraisl, D., Campbell, J., See, L., Wehn, U., Wardlaw, J., Gold, M., & Fritz, S. Mapping citizen science contributions to the UN sustainable development goals. *Sustain. Sci.* **15**, 1735–1751 (2020).
10. Walid, M., & Luetz, J. M. From education for sustainable development to education for environmental sustainability: Reconnecting the disconnected SDGs. In *Sustainability Science and Technology*, 803–826 (Springer, 2018). https://doi.org/10.1007/978-3-319- 63007-6_49
11. Setiawan, H., Surtikanti, H. K., Kusnadi, K., & Riandi, R. Sustainability awareness, engagement, and perception of Indonesian high school students during sustainability project-based learning implementation in biology education. *J. Penelitian Pendidikan IPA* **9**(6), 4227–4236 (2023).
12. Utami, I., Putra, I. L. I., Khotimah, K., & Pangestu, R. G. Maggot Black Soldier Fly sebagai agen degradasi sampah organik dan pakan ternak warga Mergangsan Yogyakarta. *LOGISTA* **4**(2), 127–135 (2020).
13. Ridwan, M. Analysis of the use of media resulting from bioconversion of organic waste in the production of maggots BSF (black soldier fly). *Jurnal Akuakultur Indonesia*, **21**(1), 1-10. (2022).
14. Sari, D. A. P., Taniwiryono, D., Andreina, R., Nursetyowati, P., Irawan, D. S., Azizi, A., & Putra, P. H. Utilization of household organic waste as solid fertilizer

with maggot black soldier fly (BSF) as a degradation agent. *Agricultural Science*, **5**(2), 82-90. (2022).
15. Putri, M. S. A., Dahlan, M., Fadlilah, A., Rusminah, S., Khitam, M. C., Yani, Y. P., ... & Haqiqi, R. Effectiveness of Organic Waste Degradation Level using the Black Soldier Fly Maggot. *Journal of Health Science and Prevention*, **7**(1), 1-9. (2023).
16. Bryman, A. *Social research methods*. (Oxford university press, 2016).

Intergrading Geological Resource Potential in Landfill Siting: A GIS-SMCE Approach in a Volcanic Landscape of Probolinggo, Indonesia

Listyo Yudha Irawan[1*], *Rizal* Sulaiman[2], *Mochammad Tri* Herwanto[1], *Agung* Suprianto[3], *Widodo Eko* Prasetyo[4], *Denisa Aulia* Pratiwi[1], *Muhammad Nurul* Huda[1], *Azril* Chairil[1]

[1]Department Geography, Faculty of Social Science, Universitas Negeri Malang, Malang, Indonesia
[2]Alumni Bachelor Geological Engineering, Gadjah Mada University, Yogyakarta, Indonesia
[3]Social Studies Education Study Program, Faculty of Social Science, Universitas Negeri Malang, Malang, Indonesia
[4]Department Geomatics Engineering, Faculty of Engineering, Institut Teknologi Sepuluh November, Surabaya, Indonesia

Abstract. The selection of suitable landfill sites is a critical challenge in rapidly urbanizing regions, where improper siting may cause long-term environmental degradation, public health risks, and land-use conflicts. This study integrates geological resource potential into a GIS-based Spatial Multi-Criteria Evaluation (SMCE) framework for landfill site selection in Lumbang District, Probolinggo Regency, East Java, Indonesia. The methodological approach combines the Indonesian National Standard for landfill siting (SNI 03-3241-1994) with the Analytical Hierarchy Process (AHP) to determine parameter weights, ensuring both regulatory compliance and scientific transparency. Four key parameters—lithology, groundwater depth, slope gradient, and rainfall intensity—were evaluated, supported by exclusion criteria such as proximity to settlements, rivers, and geological hazards. The final suitability index was calculated using a Weighted Linear Combination (WLC), producing a comprehensive landfill feasibility map. Results indicate that volcanic breccia formations with medium to deep groundwater conditions and gentle slopes (0–20%) offer the highest suitability, with Boto Village identified as the most feasible location. Importantly, the integration of geological resource potential as an exclusion criterion differentiates this study from previous approaches, preventing future conflicts over mineral and groundwater use. Despite limitations in DEM resolution, rainfall data, and resource inventories, the model demonstrates the necessity of incorporating environmental geology into landfill planning. The findings provide a robust decision-support tool for local authorities, offering policy recommendations to update the national landfill siting standard and strengthen sustainable waste management strategies in volcanic landscapes.

* Corresponding author: listyo.fis@um.ac.id

1 Introduction

The absence of environmental characterization in the government's developing program for constructing landfill sites would frequently result in future difficulties. In addition to political and social implications, landfill development plans may adversely affect public health, ecology, and the environment. The scarcity of environmental characterization for infrastructural development in East Java was particularly evident in the late 1990s and early 2000s, during which environmental awareness was limited. Although a limited number of studies have been conducted to evaluate this site's social and livestock impacts, there are no known investigations on its environmental impact. In 2010, the Government Regulation No. 2 of 2010 designated Kraksaan as the capital of Probolinggo Regency. The establishment of a new urban area will undoubtedly increase waste production. Therefore, a robust environmental and geological assessment must be conducted to select an appropriate landfill site. Based on the Unpublished report Regency Government in 2023, the Probolinggo Regency Government has identified the Lumbang sub-district as the most suitable location for the new Regional Waste Landfill, a decision that has been incorporated into the spatial and regional planning. This determination was based on a regional scale evaluation of the SNI 03-3241-1994 regulation concerning the Landfill Site Selection Procedures, which the Ministry of Public Works and Housing of Indonesia determined.

Identifying suitable landfill sites has been a widely debated issue, with many studies employing GIS-based spatial multi-criteria evaluation (SMCE) combined with the Analytical Hierarchy Process (AHP) to determine feasible locations [1], [2]. These studies generally emphasize parameters such as slope, distance from settlements, and proximity to water bodies, aligning with regulatory standards. However, despite such approaches, numerous landfill projects, particularly in Indonesia, have resulted in groundwater contamination and environmental degradation [3]. This indicates that previous research has inadequately incorporated geological aspects beyond the basic hazard criteria. In many cases, the assessment focuses on avoiding immediate risks such as unstable slopes, floodplains, or fault zones, but rarely addresses the potential geological resources that may be required for future development. Consequently, landfills are sometimes sited in areas with valuable subsurface resources, which may lead to long-term land-use conflicts and foreclose opportunities for resource utilization [4].

The novelty of this study lies in bridging that gap by explicitly integrating geological considerations into the landfill site selection process. While numerous SMCE-based landfill studies exist, few have combined compliance with national standards and a dual emphasis on both geological hazards and the pre-emptive exclusion of areas with potential mineral or groundwater resources [5]. In the context of Probolinggo Regency, this approach is particularly relevant due to its volcanic landscape, which presents both geological hazards and resource opportunities. By incorporating geological potential as an exclusion criterion, this research not only strengthens the scientific rigor of landfill siting but also ensures that areas with future economic or environmental value are preserved. Such integration provides a forward-looking framework that aligns Indonesia's national landfill siting standard (SNI 03-3241-1994) with principles of sustainable development, offering a more comprehensive decision-making tool for waste management and spatial planning.

This research addresses the critical issue of landfill site selection in Lumbang District, Probolinggo Regency, East Java, Indonesia, focusing on integrating environmental geology principles with national regulations to ensure sustainable development and minimize environmental degradation. The current plan to establish a regional waste landfill in Lumbang sub-district highlights the urgent need for a thorough environmental and geological assessment. The research hypothesizes that insufficient consideration of these principles may lead to the selection of unsuitable sites, even when national regulations are followed.

Previous studies have often treated geology in a broad sense, overlooking specific aspects such as aquifer vulnerability, slope stability, and the potential use of geological resources, which are essential for long-term sustainability [6]. Addressing these overlooked aspects constitutes the novelty of this study. While numerous SMCE-based landfill studies exist, few have explicitly integrated national regulatory frameworks with a dual focus on both geological hazards and the pre-emptive exclusion of areas with future mineral resource potential, a critical consideration for long-term sustainable development.

2 Method

This study was conducted in Lumbang District, Probolinggo Regency. The selection of this location was motivated by the absence of a landfill facility in the area, which creates an urgent need for a structured waste management system. The lack of a landfill not only poses potential environmental risks due to the accumulation of domestic waste and agricultural activities, but also has implications for public health and the surrounding ecosystem's integrity. Furthermore, Lumbang District is situated in a transitional zone between residential settlements and resource-based economic zones, making it strategically significant. Identifying a suitable landfill site in this district is thus expected to support the local government's goals for sustainable spatial development and minimize adverse environmental and socio-economic impacts.

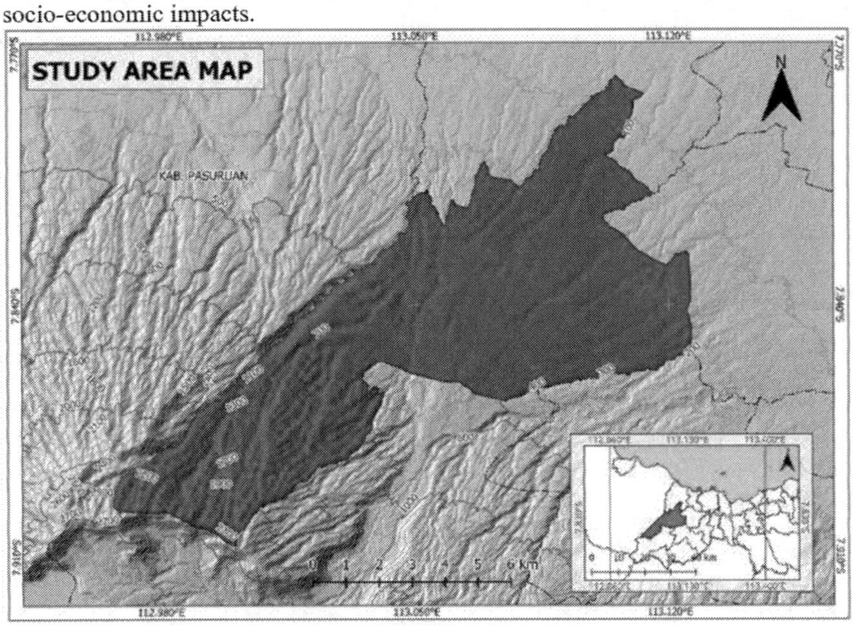

Fig 1. Location map of the study area

This research applies a conventional GIS-based Spatial Multi-Criteria Evaluation (SMCE) approach, guided by the Indonesian National Standard (SNI 03-3241-1994), which, although designed for municipal landfills, remains relevant for regional planning contexts. The GIS-based SMCE method has been extensively adopted in previous landfill siting research. A widely used technique within this framework is the integration of Analytical Hierarchy Process (AHP) to determine factor weights and enhance transparency in spatial decision-making. For instance, Rame et al. demonstrated the use of GIS and AHP in landfill site selection in Malaka Regency, emphasizing environmental geology and socio-economic considerations [1]. A raster-based overlay analysis was performed to integrate all relevant parameters into a composite suitability map. Each parameter layer was assigned a weight and standardized score, and unsuitable zones were excluded. The final suitability index was computed using the Weighted Linear Combination (WLC) method.The overall suitability score for each location was then calculated using the Weighted Linear Combination (WLC) formula:

$$Si = n\, \Sigma\, j\; wj\; \times x\, j \qquad (1)$$

Where
S = = suitability level of the landfill site
(i), wj = parameter score weighting
(j), xj = parameter score
(j), n,j = amount of parameters

The analytical hierarchy process (AHP) is a frequently employed methodology in the context of complex decision-making, whereby priorities are established for each parameter, and a pairwise comparison is conducted based on the relative importance of each parameter's ultimate objective, which in this case is landfill site selection. AHP represents an efficient approach to quantifying the qualitative factors and comparing each parameter's relative importance.

The landfill site selection criteria were formulated by harmonizing and contextualizing the procedures outlined in the national standard SNI 03-3241-1994. The national standard is primarily based on the analysis of three criteria: (1) the regional criterion, (2) the exclusion criterion, and (3) the determination criterion. Each zone is characterized by geological and hydrological conditions, slope gradient, distance from airports, proximity to nature reserves, and vulnerability to periodic flooding. These factors form the first layer of analysis, highlighting broad areas that may be suitable for landfill development. The exclusion criterion refines this process by removing areas that are deemed inappropriate due to additional parameters, such as climate, availability of utilities, biological environment, soil properties, demographic characteristics, administrative boundaries, noise and odor pollution, aesthetics, and economic constraints. These parameters create buffer zones that strictly exclude unsuitable areas, thereby narrowing the selection to only those sites that meet both environmental and regulatory standards [1], [5].

The final step is the determination criterion, which requires that the alternative sites be validated and confirmed by relevant authorities, the provincial government based on local policy frameworks and governance structures. This step ensures that the scientific analysis is consistent with practical and regulatory considerations, creating a balance between technical suitability and administrative feasibility. By integrating AHP with the three levels of criteria established in the SNI, this study creates a comprehensive decision-making hierarchy. AHP provides the mechanism for weighting the regional parameters, testing the consistency of expert judgments, and producing transparent and defensible results. At the same time, the SNI framework grounds the analysis in nationally recognized standards, ensuring that the site selection process is not only scientifically rigorous but also aligned with regulatory requirements [7]. This integration enables the identification of potential landfill sites that are environmentally sustainable, geologically stable, and institutionally acceptable.

The parameters selected here are adjusted and emphasized based on geological information. In the first stage, values and weights were assigned to shapefiles representing lithology, groundwater depth, slope gradient, and rainfall intensity. These parameters were then evaluated using the Analytical Hierarchy Process (AHP), which requires constructing a pairwise comparison matrix. The pairwise comparison matrix was developed following Saaty's 1–9 fundamental scale. In this scale, a score of 1 indicates equal importance between two parameters, while 9 indicates that one parameter is extremely more important than the other. Intermediate values (3, 5, 7) represent moderate, strong, and very strong importance, respectively, with their reciprocals (1/3, 1/5, etc.) used when the relative importance is in the opposite direction. For this study, four parameters were compared: lithology, groundwater depth, slope gradient, and rainfall intensity. Each parameter was judged against the others by a panel of experts consisting of two senior geologists, one environmental scientist, and one academic specializing in GIS-based decision support systems. For example, lithology was judged to be strongly more important than rainfall intensity due to its fundamental role in determining ground stability, and therefore was assigned a score of 5 in that comparison. Similarly, groundwater depth was considered moderately more important than slope gradient for landfill suitability, and was assigned a score of 3. The individual pairwise matrices provided by the experts were then aggregated using the geometric mean to form a consensus matrix.

After constructing the matrix, the relative weights of each parameter were calculated as the normalized principal eigenvector of the matrix. To ensure the reliability of these judgments, the Consistency Index (CI) and Consistency Ratio (CR) were computed. The CR value obtained in this study was 0.07, which is below the generally accepted threshold of 0.10. This indicates that the expert judgments were consistent and the derived weights are valid. If the CR had been greater than 0.10, it would have suggested inconsistencies in expert input, requiring a re-examination of the pairwise comparisons. The final parameter weights from the AHP were multiplied by the corresponding parameter values in the GIS environment, and the results for each location were summed to obtain a total suitability score. These scores were then combined with exclusion criteria to eliminate unsuitable areas.

Finally, all locations were grouped into three feasibility zones: not feasible (0–18), low to medium feasibility (19–55), and high feasibility (56–70). These ranges correspond to increasing levels of site suitability, from very poor (Not Feasible) up to excellent (High Feasibility). Each class represents a grouping of similar scores, making the results easier to interpret for decision-makers. Locations falling in the High Feasibility range (56–70) are considered the most suitable sites, whereas those in Not Feasible (0–18) are clearly unsuitable for development [8]. The outputs were presented as thematic maps and integrated through GIS overlay analysis, which is capable of handling diverse spatial datasets. The complete process workflow is shown in Figure 1.

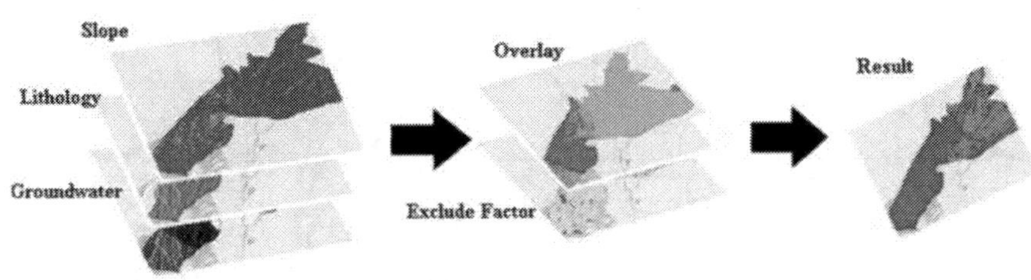

Fig. 2. Process of regional criterion which overlays the exclusion criterion leading to the elimination of unwanted candidates, thus leading to clusters of potential sites, presented in the final map.

The total scores are calculated and presented in a range, indicating the feasibility level. This study has involved the creation of a slope map, which has been developed by creating a raster layer from an 8 m digital elevation model (DEM) that categorizes the slope gradient, and a geological potential map, which has been created by combining old government documents, information gathered from local sources, and brief field observation to identify any potential geological features. A total of 14 parameters have been selected for the study, which has been categorized into two distinct analytical categories: regional analysis and exclusion analysis. The regional analysis is evaluated by expressing the values of each parameter. In contrast, the exclusion criterion establishes a buffer zone to exclude irrelevant locations. The parameters used in this study are presented in Table 1, which was modified from the Indonesian National Standard (SNI 03-3241-1994) to suit the regional context and integrate additional geological considerations.

Table 1. Environmental geology parameters encompassing regional and exclusion criteria

No	Parameters	Units	Value	Weight / Score
1	Regional Criterion Litology	Volcanic breccia Andesitic lava Volcanic sandstone No groundwater	5 4 3 0	5 / 25 5 / 20 5 / 15 5 / 0
2	Groundwater depth	Medium to deep Shallow	5 3	4 / 20 4 / 12
3	Slope gradient	0–20% 20–40% 40–100%	5 3 1	3 / 15 3 / 9 3 / 3
4	Rainfall intensity	1000–2000 mm/year 2000–3000 mm/year >3000 mm/year	5 4 3	2 / 10 2 / 8 2 / 6
5	Exclusion criterion Distance from fault zones	<100 m	-	Not feasible
6	Floods	<25 years	-	Not feasible
7	Mass movements	Low–medium	-	Not feasible
8	Distance from rivers	<150 m	-	Not feasible
9	Shoreline	<500 m	-	Not feasible
10	Protected areas	Forest, sanctuary, etc	-	Not feasible
11	Distance from settlements	<300 m	-	Not feasible
12	Distance from main roads	<150 m	-	Not feasible
13	Distance from airports	<3000 m	-	Not feasible
14	Potential Geological Resources	Water, mineral	-	Not feasible

Modified from SNI 03-3241-1994

The scoring of lithology parameters was based on the hydrogeological characteristics of each rock type, particularly their permeability, fracture density, and filtration capacity against leachate. Volcanic breccia received the highest score (5) because it consists of coarse fragments with a compacted fine matrix, functioning as an aquitard with relatively low permeability and effectively impeding leachate infiltration [9][10]. Andesitic lava was assigned a score of 4 since, although fresh andesite lava is highly impermeable (aquifuge), in field conditions it commonly develops columnar joints or cooling fractures that may increase secondary permeability and create pathways for leachate movement [11]. Volcanic sandstone was given a score of 3 because of its high intergranular porosity and relatively high permeability, which allow leachate to percolate more easily [10]. Finally, the no groundwater condition was scored 0, as the absence of groundwater implies no additional protective function from the hydrogeological system, thereby contributing nothing to site suitability [11].

 https://doi.org/10.1051/e3sconf/202565202003

3 Result and Discussion

3.1 Lithology

The lithology of an area can be utilized to minimize the contamination of surface and groundwater from leachates. Surface geological observations indicate that this area predominantly comprises volcanic sandstone, andesitic lava, and volcanic breccia. In general, volcanic breccia is observed to constitute the river's bedrock in the vicinity of the planned landfill site in the Lumbang subdistrict. Medium-sized andesite blocks are typically found overlying the volcanic breccia and tuff fragments, with the latter present in localized areas. Figure 2 illustrates that the predominant lithology in the study area is volcanic sandstone, which interfered with andesitic lava and volcanic breccia lenses. This stratigraphic condition follows the conceptual volcaniclastic facies of the volcano, which indicates that conglomerate lenses enclosed by sand material represent the dominant sequence in the medial-distal facies area, situated approximately 20 km from the summit of the mountain [5].

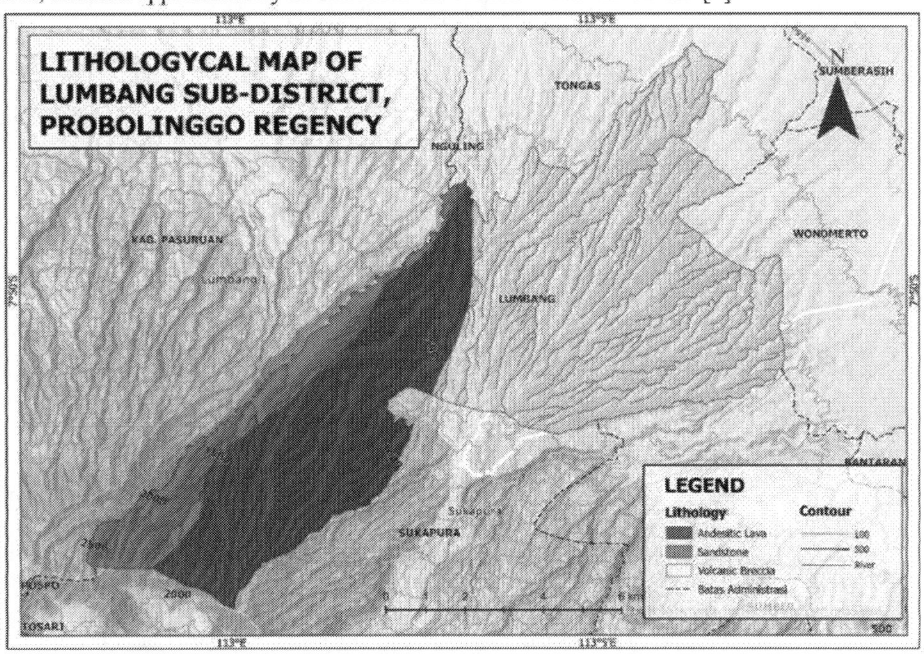

Fig. 3. Regional lithology map of Lumbang District

3.2 Groundwater Depth

The contamination of groundwater is more prevalent in shallow aquifers than in deeper ones. The lithology of the area acts as a reservoir (a groundwater-bearing formation) or as a conduit to the aquifer. Consequently, there is a correlation between these two parameters. The significant demand for groundwater in this region underscores the importance of preserving the aquifer's quality. Most environmental issues concerning landfill site selection can be attributed to this parameter. Several reports have documented groundwater contamination in landfill sites, particularly shallow aquifers. The groundwater in Lumbang is typically found in alluvial and low-lying land units (Fig. 3). The aquifer is divided into three zones: shallow groundwater, medium groundwater, and no groundwater. The regional hydrogeological map was derived from thoroughly examining the regional geology map. The shallow groundwater is found in the unconfined alluvial deposits containing sand, gravel, clay, and mud. The position of an aquifer in shallow groundwater is variable, with medium productivity. Common alluvial aquifers have a medium to high coarsegrained and low permeability. The medium to deep groundwater is located in the volcaniclastic units.

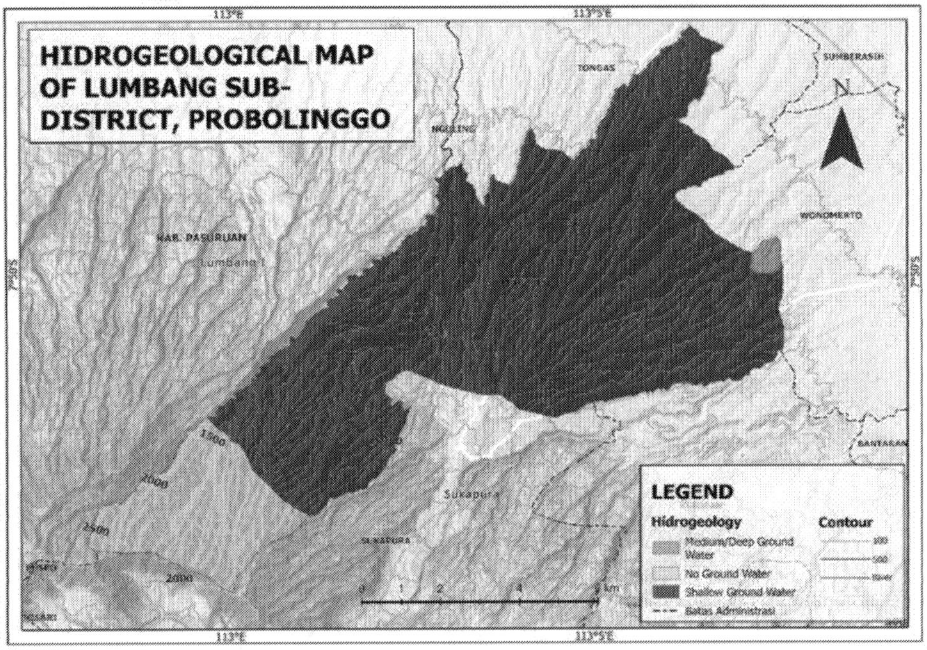

Fig. 4. Regional hydrogeological map of Lumbang District

3.3 Slope Gradient

Based on the National Standard for Landfill Site Selection (SNI 03-3241-1994), any area with a slope greater than 20% should be excluded from consideration. Although engineering techniques such as excavation could reduce steepness, these interventions demand high financial investment. The slope characteristics in Lumbang are divided into three classes derived from a 30 m Digital Elevation Model (DEM) (Fig. 4). Land with slopes between 0–20% is relatively flat and unevenly distributed, making it the most suitable for landfill placement. Meanwhile, slopes of 20–40% represent mildly hilly areas, and those between 40–100% correspond to mountainous zones. Although areas with slopes exceeding 20% are considered less appropriate, they remain as secondary alternatives with lower suitability scores. [1].

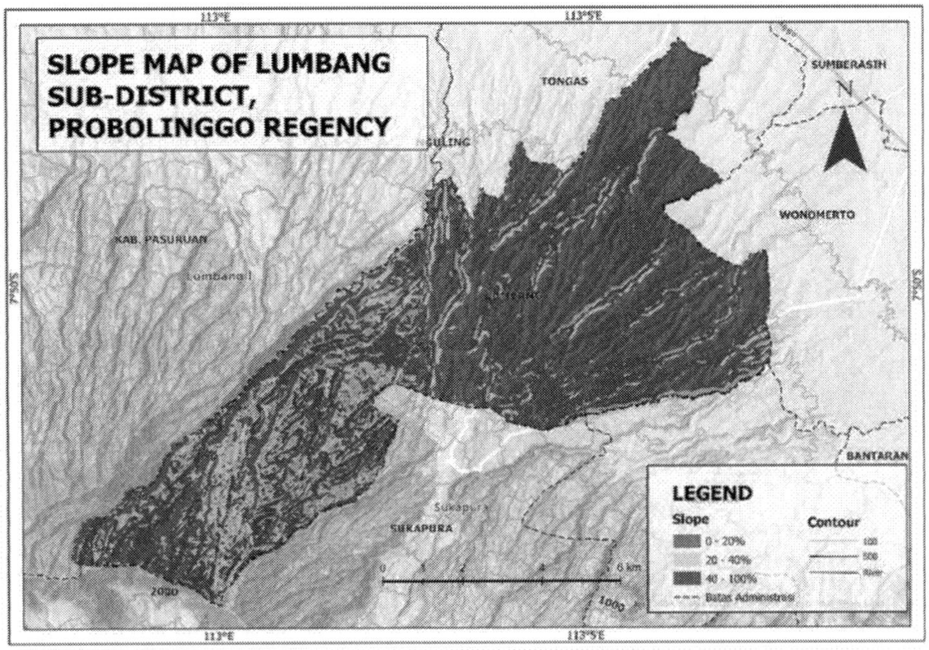

Fig. 5. Slope gradient map of Lumbang District

3.4 Rainfall Intensity

Precipitation levels significantly affect the design and maintenance complexity of landfill facilities, encompassing systems for drainage, runoff control, leachate treatment, and oxidation. This parameter also informs the selection of surface covers to minimize water infiltration and prevent ponding. Records from the East Java Climatology Class I Station indicate that Probolinggo Regency received an average of 1,900 mm of rainfall in 2023, primarily concentrated between October and March, over approximately 186 rainy days (Statistics of Probolinggo Regency, 2023). Rainfall patterns have remained relatively consistent in recent years, averaging above 1,000 mm annually. Given the absence of site-specific climatic data, an identical score for precipitation was applied across all Lumbang areas.

3.5 Exclusion Criterion

The exclusion parameters eliminate areas in the buffer zone deemed unsuitable for the landfill site. It is paramount that the site is situated at a considerable distance from rivers, lakes, or shorelines, as these parameters act as buffer zones and borders for river/coastal management. In addition to aesthetic and health concerns, organic waste readily attracts birds, which presents a potential hazard for airplanes. The nearest airport to Lumbang is Juanda International Airport, located approximately 105 km away. Therefore, landfill operation will not affect the airport's activity. To prevent the engagement of water, odor, and noise pollution, landfill sites must be at least 300 m from any residential areas. The settlement areas in Lumbang have a linear pattern that follows the main road network, mainly in the northern part of the Lumbang sub-district. Furthermore, the landfill site must be at least 150 meters from the nearest main road, as constructing a new landfill could lead to increased traffic. In their 2016 study, Anggraini and Darwati highlight the significance of investment and operational costs as a primary consideration in selecting an appropriate landfill site. These costs are related to the construction and maintenance of the landfill, the transportation of waste, and the routing of associated traffic.

3.6 Geological Hazards

Applying geological information is critical for spatial planning and environmental management, particularly when selecting landfill sites. Two aspects require close attention: geological hazards and geological resources. Hazards such as slope instability, active faults, and flooding can compromise the integrity and operation of landfills. Slope failures are common in steep terrain with highly weathered rocks or in areas close to fault zones, while seismic activity along faults poses significant risks to infrastructure. For this reason, areas with moderate to high susceptibility to slope movement should be excluded from consideration. Current regulations also prohibit landfills within protected zones, including conservation forests, nature reserves, cultural heritage sites, and designated geological reserves. In addition, candidate sites must be located at a safe distance from geological faults to reduce the likelihood of failure during earthquakes.

3.7 Geological Potential

The geological resources of a region play a vital role in supporting future economic development. For this reason, landfill construction should be avoided in areas with significant resource potential. Geological resources include water, minerals, energy, and land, yet they are often neglected in spatial planning due to limited data and the high cost of surveys. Remote sensing offers an effective approach to address this gap and should be applied more extensively. In Lumbang, preliminary investigations indicate the presence of tras sediment and small-scale mining activities, although reserve data remain limited. Such areas are not suitable for landfill sites, as they may be required for future resource development.

3.8 Potential Sites

The regional criterion map overlaps the accumulation of thematic maps of the exclusion parameters, thereby eliminating areas considered unfeasible for the landfill site. This includes areas of settlement along the main road in the north of Lumbang; areas of potential hazards such as floods, landslides, and weak zones; areas close to rivers as they may interfere with their management; and areas of potential resources such as tras sediment. The modeling results show that areas with high suitability scores are located in favorable geological, hydrogeological, and morphological patterns [1][2]. For example, the high suitability in Boto and Palangbesi Village is mainly driven by low-permeability volcanic breccia lithology and its position in a deep groundwater zone, resulting in a low risk of leachate infiltration into the aquifer. In general, locations with high suitability are determined by a combination of: fine-grained/low-permeability lithology, gentle slope gradients (0–20%), safe distance from settlements and water bodies, and not being located in a zone with potential for mass movement or a protected area [12]. The final suitability map reflects the contribution of each input map (lithology, groundwater depth, slope, land use, distance to roads/settlements), so that cells that meet technical and environmental criteria receive the highest scores [13]. Nevertheless, the area would require the approval and confirmation of the local institution/government. Consequently, a supplementary assessment of socio-economic factors or a geotechnical and hydrological study would be beneficial.

The methods employed in this assessment are suitable for determining land use. A straightforward combination of geological and environmental knowledge can be employed in land-use evaluation by calculating the requisite parameters [14]. Given the extensive reliance on geological data in environmental studies, a significant challenge would be the total investment required for surveying and data collection. Therefore, the abundance and accuracy of such datasets are crucial for characterization. The results of this study reinforce several aspects of the government's plan while highlighting differences. On the one hand, the areas recommended by the model (sloping land with impermeable lithology and far from settlements) are in line with the principles of SNI 03–3241-1994 regarding slope, buffer distance, and hydrogeological factors. On the other hand, several locations specified in the government's plan show conflicts for example, being close to potential resource zones or having a slope >20%, so this model challenges the location determination with spatial evidence. The practical recommendation is to use this suitability map as a verification and prioritization tool before final decision-making by the authorities [15].

This study has several limitations that affect its accuracy and precision. First, the 30 m DEM resolution limits the accuracy of slope analysis; local topographic inaccuracies can cause misclassification of gradients at the micro scale, especially in hilly areas. Second, the absence of complete local rainfall data forced the application of a uniform score for the entire Lumbang area, so that spatial variability in rainfall and extreme phenomena were not modeled. Third, data on geological resource reserves (e.g., tras reserves) and local-scale fault inventories are still limited, this adds uncertainty to the assessment of potential resource conflicts and seismic risk. Fourth, the weighting and parameter combination scheme uses a selected approach that can produce different results when other weighting methods (e.g., field-calibrated AHP) are used. All these limitations should be explained and reduced in further research through field surveys, higher-resolution DEMs, and local climatological data. The application of recent advancements in remote sensing may be a cost-effective and time-efficient alternative to ground surveys, potentially yielding comparable results. This assessment demonstrates the value of collaboration between geoscientists and environmental scientists with urban planners. This section introduces the significance of geoscience and environmental science in the potential landfill site selection process, which follows a generalized version of the national regulation on the physical environmental aspects (SNI 03-3241-1994). This method effectively filters out socio-economic impact, yet the connection remains irrefutable. Environmental degradation may hurt the socio-economic status of a community over time, and vice versa. A robust environmental geology characterization in urban planning minimizes natural resource degradation due to population growth and urbanization, thereby facilitating sustainable development.

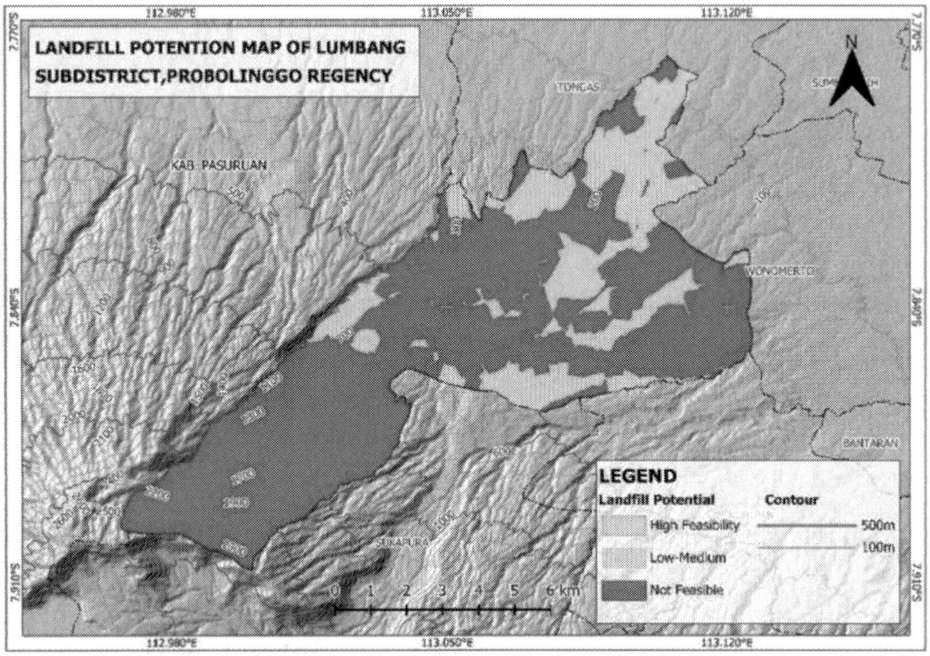

Fig. 6. Potential landfill sites map of Lumbang District

Table 2. The regional (lithology, groundwater, slope, and rainfall) and exclusion criterion analysis for the landfill site selection resulted in the feasibility level based on its total score.

No	Lithology	Groundwater	Slope	Rainfall	Exclusion criterion	Total score	Information
1	25	12	15	8	✓	60	Volcanic breccia, medium/deep groundwater, 0–20% slope, high feasibility
2	25	12	9	8	✗	54	Volcanic breccia, medium/deep groundwater, 20–40% slope, low-medium feasibility
3	25	4	15	8	✗	52	Volcanic breccia, shallow groundwater, 0–20% slope, low-medium feasibility

Based on the map and table, it is known that some areas in Boto and Palangbesi the potential to become landfills. The development and management should be carried out using the principles of sustainable development to minimize environmental damage due to the construction of landfills. These findings have several direct policy implications. First, suitability maps can be used as a tool in revising regional landfill zoning to ensure aquifer protection and reduce environmental risks, prioritizing locations with low-permeability lithology, gentle slopes, and adequate buffer distances. Second, this model can inform the update of SNI 03–3241-1994 by incorporating additional criteria based on hydrogeological risk and local resource potential (e.g., exclusion of areas with small mining potential), as well as recommendations for the use of minimum resolution DEM and climatology data for slope and drainage analysis. Third, from an operational perspective, authorities should include provisions for post-closure monitoring (groundwater quality and mass movement) at approved sites. Finally, practical policy recommendations include: (a) prioritizing hydrological and geotechnical surveys prior to construction; (b) establishing a mechanism for reviewing sites when new data (rainfall, geological reserves, or fault maps) becomes available; and (c) allocating budgets for mitigation (e.g., slope excavation/stabilization) only for candidates that truly have no alternatives. In light of these findings, the identification of Boto Village as the most suitable site is further strengthened when cross-referenced with land-use considerations. The land use map is presented in the following figure.

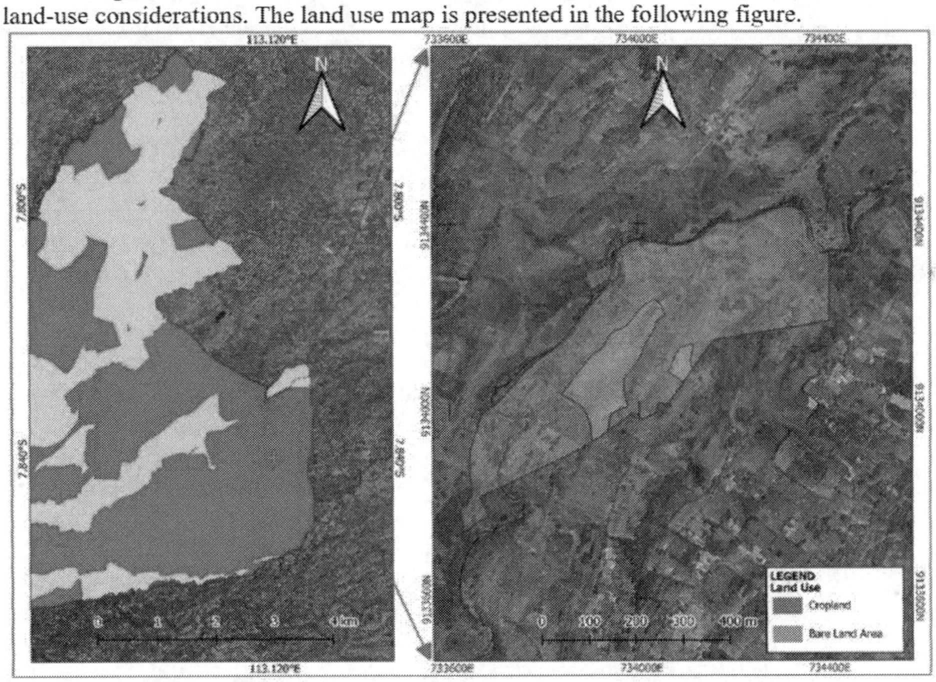

Fig. 7. Landuse Map of Potential landfill sites map of Lumbang District

The analysis indicates that areas with high feasibility are generally located within cropland, bare land, and open-pit mining zones. Nevertheless, although open-pit mining areas are spatially classified as highly feasible, their conversion into landfills is not recommended. This is due to the significant economic value associated with active mining activities, whether in the form of mineral extraction or construction materials, which far outweighs the potential benefits of using the land as a disposal site. Repurposing active mining zones as landfills may generate economic conflicts and hinder the utilization of strategic natural resources. In contrast, bare land is considered more appropriate for landfill development, as it carries the lowest economic value among the three land-use categories. Utilizing bare land not only minimizes potential conflicts with other sectors but also supports efficiency in spatial planning. Therefore, the technical recommendation of this study is to prioritize bare land for landfill construction, while ensuring that environmental factors and long-term sustainability considerations are adequately addressed.

4 Conclusions

This study identified Boto Village as the most suitable landfill site in Probolinggo, based on its favorable lithology of volcanic breccia, medium to deep groundwater conditions, and gentle slopes. The main contribution of this research is the integration of geological resource potential into the exclusion criteria, which strengthens landfill site selection by ensuring that areas with potential future resource use are avoided. However, several limitations remain, including the reliance on a 30 m DEM for slope analysis, incomplete local rainfall data, and limited information on geological resource reserves and local fault systems. Future research should incorporate higher-resolution DEMs, localized climatological records, and detailed geological surveys to reduce these uncertainties. Such improvements would not only refine the model's accuracy but also provide a stronger foundation for updating the national landfill siting regulation (SNI 03-3241-1994) to better address hydrogeological risks and support sustainable waste management planning.

References

1. Rame, L. S., Widiatmaka, W., Hartono, A., & Firmansyah, I. (2022). Multi-criteria decision making for determining landfill location in Malaka Regency, East Nusa Tenggara Province of Indonesia. *Journal of Degraded and Mining Lands Management*, *9*(2), 3405–3413. https://doi.org/10.15243/jdmlm.2022.092.3405
2. Mornya, A. A., Yola, L., & Rafee, M. (2010). Identification of Landfill Sites by Using GIS and Multi-Criteria Method in Batam, Indonesia. *International Graduate Conference on Engineering Siences and Humanities 2010, March 2010.* https://www.academia.edu/545126
3. Randazzo, L., Cusumano, A., Oliveri, G., Di Stefano, P., Renda, P., Perricone, M., & Zarcone, G. (2018). Landfill site selection for municipal solid waste by using AHP method in GIS environment: Waste management decision-support in sicily (Italy). *Detritus*, *2*(June), 78–88. https://doi.org/10.31025/2611-4135/2018.13656
4. Fagbohun, B. J., & Aladejana, O. O. (2016). Integrating knowledge-based multi-criteria evaluation techniques with GIS for landfill site selection: A case study using AHP. *Materials and Geoenvironment*, *63*(3), 169–182. https://doi.org/10.1515/rmzmag-2016-0016
5. Nugroho, A., Kusratmoko, E., & Indra, T. L. (2021). Preferred Site Selection Using GIS and AHP: Case Study in Bangka Island NPP Site. *Jurnal Pengembangan Energi Nuklir*, *23*(1), 51. https://doi.org/10.17146/jpen.2021.23.1.6404
6. Chumaidiyah, E., Dewantoro, M. D. R., Fauzi, P. M., & Kamil, A. A. (2023). Selection of Industrial Sites Using a Web-Based Geographical Information System to Minimize Risks: A Case Study in West Java, Indonesia. *Sustainability*, *15*(22), 16034. https://doi.org/10.3390/su152216034

7. Birawida, A. B., Daud, A., Mallongi, A., Silalahi, S. R., Sila, N., & Todingan, M. (2024). Boolean and Spatial Analysis Using Gis To Determine Landfill Waste on Bunaken Island With an Environmental Health Approach. *Jurnal Kesehatan Lingkungan, 16*(2), 153–165. https://doi.org/10.20473/jkl.v16i2.2024.153-165
8. Titu-Eki, A., & Kotta, H. Z. (2021). Environmental geology assessment on the regional Pitay landfill site: A case study in Kupang, Indonesia. *SN Applied Sciences, 3*(1), 92. https://doi.org/10.1007/s42452-020-03994-3
9. Suryana, D., & others. (2022). Pemodelan karakteristik aliran airtanah sistem porous. *Jurnal Geologi & Sumberdaya Mineral, 23*(1), 35–51.
10. Lubis, R. F. (2000). Hidrogeologi dan potensi resapan airtanah Cekungan Bandung. *Jurnal Riset Geologi & Pertambangan, 10*(2), 273–280.
11. Hendrayana, H., & others. (2024). Karakteristik hidrogeologi dan hidrogeokimia DAS Tempuran, Lereng Gunung Bromo. *Majalah Geografi Indonesia, 38*(1), 48–85.
12. Ayudyaningtyas, A. T., Riniwati, H., & Nugroho, A. M. (2023). Determining The Location of Regional Landfill in Kediri District Using Spatial Approach with Geographic Information System Applications. *Asian Journal of Social and Humanities, 1*(11), 961–976. https://doi.org/10.59888/ajosh.v1i11.92
13. Mukomberanwa, N. T., Chaipa, B., & Mutowo, G. (2025). GIS-Integrated Analytical Hierarchy Process (AHP) for optimal landfill site selection in Kadoma City, Zimbabwe. *Sustainable Environment, 11*(1). https://doi.org/10.1080/27658511.2025.2557727
14. Elshaboury, N., M. AlMetwaly, W., Ali, E., & E.E. Eltoukhy, A. (2025). Sustainable Landfill Site Selection for Construction and Demolition Waste Using Magnitude-Based Fuzzy AHP and GIS Technologies. *Journal of Geovisualization and Spatial Analysis, 9*(1), 27. https://doi.org/10.1007/s41651-025-00227-2
15. Mulualem, B., Ayenew, T., Tesfaw, G., Jothimani, M., & Mulualem, W. (2025). Geospatial and multi-criteria decision analysis for sustainable solid waste disposal site selection in Debre Tabor, Ethiopia. *Discover Sustainability, 6*(1), 475. https://doi.org/10.1007/s43621-025-01365-9

Sustainability Reporting in Mining Industry: Does Board Structure and Ownership Matter?

Afif Yoga Perdana[1], *Ari Kuncara* Widagdo[2*], and *Siti Rochmah* Ika[3]

[1]Magister Program in Accounting, Sebelas Maret University, Surakarta, Indonesia
[2]Department of Accounting, Sebelas Maret University, Surakarta, Indonesia
[3]Department of Accounting, Janabadra University, Yogyakarta, Indonesia

Abstract. This study examines the impact of board attributes and ownership structure on sustainability reporting disclosure (SRD) among mining companies listed on the Indonesia Stock Exchange (IDX) during the 2019–2023 period. Using a panel dataset of 265 firm-year observations, the research employs a fixed effects regression model to analyze how board size, board independence, and various types of ownership concentrated, foreign, institutional, and publica ffect the level of sustainability disclosure based on the Global Reporting Initiative (GRI) index. The findings reveal that board size does not significantly influence SRD, suggesting that the number of board members alone is insufficient to improve transparency. Conversely, board independence has a significant positive effect, indicating that independent directors enhance disclosure quality by strengthening oversight. Among ownership types, foreign ownership shows a significant negative impact on SRD, while institutional ownership is positively but weakly associated. Public ownership demonstrates a strong positive relationship, reflecting the importance of market-based transparency pressures. Concentrated ownership, however, is found to have no significant effect. These findings highlight the pivotal role of corporate governance in shaping ESG transparency in resource-intensive sectors and offer practical implications for regulators and policymakers seeking to promote sustainability accountability in emerging markets like Indonesia.

1 Introduction

Global awareness of sustainability issues has grown rapidly over the past few decades. Society, governments, and investors are increasingly demanding that companies be accountable for the social and environmental impacts of their business activities. Companies that previously focused solely on profit generation are now expected to integrate sustainability into their long-term strategic objectives. Sustainability reporting has emerged as a response to these rising expectations, enabling firms to disclose their performance in environmental, social, and governance (ESG) dimensions. According to Van et al. [1], high- quality sustainability reporting not only reflects corporate transparency but also enhances a

* Corresponding author: widagdo1998@staff.uns.ac.id

firm's perceived value among stakeholders. In this context, corporate success is no longer assessed solely by financial indicators but also by its contributions to sustainable development [2].

In Indonesia, the role of sustainability reporting has become increasingly prominent following the issuance of the Financial Services Authority Regulation (OJK) No. 51/POJK.03/2017, which mandates financial institutions, issuers, and certain public companies to publish annual sustainability reports. This regulation aims to strengthen corporate accountability and promote the integration of ESG principles into business decision-making processes [3]. However, in practice, there remains a significant gap between policy and implementation. Many companies still regard reporting obligations as mere administrative burdens rather than strategic opportunities to improve their long-term reputation and competitiveness [4]. Key challenges contributing to the low quality of sustainability disclosure include limited resources, insufficient understanding of ESG concepts, and weak market-based incentives. Dissanayake et al. [5] emphasize that sustainability reports compiled solely for regulatory compliance without real integration of sustainability values tend to be ineffective in delivering tangible impact.

The mining sector is particularly exposed to sustainability challenges due to its inherent environmental and social risks. Activities such as resource exploration and extraction directly affect ecosystems, air and water quality, and often trigger social conflicts with local communities. Nevertheless, the mining industry also plays a pivotal role in national economic development, contributing significantly to GDP and government revenue. This dual role demands that mining companies manage externalities responsibly. In this context, sustainability reporting is not only a moral obligation but also a strategic necessity to maintain social legitimacy [6], [7], [8]. From a theoretical perspective, Legitimacy Theory provides a compelling framework for understanding corporate sustainability behavior. This theory posits that firms voluntarily disclose information as a means to maintain or repair legitimacy in the eyes of stakeholders and society [9]. In high-risk industries like mining, where environmental damage and social conflict are prevalent, sustainability reporting becomes a tool to justify operations, secure stakeholder approval, and preserve a "license to operate" [10]. However, the extent to which companies adopt such disclosures is often contingent on internal governance and external stakeholder pressures.

Within corporate governance, board structure play a critical role in shaping the direction and quality of sustainability reporting. A larger board size tends to bring greater diversity of perspectives, expertise, and experience, which can facilitate more comprehensive oversight of sustainability strategies [11]. Moreover, the presence of independent directors enhances objectivity in decision-making and mitigates management dominance over information flows. Empirical findings by Aladwey et al. [12] suggest that well-structured boards significantly promote transparent and accountable ESG disclosure practices. Therefore, the influence of board characteristics on sustainability reporting is particularly relevant in high-risk sectors such as mining.

Ownership structure also significantly influences a firm's approach to sustainability disclosure. Institutional and foreign investors are generally more concerned with sustainability practices due to pressure from global financial standards, reputational concerns, and adherence to international norms [13], [14]. Public ownership, characterized by a wider distribution of shares among retail and small institutional investors, increases market pressure for information transparency, including ESG reporting. In contrast, concentrated ownership where a few dominant shareholders exert significant control can reduce public accountability, as strategic decisions may prioritize the interests of major shareholders over broader stakeholder concerns [15]. In developing countries, Shafai et al. [6] found that ESG disclosure levels are strongly influenced by the identity of the controlling parties. Thus, understanding the impact of ownership structure variations on sustainability disclosure is essential [16], [17].

Although numerous studies have investigated the individual impact of board attributes and ownership structure on sustainability reporting, there remains a notable gap in the literature regarding their simultaneous influence, particularly within the mining sector in Indonesia. This study seeks to bridge that gap by adopting an integrative approach, combining key corporate governance dimensions namely board characteristics (board size and board independence) and ownership structure (institutional, foreign, public, and concentrated ownership) into a single empirical model. Focusing on mining companies listed on the Indonesia Stock Exchange (IDX) offers a highly relevant context, given the sector's public visibility and strategic role in the national economy. The findings of this study are expected to provide both theoretical and practical contributions in strengthening sustainability governance practices in Indonesia's strategic industries.

Based on a thorough examination of existing literature and identified research gaps, this study formulates six hypotheses concerning the effects of board attributes and ownership structure on sustainability reporting disclosure.

H1 : Board size has a positive effect on sustainability reporting disclosure.
H2 : Board independence has a positive effect on sustainability reporting disclosure.
H3 : Concentrade ownership has a positive effect on sustainability reporting disclosure.
H4 : Foreign ownership has a positive effect on sustainability reporting disclosure.
H5 : Institutional ownership has a positive effect on sustainability reporting disclosure.
H6 : Public ownership has a positive effect on sustainability reporting disclosure.

2 Material and methods

2.1 Sampling and Data

This research adopts a quantitative methodological framework to examine the influence of board governance attributes on corporate sustainability disclosure, with a particular focus on three key variables: board size, board independence, and gender diversity within the board of directors. These variables were selected based on their relevance in prior corporate governance literature and their theoretical linkage to organizational transparency and accountability. The mining industry was chosen as the research context due to its substantial environmental footprint and increasing global scrutiny regarding Environmental, Social, and Governance (ESG) performance. Companies in this sector often face pressure to adopt responsible practices and report transparently, making them ideal subjects for studying sustainability disclosure behavior.

The analysis draws upon empirical data gathered from annual reports and sustainability reports of publicly listed mining companies on the Indonesia Stock Exchange (IDX), covering a five-year observation period from 2019 to 2023. The total sample consists of 265 firm-year observations, collected from 53 unique companies. This longitudinal approach allows the researcher to observe disclosure patterns and governance dynamics over time, offering more robust insights than a single-year cross-sectional design.

To ensure the credibility and validity of the analysis, multiple linear regression was applied using EViews 12 statistical software. This technique is widely used to assess causal relationships between independent and dependent variables in financial and governance research. Before conducting regression analysis, diagnostic testing was performed, including a multicollinearity test, to confirm that the independent variables did not exhibit high intercorrelation that could bias the regression results. By doing so, the study ensures that the estimated coefficients accurately reflect the unique contribution of each board structure and ownership to sustainability reporting levels.

The sampling method used was purposive sampling, a non-probability technique that allows the researcher to select units that meet certain predefined criteria. The firms included in the sample were selected based on the following specific conditions:

1. The company must have published complete and accessible annual and sustainability reports for each year from 2019 to 2023.
2. The company must be officially listed on the IDX, thereby subject to formal financial and non-financial reporting regulations in Indonesia.
3. The company must operate in the mining sector, an industry particularly relevant for studies on sustainability due to its exposure to environmental risks, community impacts, and governance scrutiny.

All data utilized in the study were extracted from publicly available company documents, including financial statements, sustainability reports, and GRI reference indices. The use of GRI-based indicators strengthens the objectivity of sustainability measurement, as GRI is the most widely recognized global standard for sustainability reporting [18]. The five-year period from 2019 to 2023 was deliberately selected to capture the most recent trends and policy developments, especially in response to regulatory changes such as POJK No. 51/POJK.03/2017, which requires financial institutions and certain public companies to disclose sustainability-related information.

This detailed research design aims not only to identify whether board characteristics have a statistically significant relationship with sustainability disclosure, but also to offer insight into how corporate governance mechanisms can be optimized to enhance transparency in ESG practices particularly in environmentally sensitive industries like mining. The results of this study are expected to contribute both theoretically and practically, providing empirical evidence for academics and actionable insights for policymakers, regulators, and corporate decision-makers in Indonesia and beyond.

2.2 Variables and data analysis

To empirically investigate the relationship between board and ownership structures and the extent of sustainability reporting disclosure, this study identifies several variables based on relevant theories and previous empirical findings. The variables consist of dependent, independent, and control variables, each of which is defined in accordance with established measurements in corporate governance and sustainability disclosure research. The operational definitions of these variables are summarized in Table 1.

Table 1. Variable of the Current Study

Variabel	Operational Definition
Dependent Variables:	
Sustainability Reporting Disclosure **(SRD)**	Total disclosure based on GRI index
Indeoendent Variabels:	
Board size **(BS)**	Total board of directors and commissioners
Board Independence **(BI)**	Precentage of independentboard members to total board members
Ownership structure:	
Concentrade Ownership **(CO)**	Percentage of shares held by the largest shareholder's owning more than 5% of total shares
Foreign Ownership **(FO)**	Percentage of shares owned by foreign investors
Institutional Ownership **(IO)**	Percentage of shares owned by institutional investors
Public Ownership **(PO)**	Percentage of shares held by the general public
Control Variables:	
Firm size **(FS)**	Log natural total asset
Leverage **(Lev)**	Total liabilities divided by total asserts
ROA	Return on assets, net income divided by total assets

To test the hypothesis in this study, panel data regression was conducted using a fixed effect model and the given equation:

$$CSD_{it} = \alpha + \beta1BS_{it} + \beta2BI_{it} + \beta3CO_{it} + \beta4FO_{it} + \beta5IO_{it} + \beta6PO_{it} + \beta7FS_{it} + \beta8LEV_{it} + \beta9$$

$$ROA_{it} + \mu_i + \epsilon_{it}$$

Description:

CSD	: Corporate Sustainability Disclosure
α	: Constant
$\beta1, \ldots, \beta9$	: Independent and Control Variable Coefficients
μ_i	: Fixed effects that capture individual company characteristics that do not change over time
ϵ_{it}	: Error term

3 Results and discussion

3.1 Descriptive statistics of CSD and the Multicollinearity test

Table 2 presents the descriptive statistics of the variables used in this study, including the mean, median, maximum, minimum, standard deviation, and the number of observations (Obs.). These statistics provide an overview of the data distribution and central tendencies for each variable, which are essential for understanding the characteristics of the sample prior to conducting further analysis

Table 2. Statistic Descriptive

	SRD	BS	BI	CO	FO	IO	PO	FSize	Lev	ROA
Mean	0.275	4.132	0.401	0.718	0.278	0.708	0.241	19.357	0.483	0.074
Median	0.259	4.000	0.400	0.775	0.173	0.806	0.206	19.069	0.451	0.040
Max	1.000	11.000	1.000	0.977	0.998	0.999	0.764	23.101	1.416	1.079
Min	0.000	1.000	0.000	0.000	0.000	0.000	0.000	15.231	0.007	-1.012
Std. Dev.	0.285	1.780	0.125	0.190	0.293	0.303	0.144	1.796	0.262	0.173
Obs	265	265	265	265	265	265	265	265	265	265

Descriptive statistics in Table 2 provide a comprehensive overview of the central tendencies and dispersion of the variables examined in this study. The mean value of Sustainability Reporting Disclosure (SRD) is 0.275, indicating that, on average, mining companies listed on the Indonesia Stock Exchange (IDX) disclose approximately 27.5% of the sustainability items based on the Global Reporting Initiative (GRI) index. This low disclosure rate may reflect a lack of regulatory pressure or limited internal capacity to implement comprehensive Environmental, Social, and Governance (ESG) practices in the mining industry—one of the sectors most exposed to environmental and social risks. The relatively high standard deviation (0.285) suggests considerable variation in disclosure levels across firms, reinforcing the view that sustainability disclosure remains voluntary and inconsistently practiced in Indonesia. Regarding board characteristics, the average board size (BS) is 4.132 members, suggesting compact governance structures that may influence the board's ability to oversee a wide range of strategic issues, including sustainability. Furthermore, Board Independence (BI) averages 40.1%, indicating that less than half of board members are independent, which might weaken board objectivity and its role in safeguarding stakeholders' interests, especially in sustainability matters. Prior research shows that independent directors are more likely to promote transparency and ESG disclosure due to their monitoring function.

In terms of ownership structure, the descriptive results show that Concentrated Ownership (CO) has a mean of 0.718, suggesting that in most firms, a single shareholder or small group of shareholders holds more than 71% of the total shares. This aligns with the typical ownership concentration pattern in emerging markets, where control is often centralized. Such ownership may lead to entrenchment issues, where controlling shareholders dominate corporate decisions, potentially limiting the firm's motivation to disclose non- financial information that does not directly benefit them. Foreign Ownership (FO) has an average of 27.8%, indicating moderate participation by international investors, who are generally more demanding in terms of transparency and sustainability performance. Institutional Ownership (IO) is relatively high at 70.8%, which is promising, as institutional investors tend to favor long-term value and good governance, including robust sustainability reporting. Conversely, Public Ownership (PO) remains low at 24.1%, suggesting limited dispersion of shares among retail investors. Regarding financial characteristics, the mean Firm Size (FSize) is 19.357 (log of total assets), with significant variation, indicating the sample covers firms of various scales. Leverage (Lev) shows a moderate mean of 0.483, suggesting that companies are partly funded by debt. Lastly, Return on Assets (ROA) has a relatively low average of 0.074, with a wide range (from -1.012 to 1.079), pointing to profitability volatility among mining firms. These patterns underscore the heterogeneous nature of the sample and provide a solid foundation for exploring how governance and ownership

https://doi.org/10.1051/e3sconf/202565203001

characteristics shape sustainability disclosure in resource-intensive sectors.

3.2 Results of the hypothesis test

To examine the influence of independent variables on the level of sustainability reporting disclosure (SRD), a multiple linear regression analysis was conducted. The results of the hypothesis testing are presented in Table 4, which includes the regression coefficients, probability values (significance levels), and supporting statistical indicators such as the adjusted R-squared, F-statistic, and the probability of the F-statistic. Asterisks (*) denote significance at the 5% and 10% levels. These findings serve as the basis for evaluating whether each independent variable has a significant effect on a firm's sustainability reporting disclosure.

Table 4. Hypothesis Test

Variable	Coefficient	Prob.
C	-3.1485	0.0006
BS	0.0213	0.1181
BI	0.5878	0.0026*
CO	-0.0931	0.3379
FO	-0.2980	0.0063*
IO	0.2399	0.0728**
PO	0.8112	0.0009*
Fsize	0.1566	0.0012
Lev	-0.3615	0.0004
ROA	0.3762	0.0003
Adjusted R-squared	0.5180	
F-statistic	5.6518	
Prob(F-statistic)	0.0000	

Note: *, ** indicate significance at the 5%, 10% levels.

The regression analysis presented in Table 4 offers critical insights into the influence of board attributes and ownership structure on the extent of corporate sustainability disclosure (SRD) among mining firms in Indonesia. The results reveal that board size (BS) does not have a statistically significant impact on SRD ($p = 0.1181$), indicating that having a larger board does not necessarily enhance sustainability transparency. This finding suggests that the effectiveness of the board in promoting ESG disclosure lies not in its size, but in the quality of its members, their expertise, and the degree of engagement in sustainability oversight. This result is consistent with Kumari et al. [19], who observed that board size had no significant effect on environmental disclosure in Indian sensitive industries, positing that larger boards may suffer from coordination inefficiencies that dilute strategic oversight.

Conversely, board independence (BI) is found to exert a significant positive influence on SRD (coefficient = 0.5878, $p = 0.0026$), thus supporting H2. The presence of independent directors appears to enhance the firm's commitment to transparent sustainability reporting, likely because such directors serve as a mechanism for monitoring management behavior and ensuring the protection of broader stakeholder interests. This finding corroborates Nguyen and Nguyen [20], who found a similar positive association between board independence and sustainability disclosure in Vietnamese listed companies. Independent board members are often perceived as more objective and less susceptible to managerial capture, thereby strengthening

https://doi.org/10.1051/e3sconf/202565203001

accountability and transparency.

Regarding ownership structure, the findings demonstrate that concentrated ownership (CO) does not significantly affect SRD ($p = 0.3379$), leading to the rejection of H3. This result suggests that dominant shareholders may not view sustainability disclosure as a priority, possibly due to their privileged access to internal information or their preference for short-term financial performance over long-term ESG strategies. This aligns with the work of Kumari et al. [19], who argued that controlling shareholders in emerging markets often lack incentives to promote voluntary ESG disclosures.

Interestingly, foreign ownership (FO) shows a statistically significant negative relationship with SRD (coefficient = -0.2980, $p = 0.0063$), contrary to the expectation outlined in H4. This unexpected outcome may reflect the behavior of foreign investors in Indonesia's mining sector who are more concerned with financial returns and regulatory efficiency than with promoting ESG transparency. This result is aligned with Zahid et al. [21], who found that foreign ownership does not always enhance sustainability reporting in developing countries, particularly when foreign investors have passive ownership or are not bound by strict home-country ESG mandates.

Institutional ownership (IO) is marginally significant at the 10% level ($p = 0.0728$), thus weakly supporting H5. This finding implies that institutional investors may exert some influence on corporate sustainability practices, though the extent of this influence may vary depending on the institutional type and their ESG engagement intensity. Similar conclusions were drawn by Bello et al. [18], who emphasized the heterogeneity in institutional investor activism regarding sustainability issues in Southeast Asia, where some investors prioritize ESG integration while others focus solely on financial metrics.

Finally, public ownership (PO) shows a strong and statistically significant positive association with SRD (coefficient = 0.8112, $p = 0.0009$), providing robust support for H6. This finding underlines the importance of public pressure and capital market discipline in shaping corporate disclosure behavior. Firms with higher public ownership are more likely to enhance transparency, possibly due to reputational considerations and the need to reduce information asymmetry with dispersed shareholders. This result is consistent with Buallay et al. [22], who reported that firms with broad public shareholding tend to disclose more sustainability information to meet the expectations of a wider and more diverse investor base, especially in high-impact sectors such as mining.

4 Conclusion

This study explores the influence of board attributes and ownership structure on sustainability reporting disclosure (SRD) in Indonesia's mining industry, using panel data from 265 firm-year observations between 2019 and 2023. The empirical findings indicate that board size does not significantly impact SRD, suggesting that the number of board members alone is not a sufficient determinant of transparency. In contrast, board independence demonstrates a significant positive effect, highlighting the importance of independent directors in strengthening sustainability oversight and improving ESG disclosure quality. These results reinforce agency and stakeholder theory, which emphasize the role of independent governance mechanisms in aligning corporate behavior with broader stakeholder interests.

Regarding ownership structure, the results show that concentrated ownership has no significant relationship with SRD, indicating that dominant shareholders may prioritize private control over public accountability. Notably, foreign ownership exhibits a significant negative association with SRD, contrary to common expectations. This suggests that foreign investors in the mining sector may be more profit-oriented and less engaged with ESG issues, especially in contexts where regulatory enforcement is weak. Institutional ownership demonstrates a marginally positive influence, reflecting the uneven ESG activism among institutional investors. On the other hand, public ownership emerges as the strongest positive predictor of SRD, suggesting that market pressures from retail investors and dispersed shareholders play a vital role in driving corporate transparency.

These findings carry meaningful theoretical and practical implications. Theoretically, they emphasize the importance of disaggregating ownership types in governance studies, as different shareholder categories exert distinct pressures on ESG behavior. Practically, regulators should consider reinforcing governance regulations by requiring higher proportions of independent directors and encouraging board accountability through targeted ESG training. Furthermore, capital market development that promotes public participation in ownership can strengthen sustainability governance. Policymakers may also consider implementing stewardship codes or disclosure mandates for institutional and foreign investors to enhance their engagement in corporate ESG practices. In sum, enhancing sustainability reporting in high-risk sectors like mining requires both internal reforms in board structure and external pressures through more inclusive and responsible ownership.

References

1. H. V. Van, N. M. Nguyen, M. Abu Afifa, D. T. T. Vo, and D. Van Bui, "Sustainability reporting quality and firm value in ASEAN+ 3: A series moderation model," *Corp. Soc. Responsib. Environ. Manag.*, vol. 32, no. 2, pp. 2207–2224, 2025, doi: https://doi.org/10.1002/csr.3059.
2. M. J. Rahman, T. Rana, H. Zhu, and Z. Dongying, "Factors Influencing Sustainability Reporting: Evidence from Medicine Manufacturing Industry in China," in *Carbon Accounting for Sustainability and Environmental Management*, Routledge, 2025, pp. 166–182.
3. A. Y. Perdana, A. K. Widagdo, and S. R. Ika, "Governance Attributes Driving Sustainability Reporting in Indonesia's Mining Industry," in *E3S Web of Conferences*, EDP Sciences, 2024, p. 15001. doi: https://doi.org/10.1051/e3sconf/202459315001.
4. T. T. Nu, N. T. T. Hai, L. T. T. Huong, and D. T. N. Lan, "Factors Influencing the Disclosure of Sustainability Reports By Listed Companies on the Vietnamese Stock Market," *Financ. Credit Act. Probl. Theory Pract.*, vol. 2, no. 61, pp. 265–281, 2025, doi: 10.55643/fcaptp.2.61.2025.4707.
5. D. Dissanayake, S. Dissanayake, and R. Ajward, "Do attainment discrepancy and slacks moderate the relationship between board characteristics and level of sustainability disclosure?," *Corp. Gov. Sustain. Rev.*, vol. 5, no. 4, pp. 28–44, 2021, doi: https://doi.org/10.22495/cgsrv5i4p3.
6. N. A. Shafai and H. Abd-Mutalib, "Ownership structures and sustainability reporting of Malaysian listed companies," *J. Pengur.*, vol. 70, pp. 1–14, 2024, doi: 10.17576/pengurusan-2024-70-2.
7. A. Osei, "Exploring the governance–digital transformation nexus: empirical evidence on sustainability reporting using 2SLS, LSDV models, heterogeneity effects, and cluster analysis," *Cogent Bus. Manag.*, vol. 12, no. 1, p. 2491686, 2025, doi: https://doi.org/10.1080/23311975.2025.2491686.
8. R. B. Adomako, P. B. Boakye, and A. Salapki, "Tech-driven governance for sustainability: enhancing environmental disclosure through corporate governance and innovation in emerging market," *SN Bus. Econ.*, vol. 5, no. 7, p. 86, 2025, doi:

https://doi.org/10.1007/s43546-025-00852-5.
9. C. Del Gesso and R. N. Lodhi, "Theories underlying environmental, social and governance (ESG) disclosure: a systematic review of accounting studies," *J. Account. Lit.*, vol. 47, no. 2, pp. 433–461, 2025, doi: https://doi.org/10.1108/JAL-08-2023-0143.
10. T. Gezgin, G. Özer, A. K. Merter, and Y. S. Balcıoğlu, "The mediating role of corporate governance in the relationship between net profit and equity and voluntary disclosure in the context of legitimacy theory," *Sustainability*, vol. 16, no. 10, p. 4097, 2024, doi: https://doi.org/10.3390/su16104097.
11. S. Alshaiba and B. Abu Khalaf, "The impact of board gender diversity on the Gulf Cooperation Council's reporting on sustainable development goals," 2024, doi: https://dx.doi.org/10.22495/cbv20i1art3.
12. L. Aladwey, A. Elgharbawy, and M. A. Ganna, "Attributes of corporate boards and assurance of corporate social responsibility reporting: evidence from the UK," *Corp. Gov. Int. J. Bus. Soc.*, vol. 22, no. 4, pp. 748–780, 2022, doi: https://doi.org/10.1108/CG-02-2021-0066.
13. M. Alta'any, V. Tauringana, A. Zalata, and L. O. Achiro, "Unpacking sustainability reporting dimensions: the impact of board characteristics," *J. Financ. Report. Account.*, no. ahead-of-print, 2024, doi: https://doi.org/10.1108/JFRA-09-2023-0568.
14. Y. Nuhu and A. Alam, "Ownership structure and sustainability reporting: emerging economies evidence," *Int. J. Account. Inf. Manag.*, vol. 32, no. 5, pp. 883–908, 2024, doi: https://doi.org/10.1108/IJAIM-09-2023-0228.
15. D. Arieftiara, S. Widyastuti, M. Masripah, M. Miftah, and S. Wijaya, "The Role of Largest Ownership Structure in ESG Reporting, Tax Avoidance, and Firm Value: Evidence from Emerging Markets," *Qubahan Acad. J.*, vol. 5, no. 2, pp. 293–321, 2025, doi: https://doi.org/10.48161/qaj.v5n2a1682.

Eco-friendly Product: Evaluating Steamed Ecoprint Training's Usefulness, Impact, and Contribution to the Green Economy

Ari Kuncara Widagdo[1], *Rahmawati* Rahmawati[1], *Siti Rochmah* Ika[2*], *M* Rudianto[3], and *Siti Nur Azizah Permata* Hati[4]

[1]Department of Accounting, Universitas Sebelas Maret, Surakarta, Indonesia
[2]Department of Accounting, Janabadra University, Yogyakarta, Indonesia
[3]Department of Fashion Design, Universitas Sebelas Maret, Surakarta, Indonesia
[4]Department of International Relations, Gadjah Mada University, Yogyakarta, Indonesia

Abstract. Steaming ecoprint, a natural fabric printing method utilizing plant-derived dyes, provides ecological benefits and income-generating opportunities for rural areas. This study seeks to evaluate the usefulness, perceived influence, and contribution to the green economy of steaming ecoprint training offered to women in Jaranan, Panggungharjo, Sewon, Bantul. The program aimed to enhance participants' production skills, promote community empowerment, and assist Ayik Ecoprint SME in fulfilling huge market demand. Twenty-three women participated in an interactive training program that imparted technical production skills, sustainable material sourcing, and business management techniques. This study employed a 5-point Likert scale grounded on empowerment and sustainability frameworks to assess three key variables before and after training: usefulness, perceived impact, and contribution to the green economy. Paired-sample t-tests with a 95% confidence level were employed to evaluate the difference in scores. Significant improvements were observed in all metrics: perceived usefulness, impact, and contribution to the green economy. Qualitative feedback revealed that participants acquired practical skills, increased confidence in product marketing, and elevated environmental awareness. The findings indicate that steaming ecoprint training markedly enhances women's skill relevance, livelihood outcomes, and environmental accountability, hence aligning Sustainable Development Goals 5, 8, and 12. This work contributes to the field by introducing a measurement approach for evaluating the effectiveness of ecoprint training in the green economy, an area that has been absent in prior research. The initiative demonstrates the need of conducting training within a Quadruple Helix cooperation involving academia, industry, government, and civil society.

* Corresponding author: ika@janabadra.ac.id

1 Introduction

The world is moving toward sustainability, and green business practices are now seen as a way to protect the environment and promote social and economic growth [1]. Creative businesses, especially those that use eco-friendly production methods, are becoming increasingly popular to balance making money with being responsible for the environment. Ecoprint, a natural dyeing method that transfers plant pigments to fabric, has become a new, eco-friendly alternative to making synthetic textiles. Ecoprint uses plant-based dyes instead of chemical-based pigments, which create toxic waste. This makes the process much less detrimental to the environment and creates unique, high-quality textile designs [2].

The development of ecoprint-oriented small and medium companies (SMEs) in Indonesia is closely aligned with the Sustainable Development Goals (SDGs), namely SDG 5 (Gender Equality), SDG 8 (Decent Work and Economic Growth), and SDG 12 (Responsible Consumption and Production) [3]. Facilitating women's involvement in green business endeavors enhances household income while promoting environmental sustainability and community resilience [4]. The active participation of women in sustainable entrepreneurship enhances socio-economic advantages, as they tend to reinvest profits into family welfare, education, and community development [5].

Ecoprint techniques illustrate this dual capacity—facilitating money generating while promoting sustainable development. Research on female-led green and inclusive MSMEs in Indonesia indicates that eco-friendly product creation might substantially enhance local green economy growth [6]. Furthermore, integrating traditional eco-friendly crafts with entrepreneurship, digital marketing, and environmental legislation can broaden market access and improve competitiveness [4]. This integration is widely regarded as vital for powerful and inclusive economic growth.

The example given is Ayik Ecoprint in Jaranan, Panggungharjo, Sewon, Bantul. With the steaming method, Ayik Ecoprint can create high-quality ecoprint fabrics and ready-to-wear clothes that are water-resistant and have vibrant, rich colors. This is achieved by utilizing plant-based dyes instead of synthetic chemicals. Still, the business has a hard time fulfilling orders during peak demand using just internal resources. Supporting the expansion of small and medium-sized enterprises (SMEs) and empowering local communities, steaming ecoprint training can help local women overcome this production constraint and provide them with a secure livelihood.

Although ecoprint training is increasingly practiced, previous research has predominantly focused on cultural preservation [7], social inclusion, eco-innovation [8], or product promotion, neglecting to systematically evaluate the training's effectiveness on socio-economic and environmental outcomes or to integrate programs within a collaborative ecosystem connecting SMEs, policy actors, and communities. This presents two deficiencies: (i) the lack of a definitive, validated measuring methodology for green-economy-focused training; and (ii) insufficient evidence about the operationalization of Quadruple Helix (QH) collaboration to convert skills into production and market access suitable for SMEs [9].

This research enhances both theoretical understanding and practical application through two contributions. We present and empirically assess a concise measurement approach for evaluating green-economy training across three outcome domains—usefulness, perceived impact, and contribution to the green economy—rooted in training evaluation, empowerment, and green-economy literature. Secondly, we incorporate a Quadruple Helix collaboration—comprising academics, industry (Ayik Ecoprint), local government, and civil society—through a community-based participatory design that aligns training content with SME production limitations and market avenues. Collectively, these contributions advance from mere descriptive community training to a theory-driven, evaluative framework that can be reproduced for enhancing local SME capacity in sustainable entrepreneurship.

The study operationalizes women's economic empowerment and green-economy models into quantifiable training outcomes and evaluates them using pre/post evidence, so advancing training assessment in community-based green entrepreneurship. The QH architecture illustrates how local training may be integrated into SME processes and marketplaces, promoting equitable livelihoods and enhancing ecologically sustainable production.

This study aligns with the UNEP's Education for Sustainable Development (ESD) framework by aiming to cultivate sustainable consumption and production practices, alongside advancing economic development and social inclusion. The research emphasizes the integration of environmental sustainability with local entrepreneurial capacity building, aligning with the broader UNEP aim of enabling communities to tackle environmental and socio-economic concerns. This study highlights the significance of training and collaboration among many sectors—government, academia, industry, and civil society—in fulfilling the global sustainability goal and promoting green entrepreneurship at the grassroots level. The QH collaboration model effectively implements UNEP/ESD's advocacy for multi-stakeholder involvement in sustainable development initiatives.

2 Material and methods

2.1 Research Design

This study utilized a community-based participatory research (CBPR) methodology to evaluate the results of steaming ecoprint training for women in Jaranan, Panggungharjo, Sewon, Bantul. The Quadruple Helix structure was used as the basis for the program. It involved cooperation between academia, business, government, and civil society to make sure that the program was both technically sound and relevant to the community.

2.2 Subjects

The training included 23 women who lived near Ayik Ecoprint. They were chosen because they were willing to take part, could attend the whole program, and could use the skills they learned to make money. The sample size of 23 women was established based on the participants' availability and willingness to engage in the complete training session. The sample was selected based on participant availability; nonetheless, it is crucial to acknowledge that it may not accurately represent the wider population of women in the region. Subsequent research should pursue larger and more randomized samples to enhance the generalizability of the results.

2.3 Training material

The training comprised three fundamental modules: (1) Technical Proficiencies - Steaming ecoprint methodologies, botanical selection, color fixation, and textile finishing. (2) Sustainable Practices — Utilization of natural dyes, minimization of waste, and ethical procurement of plants. (3) Entrepreneurship - Fundamental costing, pricing strategy, product marketing, and potential integration with Ayik Ecoprint's production processes.

2.4 Measurement indicators

The following three indicators were assessed using a structured questionnaire during the pre- and post-training surveys:
Usefulness (A1–A6): How relevant and useful the training material is.
Personal, social, and economic changes that result from training are referred to as the perceived impact (B1–B6).
C1–C6: Contribution to the green economy: sustainable production practices and environmental awareness.

Indicators were derived from the training evaluation model by Kirkpatrick and Kirkpatrick [10], empowerment measurement frameworks [11], and research on the green economy [12,13]. The Likert scale was used to rate each item on a scale of 1 to 5, with 1 representing a strong disagreement and 5 representing a strong agreement. Open-ended questions (D1–D3) were used to gather qualitative feedback. Table 1 exhibits details of the questionnaires.

The use of the Kirkpatrick model was justified based on its well-established utility in training evaluation across diverse contexts. However, future studies could explore alternative frameworks, such as the CIPP (Context, Input, Process, Product) model or the 5-Level Evaluation Model, which may provide additional insights into training outcomes and program efficiency.

2.5 Data analysis

To quantify the differences between the scores before and after training at a 95% confidence level ($\alpha = 0.05$), paired-sample t-tests were performed. However, due to the limited sample size, we perform the normality test first to decide whether to use non-parametric testing such as the Wilcoxon signed-rank test or the paired-sample t-test. Results of the normality test show that the data is normally distributed, and then we use the paired sample t-test. To enhance the quantitative findings, thematic analysis was also used to assess the qualitative input derived from field observations and additional open-ended questions. As usual, the validity and reliability of the questionnaires were also performed to ensure the questions are valid and reliable.

3 Results and discussion

3.1 Demographic of the training participants.

As stated earlier, twenty-three women took part in the steaming ecoprint training. Sixty percent (14 women) are in the age range of 40-45, and the rest, 40 percent (9 women), are age 50 or above. Eighty percent (18 women) are housewives, and the remaining 20 percent (5 women) work in the private sector, such as morning market sellers (3 persons) and kindergarten teachers (2 persons).

Table 1. The detailed questionnaire for the ecoprint training assessment

Code	Question	Indicator	Source of questionaries development
A1	The training material was useful for my daily tasks and demand.	Uselfulnes (how relevant and useful the training is)	Kirkpatrick and Kirkpatrick [10]
A2	I can use the skills I learnt to make ecoprint fabrics.		
A3	The training enhanced my capacity to create high-quality items.		
A4	I am capable of employing the steamed ecoprint procedure independently.		
A5	It was easy to understand the training materials, such as examples and demos.		
A6	I may modify the steaming ecoprint technique to accommodate various plant materials accessible in my region.		
B1	The training has enhanced my prospects for generating additional revenue.	Perceived impact	Goulart [11]
B2	I am sure that I can sell things that were produced utilizing the steamed ecoprint method.		
B3	The abilities I gain in this program will help my family's finances.		
B4	I am more inclined to establish a small enterprise that specializes in textile items.		
B5	The training helped me feel more confident while responding to consumers.		
B6	In my opinion, the production of ecoprint fabrics has the potential to serve as a sustainable source of income.		
C1	I am aware that utilizing natural dyes instead than synthetic dyes is better for the environment.	Contribution to the green economy (Awareness of the environment and sustainable practices)	Loiseau et al. [12]; SDG 12
C2	I am dedicated to employing production methods that are good for the environment.		
C3	I know more about how to cut down on wastage when making fabric.		
C4	I assert that my ecoprint items help promote eco-friendly clothing.		
C5	I am committed to advocating for environmentally sustainable practices in my community.		
C6	I am committed to sourcing plant materials in a manner that is environmentally friendly.		
D1	Which aspect of the training proved to be the most beneficial for you?	Open ended questions	DeVellis and Thorpe [14]
D2	What obstacles do you anticipate encountering when employing the steaming ecoprint method?		
D3	What recommendations do you have for enhancing this training?		

3.2 Validity and reliability test.

Validity testing is used to determine whether statements in a questionnaire are valid. The validity of a questionnaire can be determined by the correlation between the scores of each statement item and the total score. If the significance level is below 0.05 or the p-value is <0.05, the questionnaire is considered valid. Table 2 presents the results of validity test using bivariate Pearson Correlation.

Table 2. Results of validity test

Code	Indicator	R value	P value	Meaning
A1	Usefullness	0.356**	0.000	Valid
A2	Usefullness	0.304**	0.000	Valid
A3	Usefullness	0.332**	0.000	Valid
A4	Usefullness	0.333**	0.000	Valid
A5	Usefullness	0.371**	0.000	Valid
A6	Usefullness	0.325**	0.000	Valid
B1	Perceived impact	0.293**	0.000	Valid
B2	Perceived impact	0.374**	0.000	Valid
B3	Perceived impact	0.314**	0.000	Valid
B4	Perceived impact	0.323**	0.000	Valid
B5	Perceived impact	0.256**	0.002	Valid
B6	Perceived impact	0.303**	0.000	Valid
C1	Contribution to the green economy	0.261**	0.001	Valid
C2	Contribution to the green economy	0.342**	0.000	Valid
C3	Contribution to the green economy	0.400**	0.000	Valid
C4	Contribution to the green economy	0.285**	0.000	Valid
C5	Contribution to the green economy	0.272**	0.001	Valid
C6	Contribution to the green economy	0.243**	0.003	Valid

** means significant at 1 percent.

Reliability tests check the reliability of a questionnaire. A questionnaire is reliable if the answers people give to statements stay the same over time. How reliable an indicator or questionnaire is taken by its Cronbach's alpha (α) value. If the value is greater than 0.60, it means that the indicator or questionnaire is reliable. If it is less than 0.60, it means that the indicator or questionnaire is not reliable. Table 3 displays the test results for reliability tests.

Table 3. Results of reliability test

Variable	Cronbach's Alpha	Number of statements	Meaning
Usefullness	0.720	6	Reliable
Perceived impact	0.785	6	Reliable
Contribution to the Green Economy	0.723	6	Reliable

Based on Table 2 and Table 3, the questionnaires distributed to the respondents or training participants are valid and reliable.

3.3 Results of the paired t-test.

Table 4 demonstrates results of the training particpants' answer in the pre and post ecoprint training. As depicted in the table all indicators show significant improvement.

Table 4. Results of the paired t-test

Code	Pre-training mean	Post-training mean	t value	P value
A1	3.14	4.67	9.26	0.002**
A2	3.17	4.52	9.28	0.001**
A3	3.12	4,72	9.31	0.001**
A4	3.04	4.56	9.31	0.000**
A5	3.12	4.38	8.98	0.000**
A6	3.10	4.46	9.26	0.001**
B1	3.11	4.41	8.92	0.003**
B2	3.21	4.52	8.98	0.001**
B3	3.08	4.56	9.26	0.000**
B4	3.16	4.41	9.21	0.001**
B5	3.09	4.56	8.87	0.000**
B6	3.12	4.41	9.26	0.000**
C1	2.98	4.38	10.12	0.001**
C2	3.21	4.41	10.01	0.000**
C3	3.15	4.46	9.68	0.002**
C4	2.98	4.56	10.01	0.001**
C5	3.14	4.41	10.21	0.000**
C6	2.98	4.38	9.81	0.000**

** means significant at 1 percent.

A substantial improvement in Usefulness is evidence that the training was really useful in facilitating the transfer of skills [10,15]. Perceived impact has improved, which is consistent with empowerment frameworks [11] and indicates that there have been noticeable changes in self-assurance, financial possibilities, and initiative to start a business.

The significant improvements in Green Economy Contribution show that people are becoming more environmentally responsible, which is in line with Loiseau et al. [12]. Incorporating this program within a Quadruple Helix framework enhanced its effectiveness by aligning training with market demands, policy assistance, and community engagement.

This study enhances the literature by presenting a measurement approach designed to assess ecoprint training outcomes within the context of the green economy. Prior studies have addressed ecoprint training in general but have not established a systematic framework of metrics to evaluate training effectiveness across socio-economic and environmental aspects.

3.4 Qualitative feedback.

In terms of post-training participant feedback, 20 out of 23 women give comments and the summary of the comments revealed the following themes: (1) Skill enhancement, i.e. the ability to make high-quality ecoprint products on their own. (2) Economic preparedness means being sure of their ability to market and sell their products. (3) Environmental consciousness, i.e. dedication to natural dyes and waste minimization.

4 Conclusion

This study aims to assess the use, perceived impact, and contribution to the green economy of steaming ecoprint training provided to women in Jaranan, Panggungharjo, Sewon, Bantul. Meanwhile, the purpose of the program was to improve the production abilities of the participants, to encourage community empowerment, and to aid Ayik Ecoprint SME in meeting the enormous demand in the market. Participants' views on the usefulness of training, the impact it had on their lives and finances, and their sense of environmental responsibility all improved significantly after the steaming ecoprint instruction. The measurement framework that was created is a useful way to evaluate comparable programs in entities that are beneficial for the environment. The program illustrates the attainment of skill development, environmental stewardship, and SME support through collaborative, community-based activities by aligning with SDGs 5, 8, and 12. Consider the limitations of this study when interpreting the results. First, a small sample size (23 women) from a single venue limits the generalizability. Second, this study employs a short-term assessment, lacking any long-term follow-up on income or environmental behavior. Future research should undertake longitudinal studies to monitor enduring effects over a period of 6 to 12 months and replicate findings across various geographic contexts for enhanced generalizability. Subsequent research may also incorporate assessments of digital marketing and e-commerce preparedness.

References

1. M. Adamowicz, Green Deal, Green Growth and Green Economy as a Means of Support for Attaining the Sustainable Development Goals. Sustain. **14**, (2022) 10.3390/su14105901
2. E. Saptutyningsih, B. P. Kamiel, R. B. Hartarto, E. P. Johana, T. Anggoro, and S. H. B. D. M. Zailani, Eco-Friendly Product: The Role of Pounding Ecoprint Training on Green Economy. E3S Web Conf. **570**, 162 (2024) 10.1051/e3sconf/202457004003
3. United Nations, (2023) Sustainable Development Goals. https://sdgs.un.org/goals
4. S. Alzamel, Exploring the Role of E-Entrepreneurship in Fostering Future Green Economy and Environmental Policies: A Study on Saudi Women Entrepreneurs. J. Environ. Assess. Policy Manag. **26**, 2450002 (2024) https://doi.org/10.1142/S1464333224500029
5. N. A. A. Abdelwahed, A. Channa, M. A. Al Doghan, and B. A. Soomro, Building a resilient economy: green initiatives and green entrepreneurship for women's empowerment. Int. J. Dev. Issues (2025) 10.1108/IJDI-12-2024-0329
6. P. Anggahegari, R. Agustine, U. S. Azni, R. Halifu, U. B. Robby, and H. H. Setiawan, in *Green Econ. Sustain. Dev.*, edited by V. P. Gupta, A. K. Haghi, and S. Sharief (Springer Nature Switzerland, Cham, 2025)
7. I. S. Md Yusoff, Z. Hamdan, M. F. Zaini, A. S. Mohd Anuar, and A. Md Hashim,

Natural dyes and eco-print techniques elevate the art of nature and local culture. J. Pengguna Malaysia **2**, 166 (2024)

8. T. Juwariyah, D. R. Widi Ashar, A. Priyanto, D. Wahyuningsih, and E. Santoso, Ecoprint Trainning for Cadres and People with Mental Disorders Families. J. Community Serv. Heal. **5**, 049 (2024) 10.26699/jcsh.v5i2.1170
9. Y. Cai and A. Lattu, Triple Helix or Quadruple Helix: Which Model of Innovation to Choose for Empirical Studies?. Minerva **60**, 257 (2022) https://doi.org/10.1007/s11024-021-09453-6
10. D. L. Kirkpatrick and J. D. Kirkpatrick, An Excerpt From. Evaluating Training Programs The four levels. (Berrett-Koehler Publisher, California 2009)
11. C. M. Goulart, A. Purewal, H. Nakhuda, A. Ampadu, A. Giancola, J. L. Kortenaar, and D. G. Bassani, Tools for measuring gender equality and women's empowerment (GEWE) indicators in humanitarian settings. Confl. Health **15**, 1 (2021) 10.1186/s13031-021-00373-6
12. E. Loiseau, L. Saikku, R. Antikainen, N. Droste, B. Hansjürgens, K. Pitkänen, P. Leskinen, P. Kuikman, and M. Thomsen, Green economy and related concepts: An overview. J. Clean. Prod. **139**, 361 (2016) 10.1016/j.jclepro.2016.08.024
13. A. Di Vaio, R. Hassan, M. Chhabra, E. Arrigo, and R. Palladino, Sustainable entrepreneurship impact and entrepreneurial venture life cycle: A systematic literature review. J. Clean. Prod. **378**, 134469 (2022) https://doi.org/10.1016/j.jclepro.2022.134469
14. R. F. DeVellis and C. T. Thorpe, *Scale Development: Theory and Applications*, 5th ed. (Sage publications, California, 2021)
15. A. Alsalamah and C. Callinan, Adaptation of Kirkpatrick's four-level model of training criteria to evaluate training programmes for head teachers. Educ. Sci. **11**, (2021) 10.3390/educsci11030116

Fossil Energy Resilience in a Changing Climate: Risk Analysis, Strategies, and Practical Plan

Berliana Dewi Balqis[1*] and *Bagos* Armansyah[2]

[1]Alumna, Department of Industrial Engineering, Faculty of Engineering, Brawijaya University, Malang, Indonesia
[2]Alumni, Department of Civil Engineering, Faculty of Civil Engineering and Planning, Sepuluh Nopember Institute of Technology, Surabaya, Indonesia

Abstract. Climate change is a global phenomenon driven by increased greenhouse gas (GHG) emissions, resulting in various environmental impacts. Its effects include increasing global temperatures, which manifest as tangible physical risks such as sea-level rise, land subsidence, and extreme weather events. These phenomena can damage infrastructure, trigger oil spills, contribute to environmental pollution, and generate other detrimental impacts. Numerous studies emphasize that the physical risks posed by climate change have become a central focus in sustainable management within the oil and gas (O&G) industry. The O&G is projected to remain a necessary component of the energy mix for several decades, alongside the development of new and renewable energy (NRE), to ensure national energy security during the transition toward Net Zero Emissions (NZE). This study adopts a qualitative approach to examine the resilience of the O&G industry in addressing climate-related risks. The findings include a risk analysis framework, the development of physical and policy strategies, and potential adaptation scenarios implementable by the industry. These results highlight the urgency of proactive climate adaptation efforts to ensure the long-term sustainability of the O&G sector in the coming decades while supporting Indonesia's transition to NZE.

1 Introduction

Fossil energy will remain a crucial requirement in the future, making sustainable investments essential to prevent potential energy supply shortages [1]. As of 2023, fossil fuels continued to dominate global energy consumption, accounting for approximately 81.5% of the total energy mix worldwide [2]. The combustion of fossil fuels results in significant CO_2 emissions, which contribute to climate change by increasing the Earth's surface temperature. One of its primary indicators is the rise in global average surface temperature, which reached 1.1 °C higher during the 2011–2020 period compared to pre-industrial levels (1850–1900) [3]. This warming trend is driven by human activities that generate greenhouse gas emissions [3-4]. Current emission trajectories continue their upward trend, with the oil and gas sector being one of the primary contributors. Recent data reveal that oil and gas production,

* Corresponding author: berlianadewibalqis@gmail.com

transportation, and processing accounted for approximately 5.1 billion tons of greenhouse gas emissions, equivalent to roughly 15% of global energy-related emissions in 2022 [1]. This makes the energy sector one of the industries facing urgent decarbonization demands to meet the Paris Agreement targets of achieving global emission reductions.

Operational activities in the oil and gas sector are significant contributors to greenhouse gas emissions. Nevertheless, Figure 1 explains that fossil energy is still required to meet the national energy demand in the future, as outlined in the National Energy Policy (RUEN). Projections indicate fossil fuels will account for 369.5 MTOE by 2030, compared to 130.4 MTOE from renewable energy. This trend persists through 2040 (520.9 MTOE) and 2050 (696.5 MTOE), demonstrating that oil and gas will maintain vital importance for national energy security during the clean energy transition in the coming decades [5].

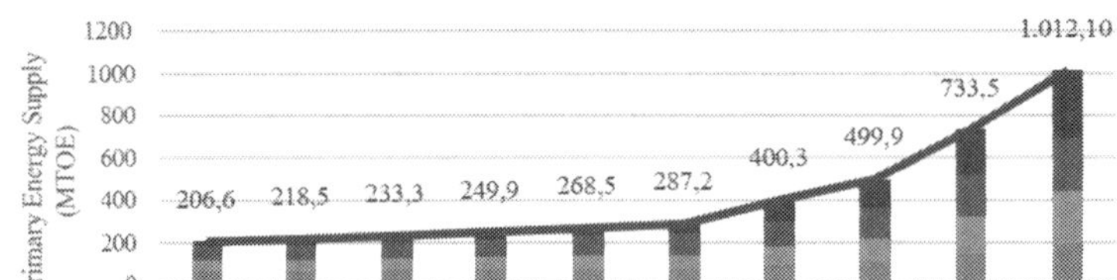

	2015	2016	2017	2018	2019	2020	2025	2030	2040	2050
RNE	20,3	22,7	25,5	29,1	33	38,6	92,3	130,4	212,6	315,6
Coal	67,6	73,5	80,7	88,4	97.6	104,8	119,8	147,5	198,4	255,9
Natural Gas	43	45,9	49,4	53,1	56,5	61	89,5	109,1	171,6	242,9
Crude Oil	75,7	76,4	77,7	79,3	81,4	82,8	98,7	112,9	150,9	197,7
Total	206.6	218,5	233,3	249,9	268,5	287,2	400,3	499,9	733,5	1.012,

Crude Oil Natural Gas Coal RNE Total

Figure 1. Indonesia's Projected Energy Mix

Fossil fuels are expected to remain essential through 2050–2060 during the transition toward renewable energy. The accumulation of CO_2 emissions over recent decades has accelerated climate change and significantly impacted oil and gas facilities, which have become increasingly vulnerable to various climate change-related risks [6]. Rising global surface temperature triggers multidimensional risks, including sea-level rise, extreme weather events, droughts, land subsidence, and intensified precipitation, which pose significant threats to oil and gas infrastructure, including both offshore and onshore facilities, compromising long-term energy security [7-8]. These climate risks pose substantial threats to operational continuity, including infrastructure degradation, production disruptions, supply chain vulnerabilities, increased spill risks, occupational health and safety hazards, and escalating financial liabilities caused by emergency maintenance costs, unplanned downtime, and climate adaptation retrofitting [7-8].

Several scientific studies have examined the effects of climate change on oil and gas infrastructure. Findings suggest that climate change may increase the risk of oil spills and lead to significant damage to large-scale infrastructure [6]. However, most existing studies remain limited in scope and therefore do not provide a comprehensive understanding of the sector's overall vulnerability. This study aims to address this gap by presenting a more comprehensive analysis and proposing more detailed strategies by offering alternative approaches for the application of climate change adaptation and mitigation measures, particularly for companies operating in the fossil energy sector.

Amid escalating climate-related disruptions, the oil and gas industry must establish comprehensive multi-decadal planning frameworks to ensure the continuity of fossil fuel operations. This study aims to assess climate-related risks to oil and gas infrastructure, formulate integrated physical and transitional strategies with corresponding policy recommendations, and propose practical implementation scenarios for industry adoption.

2 Methodology

This study conducts a qualitative analysis, synthesis, and examination of information through a desk review, related to the close relationship between climate change issues and the oil and gas industry, to provide an understanding of the projected needs of countries for fossil energy, impacts, challenges, and risks faced by the oil and gas industry in Indonesia; as well as strategies that can be implemented in the context of climate change issues.

Currently, there is no standardized methodology or official guideline for conducting a desk review. Nevertheless, this approach serves as an alternative method that allows for the integration of rigor and comprehensiveness, while also accommodating the complexity and diversity of available sources [9]. Data is collected from a variety of reliable sources, including academic databases such as ScienceDirect and Google Scholar, international and national climate-related policies, reports from leading organizations such as the IPCC and IEA, and data from the Ministry of the Republic of Indonesia.

Figure 2 illustrates the methodological flowchart adopted in this study, developed with reference to prior studies on desk review approaches [9]. The sources were selected based on two main criteria: (1) relevance to the study discussion, and (2) the credibility of the source, where only publications issued by trusted institutions or academic/industry databases are used. The data was extracted to adjust aspects of the study, such as the background of the issue of future fossil energy needs, the infrastructure of oil and gas companies that are vulnerable to the impacts of climate change, and what strategies oil and gas companies need to implement for business sustainability, while still supporting the achievement of NZE. All data were then categorized and systematically compiled to cover aspects of the study. Ethically, the study also provides appropriate acknowledgment and reference to the original sources and authors cited in the analysis process.

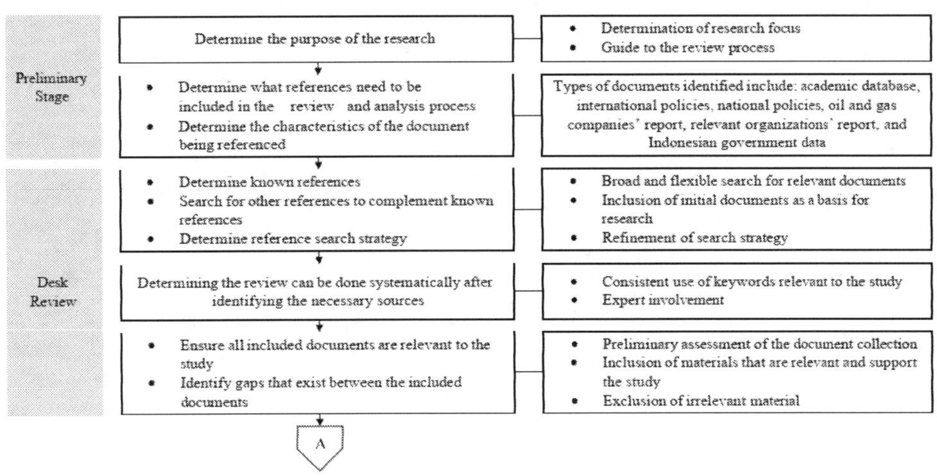

Figure 2. Flowchart for Research Methodology

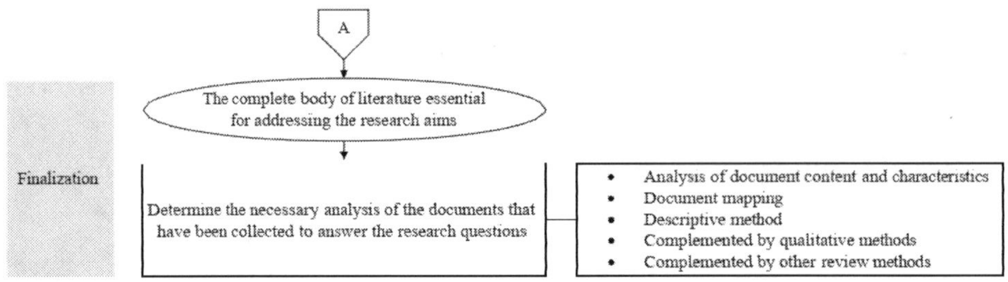

Figure 2. Flowchart for Research Methodology (Continued)

3 Results and discussion

3.1 Assessment of NZE achievements and vulnerability of oil and gas infrastructure

Following the pandemic period, Indonesia has experienced an accelerated growth in greenhouse gas emissions, with the energy sector demonstrating particularly concerning trends. In 2022, energy-related emissions surged by 21% year-on-year, reaching 715 million metric tons of CO2 equivalent (MtCO2e) [10]. The adoption of renewable energy remains significantly below target due to insufficient commitment and leadership in establishing effective policy frameworks to accelerate the transition [10]. Indonesia's energy sector has consistently represented the primary source of greenhouse gas (GHG) emissions since 2020, demonstrating a persistent upward trajectory. The Climate Action Tracker (CAT) rates Indonesia's climate targets and policies as ‘Critically Insufficient' [11]. Indonesia's commitments project an emissions trajectory consistent with a warming scenario exceeding 4°C. This assessment indicates that the ambition of Indonesia's climate policy framework remains inadequate and misaligned with the Paris Agreement's 1.5°C threshold target [11]. According to emission data reports from Indonesia's Ministry of Environment and Forestry (KLHK), the period from 2020 to 2024 demonstrates consistently high and rising emissions from the energy sector [12]. This condition has contributed to a five-year delay in achieving emission reduction targets [13]. Consequently, it can be concluded that fossil fuel energy (oil and gas) remains the dominant energy source in Indonesia's current energy mix [12, 14].

Climate change presents significant risks to the operational continuity and long-term sustainability of the oil and gas sector. Current estimates indicate that approximately 40% of global oil and gas reserves face substantial threats from the physical impacts of climate change [7]. Table 1 presents an example that shows climate change events can cause global oil and gas infrastructure to be damaged or disrupted, resulting in oil spills, company reputation issues, collapses, disruptions to the supply chain, environmental pollution, and disasters to communities [7].

Table 1. Some of the climate change events impacting the oil and gas industry

Event	Location	Period	Climate Impact
Hurricane Ida	Louisiana, United States	2021	• An oil spill occurred in the Gulf of Mexico. • 88% of offshore oil production was halted. • Over 100 oil drilling facilities were rendered inoperable. • Market losses amounted to 17.5 million barrels of oil up to September 2021. • The total estimated loss ranged from US$25 billion to US$35 billion.

Event	Location	Period	Climate Impact
Water-related issues, such as drought, extreme heat, and water scarcity	Large oil & gas companies operating in the US	2016	• Impact on oil and gas production processes due to water requirements. • Increased health and safety risks for workers in the workplace, potentially leading to decreased productivity. • 57% of the 110,000 oil and gas wells in the US are damaged. • 9 of the top 10 oil companies in the US operate 70% or more of their wells in areas with moderate or high water scarcity.
Shifting ice sheets in Russia	Norilsk, Russia	2020	• The melting of the ice sheet weakened the foundation of the storage tank, causing 20,000 tons of diesel to leak into the river. • 45% of Arctic hydrocarbon extraction fields in Russia have the potential to be damaged by 2050, and 40% of the Trans-Alaska Pipeline System (TASP) in the United States is similarly affected. • Russia's physical infrastructure in the Arctic, worth about US$250 billion, is threatened by the melting of permafrost.
Wildfires in Alberta	Alberta, Canada	2015	• 10% of oil and gas production ceased. • Canadian Natural and Cenovus suspended their operations for safety reasons. • Production losses ranged from 80,000 to 135,000 barrels per day.
Wildfires in Fort McMurray	Fort McMurray, Canada	2016	• Production declined by up to 40%, resulting in a reduction of nearly 1 million barrels per day in output. • 665 worker accommodation camp units were destroyed by fire.

In 2021, Hurricane Ida caused a record 55 oil spills in the Gulf of Mexico, severely damaging offshore oil and gas infrastructure, disrupting crude oil supply chains, and resulting in significant economic losses. Between 2011 and 2016, several oil and gas wells in the United States were situated in areas categorized as experiencing extremely high water stress. In 2020, thawing permafrost in Norilsk, Russia, compromised the structural integrity of a fuel storage tank, leading to infrastructure failure, disruption of transport networks, and considerable financial losses. Similarly, major wildfires in Alberta and Fort McMurray, Canada, in 2015 and 2016, had a substantial impact on oil sands operations, reducing production levels and temporarily halting a significant portion of the region's output [7].

Given Indonesia's geographical status as an archipelago nation rich in oil and gas resources, numerous upstream and downstream activities are highly susceptible to the impacts of climate change. Should these impacts materialize, there is a substantial risk of infrastructure collapse and disruptions to the national energy supply. Consequently, long-term strategies are imperative for predicting and analyzing the effects of climate change on both onshore and offshore infrastructure.

3.2 Risk analysis

The physical risks that occur are direct risks from climate change, including risks to facilities and infrastructure, impacts to operations, water and raw materials, and disruptions to supply chains [8]. Table 1 presents data on physical infrastructure damage resulting from climate change phenomena, which serves as empirical evidence that climate change has produced tangible impacts in the field. This study then proceeds with a risk analysis, explained in Table

2, employing a climate event parameter based approach, adapted from previous research. These climate event parameters serve as the foundation for identifying risks and assessing their potential impacts in a general context.

Table 2. Risk analysis of climate change impacts on oil and gas infrastructures

Climate Change Events	Risk Identification	Impact
Sea Level Rise, Land Subsidence, Lightning Strikes, Landslides, Storms, and Flood	Disruption of road access	• Disruption of plant operations • Risk of logistics or operational delays • Workplace accidents due to slippery roads and inundation
	Interference with electrical systems	• Potential damage to electrical systems • Operational shutdown
	Continuous exposure of facility structures to inundation	• Disruption of business activities or supply chains • Pipeline leaks/spills • Degradation of facility structures, potentially leading to collapse • Corrosion of materials or structures • Detrimental impact on the company's image among the public and globally • Compensation and/or legal issues • Fires • Increased maintenance and repair costs
	Collapse of equipment, facilities, or installations	• Disruption of business activities or supply chains • Oil spills • Detrimental impact on the company's image among the public and globally • Compensation and/or legal issues • Fires • Increased maintenance and repair costs
	Lightning strikes on equipment, facilities, or installations	• Disruption of business activities or supply chains • Oil spills • Detrimental impact on the company's image among the public and globally • Compensation and/or legal issues • Fires • Increased maintenance and repair costs
	Open access due to inundation potentially reduces plant perimeter control	• Disruption of business activities or supply chains • Theft or loss of assets • Operational disruptions • Oil spills • Fires • Detrimental impact on the company's image among the public and globally • Compensation and/or legal issues • Increased maintenance and repair costs
	Disruption of waste management facilities	• Oil spills • Ecosystem contamination • Deterioration of company image/reputation • Compensation and legal issues

From Table 2, it is evident that climate change poses significant threats to oil and gas infrastructure. This phenomenon disrupts operations through physical damage, safety hazards, environmental and legal liabilities, and security risks. Proactive adaptation and mitigation strategies, incorporating enhanced designs, advanced monitoring, robust emergency plans, and extensive stakeholder collaboration, can effectively mitigate these disruptions.

3.3 Strategies

In response to the impacts of climate change, it is essential to identify deployable strategies for corporate mitigation and adaptation efforts. These strategies represent a concrete application of risk management phases aimed at bolstering infrastructure resilience, reducing climate exposure to physical risks, and anticipating disruptions to operational and business continuity. Table 3 outlines applicable strategies for oil and gas companies, categorized into physical & transition strategies and policy strategies. Physical & transition strategies encompass physical constructions as an adjustment effort due to climate change, alongside emission prevention and reduction through technology adoption, to achieve resilience and robustness [15]. Whereas policy strategies do not involve physical construction but rather utilize knowledge, practices, or agreements to reduce risks and disaster impacts. This is achieved particularly through the implementation of policies and laws, enhanced public awareness, training, and education [15].

Table 3. Strategies and policies for climate change impacts

Types of Strategies	Strategy	Explanation
Physical & transition strategies	Reinforcing Existing Infrastructure	Strengthening infrastructure involves engineering solutions such as building embankments or retaining walls, raising elevations, additional protection of electrical panels (waterproofing), and repairing structures with materials that are more resistant to corrosion.
	Relocating Facilities	Megaprojects to move critical infrastructure (e.g., refineries, terminals) to less vulnerable zones must consider lifecycle emissions, environmental impact assessments (EIAs), and stakeholder consent. Relocation reduces exposure but entails massive Capital Expenditure (CAPEX) and decommissioning liabilities. For instance, offshore platforms in hurricane-prone regions may be relocated, but this requires a robust cost-benefit analysis that integrates climate models and asset depreciation forecasts.
	Decommissioning (Asset Retirement) and/or Well P&A	Proactive well plugging and abandonment (P&A) and asset retirement mitigate environmental liabilities (e.g., leaks, spills) in flood-prone areas. Advanced monitoring can prioritize assets for decommissioning based on risk matrices. However, this strategy demands regulatory compliance and funding mechanisms to cover abandonment costs.
	Investing in Renewables and CCS/CCUS	Transitioning to renewable energy (e.g., offshore wind, solar-powered extraction) and carbon capture (CCS/CCUS) aligns with net-zero goals while leveraging existing expertise.
Policy strategy	Emergency Response Planning	A scientifically-grounded Emergency Response Plan (ERP) must incorporate: • Climate modeling to predict region-specific hazards (e.g., 50-year flood projections) • Scenario-based contingency planning for various climate events

Types of Strategies	Strategy	Explanation
		• Resource allocation matrices
	Training and Capacity Building	• Classroom and/or safety drill training • Virtual reality simulations for high-risk scenarios • Cross-functional emergency response teams • Regular competency assessments using standardized metrics
	Expanded Coordination Networks	• Establishing public-private early warning systems • Participating in global climate initiatives (e.g., OGCI, IPIECA) • Developing shared response protocols with local communities • Creating joint research programs with academic institutions
	Sustainability-Aligned Business Policies	• Science-based targets for emission reductions • Climate risk disclosure aligned with the financial report • Renewable energy investment roadmaps
	Supply Chain Optimization	• Alternative logistics routing scenarios • Supplier climate resilience assessments

Oil and gas companies are not only responsible for reducing carbon emissions but also for protecting their infrastructure and business continuity from the impacts of climate change. Table 4 details scenarios that these companies can consider for implementing physical and transition strategies. The selection of scenarios for executing these strategies is contingent upon the specific needs and available resources of the respective oil and gas company. This table was developed based on an analysis that considered various relevant factors within the context of climate change adaptation and mitigation in the fossil energy sector.

Table 4. Practical plan scenario

Scenario	Scenario 1: Joint Venture	Scenario 2: O&M Service	Scenario 3: Full Ownership	Scenario 4: Build-Operate-Transfer
Scope of Activities	Construction, Management, Operation, Maintenance, and Financing (Partially)	• Operation and Maintenance • The provision of facilities and financing is carried out by the government or other parties	• Construction, Management, Operation, and Maintenance • Fees are charged to the state or other parties through a fee scheme	Construction, Management, Operation, and Maintenance (Specific Time Frame)
Construction Implementation	*Low Risk* The investment burden will be distributed to other companies	*Low Risk* Investment costs are borne by other parties (e.g., GOI)	*High Risk* The company bears the entire investment cost	*High Risk* The company bears the initial investment costs
Operations and Maintenance	*Medium Risk* The company is responsible for the operation and	*High Risk* The company is fully responsible for the operation and maintenance	*High Risk* The company is solely responsible for the operation	*High Risk* The company is fully responsible for the O&M for a

Scenario	Scenario 1: Joint Venture	Scenario 2: O&M Service	Scenario 3: Full Ownership	Scenario 4: Build-Operate-Transfer
	maintenance of part of the facility	of the entire facility	and maintenance of the entire facility	certain period, which is renewable
Opportunity	Full control of operations so that they get a good economy of scale	O&M cost efficiency	Owns all assets so that it can allocate other resources to more profitable projects/operations	The company can achieve project profitability by implementing pricing strategies that incorporate both cost recovery and appropriate profit margins
Risk	• Large capital expenditure on infrastructure and working capital • Highly susceptible to price changes	• Potential operational and maintenance problems that can lead to potential claims for damages (e.g., penalty)	• Construction, Management, Operation, and Maintenance • Fees are charged to the state or other parties through a fee scheme	• Initial investment risk • Risks to operating costs by the state or other parties affected by economic conditions

Table 4 presents various strategies for oil and gas companies to implement physical and adaptation measures, alongside considerations tailored to the company's needs and available resources. Based on a comparison of these scenarios, joint ventures or inter-company collaborations generally present a relatively lower risk compared to other options. However, the ultimate selection of a scenario is contingent upon the company's specific requirements and resource availability, with a thorough consideration of the risks and opportunities inherent in each.

4 Conclusion

The escalating impacts of climate change pose significant and multifaceted threats to Indonesia's O&G infrastructure, jeopardizing not only operational continuity but also national energy security and environmental sustainability. These vulnerabilities necessitate a paradigm shift from conventional risk management to a proactive, science-based framework for resilience.

To address these challenges, companies must adopt adaptation (physical) & transition strategies and policy strategies, including strengthening facilities, relocating facilities, decommissioning, investing in renewable energy sources, enhancing supply chain resilience, implementing ERP systems, providing workforce training, and fostering multi-stakeholder collaboration, such as government, institutions, and the community. Furthermore, practical plans for implementing adaptation and policy strategies can be pursued through four scenarios: full ownership, joint venture, operations and maintenance (O&M) service, and build-operate-transfer (BOT), which are selected based on organizational resources and capabilities.

Through such strategies, oil and gas companies can adapt to climate-related risks, maintain energy resilience, and advance their responsibility to reduce emissions while transitioning toward sustainable energy.

References

1. International Energy Agency (IEA), "Emissions from Oil and Gas Operations in Net Zero Transitions: A World Energy Outlook Special Report on the Oil and Gas Industry and COP28," 2023. [Online]. Available: www.iea.org
2. Energy Institute, "73rd Statistical Review of World Energy," 2024.
3. Intergovernmental Panel on Climate Change (IPCC), "Climate Change 2023 Synthesis Report: Summary for Policymakers," Geneva, Switzerland, Jul. 2023. doi: 10.59327/IPCC/AR6-9789291691647.001.
4. Intergovernmental Panel on Climate Change (IPCC), "IPCC Sixth Assessment Report, Working Group I: The Physical Science Basis," Intergovernmental Panel on Climate Change (IPCC), Jun. 2021. doi: 10.1017/9781009157896.005.
5. Republic of Indonesia, *Presidential Regulation of the Republic of Indonesia Number 22 of 2017 Concerning the National Energy General Plan*. 2017.
6. J. Dong, Z. Asif, Y. Shi, Y. Zhu, and Z. Chen, "Climate Change Impacts on Coastal and Offshore Petroleum Infrastructure and the Associated Oil Spill Risk: A Review," 2022, *MDPI*. doi: https://doi.org/10.3390/jmse10070849.
7. UN Environment Programme Finance Initiative (UNEP FI), "Climate Risks in the Oil and Gas Sector," Apr. 2023.
8. T. Katopodis and A. Sfetsos, "A Review of Climate Change Impacts to Oil Sector Critical Services and Suggested Recommendations for Industry Uptake," *Infrastructures (Basel)*, vol. 4, no. 4, Nov. 2019, doi: 10.3390/infrastructures4040074.
9. M. Barbieri *et al.*, "Desk review as a methodological approach for identifying policies and gray literature: A case study," *Nurs Outlook*, vol. 73, no. 6, Nov. 2025, doi: 10.1016/j.outlook.2025.102547.
10. IESR, "Indonesia Energy Transition Outlook 2024: Peaking Indonesia's Energy Sector Emission by 2030: The Beginning or The End of Energy Transition Promise," Jakarta, 2023. [Online]. Available: www.iesr.or.id
11. "Indonesia Climate Change Tracker," Climate Action Tracker (CAT). Accessed: Jul. 13, 2025. [Online]. Available: https://climateactiontracker.org/countries/indonesia/
12. "National Greenhouse Gas Inventory System," Directorate of Greenhouse Gas Inventory and Monitoring, Reporting and Verification. Accessed: Jul. 13, 2025. [Online]. Available: https://signsmart.menlhk.go.id/v2.1/app/
13. D. Wuryandani, "Changes to Indonesia's Renewable Energy Mix Target," 2025.
14. Ministry of Environment and Forestry of the Republic of Indonesia, "Ministry of Environment and Forestry Performance Report 2024," 2025.
15. United Nations Office for Disaster Risk Reduction (UNDRR), "The Sendai Framework Terminology on Disaster Risk Reduction 'Structural and non-structural measures.'" Accessed: Jul. 26, 2025. [Online]. Available: https://www.undrr.org/terminology/structural-and-non-structural-measures

Exploring Public Acceptance Model of Biomass as an Alternative Energy for Power Sector: a Bibliometric Analysis and Systematic Literature Review

Meiri Triani[1*], *Shochrul Rohmatul* Ajija[2], *Ahmad* Hudaifah[3], *Sri Cahyaning Umi* Salama[4], *Arivia Fikratuz* Zakia[5] and *Wahyu* Setyorini[2]

[1]PLN Puslitbang Ketenagalistrikan (Research Institute), Jakarta, Indonesia
[2]Department of Economics, Faculty of Economics and Business, Universitas Airlangga, Surabaya, Indonesia
[3]Department of Islamic Economics, Universitas Internasional Semen, Gresik, Indonesia
[4]Department of Islamic Economics, University of Muhammadiyah Malang, Malang, Indonesia
[5]Department of Development Economics, Faculty of Economics and Business, Universitas Pembangunan Nasional Veteran Jawa Timur

Abstract. The driving factor for using bioenergy is to promote a circular economy and enhance environmental sustainability. Biomass has potential as a bioenergy source to reduce the energy sector's dependence on coal and to ensure long-term energy security. Public acceptance is a key factor in shaping an effective energy strategy. Here, we analyzed global publication trends regarding public acceptance of biomass energy development and identified theoretical models using a bibliometric and content review approach. The results indicate an increasing trend in publications about the public acceptance of biomass. Biomass is generally more accepted as an alternative energy source in developing countries than in developed ones. Most models developed to explain public acceptance are based on the theoretical frameworks of the Technology Acceptance Model (TAM) and the Theory of Planned Behavior (TPB). Variables such as financial gain, information, and environmental concern were considered when constructing the biomass energy model.

1 Introduction

The increase in temperature on the Earth due to global warming is a sustainable environmental threat. The use of fossil fuels in the power sector is one of the triggers for increased emissions and causes an increase in the Earth's temperature. Fuels in the form of coal and petroleum contribute 14,766.0 $MtCO_2$ and 11,415.0 $MtCO_2$ to total carbon dioxide

* Corresponding author: meiri2mail@gmail.com

emissions [1]. The adverse effects of global warming can be mitigated through carbon reduction or decarbonization initiatives, with the energy sector playing a particularly critical role in these efforts. Renewable energy is important for mitigating climate change by decreasing dependence on fossil fuels and reducing greenhouse gas emissions. Several researchers also agreed to develop renewable energy as a step towards achieving the 7th and 13th SDG targets, realizing a sustainable environment, obtaining economic benefits such as job creation and investment possibilities, and encouraging innovation and stimulating economic development.

Currently, biomass is one of the alternative energy options for the power sector in many countries. The development of biomass energy is strongly associated with circular economy principles. The implementation of a circular economy aims to increase resource efficiency, reduce waste, and enhance environmental sustainability. Biomass can be derived from agricultural and forestry residues such as wood waste, rice husks, and corn cobs. Therefore, using biomass for electricity generation is part of Waste-to-Energy (WTE) practices. The study estimates that the global biomass potential by 2050 is between 200-500 EJ/year, potentially meeting up to one-third of the projected energy demand [2]. The maximum biomass energy production in China is estimated to be 18,833 PJ in 2030 and 24,901 PJ in 2050 [3]. India also shows substantial biomass potential from energy crops and agroforestry residues. In Australia, although woody biomass represents only 1% of total energy production but offers promising low-cost, low-emission renewable energy solutions.

Most biomass-related research focuses on the resulting impact aspect, technological readiness, and calculation of economic benefits from biomass. Social aspects must also be considered to ensure the sustainability of biomass energy development. Public acceptance is one of the keys that significantly influence energy policy. Lack of community support will hinder the implementation of policies, especially related to climate change, different backgrounds, social and economic factors. And psychological factors also affect public perception of policy implementation. Factors or obstacles such as barriers to social awareness and misunderstandings, and gaps between stakeholders, can affect the biomass industry, which will later impact the implementation of biomass energy.

Considering the importance of biomass acceptance, a study analyzing the development of literature on the topic through a review of previous literature is needed. Developing a research topic, including research methods, is crucial to determine the direction of the following research. Furthermore, the results of previous studies are used as a reference in determining the direction, indicators, or variables, and research models that will be used for future research. The study's main objective is to analyze global publication trends related to public acceptance of biomass as an alternative source for electricity generation. The methods employed in this study include bibliometric analysis and a systematic literature review. Many studies have used both methods to explore biomass-related topics. The scope of research using these two approaches includes biomass supply chains, biomass conversion, and other topics. The topic of utilizing biomass energy remains underexplored. This study also includes a comparative analysis of findings from research conducted in developing and developed countries. The objective is to propose a model for future research on biomass acceptance, drawing upon the variables and indicators identified in previous studies. Thus, there are three research questions in this study: RQ1: What trends are emerging in global publications concerning the public's acceptance of biomass energy?

RQ2: How does the acceptance of biomass energy compare in developed and developing countries?

RQ3: What is the proposed biomass acceptance model?

 https://doi.org/10.1051/e3sconf/202565203004

2 Methods

This study emphasizes the community's acceptance of biomass energy, which combines two research methods, namely the bibliometric analysis method (quantitative approach) and systematic literature review (qualitative approach). The analysis begins with a bibliometric method to search for documents and related information, which are then synthesized and mapped according to predetermined criteria to address the first research objective (RQ1). A systematic literature analysis (SLR) is conducted to provide an in-depth understanding of the selected articles based on specific criteria aligned with the research objectives. Applying SLR helps answer the second and third research objectives (RQ2 and RQ3). By combining bibliometric methods with SLR, this study can enhance the understanding of the current state of the topic being researched. In this context, the study focuses on biomass acceptance as an alternative energy source for the power sector. To develop a proposed model for biomass acceptance as an energy source, the scope of the selected article keywords was expanded to address limitations associated with the predefined keywords. An illustration of the research method is shown in Figure 1 and Figure 2.

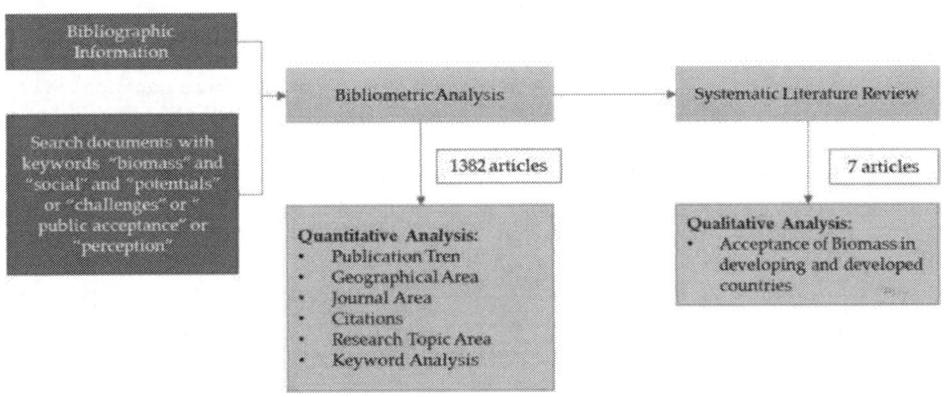

Fig. 1. Research framework for RQ1 & RQ2.

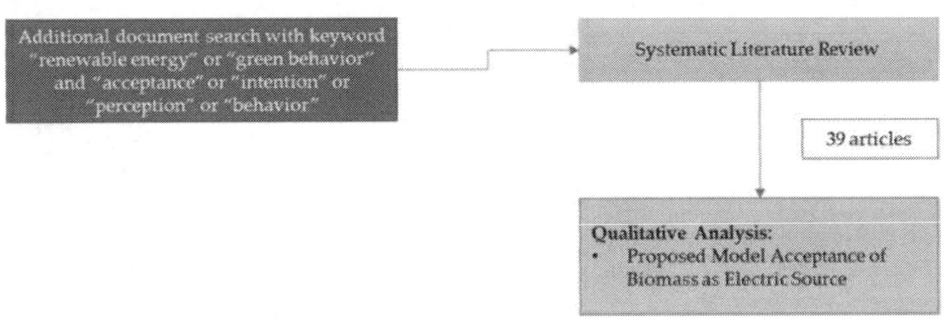

Fig. 2. Research framework for RQ3.

2.1 Bibliometric

Bibliometrics offers the advantage of conducting measurable and objective quantitative analyses based on structured knowledge—i.e., data that is measurable, objective, consistent, and easily accessible. We adopted the stages of bibliometric analysis from Donthu et al. (2021) by defining the aims and scope, choosing the appropriate techniques for conducting bibliometric research, collecting the data, and analysing the results or data from the bibliometric process. Scopus was selected as a bibliographic source because of the quality of its publications, which go through a strict curation process, and comprehensive coverage on one website and limitation settings that tend to be more comprehensive. While Google Scholar is less reliable than Scopus in terms of data accuracy. The detailed stages of bibliometric data collection are as follows:

Step 1. The keywords "biomass" AND "social" AND "potentials" OR "challenges" OR "public acceptance" OR "perception" were applied to identify research areas. From this initial step, 2858 literature were filtered.
Step 2. Selected literature is limited to literature sourced from scientific journals. At this stage, 2291 articles were selected.
Step 3. The literature must be in the final publication stage, not peer reviewed. At this stage, 2287 articles were filtered.
Step 4. The selected type of literature document is an 'article', not a conference paper or other. This stage of selection leaves 1843 articles.
Step 5. Figure 3 below shows the publication trend of biomass acceptance yearly. This study selected articles published in the 2014-2024 range as data for bibliometric analysis.

This is based on the trend of the number of publications that increased rapidly after 2014. After 2014, more than 50 articles were published every year. The number reached 200 articles in 2022 and 2023. This number will undoubtedly increase in 2024, considering the data collected is limited until April 16, 2024.

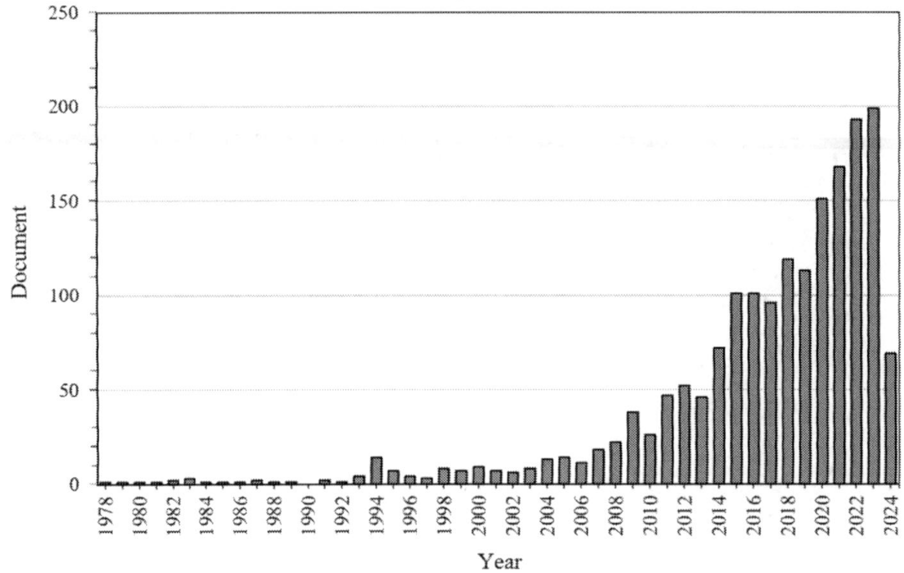

Fig. 3. Number of documents annually.

Through five stages of literature selection, 1382 articles were selected to be analyzed using bibliometrics. This study uses the VosViewer analysis tool. This analysis tool allows the use of selected bibliographic data to be visualized as a network map and calculates quantitative indicators regarding publication trends, research geographic areas, publisher journal areas, citations, research topic areas, and keyword analysis. Visual analysis enables the grouping of manuscripts into clusters based on their strong links.

2.2 Systematic Literature Review (SLR)

Systematic Literature Review (SLR) can help provide answers to research questions that have been formulated in a systematic and structured [5]. After conducting a bibliometric analysis, as many as 1382 articles were selected to determine the literature to be reviewed in depth. The process for obtaining articles suitable for deeper analysis utilized the Prisma Methodology, with specific adjustments as illustrated in Figure 4.

First, this study added the keyword 'acceptance' to the Scopus refined search, resulting in 150 selected articles. An abstract screening process enhances the selected articles' relevance to the research objectives. This process helps identify articles that will be reviewed more thoroughly. The chosen articles must include the keywords "biomass" and/or "acceptance" and "perception" in their abstracts. The screening process resulted in identifying 20 articles, all of which will be analyzed in depth. After a thorough review, seven articles aligned with the research objective of assessing the acceptance of biomass energy. The remaining articles focus on biomass in other contexts, such as its use in cooking stoves and as a substitute for fossil fuels in vehicles. A list of the selected articles can be found in Table 1. The literature review will compare the acceptance of biomass energy in both developed and developing countries.

SLR analysis was again carried out by expanding the range of keywords, as in Figure 2, to add to the reference literature of the proposed biomass model. In this second phase of SLR, the keywords "renewable energy" or "green behaviour" are added because both keywords contain studies on renewable energy (one of which is biomass). In the literature search, the keywords "acceptance", "intention", "perception", or "behaviour" are also added to select publications that focus on topics on the side to renewable energy acceptance or green behaviour.

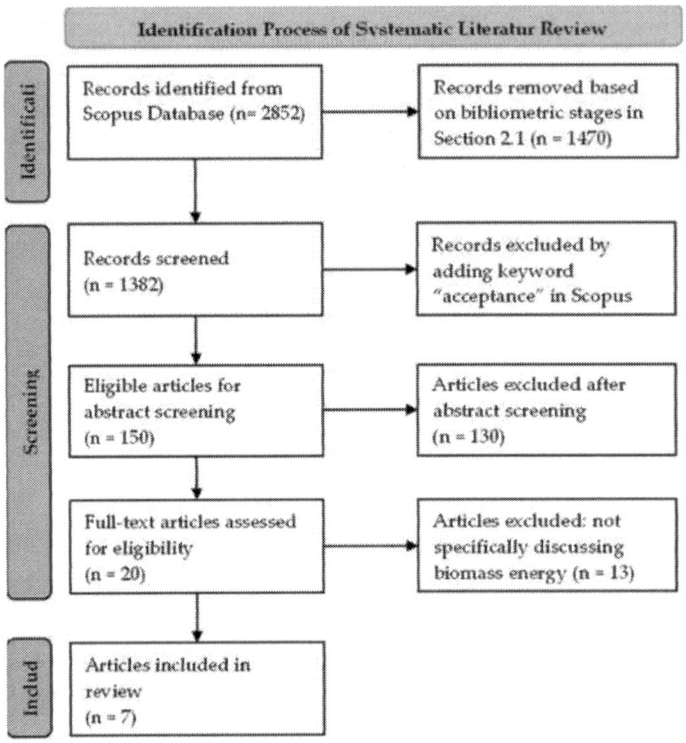

Fig. 4. Flow diagram for systematic literature review.

To comprehend and examine the large amount of secondary data being extracted from the database, the study designed a procedure that followed four stages: (1) designing the review, (2) conducting, (3) analyzing, and (4) summarizing. The selection criteria for screening the inclusion of articles being reviewed adhere to the PRISMA standard (The Preferred Reporting Items for Systematic Reviews and Meta-Analyses) to identify the relevant articles. Bibliometrics is a quantitative approach that focuses on numerical frequencies and occurrences, while the SLR method emphasizes qualitative aspects. To carry out such objectives, the study must conceptualize 4 stages in the review consisting of (1) decontextualization (familiarization with the topic and data), (2) recontextualization (determining the unit of analysis), (3) categorization (classification of all information into condensed presentation), and (4) compilation (writing a concise report).

3 Results

This section presents the findings of both the bibliometric analysis and the systematic literature review (SLR) on the selected topic.

3.1 Bibliometric Analysis

This subsection presents the results of the bibliometric analysis, focusing on several key aspects: publication trends over time, the most influential journals, and the distribution of research across subject areas.

3.1.1 Publication Over Time

The number of articles on public acceptance of biomass energy has increased significantly every time, peaking in 2022 and 2023, reaching 200 articles (Figure 5). This number then decreased drastically in 2024 due to the lack of publications in the current year. Thus, the total publications in 2024 cannot be concluded predictively.

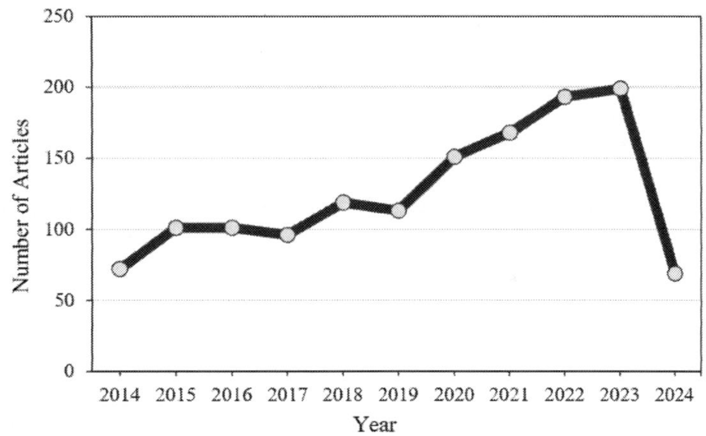

Fig. 5. Development of publications related to the public acceptance of biomass every year.

3.1.2 Geographical Publication Area and Journal Distribution

Research on public acceptance of biomass energy is widely carried out in developed countries. Based on Figure 6, the countries or locations where research studies on this topic are carried out are mostly in the United States and China. The number of studies conducted in the top four countries is over 100 articles/publications. The main journals for the public acceptance of biomass are Journal of Cleaner Production, Science of the Total Environment, Energies, Biomass and Bioenergy, and Sustainability Switzerland (Figure 7).

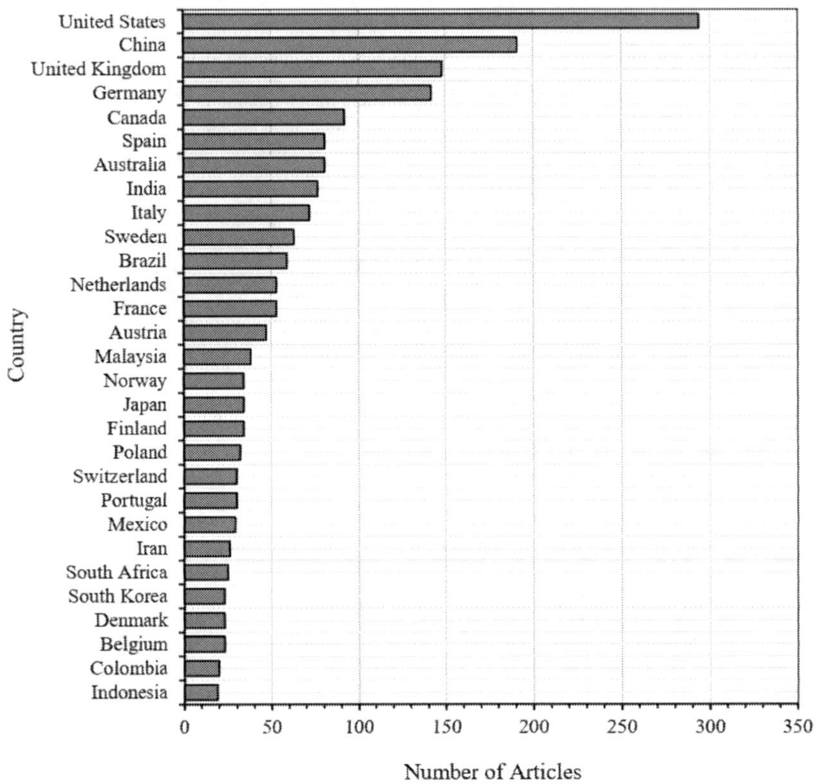

Fig. 6. Country/location of study sites related to public acceptance of biomass.

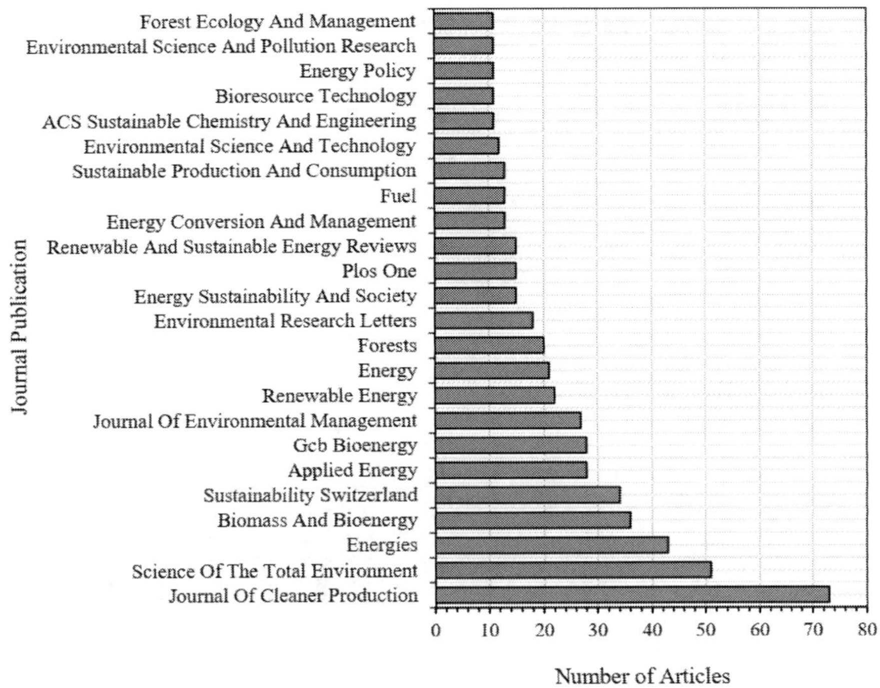

Fig. 7. Journal of publications related to public acceptance of biomass.

3.1.3 Research Topic Area

According to the classification of research topic areas, most journals and articles related to public acceptance of biomass are found in environmental science, energy, agricultural and biological sciences, and engineering, as illustrated in Figure 8.

The analysis of metadata and raw data (tabulation data) from the Scopus database, processed using Vosviewer software, reveals four distinct clusters of keyword networks (Figure 9): 1. Green Group: This cluster focuses on topics related to biomass derived from forest products; 2. Red Group: This group highlights keywords associated with waste, environmental impact, and fossil fuels; 3. Blue Group: This cluster emphasizes themes surrounding community, knowledge, household dynamics, perception, and behavior. It connects various issues concerning public acceptance and the implementation of biomass; 4. Yellow Group: This group contains the least significant themes, including mitigation strategies and bioenergy with carbon capture and storage (BECCS). The yellow group is considered less significant than the other three themes.

 https://doi.org/10.1051/e3sconf/202565203004

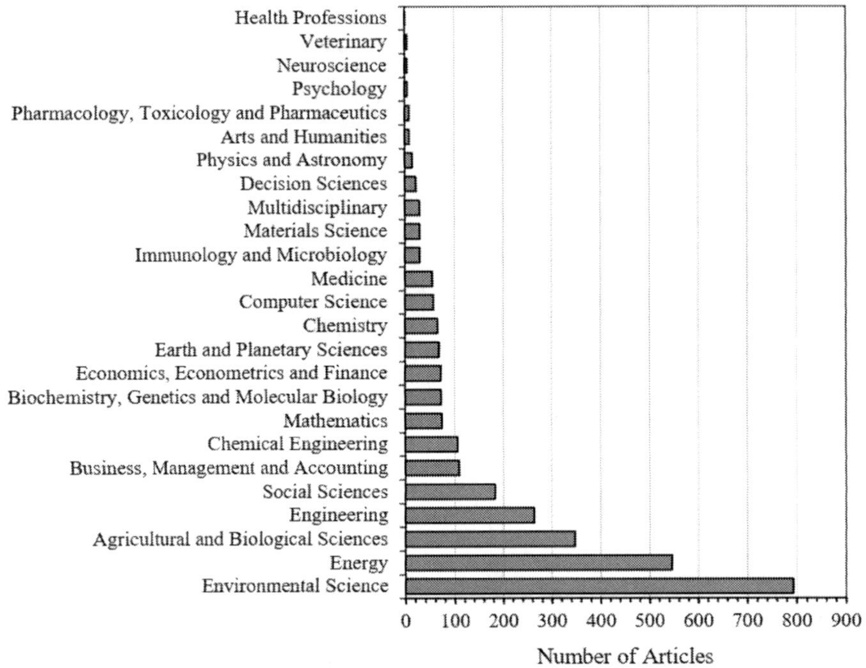

Fig. 8. Research topic areas related to public acceptance of biomass.

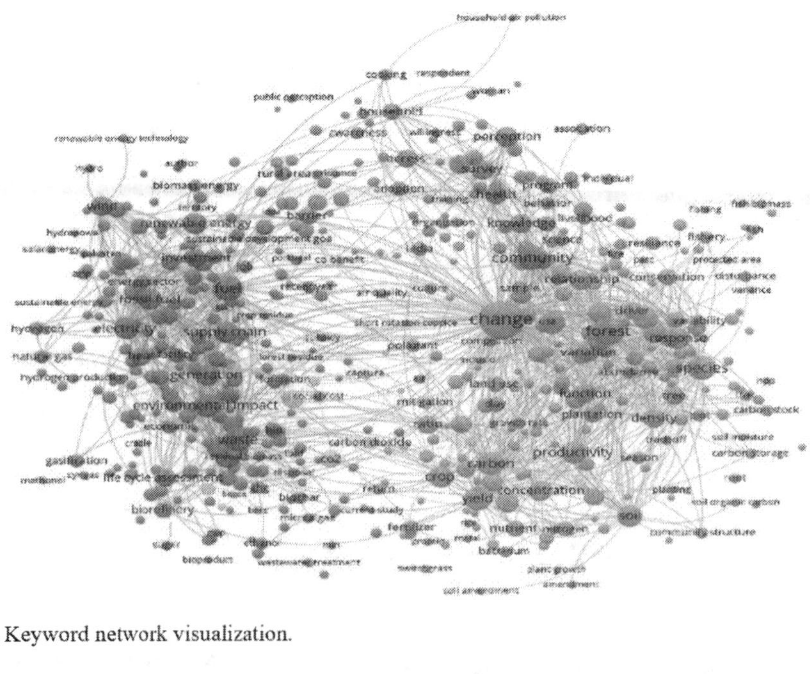

Fig. 9. Keyword network visualization.

3.2 Synthesis of The Literature Review Analysis

This subsection presents the results of the systematic literature review (SLR), focusing on methodological approaches and the contextual differences in admission conditions between developed and developing countries.

3.2.1 Admission Conditions in Developed and Developing Countries

Selected studies on biomass energy were conducted in developed European countries. The public acceptance of various energy source options, both fossil and non-fossil fuels. The study concluded that biomass is one of the energy sources with low acceptance. Additionally, the public is concerned about its application, which is considered to have a higher risk than other renewable energy sources.

The study about public acceptance in Michigan revealed that most respondents opposed using wood biomass as a raw material for power plants around them [6]. However, the same results were revealed in this study. People are more familiar with using biomass as a heating fuel (for example, for the use of stoves, barbecues, and the like) compared to its utilization in the power sector. Research focusing on biomass energy acceptance in developing countries is still very limited. A relevant study on this topic, focuses on Malaysia [7]. The research indicates that biomass is the most effective renewable energy source for enhancing public acceptance of sustainable energy compared to wind, water, or solar power.

The literature review results produced a mixed picture of the community's acceptance of biomass energy. People in developed countries tend to have positive sentiments toward biomass. However, in developed countries, biomass is not yet the preferred choice for electricity generation. In regions such as Australia, Europe, and the Americas, people tend to favor wind and solar energy as the primary alternatives to fossil fuels. In contrast, developing countries exhibit different trends. In Malaysia, biomass is viewed as a promising and optimal alternative among renewable energy sources, primarily because it is believed to create more job opportunities. Additionally, the implementation costs for this energy source are lower and more efficient, requiring less land compared to the other three renewable energies. To enhance the existing literature on developing countries, this study extends the years covered in previous research. Studies in both developing and developed countries agree that wind energy is a viable substitute for fossil fuels. However, while individuals in developed countries tend to prefer solar energy or solar photovoltaics, those in developing countries favor biomass energy.

4 Discussion

According to the results of a systematic literature review on public acceptance of biomass energy, the most frequently employed method is SEM-PLS, which is often based on theoretical frameworks such as the Theory of Planned Behavior (TPB) or the Technology Acceptance Model (TAM).

4.1.1 Theoretical Framework

The framework of TPB has been utilized in several studies. TPB is commonly applied to understand and estimate considerations influencing human behavior. According to Ajzen (1991), the nature of human behavior can be considered and planned. TPB has been successfully used to explain and predict behavior in many behavioral domains, financial behavior, the behavior of accepting treatment for health, and many others. TPB has five main variables, including attitude, subjective norms, perceived behavioral control, intention, and behavior.

Attitudes are considered antecedents of behavioral intentions in the TPB model. Attitude towards behavior affects a person's intention to perform an action. In the TPB, attitude refers to the evaluation of individuals who are positive or negative towards the implementation of certain behaviors. In addition, TPB states that attitudes towards a behavior do not stand alone but interact with two other factors, namely subjective norms and perceptions of behavioral control, in shaping behavioral intentions Subjective norms are an individual's perception of social expectations from those closest to them, such as family, friends, and colleagues, towards the implementation or non-implementation of a behavior Perceived Behavioral Control refers to individuals' beliefs about a certain behavior. Perceived behavioral also directly affects the behavior itself, especially in situations where the individual has a high degree of control over the behavior.

In the TPB, intention is the predictor of actual behavior. Intention reflects an individual's motivation or determination to perform a behavior and is considered a strong indicator of how much effort they will put into carrying out the behavior While behavior is the actual action taken by an individual due to a decision process influenced by intentions and other factors.

Another theory that is usually used is the Technology Acceptance Model (TAM). TAM includes two exogenous/independent variables that influence an individual's decision, including perceived usefulness/PU and perceived ease of use/PEU [9]. PU is defined as the level of confidence that an individual's use of a particular system will improve his or her job performance. Meanwhile, PEU is defined as the individuals' confidence level that the technology offered is easy to use. PEU is also intended to have a direct influence on PU. There have been many studies that use TAM to analyze technology users, such as in the fields of economics and finance, in the field of energy, Health, and many others.

4.1.2 Proposed Model of Public Acceptance of Biomass Energy

Due to the limited literature on public acceptance of biomass, this study also considers community acceptance of other renewable energy. The proposed model integrated the TAM and TPB variables with three additional variables (financial benefits, information, and environmental concerns), as illustrated in Figure 10. Based on the model formed in Figure 10, 14 hypotheses of the TAM and TPB model of community acceptance of biomass are present.

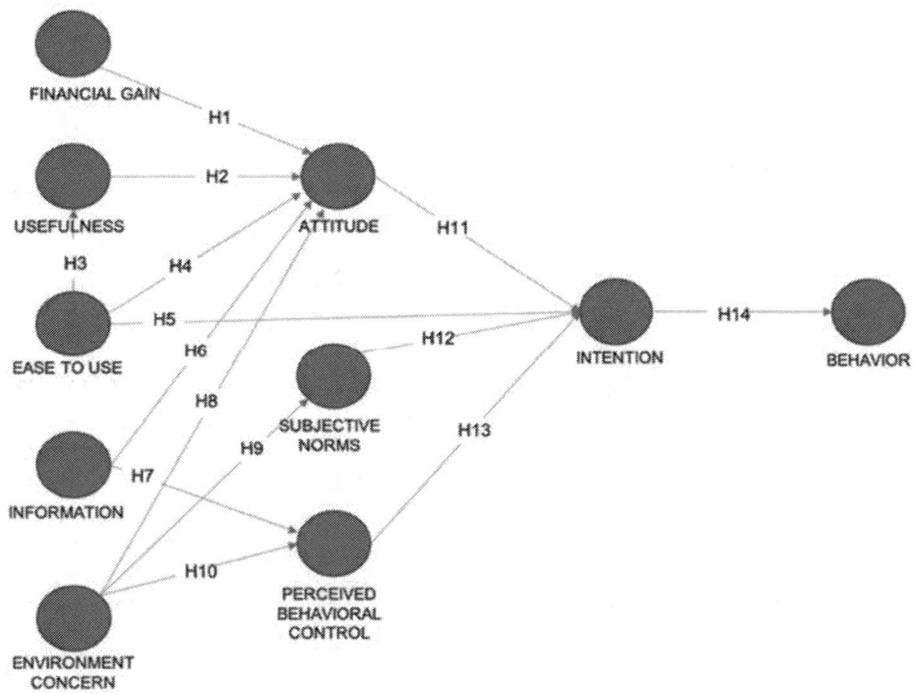

Fig. 10. Proposed model.

H1: There is an influence of Financial Gain on Attitude

Financial gain serves as an antecedent variable in this context. In energy-related research, individuals with a stronger environmental concern are more likely to engage in pro-environmental behavior if they perceive higher benefits and lower costs. Research involving energy, particularly biomass, that incorporates financial gain as a variable is still quite limited and requires further exploration. Existing studies have utilized the financial benefit variable to analyze pro-environmental behavioral factors. For example, Employed mixed-effects logistic regression with respondents from Amazon Mechanical Turk; conducted descriptive analysis methods, ANOVA, t-tests, and logit analyses to assess social acceptance of agricultural biomass across 22 European countries; and analyzed participation factors in renewable energy in Belgium using data from prior research respondents.

H2: There is an influence of Perceived Usefulness (PU) on Attitude Toward Using (ATU)

The relationship of PU to ATU is in the TAM, carried out by Davis & Granić. Previous studies analyzed the influence of PU variables on ATU to analyze the main factors that shape a person's intention to utilize renewable energy in Malaysia and Ghana. Both studies used TAM with the SEM-PLS method, which only measured intentions, not behavior. When individuals believe that alternative energy sources, such as biomass technology, can offer real benefits and meet their needs, their attitude towards adopting such technologies becomes more positive [10].

H3: There is an influence of Perceived Ease to Use (PEU) on Perceived Usefulness (PU)

The PEU is a variable that impacts PU (Perceived Usefulness) within the TAM theory. When users feel comfortable, they are more likely to appreciate its benefits. Several studies have employed the PEU and PU variables in their hypotheses. For instance, research indicates a significant influence of PU on the use of renewable energy in Ghana and Malaysia, as well as the adoption of electric vehicles in China. PEU is shown to be directly related to both Usefulness and Intention.

H4: There is an influence of Perceived Ease to Use (PEU) on Attitude Toward Using (ATU)

The PEU is important in forming a positive attitude. When users find the technology easy to learn and use, they tend to develop a more favorable attitude. Conversely, if a technology feels complicated and confusing, it can lead to frustration and negative attitudes towards its use. Several studies in the energy sector have indicated that PEU significantly affects ATU. The findings suggest that PEU can also enhance users' perception of the value associated with alternative or renewable energy technologies. When users perceive technology as easy to use, they are more likely to recognize its added benefits, such as energy efficiency and environmental advantages. A positive perception of these values is likely to lead to a better attitude toward adopting technology. However, the application of biomass as an alternative energy source remains limited.

H5: There is an influence of Ease to Use (PEU) on Intention (INT)

Unlike the previous hypothesis, the relationship between PEU and INT is not depicted in TAM or TPB (no arrow connects the two). However, recent studies also try to connect the two by examining the intentions of farmers in Italy in adopting technology. The rest of the pre-existing studies used PEU as an antecedent or factor that indirectly affects INT.

H6: There is an influence of Information Publicity (IF) on Attitude Toward Using (ATU)

Understanding the environment can indirectly lead to behavioral changes by shaping intentions to participate in conservation efforts. This understanding can be used to predict both behavioral intentions and actual environmental behaviors, such as the purchase of eco-friendly products. Additionally, if individuals lack sufficient knowledge about environmental issues, they are unable to accurately assess environmental risks [11].

Information variables, which are one of the antecedents of intention and behavior in the energy sector, are still very rare. Previous research used information as an antecedent of the variable intention to use renewable energy and reduce waste in Pakistani society using the SEM method. Unlike other previous studies, a variable of information or knowledge about biomass was used in his research in Hungary, but the method used was multinomial logistic regression which resulted in the information variable had a significant influence on biomass support. It can be concluded that using information variables as direct or indirect factors is rarely used to research biomass, especially in Southeast Asian countries.

H7: There is an influence of Information (IF) on Perceived Behaviour Control (PBC)

Information or knowledge has an indirect influence on the TPB [12]. Sufficient knowledge or information plays a significant role in improving this sense of control. Thus, knowledge not only provides an understanding of how an action can be performed but also increases individual confidence in managing various aspects of the action. Further, good knowledge can help individuals identify the resources and strategies needed to achieve their goals. Knowledge also plays a role in reducing uncertainty and improving predictability. With accurate and comprehensive information, individuals can make better decisions and prepare themselves for a wide range of possibilities. This is especially important in improving Perceived Behavioral Control, as uncertainty is often a barrier to feeling in control. In previous research, information that has a significant influence on PBC is measured using SEM in Pakistani people who use alternative energy in the form of solar panels and environmentally friendly products. The limited number of studies that use information variables on PBC is the need for more research in using these two variables, especially in research related to alternative energy.

H8: There is an influence of Environmental Concern (EC) on Attitude Toward Using (ATU)

The EC and ATU are rooted in how an individual's awareness and concern for environmental issues affect their attitudes towards the use of products or technologies. Individuals who have a high level of concern for the environment tend to develop a positive attitude towards products and technologies that are considered environmentally friendly. Furthermore, high levels of Environmental Concern motivate individuals to seek information and assess products based on the environmental benefits offered. Products that offer real environmental benefits are often seen as more profitable and ethical, thus increasing Attitude Toward Using. Thus, Environmental Concern serves as the main driver in shaping and strengthening a positive attitude towards the use of products that can reduce environmental impacts.

H9: There is an influence of Environmental Concern on Subjective Norms

The EC can be considered as an evaluation of one's view of environmental impacts, includes both cognitive (thought and knowledge-based) and affective (emotion and value-based) evaluations of environmental protection. EC significantly influences social norms [13]. When individuals within a community prioritize environmental issues, they are more likely to promote social norms that advocate for the adoption of technologies, such as biomass technology, which is a form of renewable energy. By raising environmental awareness in the community, these norms supporting biomass adoption can be reinforced, leading to broader and more effective use of this technology. This, in turn, will contribute to achieving sustainability goals and reducing negative environmental impacts.

H10: There is an influence of Environmental Concern (EC) on Perceived Behavioral Control (PBC)

A person who has a concern for the environment will have better control of behavior in technologies that protect the environment, including alternative energy technologies such as biomass. In the Philippines, environmental concern also affects the behavior of Filipino people in using renewable energy technology [13]. Several studies have shown that people who have a higher level of environmental concern are more likely to adopt renewable energy technologies such as solar panels, wind turbines, and electric vehicles. Additionally, they tend to engage in environmentally conscious behaviors, such as conserving energy, recycling, and reducing waste. Environmental concerns can also influence policy decisions related to renewable energy.

PBC in the TPB is explained as one of the factors that affect a person's intention to take action. PBC in TPB refers to the extent to which a person has the ability in the form of knowledge, financial, and previous experience resources to intend to take action, which in this study is the intention to utilize biomass in his environment. Based on the explanation above, it can be concluded that the Environmental Concern (EC) variable influences PBC.

H11: There is an influence of Attitude Toward Using (ATU) on Intention (INT)

In the context of ATU, it evaluates whether individuals have a positive attitude toward a technology. This variable helps identify the factors that influence a person's intention to use energy-related technologies, such as biomass. Positive attitudes toward these technologies significantly affect individuals' intentions to adopt them. By fostering and maintaining a positive attitude toward biomass technology, developers and providers can encourage broader adoption, thus supporting sustainability and energy efficiency goals. A favorable attitude toward biomass technology can serve as a crucial driver for individuals' intentions to adopt it. This positive perception generates an intrinsic motivation to explore and integrate new technologies, reinforcing the decision to commit to their use. Such motivation is vital in stimulating a strong and consistent intention to utilize biomass.

H12: There is an influence of Subjective Norms (SN) on Intention (INT)
Subjective norms influence an individual's intention to adopt technology. Subjective norms can influence an individual's intention to adopt biomass technology through social pressure. When a person feels that the people around them (such as family and friends) support or expect them to use biomass technology, they tend to have a stronger intention to do so. Social pressure can motivate individuals to follow social expectations and adopt behaviors that are perceived positively by their environment. Subjective norms (SN) are used to determine whether they influence intention in utilizing biomass. When the circumstances are supportive, it can influence a person to have the intention to do the same thing as others [14].

H13: There is an influence of Perceived Behavioral Control (PBC) on Intention (INT)
The PBC has a significant influence on an individual's intention to adopt energy-related technologies. When individuals feel they have the knowledge to use technology, they are more likely to have a strong intention to adopt it. By increasing support and providing the necessary resources, technology developers and providers can strengthen the perception of behavioral control, driving the intention and wider adoption of biomass technology. This will help accelerate the transition to more efficient and sustainable energy solutions. Several previous studies have been conducted in several countries, such as Lithuania and countries in Eastern Europe, using variables in the TPB to determine the intention of Turkiye youth in using renewable energy. The methods used by these studies are similar in that they use SEM and path analysis.

H14: There is an influence of Intention (INT) on Behavior
Intentions significantly affect behavioral intentions in the TPB [12,15]. A strong intention to adopt technology is usually followed by a behavioral intention to use it, which in turn drives real action in using the technology. Several studies have used this variable to determine a person's behavior in utilizing or adopting a technology, ranging from research in the energy sector, as carried out in this study, to the financial sector. A clear and planned intention is often followed by a behavioral intention to use the technology. When individuals have established their intention to adopt technology, they are likely to make the necessary plans and preparations to realize those intentions.

5 Conclusions

The significant potential of biomass energy necessitates a comprehensive examination from all perspectives, including the social aspect. Research on public acceptance of biomass energy has surged, driven by global efforts for energy transition. Common keywords in scientific research related to this topic include biomass, social and economic effects, and climate change. Most studies have been conducted in the United States and China. Studies indicate that the acceptance of biomass as an electrical energy source varies between developed and developing countries. Biomass is generally less accepted in developed nations than solar and wind energy. While solar energy is preferred in Europe and the United States, biomass still receives some positive feedback. However, in Denmark, many respondents remain sceptical about wood biomass, whereas in Michigan, biomass is more commonly recognized as a heating fuel. Conversely, in developing countries, biomass tends to be more accepted due to its lower costs and the job opportunities it creates, as seen in Malaysia and Pakistan. As demonstrated in Hungary, such as knowledge, trust in authorities, and education level influence public acceptance. Moreover, the differences in methodological approaches, including the use of SEM, also impacted the research outcomes. To understand public acceptance, a psychological perspective is essential, which can be effectively analyzed using the Technology Acceptance Model (TAM) and the Theory of Planned (TPB) by integrating the five main variables of the TPB model and the independent variables from the TAM model with three additional variables (financial benefits, information, and environmental concerns), various factors influencing the acceptance of biomass can be assessed.

References

1. IEA Data and Statistics; 2020;
2. Dornburg, V.; van Vuuren, D.; van de Ven, G.; Langeveld, H.; Meeusen, M.; Banse, M.; van Oorschot, M.; Ros, J.; Jan van den Born, G.; Aiking, H.; et al. Bioenergy Revisited: Key Factors in Global Potentials of Bioenergy. Energy Environ Sci 2010, 3, 258, doi:10.1039/b922422j.
3. Zhao, G. Assessment of Potential Biomass Energy Production in China towards 2030 and 2050. International Journal of Sustainable Energy 2018, 37, 47–66, doi:10.1080/14786451.2016.1231677.
4. Donthu, N.; Kumar, S.; Mukherjee, D.; Pandey, N.; Lim, W.M. How to Conduct a Bibliometric Analysis: An Overview and Guidelines. J Bus Res 2021, 133, 285–296, doi:10.1016/j.jbusres.2021.04.070.
5. Mengist, W.; Soromessa, T.; Legese, G. Ecosystem Services Research in Mountainous Regions: A Systematic Literature Review on Current Knowledge and Research Gaps. Science of the Total Environment 2020, 702.
6. Huff, E.S.; Mittlefehldt, S.; Bunting, E.; Welsh, J. Attitudes and Perceptions of Wood Energy Technologies in the Great Lakes Region, USA. Biomass Bioenergy 2023, 176, 106897, doi:10.1016/j.biombioe.2023.106897.
7. Qazi, A.; Bhowmik, C.; Hussain, F.; Yang, S.; Naseem, U.; Adebayo, A.A.; Gumaei, A.; Al-Rakhami, M. Analyzing the Public Opinion as a Guide for Renewable-Energy Status in Malaysia: A Case Study. IEEE Trans Eng Manag 2023, 70, 371–385, doi:10.1109/TEM.2020.3046749.
8. Ajzen, I. The Theory of Planned Behavior. Organ Behav Hum Decis Process 1991, 50, 179–211.
9. Davis, F.D. Perceived Usefulness, Perceived Ease of Use, and User Acceptance of Information Technology. MIS Q 1989, 13, 319–340, doi:10.5962/bhl.title.33621.
10. Davis, F.D.; Granić, A. Evolution of TAM. 2024, 19–57, doi:10.1007/978-3-030-45274-2_2.
11. Saari, U.A.; Damberg, S.; Frömbling, L.; Ringle, C.M. Sustainable Consumption Behavior of Europeans: The Influence of Environmental Knowledge and Risk Perception on Environmental Concern and Behavioral Intention. Ecological Economics 2021, 189, doi:10.1016/j.ecolecon.2021.107155.
12. Ajzen, I. The Theory of Planned Behavior: Frequently Asked Questions. Hum Behav Emerg Technol 2020, 2, 314–324, doi:10.1002/hbe2.195.

13. Gumasing, Ma.J.J.; Bayola, A.; Bugayong, S.L.; Cantona, K.R. Determining the Factors Affecting Filipinos' Acceptance of the Use of Renewable Energies: A Pro-Environmental Planned Behavior Model. Sustainability 2023, 15, 7702, doi:10.3390/su15097702.
14. Liobikienė, G.; Dagiliūtė, R.; Juknys, R. The Determinants of Renewable Energy Usage Intentions Using Theory of Planned Behaviour Approach. Renew Energy 2021, 170, 587–594, doi:10.1016/j.renene.2021.01.152.
15. Bosnjak, M.; Ajzen, I.; Schmidt, P. The Theory of Planned Behavior: Selected Recent Advances and Applications. Eur J Psychol 2020, 16, 352–356, doi:10.5964/ejop.v16i3.3107.

Innovative APTA (Air Purification Toxic Adsorbent) for Reducing Heavy Metals and Particulate Matter in Incense Stick Manufacturing

Mika Vernicia Humairo*[1], Tika Dwi Tama[2], *Dian* Puspitaningtyas Laksana[3], Alfian Ihza Mahendra[4], Nadya Nur Hikmah[5], Siti Aisah binti Mokhtar[6]

[1,2,3,4,5] Public Health Department, Faculty of Sport Science, Indonesia, Universitas Negeri Malang, District Semarang 5, 65145, Malang, Indonesia
[6] Department of Community Health, Faculty of Medicine and Health Science, Malaysia, Universiti Putra Malaysia, District Universiti 1, 43400 Serdang, Selangor, Malaysia

Abstract. Small-scale industries significantly contribute to air pollution, causing occupational health hazards and environmental damage, including effects on vegetation and heritage sites. This pre-experimental study was conducted in an incense stick home industry in Dalisodo village, Malang Regency, East Java, Indonesia, from July to October 2024. The APTA (Air Purification Toxic Adsorbent), an air purification system designed to improve workplace air quality, was installed. The system includes five main components: a HEPA filter, zeolite, silica gel, SAP sand, and a dust cotton filter. The Wilcoxon Signed Ranks Test compared concentrations of PM2.5, PM20, Pb, and Mn at 10 locations before and after APTA installation, using a significance level of 0.05. Results showed a significant reduction in manganese levels, with a Z score of -2.703 and p-value of 0.007, indicating effective filtration of Mn particulates. This reduction is important given manganese's known respiratory toxicity, highlighting the potential of APTA to reduce harmful airborne contaminants in small-scale industrial settings.

*Corresponding author: vernicia.humairo.fik@um.ac.id

https://doi.org/10.1051/e3sconf/202565203005

1 Introduction

Small-scale industries significantly contribute to air pollution, posing serious occupational health hazards and environmental damage, including to vegetation and heritage sites. Among these, incense stick manufacturing is a common activity in developing countries, often operating without adequate environmental controls. This industry emits a complex mixture of pollutants such as suspended particulate matter (PM2.5 and PM10), heavy metals (notably manganese [Mn] and lead [Pb]), volatile organic compounds (VOCs), and gases (carbon monoxide, nitrogen dioxide, sulfur dioxide, and polycyclic aromatic hydrocarbons) [11]. The composition of an incense stick consists of herbal and wood powder, fragrance material, staining matter, adhesive powder, and a bamboo stick and the production involves several steps. First, powdered aromatic woods, herbs, spices, and other ingredients are accurately measured and mixed. The mixed powders are sifted to ensure even mixing and removal of unwanted particles. Warm water and mineral colors are gradually added to knead the powders into a dough-like material. This dough is then extruded or pressed through molds to form cylindrical shapes resembling sticks. These pollutants arise during both production and burning processes, largely due to incomplete combustion and handling of raw materials.

Exposure to these airborne contaminants is associated with a range of adverse health effects. PM can penetrate deep into the respiratory system, causing oxidative stress, respiratory diseases, and cardiovascular issues, while Mn and Pb are neurotoxic metals linked to neurological impairments and cognitive dysfunction. VOCs and gaseous pollutants further contribute to respiratory and systemic health risks. Recent data suggest that workers in small-scale incense industries exhibit elevated prevalence of respiratory symptoms and reduced lung function, underscoring the urgent need for effective air quality management [13]. Due to these pollutants, assessing and controlling air quality in incense stick industry environments is crucial because can give impact to air and health quality. Incense stick industry poses significant health risks due to emissions of manganese (Mn), lead (Pb), and particulate matter ($PM_{2.5}$ and PM_{10}). These pollutants are linked to neurotoxicity, respiratory diseases, and cardiovascular disorders among workers and nearby communities.

Recent studies highlight the effectiveness of hybrid filtration systems in industrial applications using zeolite-based filtration that has been proven to remove lead and manganese from contaminated air streams, with clinoptilolite zeolite showing high adsorption capacity for heavy metals, combination HEPA with activated carbon composites significantly reduce PM and VOC emissions in incense production facilities, where traditional fabric filters fail and biofiltration enhancements can further degrade organic pollutants, though they are less effective for inorganic metals like Mn and Pb [7]. This multi-layer filter approach exemplified by the APTA system demonstrates how combining physical filtration with adsorbent materials leads to improved air purification effectiveness, crucial for controlling industrial emissions of harmful substances. Occupational exposure limits and regulatory standards are established worldwide to protect workers from harmful airborne pollutants. These standards are vital in industries with potential exposure to toxic substances and fine particles, including small-scale manufacturing sectors like incense stick production. The recommended occupational exposure limit for respirable manganese dust is generally around 0.1 mg/m^3. Lead exposure is tightly regulated due to its high toxicity even at low levels. Regulatory agencies like OSHA set enforceable limits to minimize worker lead exposure. Air quality standards for particulate matter by organizations such as the World Health Organization (WHO) set annual average limits of 5 $\mu g/m^3$ for $PM_{2.5}$ and 15 $\mu g/m^3$ for PM_{10}, with higher short-term limits [5].

https://doi.org/10.1051/e3sconf/202565203005

Innovative filtration systems such as the APTA filter system can play a vital role in reducing these harmful emissions by using multi-layer filters including HEPA filters, zeolite, silica gel, SAP sand, and dust cotton filters to effectively capture manganese, lead, PM2.5, and PM10 particles. HEPA filters capture 99.97% of $PM_{2.5}$ and PM_{10} particles down to 0.3 microns, preventing deep lung penetratio, and then zeolite and silica gel can adsorb heavy metals especially Mn, Pb, and VOC through ion exchange and surface binding. SAP (Super Absorbent Polymer) layer enhances moisture retention and improving the adsorption efficiency of gaseous pollutants. And the last layer dust cotton filters act as a pre-filter, trapping larger particles and extending the lifespan of HEPA and chemical adsorbents [14]. Multi-layered filtration systems offer significant advantages over single-layer filters, particularly in capturing and adsorbing complex manganese, lead, $PM_{2.5}$, and PM_{10}, which are common in industrial emissions like those from incense stick industry. Occupational exposure limits for respirable Mn dust are generally around 0.1 mg/m^3, with stringent regulations for Pb due to its high toxicity. WHO recommends annual average limits of 5 $\mu g/m^3$ for PM2.5 and 15 $\mu g/m^3$ for PM10. However, these thresholds are frequently exceeded in small-scale industrial settings, hopefully APTA can increased adsorption capacity from multi-layer filters, improved flow distribution and reduced changeling by offsetting the alignment of pores or channels between consecutive layers, enhanced removal of multiple pollutant types with combining physical filtration elements, and extended filter lifespan with dust cotton as a preliminary filter [10]. So, this research aims to analyze APTA's performance and quantify reduction rates for Mn, Pb, $PM_{2.5}$, and PM_{10} in incense stick industry environments using lab and field tests.

2 Method

2.1. Research Design

This pre-experimental study was conducted in an incense stick home industry, Dalisodo village, Malang Regency, East Java, Indonesia, from July to October 2024. The APTA (Air Purification Toxic Adsorbent), an air purification system designed to enhance the air quality, was installed in the workplace area. It consisted of five main components, including a HEPA filter, zeolite, silica gel, SAP sand, and a dust cotton filter.

In this study, we focused on measuring the pollutant concentrations, such as PM2.5, PM10, Pb, and Mn, to assess the air quality of the indoor workplace environment of incense stick production. PM2.5 and PM10 concentrations were measured by the HTI HT-9600 Particle Counter (Dongguan Xintai Instrument Co Ltd, Dongguan City, PR China). Meanwhile, the measurement of Pb and Mn concentrations in the workplace air was carried out by collecting particulate samples, which were then sent to the Faculty of Mathematics and Natural Sciences laboratory, State University of Malang, for testing using the Atomic Absorption Spectrophotometry (AAS) technique. The data on those parameters were collected from 10 areas of the workstations over a working day. Differences in M2.5, PM10, Pb, and Mn concentrations before and after the APTA installation were analyzed using a Wilcoxon Signed Ranks Test, with significance set at $p < 0.05$.

A single workday experimental intervention was assess the effectiveness of the APTA filter system in reducing airborne pollutants (Mn, Pb, $PM_{2.5}$, PM_{10}) within an active incense stick manufacturing area. The approach is informed by validated air filtration intervention methods reported in journal literature for industrial and indoor air quality assessment.

2.2. Sampling Site and Conditions

Sampling was conducted within a small-scale incense stick production facility characterized by manual and semi-mechanical operations. The workplace environment consisted of a production area where workers perform sequential processes including powder mixing, dough preparation, shaping, drying, and packaging incense sticks. The area is typically enclosed but allows some natural ventilation through doors and windows, with frequent worker presence during production shifts.

The facility's ambient temperature ranged from 25 to 30°C, with relative humidity between 50% and 65%. Ventilation was limited, contributing to possible accumulation of airborne pollutants. Sampling locations were selected at workers' breathing zones and central points within the production area to represent exposure levels during active production.

2.3. Data Analysis

The reduction efficiency of the APTA filtration system was calculated by comparing pollutant concentrations measured before and after the filter installation. Evaluating the statistical significance of pollutant reductions used Wilcoxon signed-rank test, a non-parametric test for paired data. This test assessed whether the median pollutant concentrations significantly differed between pre- and post-intervention measurements, accounting for potential non-normality of the data distribution. This analysis used significance level of $p < 0.05$. $P<0.05$ was used to determine statistically meaningful reductions attributable to the APTA filter system. Data analysis was conducted using statistical software SPSS 21.

3 Result and Discussion

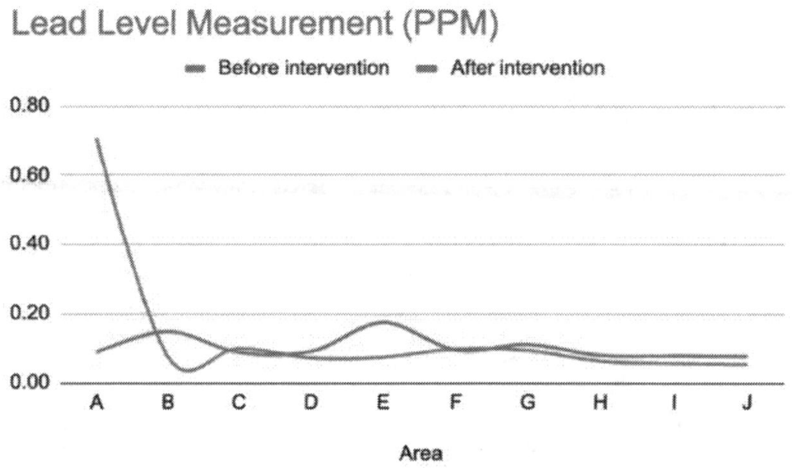

Fig. 1. Pb concentration before and after the installation of APTA

Illustrates the concentration levels of lead (Pb) measured at 10 different locations within the incense stick production site. The x-axis indicates the sampling locations numbered 1 to 10, while the y-axis represents lead concentration levels, presumably in ppm. The graph reveals variability in lead concentrations across these locations, with some points exhibiting higher levels, potentially near active production or material processing areas. Location number 1 identified as the entrance area of the industry where production activities take place shows the highest lead concentration before the intervention. This reflects concentrated emissions or airflow patterns that bring higher pollutant levels near the entry point during active manufacturing. After the APTA filter installation, lead levels decreased across most locations, with particularly notable reductions seen at location 1 and surrounding areas, indicating the filter system's impact on controlling airborne lead dust near the pollution source.

The highest lead levels at the entrance suggest that this area be a major ingress and egress point for airborne contaminants during production. Workers passing through or stationed in this zone are potentially exposed to elevated lead levels, posing health risks including neurotoxicity and kidney damage [10]. The post-installation APTA, data demonstrate a clear trend of lead reduction throughout the facility, especially near the source. This indicates that the APTA system's multi-layer filtration components (HEPA, zeolite, silica gel, SAP sand, dust cotton filters) are effective in trapping lead particles and limiting their dispersion. The physical trapping of particulates combined with zeolite's ion-exchange properties contributes to this performance [2]. Reducing lead exposure in critical zones such as location 1 is essential to minimize chronic health effects among workers. The positive impact seen underscores the importance of localized filtration solutions directly addressing emission hotspots in small-scale industries.

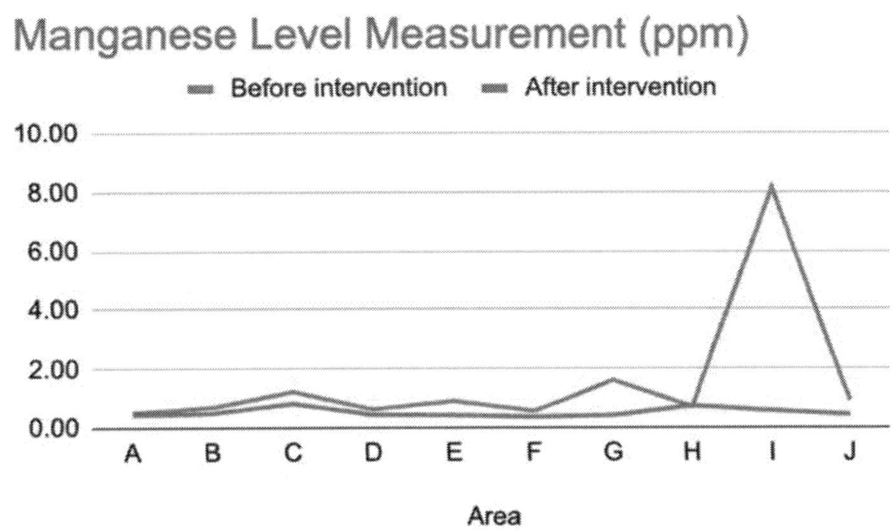

Fig. 2. Manganese concentration before and after the installation of APTA. Below the figure

This line graph Figure 2. Manganese concentration before and after the installation of APTA illustrates airborne manganese (Mn) concentration levels measured at 10 specified locations within the incense stick production facility, both before and after the installation of the APTA filtration system. The x-axis shows the measurements taken at the 10 locations distributed across the workplace, while the y-axis represents manganese concentrations.

Location 9, identified as the drying area for incense sticks after production, shows the highest manganese concentration before the APTA intervention. This suggests that manganese particulates accumulate notably in this specific area, likely due to dust and residue settling during the drying process. This elevated presence of manganese during drying can be attributed to the resuspension of fine particulate matter laden with metal residues, as the

 https://doi.org/10.1051/e3sconf/202565203005

drying area typically involves handling large quantities of incense sticks with surface residues and combustion remnants [10].

The peak of manganese concentration at drying area shows that that area is a critical exposure hotspot for Mn, possibly due to particulate dispersion from both the manufacturing and drying stages. This aligns with studies indicating that dust and metal emissions tend to concentrate in specific production phases, increasing occupational exposure risks. Following the implementation of the APTA system, manganese concentrations decreased substantially across most measurement sites, including a marked reduction at the drying area. This reflects the filtration system's capacity to effectively reduce airborne manganese particles in zones with high dust and particulate generation. The reduction of manganese levels across all locations, especially the highest concentration site, underscores the multi-layer filtration system's efficiency. Components like zeolite and HEPA filters work synergistically to adsorb manganese ions and physically trap fine particles, consistent with previous research demonstrating zeolite's effectiveness in metal adsorption and HEPA's in particulate removal [13].

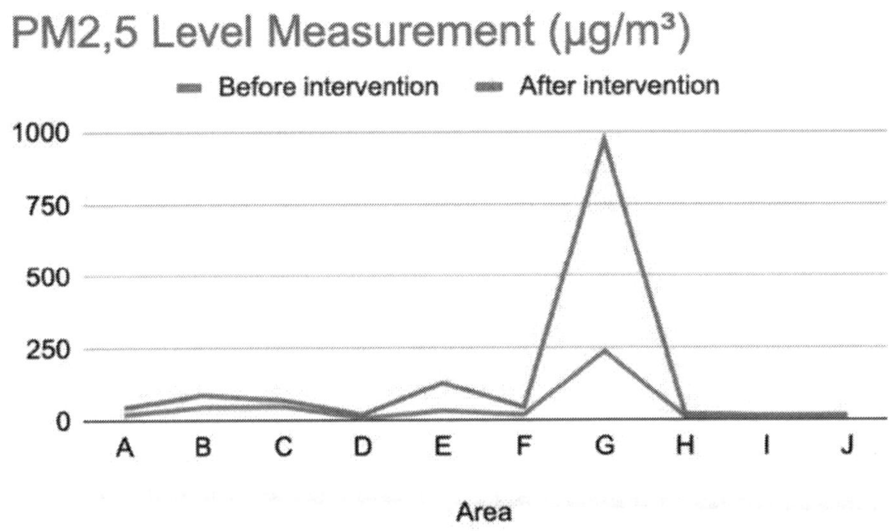

Fig. 3. PM2,5 concentration before and after the installation of APTA. Below the figure

Based on Figure 3. PM2,5 concentration before and after the installation of APTA The graph depicts the PM2.5 levels measured at 10 different locations before and after the installation of an Air Pollution Treatment Apparatus (APTA). The x-axis represents the location numbers, and the y-axis shows the PM2.5 concentration levels. Location number 6 stands out with the highest lead levels, identified as the area where raw material mixing for incense sticks occurs. Before APTA installation, the PM2.5 levels at several locations, especially location 6, were the highest, indicating significant particulate pollution likely due to the mixing activities in the incense stick production process. After the installation of APTA, a substantial reduction in PM2.5 levels is observed across all 10 locations, demonstrating the device's efficiency in lowering particulate matter in the air. Location 6, with the lead highest PM2.5 levels, showed a remarkable decrease, underscoring the critical role of emissions control in areas with intensive raw material handling [15].

https://doi.org/10.1051/e3sconf/202565203005

The spatial distribution of PM2.5 concentration before APTA aligns with typical emission patterns in industrial settings related to incense production, where particulate matter is primarily released during material mixing and combustion processes. The reduction achieved post-APTA aligns with studies showing that air purification and treatment technologies can reduce indoor and localized particulate pollution significantly, improving air quality and reducing health risks associated with PM2.5 exposure. This reinforces the importance of controlling particulate emissions during incense production processes, especially in raw material mixing zones where PM2.5 can be concentrated. Another supporting reference discusses the improvements in indoor air quality following the installation of ventilation and pollution control systems like APTA, which can achieve up to 44% reduction in PM2.5 levels, consistent with the observed data in the graph.

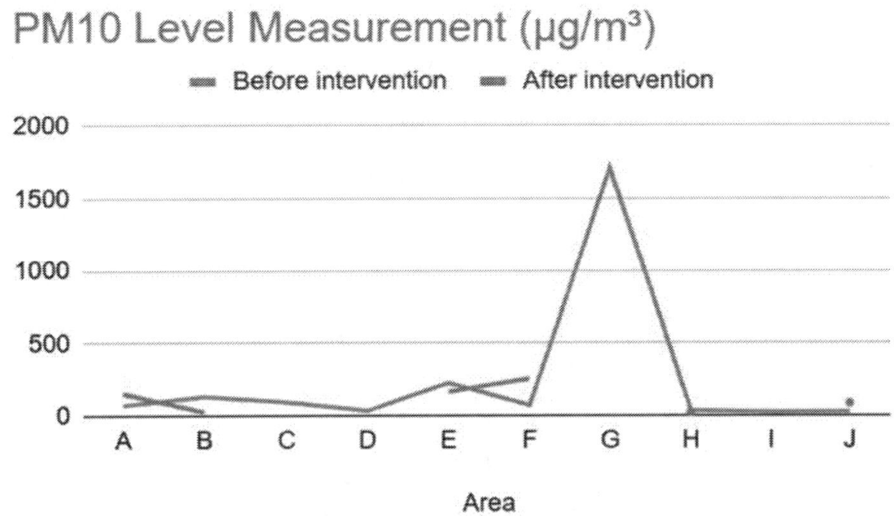

Fig. 4. PM10 concentration before and after the installation of APTA. Below the figure

Figure 4. PM10 concentration before and after the installation of APTA illustrates the PM 10 concentration in each working area before and after the APTA installation. Before intervention, the PM10 level ranged approximately between 20 and 1700 μg/m³, with area G (location 6) having the highest PM10 concentration. After the intervention, the levels of PM10 dropped significantly, stabilizing close to 0 μg/m³ across the areas. The G point measurement was located in the mixing area. The highest concentration of PM10 in the mixing area is in line with previous study conducted in Thailand [11]. However, the PM10 concentrations found in Thai factories were lower than those found in this study. The dust concentration in the mixing process area was 0.31 ± 0.18 mg/m³. This is 5 times lower than the findings in this study, where the mixing area reached 1.704 ± 0.25 mg/m³.

The principal activity in the G area is mixing all raw ingredients for making incense sticks, including resin, wood powder, sticky powder, or binding agents. The mixing process is carried out by using a mechanical machine. Workers in this area typically measure, weigh, and arrange ingredients, which are then mixed and fed into machines. Since most of the ingredients are dry, this process can produce significant amounts of particulate matter [13].

The installation of APTA in the workstation had an impact on reducing particulates in the work area. The decrease in PM10 concentration in the work area is due to the contribution of each filter component that builds up the APTA. A HEPA filter can filter over 99% particulate matter with a size of 0.3 μm, making it very effective in reducing PM10 levels. [8]. With their microporous structure, zeolites accommodate a variety of cations, water molecules, and other compounds. This structure enhances zeolites' ability to capture small particles, such as PM10 [9]. The cotton dust filter contributes to filtering up to 99.8% of particulate matter [12].

Table 1. Wilcoxon Signed Ranks Test results before and after the installation of APTA for the parameters Pb and Mn. Below the table

Variable	Ranks	n	Mean Rank	Z score	P-value
Pb after - Pb before	Negative ranks Positive ranks Ties Total	3 7 0 10	4.33 6.00	-1.489	0.137
Mn after - Mn before	Negative ranks Positive ranks Ties Total	9 1 0 10	6.00 1.00	-2.703	0.007*

Table 2. Wilcoxon Signed Ranks Test results before and after the installation of APTA for the parameters $PM_{2.5}$ and PM_{10}. Below the table

Variable	Ranks	n	Mean Rank	Z score	P-value
$PM_{2.5}$ after - $PM_{2.5}$ before	Negative ranks Positive ranks Ties Total	7 3 0 10	6.14 4.00	-1.581	0.114
PM_{10} after - PM_{10} before	Negative ranks Positive ranks Ties Total	6 4 0 10	4.83 6.50	-0.153	0.878

Based on the Table 1. Wilcoxon Signed Ranks Test results before and after the installation of APTA for the parameters Pb and Mn using significance level 0.05 shows in the table compare concentrations of lead (Pb) and manganese (Mn) before and after the installation of the APTA at 10 locations. Manganese level has Z score -2.703 and p-value is 0.007, statistically significant because less than 0.05 level. From 4 parameters only manganese that significantly there is a different before and after installation of APTA.

The wilcoxon signed ranks test is a non-parametric test suitable for comparing paired samples without assuming normal distribution, was used to analyze 4 pollutant concentration (Mn, Pb, PM2,5, and PM10) changes before and after APTA installation. The statistical results demonstrate a significant decrease in manganese concentrations after APTA implementation, indicating effective filtration or reduction in Mn particulate emissions. Manganese's significant reduction can be particularly relevant due to its known respiratory toxicity when bound to PM2.5 [6].

In contrast, lead (Pb) concentration reductions were not statistically significant. This may suggest that either Pb levels were less affected by the treatment or that the initial Pb concentrations were more variable or influenced by other sources beyond the APTA's control. Despite this, numerical decreases observed in Pb and the general reduction pattern should not be disregarded as they contribute to overall air quality improvement. This aligns with literature indicating the variability in effectiveness of air cleaning systems on specific heavy metals in PM2.5, emphasizing the need for targeted mitigation strategies in industrial settings such as incense raw material mixing areas where high heavy metal emissions occur.

Various types of filters and adsorbents play distinct roles in mitigating air pollutants including heavy metals (Mn, Pb) and particulate matter ($PM_{2.5}$ and PM_{10}). This discussion synthesizes evidence on the performance contributions of HEPA filters, zeolite, silica gel, super absorbent polymer (SAP) sand, and cotton filters in pollutant reduction. High-Efficiency Particulate Air (HEPA) filters use dense mats of fibers to physically trap particles primarily through interception, diffusion, and impaction. Their efficiency against particulate matter, especially sizes of $PM_{2.5}$ and PM_{10}, is typically above 99.97% for particles ≥ 0.3 µm. Studies confirm HEPA filters significantly reduce indoor $PM_{2.5}$ concentrations, resulting in measurable health benefits. However, HEPA filters primarily act as physical barriers and have limited direct chemical adsorption for heavy metal ions like Mn and Pb, functioning more as particulate traps for metal-containing aerosol [4].

Zeolites are aluminosilicate crystalline materials with a microporous structure that enables strong ion-exchange and adsorption capabilities. When combined with cotton fibers (zeolite-cotton composites), they provide a low-cost, high-efficiency filter for aqueous heavy metal removal. For instance, zeolite-cotton filters reduce Pb^{2+} and other heavy metals concentrations from contaminated water to below 5 ppb, demonstrating rapid and selective heavy metal ion capture. Their application in air purification benefits from ion exchange and adsorptive capture of metal ions and trace gases but requires integration into composite filter media for particulate removal [13].

Primarily used for moisture control, silica gel is a highly porous adsorbent with some affinity for trapping volatile and polar compounds. Its role in removing Mn, Pb, or particulate matter from air is less direct and more limited than zeolites or HEPA filters. Silica gel can contribute to pre-conditioning air by reducing humidity, potentially enhancing the performance of other filters. Super Absorbent Polymer (SAP) sand is designed to absorb large amounts of liquids and retains moisture, typically applied in water and wastewater treatment. Its capacity to adsorb heavy metals arises from physical entrapment and weak chemical interactions. While less common in airborne pollutant filtration, SAP sand may assist in localized pollutant capture in humid environments or serve as a substrate for immobilizing functional adsorbents but is not a standalone solution for Mn, Pb, $PM_{2.5}$, or PM_{10} removal [3].

Cotton fibers provide a natural, porous matrix suitable for physical filtration and can be functionalized with adsorbents like zeolite or metal oxides. Zeolite-cotton composites have exhibited high adsorption efficiencies of heavy metals in water treatment and can be adapted for air filtration to capture metal particulates and gaseous pollutants. Pure cotton filters alone provide limited particulate removal efficiency compared to specialized synthetic filters [12].

Air purification technologies employing toxic adsorbents offer significant benefits for occupational health by effectively reducing worker exposure to hazardous pollutants like manganese (Mn), lead (Pb), $PM_{2.5}$, and PM_{10}. Understanding their impact on exposure risk reduction and sustainability is key for long-term workplace safety. Air purification toxic adsorbents have significant potential to reduce worker exposure risks to heavy metals and particulate matter in incense stick manufacturing environments, where emissions of fine particulates and metals like manganese and lead are prevalent. By capturing and chemically transforming airborne contaminants, these adsorbents lower the inhalation doses of harmful substances, thereby mitigating risks of respiratory, neurological, and systemic toxicity commonly reported among such workers. Studies in occupational settings emphasize that

https://doi.org/10.1051/e3sconf/202565203005

reducing indoor particulate levels by over 90% through advanced filtration correlates with decreased biomarkers of exposure and improved respiratory health outcomes. Moreover, combining toxic adsorbents with routine environmental monitoring and personal protective equipment can further safeguard workers from chronic exposures inherent to incense production. These interventions are essential given the documented high dust and metal aerosol concentrations that exceed safety thresholds without adequate filtration, posing long-term occupational health hazards. Thus, air purification systems utilizing toxic adsorbents represent a promising strategy to substantially reduce hazardous exposures and improve workplace health in incense manufacturing facilities.

In incense stick manufacturing environments, long-term sustainability and maintenance of air purification toxic adsorbents for reducing heavy metals and particulate matter present important considerations. Toxic adsorbents like metal-oxide-functionalized materials offer high adsorption capacity and catalytic degradation of pollutants, making them effective for capturing hazardous emissions from incense burning, which is known for generating significant particulate matter and heavy metals. Sustainable implementation requires selecting adsorbents synthesized from abundant, low-cost, and eco-friendly precursors such as biomass- or plant-derived carbons to minimize environmental impact during production and disposal. For example, nanoporous carbon foams derived from edible feedstocks via salt-assisted pyrolysis exhibit not only outstanding PM2.5 filtration efficiency (>99%) but also a simple, green fabrication process that reduces secondary pollution risks, critical for sustainable manufacturing environments. Moreover, periodic regeneration through thermal or chemical treatments is vital to restore adsorbent capacity, prolong operational life, and reduce material waste. Incense manufacturing facilities can benefit from integrating such regenerative adsorbent filters with routine maintenance schedules to maintain consistent air quality while minimizing downtime and operational costs. Overall, leveraging sustainable, regenerable toxic adsorbents aligns with circular economy principles, fostering both environmental and occupational health goals in incense production settings [1].

4 Conclusion

This study demonstrated that the APTA multi-layer filtration system significantly reduced manganese concentrations in the incense stick manufacturing environment, especially in high-exposure areas such as the drying zone. This finding confirms the system's strong capabilities in adsorbing hazardous metals and filtering particulate matter. Although reductions in lead and particulate matter (PM2.5 and PM10) were observed, these changes were not statistically significant, indicating a need for further optimization or longer-term application to improve overall pollutant removal. The results validate the effectiveness of a multi-layer filtration approach tailored to the complex emission profile of small-scale incense production, which often lacks adequate pollution controls. By significantly lowering neurotoxic manganese exposure, the APTA system demonstrates clear potential to enhance occupational health and safety in artisanal manufacturing settings. These findings support targeted implementation of such filtration technologies to mitigate respiratory and neurological risks among workers. Adoption of APTA and similar systems could guide policy development and improve regulatory standards for air quality in small industries. Future research should focus on extended field trials, system refinement, and assessment of long-term health outcomes to solidify guidelines for broader application in incense and related manufacturing sectors.

Acknowledgements

We would like to express our sincere gratitude to Universitas Negeri Malang for their generous funding and support throughout the course of this research project. Their commitment to advancing academic research has been instrumental in enabling us to explore innovative solutions for air purification in the incense stick industry. Additionally, we extend our heartfelt thanks to the incense stick industry in Dalisodo Village in Malang Regency, East Jave who participated in this study. Their cooperation and insights were invaluable in understanding the specific challenges related to manganese, lead, and particulate matter emissions. This collaborative effort is crucial for developing effective remediation strategies that not only enhance air quality but also ensure the health and safety of workers within the industry.

References

1. Akhtar, M. S., Ali, S., & Zaman, W. (2024). Innovative Adsorbents for Pollutant Removal: Exploring the Latest Research and Applications. *Molecules*, *29*(18), 1–37. https://doi.org/10.3390/molecules29184317
2. Al-Ghouti, M. A., Ashfaq, M. Y., Khan, M., Al Disi, Z., Da'na, D. A., & Shoshaa, R. (2023). State-of-the-art adsorption and adsorptive filtration based technologies for the removal of trace elements: A critical review. *Science of the Total Environment*, *895*(March), 164854. https://doi.org/10.1016/j.scitotenv.2023.164854
3. Ayach, J., Duma, L., Badran, A., Hijazi, A., Martinez, A., Bechelany, M., ... & Hamad, H. (2024). Enhancing Wastewater Depollution: Sustainable Biosorption Using Chemically Modified Chitosan Derivatives for Efficient Removal of Heavy Metals and Dyes. Materials, 17(11), 2724.
4. Djouider, F. (2023). Performance Evaluation of High Efficiency Particulate Filters for the Removal of Sub-Micrometer Radioactive Aerosols in Nuclear Power Reactors. Simulation Study. Journal of King Abdulaziz University: Engineering Sciences, 33(1).
5. Fang, X., Chang, R., Zhang, Y., Zuo, J., Zou, Y., & Han, Y. (2024). Monitoring airborne particulate matter from building construction: A systematic review. *Journal of Building Engineering*, *86*(January), 108708. https://doi.org/10.1016/j.jobe.2024.108708
6. Kermani, M., Jonidi Jafari, A., Gholami, M., Arfaeinia, H., Shahsavani, A., & Fanaei, F. (2021). Characterization, possible sources and health risk assessment of PM2.5-bound Heavy Metals in the most industrial city of Iran. *Journal of Environmental Health Science and Engineering*, *19*(1), 151–163. https://doi.org/10.1007/s40201-020-00589-3
7. Li, J., Xu, H., Song, D., Wang, Z., Zhang, B., Feng, R., Gu, Y., Jiang, Z., Ji, X., Liu, S., Ho, S. S. H., Qu, L., & Sun, F. (2022). Emission characteristics and assessment of potential health risks on PM2.5-bound organics from incense burning. *Atmospheric Pollution Research*, *13*(3), 1–9. https://doi.org/10.1016/j.apr.2022.101326.
8. Mata, T. M., Martins, A. A., Calheiros, C. S. C., Villanueva, F., Alonso-Cuevilla, N. P., Gabriel, M. F., & Silva, G. V. (2022). Indoor Air Quality: A Review of Cleaning Technologies. *Environments*, *9*(9), 118. https://doi.org/10.3390/environments9090118
9. Nisamaneenate, J., Worakitchareonpol, P., Idris, I. A., Tocharoen, S., Atong, D., & Sricharoenchaikul, V. (2025). Enhancing particulate matter removal by PES filter coated with zeolite via a coupling agent. Process Safety and Environmental Protection, 107491.
10. Onur, A., Ng, A., Batchelor, W., & Garnier, G. (2018). Multi-Layer Filters: Adsorption and Filtration Mechanisms for Improved Separation. *Frontiers in Chemistry*, *6*(September), 1–11. https://doi.org/10.3389/fchem.2018.00417
11. Siripanich, S., Siriwong, W., Keawrueang, P., Borjan, M., & Robson, M. (2014). Incense and Joss Stick making in small household factories, Thailand. *International Journal of*

Occupational and Environmental Medicine, *5*(3), 137–145.
12. Rana, A. K., Mostafavi, E., Alsanie, W. F., Siwal, S. S., & Thakur, V. K. (2023). Cellulose-based materials for air purification: A review. Industrial Crops and Products, 194, 116331.
13. Trinh, T. H., Pham, C., Reiminger, N., Nhut, J. M., & Pham-Huu, C. (2025). Filtration systems for particulate matter reduction in outdoor air: a review. *Journal of Environmental Management*, *390*(June), 126263. https://doi.org/10.1016/j.jenvman.2025.126263
14. Yadav, V. K., Bijekar, S., Gacem, A., Alkahtani, A. M., Yadav, K. K., Alreshidi, M. A., Kumar, P., Ghosh, T., Verma, R. K., Mishra, S., Patel, A., & Choudhary, N. (2024). The Impact of Fine Particulate Matters (PM10, PM2.5) from Incense Smokes on the Various Organ Systems: A Review of an Invisible Killer. *Particle and Particle Systems Characterization*, *41*(5). https://doi.org/10.1002/ppsc.202300157
15. Yadav, V. K., Choudhary, N., Heena Khan, S., Khayal, A., Ravi, R. K., Kumar, P., ... & Gnanamoorthy, G. (2020). Incense and incense sticks: types, components, origin and their religious beliefs and importance among different religions. *J. Bio Innov*, *9*, 1420-1439.

Analysis of Flood Vulnerability and Evacuation Sites Based on ArcGIS StoryMap in Wonosegoro Sub-district, Boyolali District

Dika Kurniawan[1], and *Yuli Priyana*[2*]

[1,2] Faculty of Geography, Universitas Muhammadiyah Surakarta, Surakarta 57162, Central Java, Indonesia

Abstract. Wonosegoro Subdistrict, Boyolali District, Central Java, is an area with a high level of vulnerability to flooding due to the overflow of the Serang River. Based on data from the Regional Disaster Management Agency (BPBD) of Boyolali Regency, flooding occurred 3 times in 2020, 2021 and 2022. The data makes this research has objectives. Analyzing potential flood areas, determining the fastest evacuation route to the evacuation site, and presenting a web-based interactive map through the ArcGIS Story Map platform. Flood vulnerability analysis was conducted using the overlay scoring weighting method in the Geographic Information System (GIS) by considering the variables that cause flooding. Determination of evacuation routes is carried out at safe points when flooding occurs. Using Network Analyst, which resulted in 18 evacuation routes from 9 vulnerable villages to the nearest evacuation location. The evacuation points used are public facilities such as village halls, schools and places of worship, with distances ranging from 0.197 km to 3.707 km. The results showed that the highly vulnerable class covered 436.749 ha (4%), the vulnerable class covered 1,158.62 ha (44%), the quite safe class covered 1,511.47 ha (25%), the safe class covered 2,644.669 ha (19%), and the very safe class covered 215.504 ha (7%).

1 Introduction

Indonesia is a tropical country with diverse geomorphological characteristics where three major tectonic plates meet, namely the Indo-Australian Plate, the Eurasian Plate, and the Pacific Plate. The convergence of these plates results in Indonesia having a high risk of natural disasters [1]. This is also reinforced by several factors such as geological, geohydrological, geomorphological, and various other factors. According to data from INARISK BNPB (National Disaster Management Agency), floods are a frequent disaster in Indonesia from 2020 to 2024, with a total of 7,473 cases. [2]

There are various types of floods, such as tidal floods, lava floods, and flash floods. Floods generally occur in low-lying areas, but they can also be caused by various factors such as high rainfall, proximity to rivers, soil type, and land use [3]. One region in Indonesia that

* Corresponding author: yp121@ums.ac.id

https://doi.org/10.1051/e3sconf/202565204001

has diverse characteristics is Boyolali Regency. Boyolali Regency is one of the regencies located in Central Java. Boyolali has diverse geographical characteristics, with highlands located in the subdistricts of Selo, Cepogo, Musuk, and part of the Ampel subdistrict. The subdistricts in Boyolali Regency that are classified as lowlands include Teras, Nogosari, Sambi, Sawit, Wonosegoro, and so on. [4].

Wonosegoro Subdistrict consists of 11 villages. The most common problem in Wonosegoro Subdistrict is flooding, because Wonosegoro Subdistrict is a low-lying area and is crossed by the Serang River. Based on data from the Regional Disaster Management Agency (BPBD), Wonosegoro Subdistrict is an area that frequently experiences flooding. From 2020 to 2022, Wonosegoro Subdistrict experienced three cases of flooding each year [5]. This incident has certainly caused various detrimental impacts on the community of Wonosegoro subdistrict.

Previous research in Boyolali Regency has mostly focused on landslide hazards and vulnerability, especially in mountainous areas such as Selo, Cepogo, and Musuk subdistricts [6]. Research on flood vulnerability in lowland areas, particularly in Wonosegoro subdistrict, is still very limited. Furthermore, there has been no research that integrates flood vulnerability analysis with evacuation route planning and presents the results through ArcGIS StoryMap. This study aims to fill this gap by combining flood vulnerability mapping and evacuation planning in an interactive web-based format.

Flood vulnerability mapping alone is not sufficient for disaster mitigation, as communities also need practical guidance on evacuation locations and the safest routes for evacuation during emergencies. Without clear evacuation planning, residents in flood-prone areas risk delayed responses, which can increase casualties and economic losses. Therefore, this study not only analyzes flood vulnerability but also identifies evacuation routes to strengthen flood disaster preparedness in Wonosegoro District.

Therefore, it is necessary to take steps to minimize the impact of flooding, which is an annual occurrence in the Wonosegoro subdistrict. The steps that can be taken to minimize the impact of disasters are mitigation measures. The objectives of this study are: 1. To identify flood-prone areas in Wonosegoro sub-district, 2. To provide the fastest evacuation routes when flooding occurs, and 3. To present evacuation route information in the form of an interactive map using ArcGIS Story Map.

2 Method

The method used to determine flood vulnerability is weighted scoring. The scoring and weighting method is a method used to assign values or scores to various variables or criteria based on their level of importance in an analysis. Scoring refers to the process of assigning numerical values to a parameter. Meanwhile, weighting is the process of determining the level of influence or contribution of each variable to the final result, expressed in the form of weights. The reference for determining the score can refer to previous research [7].

The selection of evacuation sites was carried out after safe locations were identified using a scoring method. Safe evacuation sites were determined based on guidelines from the National Disaster Management Agency (BNPB) [2].

Including those provided through the InaRISK platform. The selected sites consisted of public facilities such as village offices, sub-district offices, and educational facilities that meet safety and accessibility criteria

Determining the fastest evacuation route using Network Analyst, a feature in geographic information system (GIS) software used for road network-based analysis. This feature allows users to model the shortest route.

2.1. Study Location

The location of this research is Wonosegoro Sub-district, one of the sub-districts in Boyolali Regency. Wonosegoro District has an area of approximately 5752.401 hectares and consists of 11 villages namely Karangjati, Ketoyan, Bandung, Kauman, Lemahireng, Guwo, Banyusri, Bojong, Bolo, Gosono and Wonosegoro with Coordinates: 7°19′8″S 110°36′31″E / 7.31889°S 110.60861°E / -7.31889; 110.60861. The boundaries of the study area can be described as follows: in the north, the area is bordered by Juwangi Subdistrict, the east is bordered by Kemusu Subdistrict, the south is bordered by Klego Subdistrict and Karanggede Subdistrict and the west is bordered by Wonosamodro Subdistrict. Location details are depicted in Figure 1.

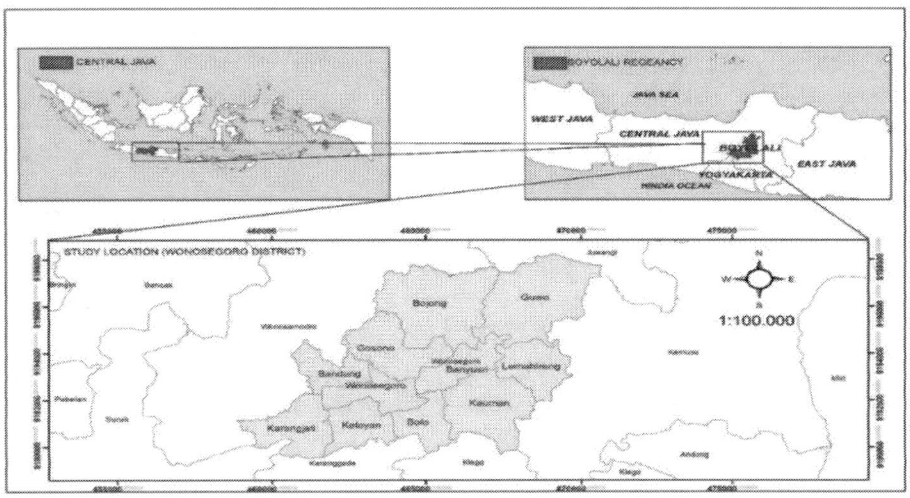

Fig.1. Research Location Map

2.2. Material and Method

This study uses Geographic Information System (GIS) software to analyze the fastest evacuation routes during floods. In determining the level of vulnerability to flooding, this study refers to several spatial variables, such as slope gradient, elevation, distance from rivers, soil type, land cover, and average annual rainfall. Meanwhile, the analysis of the fastest evacuation routes is carried out by utilizing road network data and coordinates identified as safe areas from flooding.

Tabel 1. Tools and Materials

Data	Source
Slope	INA Geoportal
Elevation	INA Geoportal
Soil type	Fao Network
Land Use	Geospatial Information Agency (BIG)
Annual Rainfall	Climate Hazards Center (CHIRPS)
River shape file	Geospatial Information Agency (BIG)
Point coordinate	Google Earth Pro
Road network Indonesia Shapefile	Geospatial Information Agency (BIG)

2.3. Flood Vulnerability Determination Process

Flood vulnerability in Wonosegoro Subdistrict was determined using a weighted overlay method in ArcGIS 10.8, based on the parameters listed in Table 1. Slope and elevation data were obtained from a 30 m resolution DEM obtained from the INA Geoportal. Slope was classified into five classes according to BNPB (2012) guidelines, where gentler slopes are considered more vulnerable to flooding. Elevation data was used to distinguish between lowland and highland areas, with lower elevations given a higher vulnerability score.

Annual rainfall data for the period 2020–2022 was obtained from the CHIRPS database and averaged to produce a representative rainfall map. Rainfall intensity was then classified into three categories: high, medium, and low, with areas with higher rainfall contributing to higher vulnerability.

Soil type data from the FAO and land use maps from BIG were reclassified based on vulnerability to flooding. For example, clay and alluvial soils were categorized as highly vulnerable, while sandy soils were less vulnerable. Similarly, land uses such as settlements and rice fields were classified as more vulnerable, while forests and mixed gardens were considered less vulnerable.

River buffer zones were generated at distances of 25 m, 50 m, 75 m, 100 m, and 125 m from the river network using the Buffer tool in ArcGIS. Areas closer to the river were given a higher vulnerability score, while areas further away were given a lower score.

Finally, all spatial layers were reclassified to the same scale and combined through a weighted overlay process to produce a flood vulnerability map. The weights for each parameter were adopted from previous studies [7], [8] and adjusted according to local conditions in Wonosegoro. The resulting flood vulnerability map was classified into five categories: Very Vulnerable, Prone, Quite Safe, Safe, and Very Safe.

Mapping of flood-prone areas in Wonosegoro District using an assessment method in the Geographic Information System (GIS). This method assigns weights and scores to each parameter to determine the flood potential of an area. Based on previous studies [9], [11], and referring to the national guideline SNI 8197:2015 [9] on Flood Hazard Mapping published by BSN, six parameters were used in this study, as summarized in Tables 2–7.

Table 2. Slope Classification %

Parameters	Weight	Skor
0 - 8		9
9- 15		7
16- 25	10%	5
26- 40		3
> 40		1

Table 3. Elevation Classification Meters above sea level

Parameters	Weight	Skor
0 -20		9
21-50		7
51-100	20%	5
101-300		3
>300		1

Table 4. Soil Type Classification

Parameters	Weight	Skor
Aluvial		9
Grumusol		7
Mediteran and latosol	10%	5
Regosol		3
Litosol		1

Table 5. Land Use Classification

Parameters	Weight	Skor
Water Body		9
Settlement		7
Rice fields	25%	5
Grass and Garden		3
Bush		1

Table 6. Average Rainfall (mm/year)

Parameters	Weight	Skor
> 3000		9
2.501 - 3.000		7
2001 - 2500	15%	5
1501 – 2000		3
< 1500		1

Table 7. River Buffer Classification (Meter)

Parameters	Weight	Skor
25		9
50		7
75	20%	5
100		3
125		1

Map processing using (GIS), with reference to previous research [7], [8], allows for the mapping of flood vulnerability. This process involves analyzing the various parameters that contribute to an area's vulnerability. To determine the total vulnerability score, calculations are made using a specific formula that considers the weight and value of each parameter

Total Skor = (10% * Slope) + (20% * Skor Elevation) + (10% * Skor Soil Type) + (25% * Skor Land Use) + (15% * Skor Average Rainfall) + (20% * Skor River Buffer)

To determine the categories of flood vulnerability levels, the interval calculation method was used to systematically assign vulnerabilities between classes. This approach allows the division of vulnerability levels based on the distribution of values obtained, so that each category has clear and measurable boundaries.

$$Interval = \frac{Highest\ score\ total - Lowest\ score\ total}{Number\ of\ classes}$$

$$Interval = \frac{8{,}4 - 0{,}2}{5}$$

$$Interval = 1{,}64$$

After the relevant parameters have been combined and the interval classes have been determined, a classification of flood vulnerability levels is obtained, which serves as the basis for determining evacuation sites. In this process, areas with safe to Quite Safe vulnerability levels were selected as evacuation site locations, given their stability and minimal disaster risk. Meanwhile, settlements in the highly vulnerable category are identified as priority areas that require evacuation measures. This approach aims to ensure the safety of residents by placing them in safer locations and optimizing disaster risk mitigation strategies

The overall vulnerability score was classified into five levels to determine the flood vulnerability class. Determining the level of vulnerability is done by calculating the value in the lowest class, then adding intervals until the highest value is reached. This approach allows for systematic grouping of areas by risk level. More detailed information on the classification can be seen in table 8 below

Table 8. Total flood vulnerability

Flood Prone Level	Flood Total Score Range
Very Vulnerable	6,77 - 8,40
Prone	5,13 - 6,76
Quite Safe	3,49 - 5,12
Safe	1,85 - 3,48
Very Safe	0,2 - 1,84

2.4. Evacuation Route Determination Process and ArcGIS Story Map Usage

Determination of the fastest evacuation route using one of the tools available in ArcGIS, namely Network Analyst. Network Analyst is one of the tools in the geographic information system to determine the fastest evacuation route [10]. Determination of the fastest route is useful when a disaster occurs. The existence used is the closest facility where points prone to flooding will be directed to a safe evacuation site. The chosen evacuation site adapts the standard guidelines of the National Disaster Management Agency where the evacuation site must have access to clean water, electricity and bathrooms and also considers certain criteria such as in the research [11]. The selected evacuation sites include 11 village halls, 1 sub-district office, and 7 schools.

After the evacuation routes and safe locations were determined using the Network Analyst tool, the results were then published through ArcGIS Story Map. Story Map was used as a medium to integrate flood vulnerability maps, evacuation routes, and evacuation facility locations into an interactive web format. This platform allows users to visualize maps online, navigate evacuation routes connected to Google Maps, and access detailed information about evacuation facilities. By using Story Map, the results of this study are not only analytical but also practical, as they can be easily accessed and utilized by local governments and communities in Wonosegoro District.

2.5. Flood Vulnerability Map Validation

To ensure the validity of the flood vulnerability map, the validation process was carried out by comparing the classification results with actual flood events in Wonosegoro District during the 2020–2022 period. Secondary data was obtained from Boyolali Regional Disaster Management Agency (BPBD) reports and local news sources. Validation was carried out by checking whether the observed flood events matched the vulnerability categories listed on the map. The accuracy rate was then calculated using the following formula:

$$Overall\ Accuracy = \frac{Number\ Of\ Matches}{Total\ Cases} \times 100\%$$

2.6. Research Flow Chart

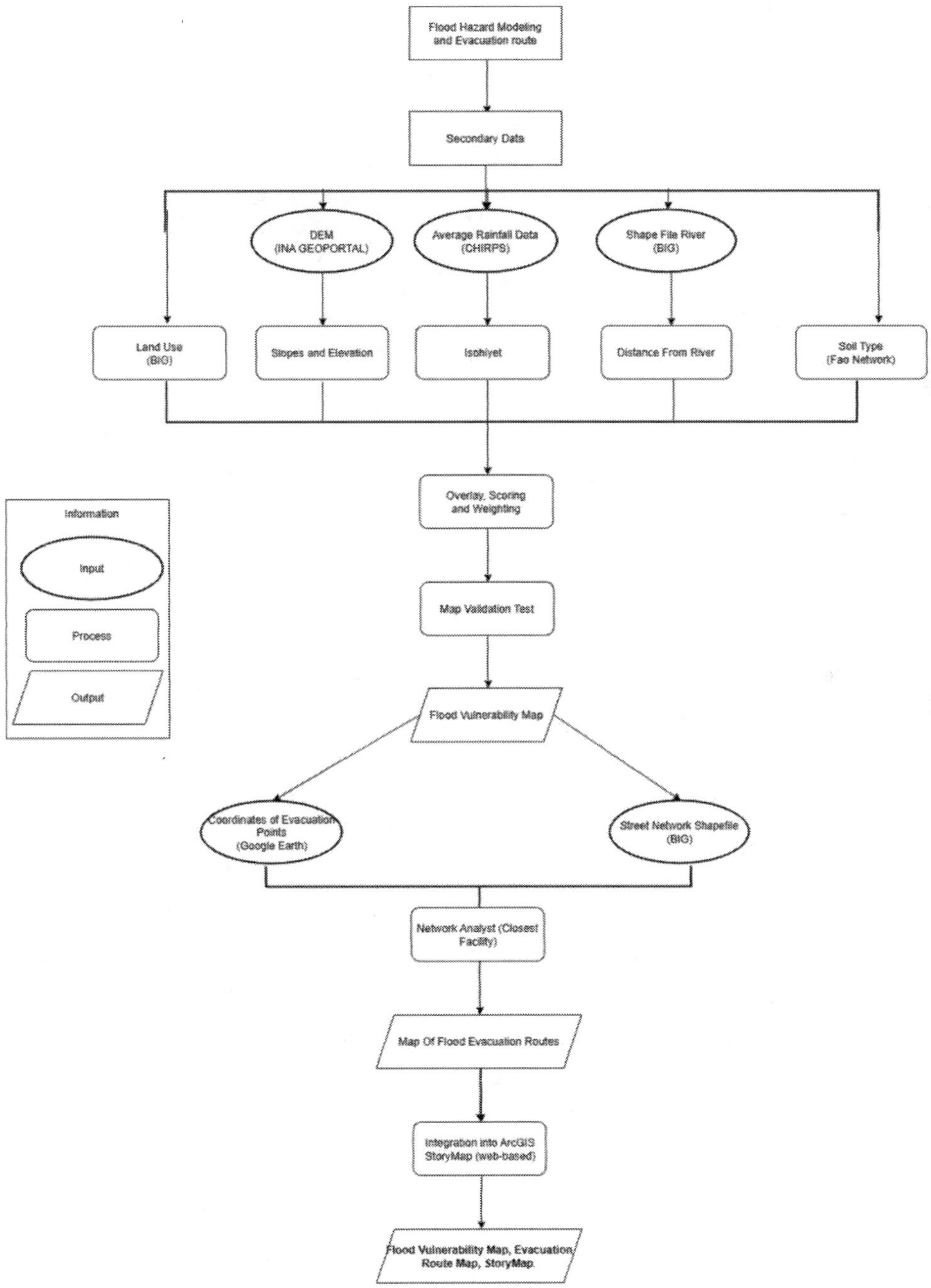

Fig 2. Research flow chart of flood vulnerability and evacuation route analysis in Wonosegoro Sub-district, Boyolali Regency

 https://doi.org/10.1051/e3sconf/202565204001

3 Results

3.1. Map analysis of flood hazard levels

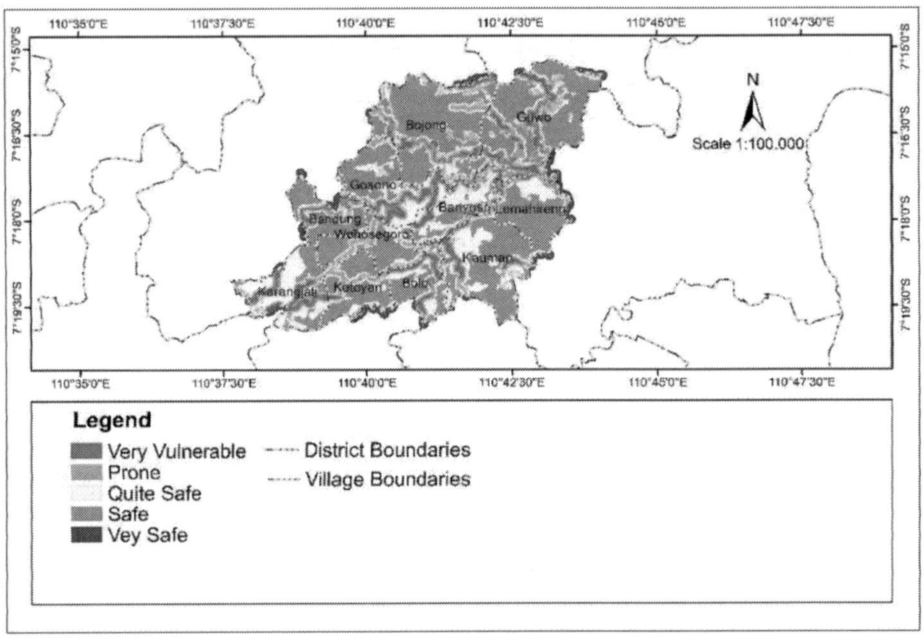

Fig.3. Flood Vulnerability Map

Table 9. Extent and Class of Flood Prone Area

No	Class	Area (Ha)	Persentase
1	Very Vulnerable	436,749	4%
2	Prone	1158,620	44%
3	Quite Safe	1511,470	25%
4	Safe	2644,660	19%
5	Very Safe	215,504	7%
Total		5967,01	100%

Based on Figure 2 and Table 9, areas with potential flood vulnerability in Wonosegoro Subdistrict are classified into five levels of vulnerability. The highly vulnerable category covers 436.749 hectares (4%) and is concentrated along the Serang River, especially in Karangjati and Badung villages, where low elevation and predominantly clay soil increase vulnerability to flooding. The vulnerable category dominates with 1,158.62 hectares (44%), including Wonosegoro and Guwo villages, where intensive land use such as settlements and rice fields contributes to higher risks. The Quite Safe category covers 1,511.47 hectares (25%), while the safe category covers 2,644.677 hectares (19%), mostly located in northern mountain villages such as Gunungsari and Kenteng, which benefit from better drainage

 https://doi.org/10.1051/e3sconf/202565204001

systems. The very safe category, covering 215,504 hectares (7%), is limited to areas with higher elevations and minimal exposure to flood hazards.

These findings are consistent with BPBD records of recurrent flooding in Karangjati and Wonosegoro during the 2020–2022 period, confirming the reliability of the vulnerability mapping. Therefore, mitigation strategies and evacuation planning should prioritize villages categorized as highly vulnerable and vulnerable to minimize disaster risk.

3.2. Map Analysis Wonosegoro District Flood Disaster Evacuation Route

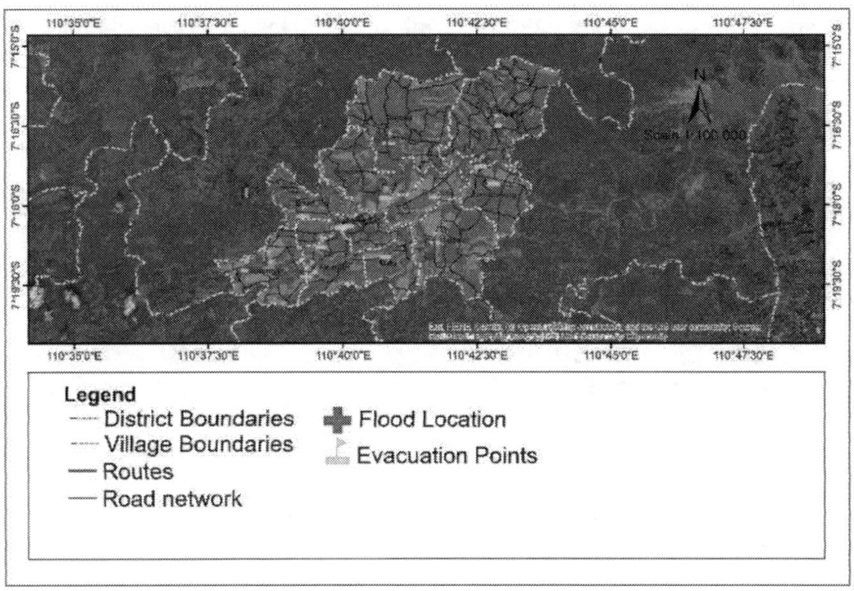

Fig.4. Evacuation Route Map of Wonosegoro Sub-district

Tabel 10. Flood Disaster Evacuation Routes and Places in Wonosegoro District

No	Flood Danger Locations in the Village	Evacuation Points	Distance (KM)
1	Guwo	SMK Bina Karya Kemusu	3,707
2	Guwo	SMK Bina Karya Kemusu	1,473
3	Lemah Ireng	Balai Desa Lemah Ireng	0,562
4	Banyusri	Mts Al Islam Banyu Sri	1,486
5	Banyusri	Banyusri Village Hall	1,056
6	Lemah Ireng	Lemah Ireng Village Hall	0,630
7	Bojong	Bojong Village Hall	1,862
8	Bojong	Bojong Village Hall	3,249
9	Kauman	Kauman Village Hall	1,294
10	Bolo	Bolo Village Hall	0,312
11	Karangjati	Smp 2 Wonosegoro	0,197
12	Guwo	Guwo Village Hall	0,518
13	Banyusri	Gosono Village Hall	1,456

No	Flood Danger Locations in the Village	Evacuation Points	Distance (KM)
14	Bandung	Bandung Village Hall	1,014
15	Bandung	SMP 1 Wonosegoro	1,573
16	Bolo	SMK 1 Wonosegoro	2,296
17	Karangjati	Karang Jati Village Hall	0,422
18	Karangjati	Ketoyan Village Hall	0,957

In this study, evacuation routes were determined using a Network Analysis approach in ArcGIS, taking into account the road network and the locations of nearby public facilities as potential evacuation points. The analysis identified 18 optimal evacuation routes leading to 19 evacuation points, spread across 11 villages in Wonosegoro Subdistrict (Figure 4 and Table 10). The shortest evacuation route was found in Karangjati Village, where residents can reach Karangjati 1 Elementary School at a distance of only 0.197 km. In contrast, the longest evacuation route was recorded in Guwo Village, where residents must travel a distance of 3.707 km to reach Bina Karya Kemusu Vocational School. Despite these variations, the distribution of evacuation points is relatively even throughout the subdistrict, ensuring that every vulnerable village has access to a safe location within a reasonable distance. These findings indicate that evacuation planning in Wonosegoro is feasible, but villages with longer evacuation distances, such as Guwo, require special attention in terms of preparation and resource allocation.

3.3. Flood Vulnerability Map Validation

To validate the accuracy of the flood vulnerability map, an accuracy test was conducted by comparing the vulnerability classification with flood events recorded by the Boyolali Regional Disaster Management Agency (BPBD) and local news reports from 2020 to 2022. The results of this comparison are presented in Table 11.

Tabel 11. Flood Vulnerability Map Validation

Year	Village Locations	Reported by	Vulnerability Class	Match with model
2020	Karangjati	BPBD/Local news	Very Vulnerable	Yes
2020	Wonosegoro	BPBD/Local news	Prone	Yes
2020	Guwo	Local news	Prone	Yes
2021	Gunungsari	Local news	Quite Safe	Not
2021	Badung	BPBD	Very Vulnerable	Yes
2021	Tanjung	Local news	Prone	Yes
2022	Karangjati	BPBD/Local news	Very Vulnerable	Yes
2022	Guwo	BPBD	Prone	Yes
2022	Wonosegoro	BPBD	Prone	Yes

As shown in Table 11, most of the flood events reported by BPBD and local media from 2020 to 2022 occurred in areas classified as Very Vulnerable and Vulnerable on the flood vulnerability map. This shows a strong correlation between model predictions and actual flood events. Of the nine cases recorded, eight events corresponded to the vulnerability class on the map, while only one case in Gunungsari Village was classified as Quite Safe but still

 https://doi.org/10.1051/e3sconf/202565204001

experienced flooding. This discrepancy may have been influenced by local factors such as inadequate drainage or land use changes that were not fully covered in the model. Based on these results, the overall accuracy of the flood vulnerability map reached 88.9%, indicating that the model is reliable enough to represent the actual flood conditions in Wonosegoro District. Therefore, this map can be used as a valid basis for evacuation planning and disaster mitigation strategies.

3.4. General interface about ArcGIS Story Map

After identifying flood vulnerability maps, evacuation locations, and evacuation routes, the results were integrated into ArcGIS StoryMap. The use of StoryMap is in line with the objectives of this study, which is to present evacuation route information in an interactive web-based format. StoryMap allows the analysis results to be converted into an easily accessible platform, where users can interactively explore maps, visualize evacuation routes, and identify safe locations linked to Google Maps. This approach ensures that the research results are not only academic but also practical for local governments and communities in Wonosegoro District. See the image below for the interface display.

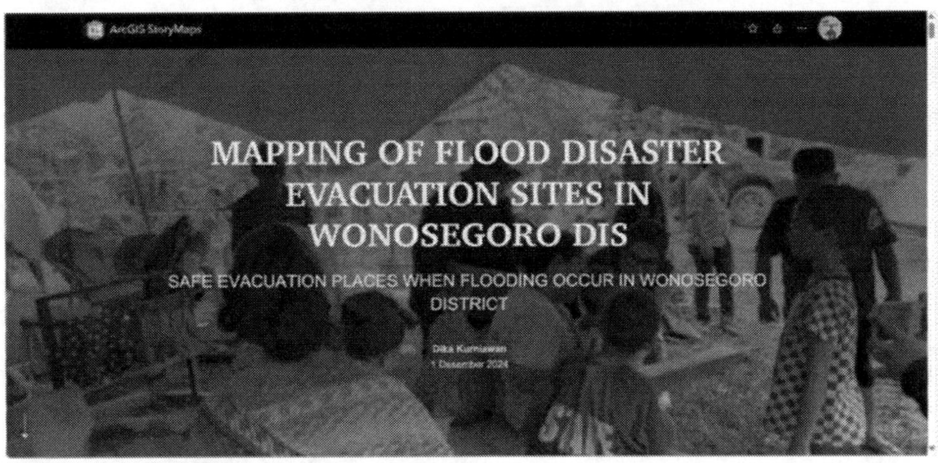

Fig.5. Story Map Interface

As shown in Figure 4, the Story Map interface provides interactive visualizations of flood-prone areas and evacuation routes. This platform allows users to view evacuation points, measure distances, and plan safe movements during flood events. By integrating research results into StoryMap, this study strengthens its contribution to disaster preparedness by ensuring that spatial information is easily accessible and understandable to non-technical users For links please visit: https://storymaps.arcgis.com/stories/6b723dd899874732b682d964ea63d769

4 Discussion

Based on the results of the study, the flood vulnerability parameters in Wonosegoro Sub-district include rainfall, soil type, slope, land cover, and distance from the river with 5 vulnerability classes. The highly vulnerable class has an area of 436.749Ha out of 4% of the total area. The vulnerable area has an area of 1158.620Ha out of 44% of the total area. Quite Safe has an area of 1511.470Ha out of 25% of the total area. Safe areas account for

2644.660Ha of 19% of the total area, and very safe areas account for 215.504 of 7% of the that are prone to flooding flooding, based on the results in Figure 8, there is also concrete evidence as reported by [12], [13] and also recorded in the Indonesian Disaster Information Data [14] that the flood caused 29 houses to be submerged and 265 casualties. The areas most prone to flooding in Wonosegoro Sub-district include Karangjati Village, Guwo Village and Banyusri Village. Based on the results in Table 9, each of these villages has 3 settlement points that are prone to flooding. These findings are consistent with [12] and [13], which also report frequent flooding in the villages of Karangjati and Guwo. Similarly, [14] confirms recurring flooding in Wonosegoro Subdistrict, reinforcing the validity of the flood vulnerability mapping results in this study.

The results show that half of Wonosegoro Sub-district is vulnerable to flooding. This indicates that the disaster risk in this area is quite high. The impact of flooding threatens infrastructure, settlements and the lives of local communities. Flooding conditions can be exacerbated by land use changes that do not pay attention to the drainage system, because areas prone to flooding are areas with low elevation and close to rivers. Determination of evacuation routes is prioritized at a distance of < 4 Km because it is for time efficiency and facilitates the need for supplies, but several places may experience excess capacity of capacity, alternatives when flooding occurs but the capacity of an evacuation site exceeds capacity can be made shelter tents by choosing a location that is safe from flooding around the evacuation site. The presentation of maps with ArcGIS Story Map also makes it easier to evacuate because the evacuation points are connected to google maps as done by [15] This shows that the information presented through the website is highly approved by respondents, but evacuation sites have limited capacity such as the Bojong Village hall which must accommodate from 2 points at once. This challenge is in line with the findings [15], which also reinforce the importance of spatial decision-making tools such as ArcGIS StoryMap in improving community preparedness.

5 Conclusion

This study concludes that Wonosegoro Subdistrict has a high level of flood vulnerability, with the vulnerable category covering 1,158.62 ha or 44% of the total area. The main factors affecting flood vulnerability include rainfall, soil type, slope, land cover, distance from rivers, elevation, and slope gradient. In line with the research objectives, three results were achieved: (1) identification of flood-prone areas with 5 classes, (2) determination of 18 evacuation routes connected to 16 evacuation locations, and (3) presentation of results through ArcGIS Story Map as an interactive platform. Although the distance to evacuation sites (< 4 km) supports a rapid response, the limited capacity of these sites remains a challenge that requires alternatives such as emergency tents. Overall, the integration of spatial analysis and Story Map technology improves disaster preparedness in Wonosegoro by providing scientifically valid and practically accessible information. Strengthening collaboration between the government, community, and technology utilization is key to reducing flood risk and building regional resilience.

References

1. M. Mustikaningrum et al., "Multi-Hazard Analysis in Gunungkidul Regency Using Spatial Multi-Criteria Evaluation," Forum Geografi, vol. **37**, no. 1, pp. 35–45, Jul. 2023, doi: 10.23917/forgeo.v37i1.19041.
2. "inaRISK." Accessed: Jun. 24, 2025. [Online]. Available: https://inarisk2.bnpb.go.id/

3. D. N. Sari, A. N. Anna, T. Taryono, M. F. Maulana, and D. N. F. Khumaeroh, "DETECTION OF FLOOD HAZARD POTENTIAL ZONES BY USING ANALYTICAL HIERARCHY PROCESS IN TUNTANG WATERSHED AREA, INDONESIA," Geographia Technica, vol. **19**, no. 1, pp. 61–78, Mar. 2024, doi: 10.21163/GT_2024.191.05.
4. M. Abdul Aziz, W. Sanjaya, and Rama Pratama Zulkifli Putra, "Pengelompokkan daerah rawan bencana di kabupaten Boyolali menggunakan algoritma k-means," JITU : Journal Informatic Technology And Communication, vol. **8**, no. 2, pp. 61–70, Nov. 2024, doi: 10.36596/jitu.v8i2.1726.
5. "Jumlah Kejadian Bencana Menurut Kecamatan dan Jenis Bencana di Kabupaten Boyolali 2021 - Kumpulan data - Portal Satu Data Kabupaten Boyolali." Accessed: Jun. 24, 2025. [Online]. Available: https://data.boyolali.go.id/dataset/jumlah-kejadian-bencana-menurut-kecamatan-dan-jenis-bencana-di-kab-boyolali-2021
6. A. Aminudin, A. P. Wijaya, and F. Hadi, "ANALISIS ZONA RAWAN TANAH LONGSOR MENGGUNAKAN METODE FUZZY ANALYTICAL HIERARCHY PROCESS (FAHP) (Studi Kasus: Kabupaten Boyolali)," Jurnal Geodesi Undip Juli, 2023.
7. H. Rakuasa, D. A. Sihasale, M. C. Mehdila, and A. P. Wlary, "Analisis Spasial Tingkat Kerawanan Banjir di Kecamatan Teluk Ambon Baguala, Kota Ambon," Jurnal Geosains dan Remote Sensing, vol. **3**, no. 2, pp. 60–69, Nov. 2022, doi: 10.23960/jgrs.2022.v3i2.80.
8. A. Musfida, M. Manaf, A. G. Tantu, H. Hadijah, S. Syafri, and K. Kastono, "Kajian Lokasi Rawan Bencana Banjir Pada Daerah Aliran Sungai (DAS) Walanae Kecamatan Dua Boccoe Kabupaten Bone," Jurnal Ilmiah Ecosystem, vol. **21**, no. 2, pp. 348–357, Aug. 2021, doi: 10.35965/eco.v21i2.1111.
9. Badan Standardisasi Nasional (BSN), "Metode Pemetaan Rawan Banjir (SNI 8197:2015)," Jakarta, 2015.
10. "ArcGIS Network Analyst | Vehicle Routing Problem & Spatial Network Analysis." Accessed: Jun. 24, 2025. [Online]. Available: https://www.esri.com/en-us/arcgis/products/arcgis-network-analyst/overview
11. J. Andretha, J. Lumban Batu, C. Fibriani, and K. S. Wacana, "ANALISIS PENENTUAN LOKASI EVAKUASI BENCANA BANJIR DENGAN PEMANFAATAN SISTEM INFORMASI GEOGRAFIS DAN METODE SIMPLE ADDITIVE WEIGHTING (Studi Kasus : Kota Surakarta)," vol. **4**, no. 2, pp. 127–135, 2017, [Online]. Available: www.surakarta.go.id
12. "Tergerus Kali Serang, Empat Rumah di Wonosegoro Direlokasi - Suara Merdeka Solo." Accessed: Jun. 24, 2025. [Online]. Available: https://solo.suaramerdeka.com/solo-raya/pr-054936103/tergerus-kali-serang-empat-rumah-di-wonosegoro-direlokasi
13. "Awas Sungai Serang Boyolali Meluap saat Hujan! Ini 3 Lokasi yang Rawan Banjir - Espos.id | Espos Indonesia dari Solo untuk Indonesia." Accessed: Jun. 24, 2025. [Online]. Available: https://solopos.espos.id/awas-sungai-serang-boyolali-meluap-saat-hujan-ini-3-lokasi-yang-rawan-banjir-1444518
14. "Data Kejadian Bencana." Accessed: Jun. 24, 2025. [Online]. Available: https://inarisk.bnpb.go.id/databencana/
15. A. Saputra et al., "A LOW-COST DRONE MAPPING AND SIMPLE PARTICIPATORY GIS TO SUPPORT THE URBAN FLOOD MODELLING," Geographia Technica, vol. **17**, p. 46, 2022, doi: 10.21163/GT_.

 https://doi.org/10.1051/e3sconf/202565204002

Detection of Sinkhole Potential Using Very Low Frequency (VLF) Method: A Case Study in Kedungsalam Village, Malang Regency

Rahmat Faizal[1,2], *Adi* Susilo[1,6*], *Alamsyah M* Juwono[1], *Emma* Yuliani[3], *Muwardi* Sutasoma[1], Muhardi[1,4], *Mohammad Habibiy* Idmi[1], & *Muhammad Fathur Rouf* Hasan[5,6]

[1]Department of Physics, Faculty of Mathematics and Natural Sciences, Brawijaya University, Malang 65145, Indonesia
[2]Department of Civil Engineering, Faculty of Engineering, University of Borneo Tarakan, Tarakan, Indonesia
[3]Department of Water Resource Engineering, Faculty of Engineering, Brawijaya University, Malang 65145, Indonesia
[4]Department of Geophysics, Tanjungpura University, Pontianak 78124, Indonesia
[5]Graduate School, Brawijaya University, Malang 65145, Indonesia
[6]Center Study on Geosciences and Hazard Mitigation, Universitas Brawijaya, 65145, Malang, Indonesia

Abstract. Several caves and sinkholes in Kedungsalam Village, Malang Regency, indicate an intensive and active carbonate rock dissolution process beneath the surface. This condition poses a risk of subsidence disaster that could threaten the safety of residents. This study aims to determine the potential for subsidence using the Very Low Frequency (VLF) method. The VLF method detects the presence of cavities based on the conductivity properties of the rock. Data collection was carried out on four routes adapted to field conditions. Data processing was carried out using the Fraser Filter and Karous Hjlet. The interpretation results identified several zones with high conductivity values, indicating the presence of cavities that could cause soil structural instability. The interpretation results from line 1 showed four anomalies at distances of 20–50 m, 125–150 m, 230–250 m, and 410–440 m. Line 2 presented two anomalies at distances of 280–340 m and 450–500 m. Line 3 shows three anomalies that are likely cavities at distances of 95–125 m, 180–230 m, and 320–370 m. Line 4 shows four anomalies at distances of 80–150 m, 200–240 m, 350–400 m, and 450–480 m. The results of this study can be used to inform policymaking to mitigate the risk of land subsidence and ensure safe spatial planning.

1 Introduction

Karst landscapes, characterised by unique and complex geology, have the potential for natural resources, ecology, and tourism that can benefit surrounding communities

* Corresponding author: adisusilo@ub.ac.id

economically. However, karst is also vulnerable to environmental degradation and disasters such as sinkholes [1].

A sinkhole is a ground surface collapse caused by a subsurface cavity formed by the dissolution of carbonate rocks such as limestone, dolomite, or gypsum over a long period [2]. This phenomenon often occurs in karst areas and can cause significant losses, especially in densely populated areas and critical infrastructure.

Kedungsalam Village in Donomulyo Subdistrict is a karst area in southern Malang (Figure 2). Based on the Blitar Geological Map, this area is part of the Wonosari Formation, as indicated on the Blitar Geological Map, composed of various types of limestone such as coral limestone, dolomitic limestone, tuffaceous limestone, sandy limestone, as well as other sedimentary rocks like shale, black peat clay, and calcareous mudstone [3, 4]. The presence of soluble stones leads to cavities whose locations are difficult to predict accurately [5]. These subsurface cavities pose a risk of sudden ground collapse, which can cause damage to the environment and infrastructure above. For example, ground subsidence on highways and the collapse of building foundations could endanger the safety of residents [5].

In Indonesia, several studies have been conducted to identify potential sinkholes, including using the Analytical Hierarchy Process (AHP) [6] and Geographic Information System (GIS) methods [7]. However, these two methods only provide information based on surface parameters and cannot directly detect the subsurface conditions that cause sinkhole formation.

Several karst features visible on the surface in Kedungsalam Village (Figure 3) indicate the presence of many underground cavities, which need to be detected to support rapid mitigation. The presence of these cavities is essential for identification because they can affect the stability of the land and infrastructure above it [8].

One effective geophysical method for detecting subsurface cavities is the VLF method. This method operates based on the principle of passive electromagnetism, utilising very low-frequency waves emitted from a transmission station [9,10]. VLF is highly sensitive to changes in electrical conductivity beneath the surface, which are associated with cavities, fractures, or dissolution zones that could potentially lead to the formation of sinkholes.

2 Methods

2.1 Principles of the VLF Method

The basic theory of the VLF method uses Maxwell's equations, which describe the propagation of electromagnetic waves related to electric and magnetic fields. This principle is used to detect disturbances to the electromagnetic field emitted by VLF sources, which indicate the presence of geological structures such as cavities, fractures, or dissolution zones [11,12].

$$\nabla \times \boldsymbol{E} = \frac{-\partial \mathbf{B}}{\partial t} \quad (1)$$

$$\nabla \times \boldsymbol{H} = \boldsymbol{J} + \frac{-\partial \mathrm{D}}{\partial t} \quad (2)$$

If the magnetic field (**H**) and electric field (**E**) are assumed to be exponential functions of time, then the vector equation is,

$$\nabla^2 \boldsymbol{H} = (i\omega\mu\sigma\boldsymbol{H}) - (\omega^2 \varepsilon\mu\boldsymbol{H}) \quad (3)$$

Working principle of the VLF method (Figure 1). Very low frequency electromagnetic waves emitted by a VLF transmitter have a magnetic field component called the primary field. This primary field propagates until it reaches conductive subsurface rocks, inducing

 https://doi.org/10.1051/e3sconf/202565204002

eddy currents (induced currents). The eddy currents formed generate a secondary magnetic field. The VLF receiver on the surface then captures the primary field directly from the transmitter and the secondary field from below the surface. By analysing the phase and amplitude differences between the primary and secondary fields, conductive subsurface structures such as fracture zones, faults, or rock layers containing metallic minerals or groundwater can be identified [12,13].

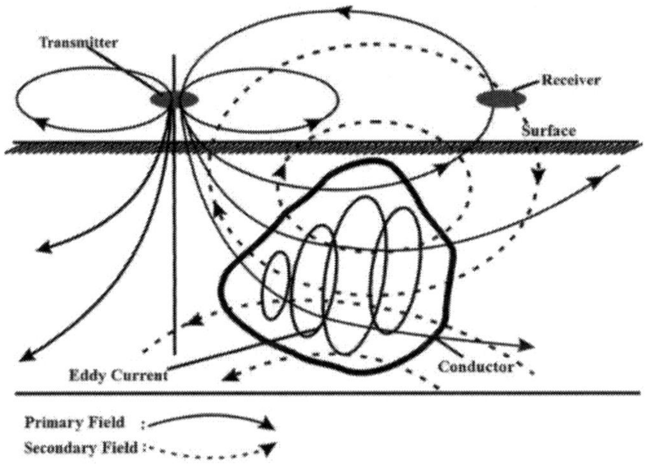

Fig. 1. The working principle of the VLF method [12]

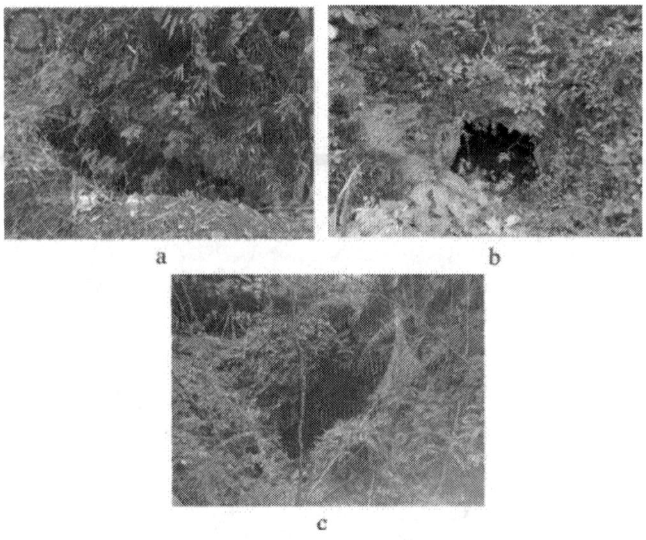

Fig. 2. Surface Karst Features in Kedungsalam Village Indicate Dissolution Activity and Potential Subsurface Cavity Formation

2.2 Research methodology

Data collection was conducted in Kedungsalam Village, Donomulyo District, South Malang Regency, East Java, using a VLF receiver that records magnetic field components from an 18.2 kHz transmitter station located in southern India. Four lines, ranging in length from 400 to 650 meters, were used to match the topography (Figure 3). Data were collected at 10-meter intervals for each line to ensure clear target anomaly detection. Data measured in the VLF-EM acquisition included in-phase, quadrature, tilt, and total field.

Fig. 3. Map of research location

VLF data were filtered using Noise-Assisted Multivariate Empirical Mode Decomposition (NA-MEMD) to reduce noise [14], then processed with Fraser and Karous-Hjelt filters to produce In-Phase, Quadrature, and subsurface conductivity graphs. Zones with high conductivity are interpreted as indications of voids, fractures, or dissolution zones [15]. The final interpretation is reinforced with surface geological data and field observation results to obtain more accurate subsurface information.

3 Results and discussions

Measurements were taken at four lines in several locations in Kedungsalam Village, Donomulyo Subdistrict, Malang Regency. The location of the lines was determined based on the presence of significant surface karst features, such as active caves (Lo Cave and Bangi Cave), dolines (dissolution depressions), fractures, and their proximity to critical infrastructure such as residents' houses and village roads.

The data was processed using the NA-MEMD filter, then processed using the Fraser filter, which aims to clarify the conductive anomalies detected in the VLF data, thereby facilitating the identification of conductive subsurface areas. The results of data processing using the Fraser filter can be seen in Figure 4.

The results of VLF data processing using the Fraser filter (Figure 4) show strong indications of subsurface cavities at several points along the survey line. The In-Phase component (blue line) shows positive and significant fluctuations at several points, indicating the presence of conductive anomalies, such as cavities or zones of rock weathering, where

https://doi.org/10.1051/e3sconf/202565204002

the Fraser filter amplifies changes in the electromagnetic field gradient, making subsurface anomalies more clearly visible.

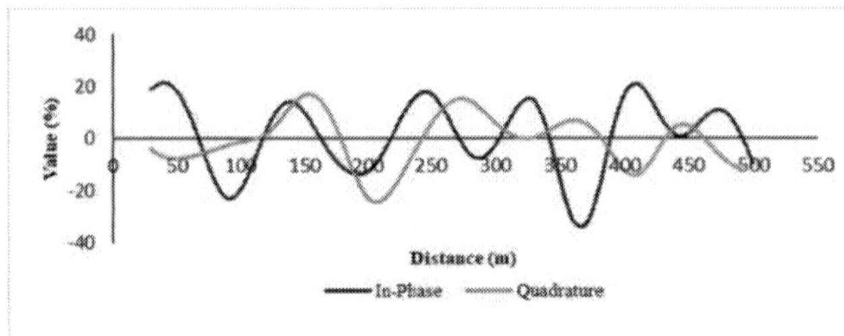

a

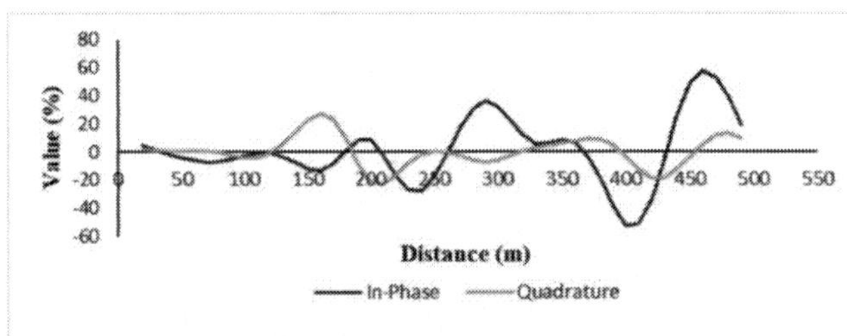

b

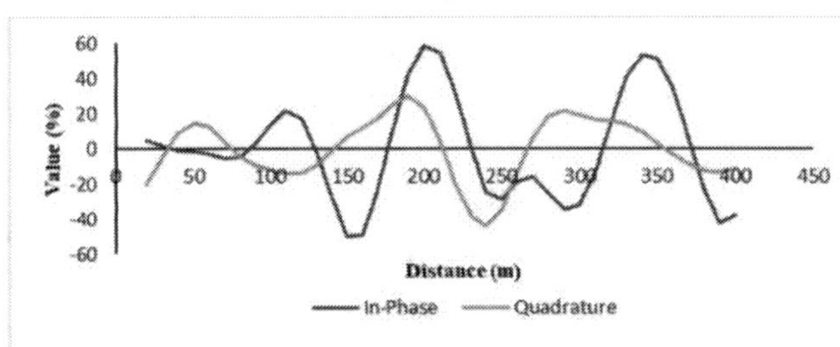

c

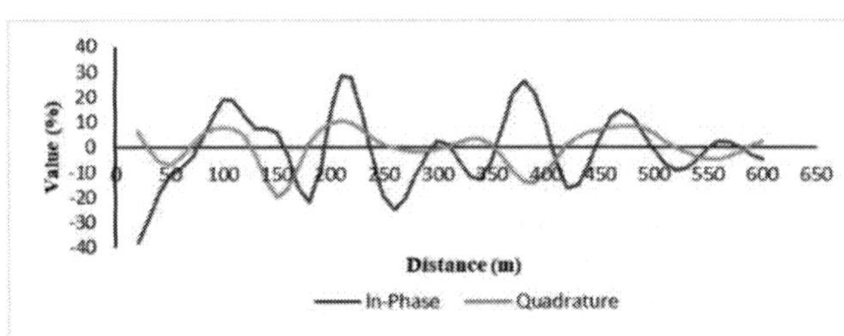

d

Fig. 1. Results of VLF data processing with Fraser Filter on (a) Line 1, (b) Line 2, (c) Line 3, and (d) Line 4

The data processing results using Karous-Hjelt (Figure 5) show several conductive zones, represented by a gradient of red to yellow colors. These zones indicate the potential for

https://doi.org/10.1051/e3sconf/202565204002

subsurface cavities, fractures, or underground rivers. Conversely, blue zones indicate low conductivity (resistive areas), which may be associated with intact limestone or solid material.

Figure 5 shows conductive anomalies on four VLF measurement lines. Line 1 shows four conductive anomalies, namely at distances of 20–50 m, 125–150 m, 230–250 m, and 410–440 m. Line 2 has two conductive anomalies at distances of 280–340 m and 450–500 m. Line 3 has three conductive anomalies, which are suspected to be underground cavities filled with water, at distances of 95–125 m, 180–230 m, and 320–370 m. Line 4 has four conductive anomalies, located at distances of 80–150 m, 200–240 m, 350–400 m, and 450–480 m.

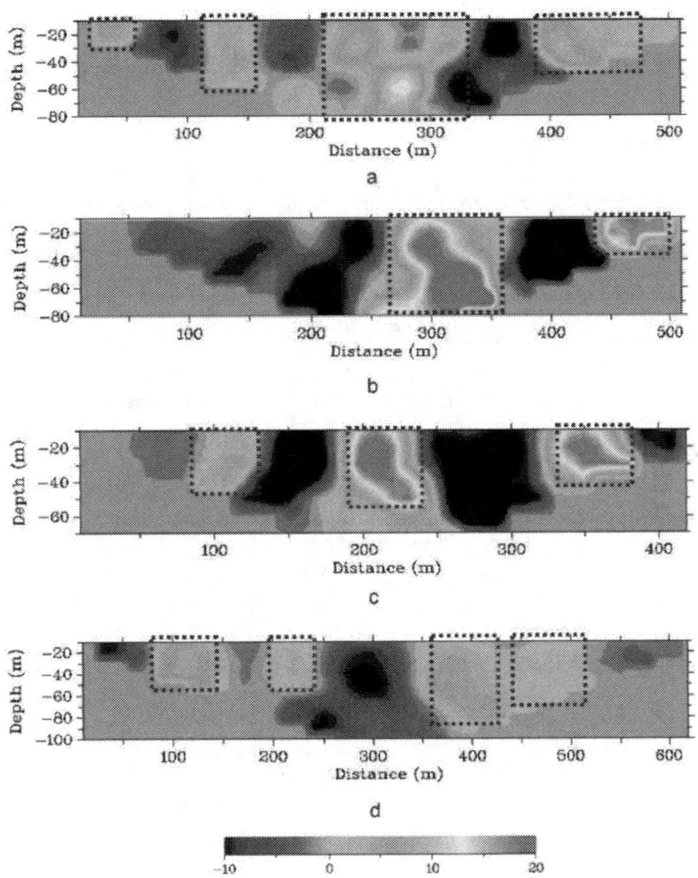

Fig. 2. Results of VLF data processing with the Karous-Hjelt filter on (a) Line 1, (b) Line 2, (c) Line 3, and (d) Line 4

The results of VLF data interpretation show that conductive anomalies along the entire traverse (Figure 5) are consistently located at the same position as the positive peaks on the in-phase graph (Figure 4). This indicates that the lateral in-phase peaks represent the location of subsurface conductive zones, which are strongly suspected to be cavities, fractures, or dissolution zones.

4 Conclusion

This study confirms that the VLF geophysical method is an effective tool for detecting potential sinkhole zones in the karst area of Kedungsalam Village, Malang Regency, East Java. VLF measurements successfully revealed some subsurface conductive anomalies interpreted as cavities, dissolution zones, or rock fractures—features commonly associated with sinkhole formation.

Based on the VLF data processing results, the importance of using non-invasive geophysical methods such as VLF in early detection and monitoring of geological hazards, particularly in karst areas, as a form of mitigation, is highlighted.

This study is an initial identification of subsurface geology in karst areas. Applying the VLF method should be integrated with other geophysical techniques and field geological observations. This integrated approach will strengthen efforts to identify and mitigate disasters in areas vulnerable to karst hazards.

References

1. R. Arjwech and M. E. Everett, Sinkhole formation induced by descending groundwater in a karst aquifer near a limestone quarry. Earth Surf. Process. Landf. 49, 5027–5037 (2024). https://doi.org/10.1002/esp.6010
2. J.K. Lee and J. Kim, Stability charts for sustainable infrastructure: Collapse loads of footings on sandy soil with voids. Sustainability. 11, 3966 (2019). https://doi.org/10.3390/su11143966
3. M. A. Labib, E. Haryono, and Sunarto, The development of cave passage in Donomulyo, Malang-Indonesia, in Proceedings of the 4th International Conference on Science and Technology (ICST 2018), Yogyakarta, Indonesia, August 7-8 (2018), 04010.
4. Y. Boimau, S. Ryo, and A. Susilo, Identification of underground river flow in karst area of Sumber Bening-Malang, Indonesia based on geoelectrical self-potential and resistivity data. Int. J. Appl. Phys., **5**, pp. 1–7, (2018). https://doi.org/10.14445/23500301/IJAP-V5I3P101
5. Y. Hussain, M. F. C. Alves, R. A. C. Menezes, A. F. Bezerra, F. H. R. Bezerra, and A. M. R. de Oliveira, Detection of cover collapse doline and other epikarst features by multiple geophysical techniques, case study of Tarimba Cave, Brazil. Water (Switz.). 12 (10) 2020. https://doi.org/10.3390/w12102835
6. R. R. Fikrillah, S. Say, and D. P. E. Putra, Land Subsidence Hazard Mapping in relation with Sinkholes in Saptosari District, Gunung Kidul, Indonesia, in Proceedings of the 13th AUN/SEED-Net Regional Conference on Geological and Geo-Resource Engineering (RCGeoE 2021), Yogyakarta, Indonesia, December 20-21 (2021), 012004
7. Damayanti, F. Riadini, and F. D. Pamungkas, Potential areas of land subsidence in karst landscape: Case study in Ponjong and Semanu District, Gunungkidul Regency, Yogyakarta, Indonesia. Int. J. Adv. Sci. Eng. Inf. Technol.11(2), 712–719 (2021). https://doi.org/10.18517/ijaseit.11.2.6754
8. A. Nezari, R. Boufarh, T. Mansouri, Numerical investigation of bearing capacity of centrally and eccentrically loaded surface footing on sand-clay with void using Plaxis 2D. J. Eng. Exact Sci., 10, 17506 (2024). https://doi.org/10.18540/jcecvl10iss3pp17506
9. R. Faizal, S. Sismanto, R. Handayani, and A. Asta, Pendugaan aliran sungai bawah tanah dalam pemenuhan kebutuhan air masyarakat Desa Hargosari Gunungkidul berdasarkan data VLF-EM terkoreksi topografi. Borneo Eng. J. Tek. Sipil, 1, 44-55 (2017). https://doi.org/10.35334/be.v1i2.601

10. Y. Hussain, et al., The potential use of geophysical methods to identify cavities, sinkholes and pathways for water infiltration. Water, 12(8), 2289 (2020). https://doi.org/10.3390/w12082289
11. S. P. Sharma, A. Biswas, and V. C. Baranwal, Very low-frequency electromagnetic method: A shallow subsurface investigation technique for geophysical applications. in Recent Trends in Modelling of Environmental Contaminants, New Delhi: Springer. pp. 119–141, (2014). https://doi.org/10.1007/978-81-322-1783-1_5
12. E. D. Nugraha, H. Susanto, and Khumaedi, Application of VLF method to identify flow patterns of underground river in the karst area Ngargoharjo, Wonogiri. J. Phys. Conf. Ser. 1170, 012071 (2019). https://doi.org/10.1088/1742-6596/1170/1/012071
13. J. J. Simpson, An established numerical method applied to geophysics. Eos Trans. AGU. 93, 265–266 (2012). https://doi.org/10.1029/2012EO290001
14. B. Sungkono, B.J. Santosa, A.S. Bahri, F.M. Santos, and A. Iswahyudi, Application of noise-assisted multivariate empirical mode decomposition in VLF-EM data to identify underground river. Adv. Data Sci. Adapt. Anal. 9, 1650011 (2017). https://doi.org/10.1142/S2424922X1650011X
15. S. Sh. Abbas, W. D. Al-Mahemmdi, and H. A. Al-Saady, Caves and fractured zones investigation using 2D electrical resistivity imaging in an engineering site, Haqlaniyah, western Iraq. Iraq. Geol. J., 57, pp. 33–40 (2024). https://doi.org/10.46717/igj.57.2E.3ms-2024-11-12

Identification of Priority Areas Vulnerable to Tsunamis Along the Coastal Zone of Bandar Lampung City

Sugeng Widodo[1*], *Abdurrahman* Abdurrahman[2], *Rahma* Kurnia Sri Utami[3], *Suhendro* Suhendro[2], and *Daska* Azis[3]

[1,2,3]Departement of Geography Education, Universitas Lampung, 35141 Bandar Lampung, Indonesia
[4]Departement of Geography Education, Universitas Pendidikan Indonesia, 40154 Bandung, Indonesia
[5]Departement of Geography Education, Universitas Syiah Kuala, 23111 Banda Aceh, Indonesia

Abstract. The coastal area of Bandar Lampung City is at risk of tsunami disasters due to the morphological characteristics of its shoreline. Major threats such as the eruption of Anak Krakatau and megathrust earthquake activity are the primary sources of this risk. Through the first study that integrated data from various sources at the village level for Bandar Lampung. The objective of this study is to identify priority areas at the village level along the coast of Bandar Lampung City. This study employs a geospatial approach combined with quantitative analysis using percentage-based calculations. The data analyzed include InaRISK data, Probabilistic Tsunami Hazard Assessment (PTHA) data, and historical tsunami records, which were processed using GIS. The findings reveal several priority villages with the highest levels of vulnerability, including Sukaraja, Karang Maritim, Panjang Selatan, Panjang Utara, Pidada, Srengsem, Way Lunik, Keteguhan, Sukamaju, and Way Tataan. Among these ten villages, six are located along the eastern coast and four along the western coast. These findings have limited but important implications as a foundational reference for further empirical investigations into the conditions of priority coastal areas in Bandar Lampung City. Such investigations are essential to explicitly construct localized disaster mitigation strategies and supporting risk-oriented evacuation planning and spatial development.

1 Introduction

The coastal zone of Bandar Lampung City is a highly vulnerable area to natural hazards due to its location at the intersection of tectonic plates and the presence of the active Anak Krakatau volcano [1]. This vulnerability is further exacerbated by its low-lying coastal morphology, bay-like geography, and dense urban population, placing the region in a high-risk zone [2, 3, 4, 5]. Several previous studies have noted that tsunami risk along coastal regions tends to increase in the absence of public awareness and the lack of integration of mitigation measures into spatial planning [6, 7, 8, 9]. Risk perception plays a key role in

* Corresponding author: sugeng.widodo@fkip.unila.ac.id

https://doi.org/10.1051/e3sconf/202565204003

shaping mitigation efforts, including the design of preparedness programs and tools such as drills, community-based activities, emergency plans, and more [7, 8].

The geospatial approach has become a widely adopted method in tsunami risk assessment studies, particularly for identifying priority impact zones [10, 11, 12]. Previous research has demonstrated that the use of quantitative data such as vulnerability scores, capacity indices, and geospatial analysis provides a scientific basis for identifying areas requiring immediate and systematic intervention These studies have consistently shown that areas with limited infrastructure and high population density are often ranked as top priorities in tsunami hazard risk maps [13, 14, 15].

Tsunami risk mapping can serve as a critical reference in disaster-mitigation-based spatial planning, especially in coastal urban areas such as those in Chile [2], the southwestern coast of Portugal [13], Banda Aceh, Indonesia [14] and the Andaman region [15]. Previous studies have emphasized the importance of integrating multiple data sources such as vulnerability indicators [8], Probabilistic Tsunami Hazard Assessment (PTHA) data [9] and historical earthquake and tsunami records [10] into a unified risk analysis to support geospatially oriented disaster management policies. Therefore, identifying priority areas at the village level must be conducted systematically and based on quantitative data to ensure that mitigation programs are focused on the highest-risk zones such as the bay-shaped, tsunami-prone coastal areas of Bandar Lampung City.

2 Method

2.1 Overview Method

This study employs a quantitative, geospatial-based approach to identify the highest-priority areas of tsunami vulnerability along the coastal region of Bandar Lampung City, Indonesia (Figure 1). The rationale for selecting this location lies in its distinctive coastal morphology, particularly its bay-like formation, which increases exposure to tsunami impacts. This morphological characteristic underpins the need to systematically identify severely affected areas at the village level using geospatial analysis.

2.2 Data Collection

This method involves the utilization of Geographic Information Systems (GIS) integrated with disaster risk data from multiple sources, including InaRISK, Probabilistic Tsunami Hazard Assessment (PTHA) data, and historical records of tsunami events. The analysis was conducted using spatial overlay techniques and weighted scoring across these three data models. The data used in this study were secondary in nature and are specifically detailed in Table 1.

Table 1. Research Data Collection

No	Data Name	Year	Function	References
1	InaRISK Data	2025	Tsunami Disaster Vulnerability Based on High, Medium, and Low Categories	[12]
2	PTHA Data	2014	Vulnerability of Areas with Probability of Tsunami Heights > 3 m and < 5 m	[13]
3	Historical Data	2025	History of Earthquakes and Tsunamis	[14]

 https://doi.org/10.1051/e3sconf/202565204003

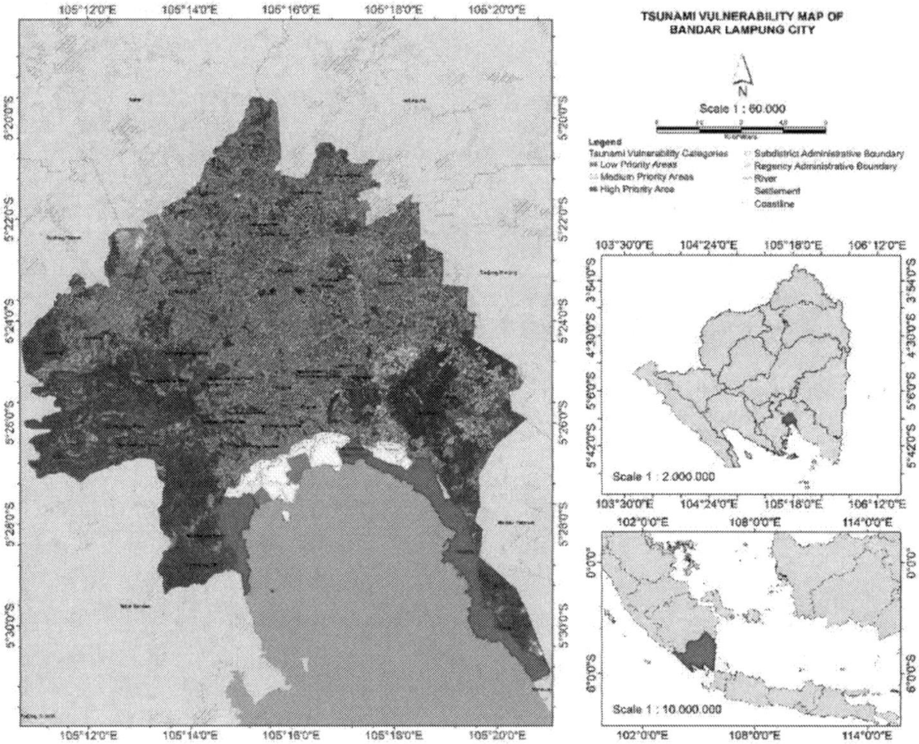

Fig 1. Study Area

2.3 Data Analysis

The analytical technique employed in this study adopts a quantitative approach supported by geospatial methods based on GIS. The scoring process was conducted using three primary datasets, detailed as follows:

1) InaRISK Data: Areas were categorized into high, medium, and low vulnerability zones. The score was calculated by summing the area of each vulnerability class within a region, dividing it by the total area of the region, and multiplying the result by 10.
2) PTHA Data: Two parameters were considered wave height probabilities of >0,5 meters and greater than >3 meters. Each region exposed to these conditions received a score of 10, while areas with no risk were assigned a score of 0.
3) Historical Earthquake and Tsunami Records: This dataset served as a supplementary indicator, providing a historical footprint of tsunami events in the area.

Formulation contructs:

$$a_{total} = \frac{a_2}{a_3} x\ 100 \quad (1)$$

$$b_{total} = b_1 + b_2 \quad (2)$$

$$b_1 = \begin{cases} 10 & Jika\ nilai\ PTHA\ dengan\ tinggi\ gelombang\ >\ 0{,}5\ meter \\ 0 & Jika\ nilai\ PTHA\ dengan\ tinggi\ gelombang\ >\ 3\ meter \end{cases} \quad (2a)$$

$$b_2 = \begin{cases} 10 & Jika\ nilai\ PTHA\ dengan\ tinggi\ gelombang\ >\ 0{,}5\ meter \\ 0 & Jika\ nilai\ PTHA\ dengan\ tinggi\ gelombang\ >\ 3\ meter \end{cases} \quad (2b)$$

$$\text{Nilai } T_1 = a_{total} + b_{total} \quad (3)$$
$$\text{Nilai } T_2 = \frac{\text{Nilai perolehan dari } T_1}{Skor\ Maksimal\ dari\ Nilai\ T_1} \times 100 \quad (4)$$

The final vulnerability score for each area was obtained by summing the individual scores from these three data sources. Areas with the highest total scores were classified as having the highest tsunami vulnerability, and vice versa.

3 Result and Discussion

3.1 Result

The results of the study indicate that among all coastal villages in Bandar Lampung City, ten villages exhibit the highest levels of tsunami vulnerability based on the geospatial scoring analysis. These villages include Sukaraja, Karang Maritim, Panjang Selatan, Panjang Utara, Pidada, Srengsem, Way Lunik, Keteguhan, Sukamaju, and Way Tataan. These findings were derived using a novel methodology that integrates multiple data sources (Table 2).

Table 2. A Methodology for Assessing Tsunami Vulnerability in Priority Coastal Areas of Bandar Lampung City

Village	inaRISK (*a*)				Probabilistic Tsunami Hazard Assessment (*b*)				c_{total}
	a_1	a_2	a_3	a_{total}	b_1	$b1_{total}$	b_2	$b2_{total}$	
Bumi Raya	High	17,440	36,056	4,84	0	0	0	0	0
	Low	7,499	36,056	2,08	0	0	0	0	0
	Medium	11,118	36,056	3,08	0	0	0	0	0
Bumi Waras Garuntang	High	92,576	93,031	9,95	0	0	< 3m	10	0
	Medium	0,454	93,031	0,05	0	0	< 3m	10	0
Kangkung	High	36,860	48,897	7,54	0	0	0	0	0
	Low	5,279	48,897	1,08	0	0	0	0	0
	Medium	6,758	48,897	1,38	0	0	0	0	0
	High	39,740	39,740	10,00	0	0	< 3m	10	0
Sukaraja	High	103,566	109,535	9,46	0	0	< 3m	10	0
	Low	2,457	109,535	0,22	0	0	< 3m	10	0
	Medium	3,513	109,535	0,32	0	0	< 3m	10	0
Karang Maritim	High	37,833	51,727	7,31	0	0	< 3m	10	0
	Low	6,748	51,727	1,30	0	0	< 3m	10	0
	Medium	7,145	51,727	1,38	0	0	< 3m	10	0

Village	**inaRISK (*a*)**				**Probabilistic Tsunami Hazard Assessment (*b*)**				c_{total}
	a_1	a_2	a_3	a_{total}	b_1	$b1_{total}$	b_2	$b2_{total}$	
Ketapang	High	0,046	16,973	0,03	0	0	0	0	0
	Low	8,526	16,973	5,02	0	0	0	0	0
	Medium	8,400	16,973	4,95	0	0	0	0	0
Ketapang Kuala	High	18,974	31,952	5,94	0	0	0	0	0
	Low	5,103	31,952	1,60	0	0	0	0	0
	Medium	7,875	31,952	2,46	0	0	0	0	0
Panjang Selatan	High	42,041	48,124	8,74	0	0	< 3m	10	0
	Low	2,560	48,124	0,53	0	0	< 3m	10	0
	Medium	3,524	48,124	0,73	0	0	< 3m	10	0
Panjang Utara	High	97,648	104,905	9,31	0	0	< 3m	10	0
	Low	3,125	104,905	0,30	0	0	< 3m	10	0
	Medium	4,131	104,905	0,39	0	0	< 3m	10	0
Pidada	High	44,907	67,663	6,64	0	0	< 3m	10	0
	Low	10,081	67,663	1,49	0	0	< 3m	10	0
	Medium	12,676	67,663	1,87	0	0	< 3m	10	0
Srengsem	High	209,443	256,907	8,15	0	0	< 3m	10	0
	Low	24,583	256,907	0,96	0	0	< 3m	10	0
	Medium	22,881	256,907	0,89	0	0	< 3m	10	0
Way Lunik	High	126,012	159,874	7,88	0	0	< 3m	10	0
	Low	13,530	159,874	0,85	0	0	< 3m	10	0
	Medium	20,332	159,874	1,27	0	0	< 3m	10	0
Way Laga	High	1,717	10,711	1,60	0	0	0	0	0
	Low	5,839	10,711	5,45	0	0	0	0	0
	Medium	3,156	10,711	2,95	0	0	0	0	0
Bakung	High	10,253	43,970	2,33	0	0	0	0	0
	Low	14,405	43,970	3,28	0	0	0	0	0
	Medium	19,312	43,970	4,39	0	0	0	0	0
Kuripan	High	2,293	23,866	0,96	0	0	0	0	0

Village	**inaRISK (*a*)**				**Probabilistic Tsunami Hazard Assessment (*b*)**				c_{total}
	a_1	a_2	a_3	a_{total}	b_1	$b1_{total}$	b_2	$b2_{total}$	
	Low	4,238	23,866	1,78	0	0	0	0	0
	Medium	17,335	23,866	7,26	0	0	0	0	0
Negeri Olok Gading	Low	6,089	6,432	9,47	0	0	0	0	0
	Medium	0,342	6,432	0,53	0	0	0	0	0
Gedong Pakuon	High	2,036	6,147	3,31	0	0	0	0	0
	Low	2,529	6,147	4,11	0	0	0	0	0
	Medium	1,582	6,147	2,57	0	0	0	0	0
Gunungmas	High	10,832	17,328	6,25	0	0	0	0	0
	Low	1,827	17,328	1,05	0	0	0	0	0
	Medium	4,668	17,328	2,69	0	0	0	0	0
Pesawahan	High	89,214	94,028	9,49	0	0	< 3m	10	0
	Medium	4,814	94,028	0,51	0	0	< 3m	10	0
Talang	High	3,686	18,397	2,00	0	0	0	0	0
	Low	4,945	18,397	2,69	0	0	0	0	0
	Medium	9,766	18,397	5,31	0	0	0	0	0
Telukbetung	High	18,470	18,470	10,00	0	0	0	0	0
Keteguhan	High	88,711	106,368	8,34	0	0	< 3m	10	0
	Low	7,061	106,368	0,66	0	0	< 3m	10	0
	Medium	10,596	106,368	1,00	0	0	< 3m	10	0
Kota Karang	High	48,025	49,898	9,62	0	0	< 3m	10	0
	Medium	1,873	49,898	0,38	0	0	< 3m	10	0
Kota Karang Raya	High	25,415	25,415	10,00	0	0	< 3m	10	0
Parwata	High	15,034	16,238	9,26	0	0	< 3m	10	0
	Medium	1,204	16,238	0,74	0	0	< 3m	10	0
Sukamaju	High	79,002	98,900	7,99	0	0	< 3m	10	0
	Low	9,866	98,900	1,00	0	0	< 3m	10	0
	Medium	10,033	98,900	1,01	0	0	< 3m	10	0
Way Tataan	High	56,388	77,522	7,27	0	0	< 3m	10	0

 https://doi.org/10.1051/e3sconf/202565204003

Village	inaRISK (a)				Probabilistic Tsunami Hazard Assessment (b)				c_{total}
	a_1	a_2	a_3	a_{total}	b_1	$b1_{total}$	b_2	$b2_{total}$	
	Low	8,963	77,522	1,16	0	0	< 3m	10	0
	Medium	12,172	77,522	1,57	0	0	< 3m	10	0
Kupang Kota	High	1,071	8,725	1,23	0	0	0	0	0
	Low	3,826	8,725	4,38	0	0	0	0	0
	Medium	3,829	8,725	4,39	0	0	0	0	0
Kupang Raya	High	5,563	14,586	3,81	0	0	0	0	0
	Low	3,683	14,586	2,52	0	0	0	0	0
	Medium	5,340	14,586	3,66	0	0	0	0	0
Kupang Teba	High	7,344	25,945	2,83	0	0	0	0	0
	Low	7,908	25,945	3,05	0	0	0	0	0
	Medium	10,693	25,945	4,12	0	0	0	0	0

Describe: Vulnerability category (a_1); Area size (a_2); Total area score (a_3); Score result (a_{total}); PTHA Wave Height Probability >0,5 meters (b_1); Score result ($b1_{total}$); PTHA Wave Height Probability >3 meters (b_2); Score result ($b2_{Total}$); Historical earthquake and tsunami data (c_{total}).

The resulting T1 (Total Score) was further processed by applying a weighting technique. This involved dividing each village's T1 score by the maximum score obtained among all assessed villages. This normalization step enables the relative comparison of tsunami vulnerability across locations. Several categorization processes involve the following stages and results:

1) Categorization process 1 (K_1):

Table 3. Categorization process K_1

Mean value	43,29
SD value	30,50
X < M – 1SD	12,79
X > M + 1 SD	73,79

Table 4. Categorization results K_1

Categories	Formulation	Interval
Low Priority Area	X < M – 1SD	X < 12,79
Medium Priority Area	M – 1 SD ≤ X ≤ M + 1 SD	12,79 ≤ X ≤ 73,79
High Priority Area	X > M + 1 SD	X > 73,79

2) Categorization process 2 (K_2):

Table 5. Categorization process K_2

Mean value	5,32
SD value	1,42
X < M – 1SD	3,90
X > M + 1 SD	6,75

Table 6. Categorization results K_2

Categories	Formulation	Interval
Low Priority Area	X < M – 1SD	X < 3,90
Medium Priority Area	M – 1 SD ≤ X ≤ M + 1 SD	3,90 ≤ X ≤ 6,75
High Priority Area	X > M + 1 SD	X > 6,75

The weighted scores are presented in Table 7.

Table 7. Categorization of Tsunami-Prone Priority Villages in Bandar Lampung City

Village	**T1**	**Decision Making (*d*)**			
		T2	**K1**	**K2**	**Decision**
Bumi Raya	4,84	24,18	2,00	5,00	Medium Priority Areas
	2,08	10,40	1,00		
	3,08	15,42	2,00		
Bumi Waras	19,95	99,76	3,00	5,00	Medium Priority Areas
	10,05	50,24	2,00		
Garuntang	7,54	37,69	2,00	4,00	Medium Priority Areas
	1,08	5,40	1,00		
	1,38	6,91	1,00		
Kangkung	20,00	100,00	3,00	3,00	Low Priority Areas
Sukaraja	19,46	97,28	3,00	7,00	High Priority Area
	10,22	51,12	2,00		
	10,32	51,60	2,00		
Karang Maritim	17,31	86,57	3,00	7,00	High Priority Area
	11,30	56,52	2,00		
	11,38	56,91	2,00		
Ketapang	0,03	0,14	1,00	5,00	Medium Priority Areas
	5,02	25,12	2,00		
	4,95	24,75	2,00		
Ketapang Kuala	5,94	29,69	2,00	4,00	Medium Priority Areas
	1,60	7,99	1,00		

 https://doi.org/10.1051/e3sconf/202565204003

Village	T_1	Decision Making (*d*)			
		T_2	K_1	K_2	Decision
	2,46	12,32	1,00		
Panjang Selatan	18,74	93,68	3,00	7,00	High Priority Area
	10,53	52,66	2,00		
	10,73	53,66	2,00		
Panjang Utara	19,31	96,54	3,00	7,00	High Priority Area
	10,30	51,49	2,00		
	10,39	51,97	2,00		
Pidada	16,64	83,18	3,00	7,00	High Priority Area
	11,49	57,45	2,00		
	11,87	59,37	2,00		
Srengsem	18,15	90,76	3,00	7,00	High Priority Area
	10,96	54,78	2,00		
	10,89	54,45	2,00		
Way Lunik	17,88	89,41	3,00	7,00	High Priority Area
	10,85	54,23	2,00		
	11,27	56,36	2,00		
Way Laga	1,60	8,01	1,00	5,00	Medium Priority Areas
	5,45	27,26	2,00		
	2,95	14,73	2,00		
Bakung	2,33	11,66	1,00	5,00	Medium Priority Areas
	3,28	16,38	2,00		
	4,39	21,96	2,00		
Kuripan	0,96	4,80	1,00	4,00	Medium Priority Areas
	1,78	8,88	1,00		
	7,26	36,32	2,00		
Negeri Olok Gading	9,47	47,34	2,00	3,00	Low Priority Areas
	0,53	2,66	1,00		
Gedong Pakuon	3,31	16,56	2,00	6,00	Medium Priority Areas
	4,11	20,57	2,00		
	2,57	12,87	2,00		
Gunungmas	6,25	31,26	2,00	5,00	Medium Priority Areas
	1,05	5,27	1,00		
	2,69	13,47	2,00		
Pesawahan	19,49	97,44	3,00	5,00	Medium Priority Areas
	10,51	52,56	2,00		
Talang	2,00	10,02	1,00	5,00	Medium Priority Areas

Village	T1	Decision Making (*d*)			
		T2	K1	K2	Decision
	2,69	13,44	2,00		
	5,31	26,54	2,00		
Telukbetung	10,00	50,00	2,00	2,00	Low Priority Areas
Keteguhan	18,34	91,70	3,00	7,00	High Priority Area
	10,66	53,32	2,00		
	11,00	54,98	2,00		
Kota Karang	19,62	98,12	3,00	5,00	Medium Priority Areas
	10,38	51,88	2,00		
Kota Karang Raya	20,00	100,00	3,00	3,00	Low Priority Areas
Parwata	19,26	96,29	3,00	5,00	Medium Priority Areas
	10,74	53,71	2,00		
Sukamaju	17,99	89,94	3,00	7,00	High Priority Area
	11,00	54,99	2,00		
	11,01	55,07	2,00		
Way Tataan	17,27	86,37	3,00	7,00	High Priority Area
	11,16	55,78	2,00		
	11,57	57,85	2,00		
Kupang Kota	1,23	6,14	1,00	5,00	Medium Priority Areas
	4,38	21,92	2,00		
	4,39	21,94	2,00		
Kupang Raya	3,81	19,07	2,00	5,00	Medium Priority Areas
	2,52	12,62	1,00		
	3,66	18,30	2,00		
Kupang Teba	2,83	14,15	2,00	6,00	Medium Priority Areas
	3,05	15,24	2,00		
	4,12	20,61	2,00		

Describe: Vulnerability category (a_1); Area size (a_2); Total area score (a_3); Score result (a_{total}); PTHA Wave Height Probability >0,5 meters (b_1); Score result ($b1_{total}$); PTHA Wave Height Probability >3 meters (b_2); Score result ($b2_{Total}$); Historical earthquake and tsunami data (c_{total}); Total 1 is INARisk + PTHA (T_1); Total 2 is T1/ Skor Maksimal (T_2); Hasil kategorisasi 1 (K_1); Jumlah hasil kategorisasi (K_2); Decision Making (*d*).

The results of the study show that of all coastal villages in Bandar Lampung City, there are 15 villages with low priority status, 5 villages with medium priority status, and 10 villages with high priority status. We also visualized the locations of tsunami vulnerability studies along the coast of Bandar Lampung City (Figure 2).

https://doi.org/10.1051/e3sconf/202565204003

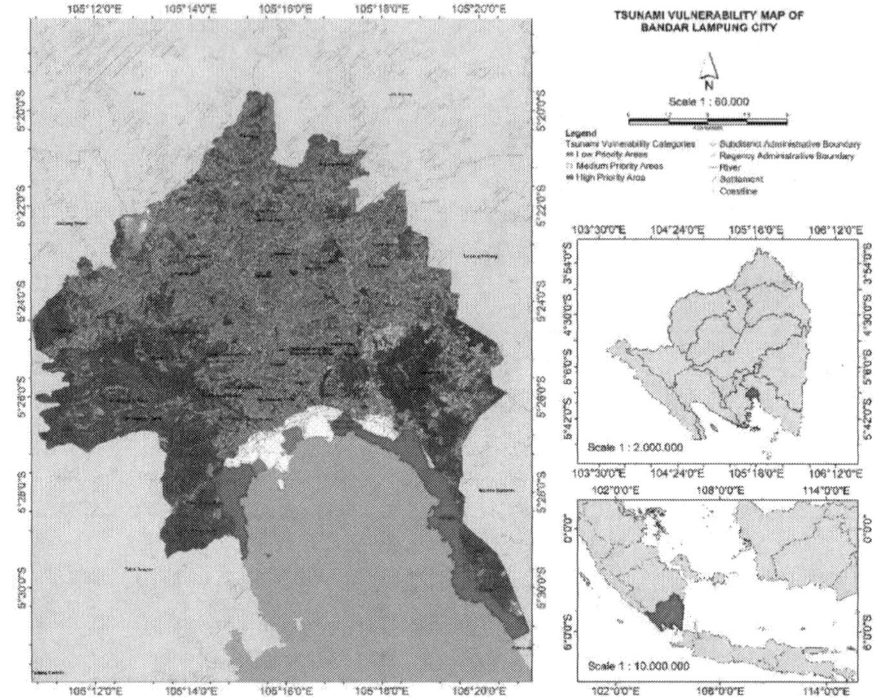

Fig 2. Tsunami Vulnerability Map of the Bandar Lampung Coast

An observable indication from the scoring results of the prioritized villages reveals a clear dominance of coastal areas, particularly along the western and eastern coastal zones of Bandar Lampung City. This highlights the spatial concentration of tsunami vulnerability, warranting a more refined vulnerability analysis in these regions.

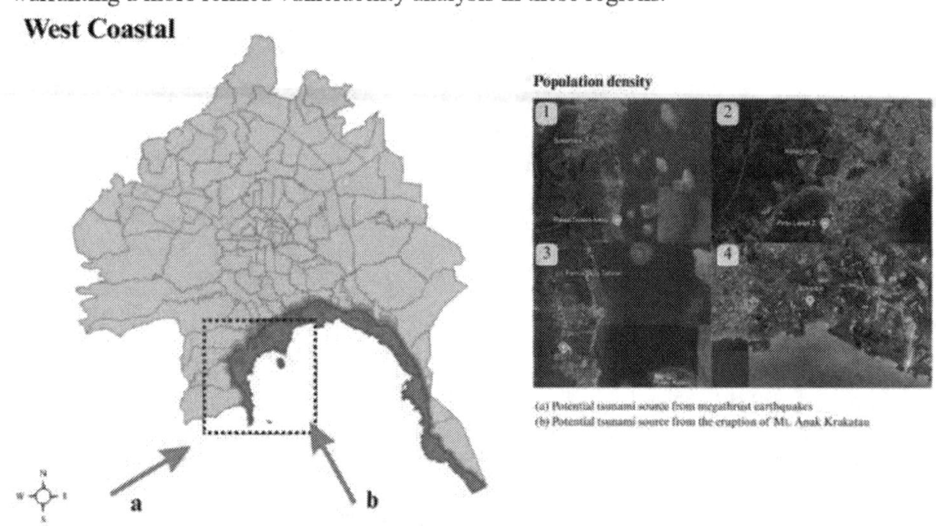

Fig 3. West Coastal of Bandar Lampung - Describe: Sukamaju Village (1); Keteguhan Village (2); Way Tataan Village (3); Sukaraja Village (4)

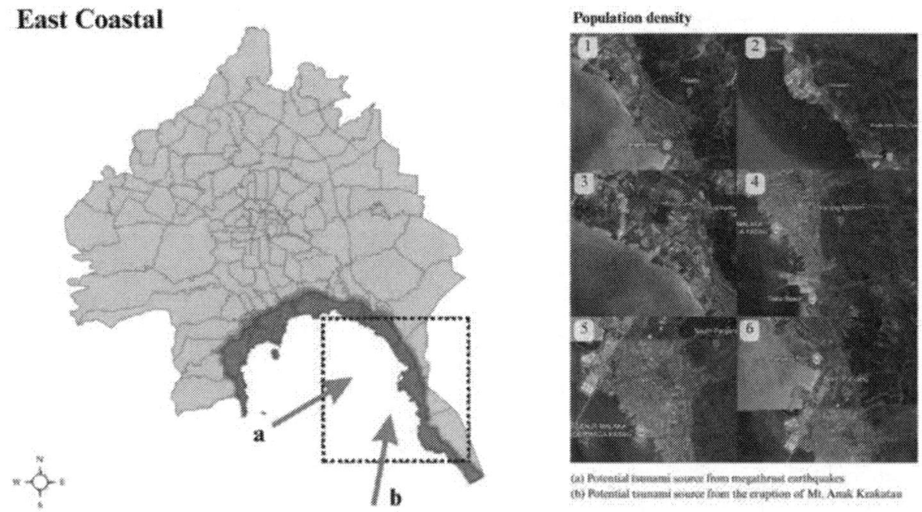

Fig 4. East Coastal of Bandar Lampung - Describe: (1) Pidada Village; Srengsem Village (2); Way Lunik Village (3); Karang Maritim Village (4); Panjang Selatan Village (5); Panjang Utara Village (6).

We identified that the western coastal zone has a high tsunami hazard potential, although it appears to be less vulnerable compared to the eastern coast. This observation is based on the dual tsunami risk sources: (a) the potential impact of the Anak Krakatau volcanic eruption, and (b) a megathrust earthquake originating from the western offshore of Lampung. It is important to note that the western region is more susceptible to volcanic-induced tsunamis from Anak Krakatau. Nevertheless, the morphological characteristics of the coastal terrain and the possible influence of wave height must also be taken into account when assessing the tsunami risk potentially triggered by megathrust earthquakes in the western offshore area.

Furthermore, the analysis of the eastern coastal area reveals that its bay-shaped morphology presents a higher potential for tsunami impact. This heightened vulnerability is likely influenced by two primary hazard sources: (a) the proximity to Anak Krakatau and (b) the subduction zone of the Sunda Plate. Although these indications warrant further verification through numerical modeling and detailed field observation, they highlight the importance of spatial dynamics in hazard assessment. In addition, this vulnerability is exacerbated by high population density, which complicates spatial planning efforts as part of disaster mitigation strategies along the coastal zones of Bandar Lampung.

3.2 Discussion

Previous studies have emphasized that the spatial pattern of tsunami vulnerability is strongly influenced by land-use planning and the distribution of coastal populations [2, 14, 15]. In many cases, high-density settlements have developed within red tsunami hazard zones without risk-based planning, significantly constraining mitigation strategies [10]. Moreover, the bay-shaped morphology contributes significantly to supporting a greater volume of seawater flowing toward the land with strong energy. The powerful energy of a tsunami can cause loss of life and damage to public facilities [12]. Several studies support this statement, such as studies along the coast of Ancon Bay, Peru [5], Pacitan Bay, Indonesia [6], Pandeglang Bay, Indonesia [7], along the western Makran coast, Iran [8], Cádiz Bay in Spain [9], Cartagena Bay in Chile [10], and the Hawkes Bay region, New Zealand [11].

 https://doi.org/10.1051/e3sconf/202565204003

In response to this, an implementation of effective mitigation strategies has led to a substantial reduction in fatalities [3]. These findings align with the results of the present study, reinforcing that tsunami mitigation strategies must not only be spatially grounded but also incorporate social dimensions seperti eksistensi kearifan lokal [4], evacuation capacity, and the integration of risk-informed spatial planning by GIS [6,7,8]. It is important to note that tsunami vulnerability in Bandar Lampung has also been addressed in prior studies. For example, Susilorini et al. investigated the levels of knowledge, awareness, and disaster resilience among coastal communities in the city in the context of earthquake and tsunami risk reduction [10]. Similarly, Febrina et al. explored the integration of geospatial technology with evacuation route planning for tsunami mitigation [12].

The present findings should not be interpreted as an effort to induce public fear, but rather as a call for heightened awareness and continuous adaptation in regions exposed to natural hazards such as the eruption of Mount Anak Krakatau and potential megathrust earthquakes along the southern Sunda subduction zone in Lampung. Ending the discussion on this study, we provide concrete recommendations for future mitigation efforts, such as evacuation route design and risk-based spatial planning using our findings in order of priority, as well as ongoing public awareness activities on the dangers of tsunamis and adaptation measures before and after, especially in priority villages along the coast of Bandar Lampung.

4 Conclusion

The coastal area of Bandar Lampung City, Indonesia, is characterized by a bay-shaped morphology, which suggests a high vulnerability to tsunami hazards. The active presence of Mount Anak Krakatau's volcanic eruptions and the predicted megathrust earthquakes along the Sunda subduction zone emphasize the urgent need for both structural and non-structural disaster mitigation efforts. Utilizing a geospatial approach and integrating data from multiple perspectives, this study introduces a novel methodological framework to identify high-impact tsunami-prone areas along the coastal zone of Bandar Lampung. Based on this analysis, ten villages were identified as high-priority areas: Sukaraja, Karang Maritim, Panjang Selatan, Panjang Utara, Pidada, Srengsem, Way Lunik, Keteguhan, Sukamaju, and Way Tataan. Most of these priority areas are concentrated along the eastern coast, rather than the central or western zones. This spatial pattern is primarily influenced by population density, low-lying bay morphology, tsunami hazard sources, and varying levels of community awareness and preparedness.

We would like to emphasize that this study contributes scientifically to studies within the multi-source geospatial framework for village-level tsunami vulnerability. However, we realize that, we acknowledge the empirical limitations of this study, particularly the absence of explicit tsunami modeling, which ideally requires not only a geospatial approach but also an integrated elaboration with numerical and empirical simulations. Therefore, we recommend that future research undertake detailed numerical and empirical modeling of tsunami hazards using advanced and emerging methodologies. Such approaches will be essential to strengthen the recommendations and formulate effective disaster mitigation strategies both structural and non-structural across the coastal zones such as urban spatial planning or community evacuation drills of Bandar Lampung City, Indonesia.

The Acknowledgements, We express our sincere gratitude to the Ministry of Higher Education, Science, and Technology of the Republic of Indonesia for supporting this research through the BIMA program under the Regular Fundamental Scheme, Fiscal Year 2025.

Fundings, This research was funded by the Ministry of Higher Education, Science, and Technology of the Republic of Indonesia under the BIMA Program, Regular Fundamental Scheme 2025, through contract number 076/C3/DT.05.00/PL/2025 (Master Contract Date: May 28, 2025).

Data availability statement, All data in this study are available upon request to the author.

Author contribution statement, All authors have contributed substantially to the conception, design, analysis, and interpretation of data for this study. All authors have read and approved the final version of the manuscript submitted for publication.

References

1. A. M. Asbi, D. Mardiatno, and D. Ruslanjari, "Tsunami hazard area zonation using run-up modeling in the coastal area of Bandar Lampung City," IOP Conf. Ser.: Earth Environ. Sci., vol. 1479, no. 1, p. 012007, 2025, doi: 10.1088/1755-1315/1479/1/012007.
2. M. Vicuña, J. León, and S. Guzmán, "Urban form planning and tsunami risk vulnerability: Analysis of 12 Chilean coastal cities," Environ. Plann. B: Urban Anal. City Sci., vol. 49, no. 7, pp. 1967–1979, 2022, https://doi.org/10.1177/23998083221075635
3. J. Woessner and R. J. Farahani, "Tsunami inundation hazard across Japan," Int. J. Disaster Risk Reduct., vol. 49, p. 101654, 2020, https://doi.org/10.1016/j.ijdrr.2020.101654
4. H. A. Vardar et al., "Scenario-based tsunami hazard assessment for the northwestern coast of Marmara Sea," Geo-Mar. Lett., vol. 43, no. 3, p. 12, 2023, https://doi.org/10.1007/s00367-023-00752-3
5. A. K. Kondarathi et al., "Tsunami modeling for a possible tsunamigenic earthquake from the 1762 Arakan source region and its effect along the South Andaman Coast," J. Geol. Soc. India, vol. 101, no. 6, pp. 866–871, 2025, https://doi.org/10.17491/jgsi/2025/174175
6. D. Tufekci-Enginar, M. L. Suzen, and A. C. Yalciner, "The evaluation of public awareness and community preparedness parameter in GIS-based spatial tsunami human vulnerability assessment (MeTHuVA)," Nat. Hazards, vol. 105, no. 3, pp. 2639–2658, 2021, https://doi.org/10.1007/s11069-020-04416-8
7. L. Cugliari, A. Cerase, and A. Amato, "Tsunami risk perception, a state-of-the-art review with a focus in the NEAM region," Front. Earth Sci., vol. 10, p. 995536, 2022, https://doi.org/10.3389/feart.2022.995536
8. P. Salah and J. Sasaki, "Knowledge, awareness, and attitudes toward tsunamis: A local survey in the southern coast of Iran," Sustainability, vol. 13, no. 2, p. 449, 2021, https://doi.org/10.3390/su13020449
9. A. Dhellemmes et al., "Tsunami awareness and preparedness in Aotearoa New Zealand: The evolution of community understanding," Int. J. Disaster Risk Reduct., vol. 65, p. 102576, 2021, https://doi.org/10.1016/j.ijdrr.2021.102576
10. N. Wood et al., "Multi-hazard risk analysis for the US Department of the Interior: an integration of expert elicitation, planning priorities, and geospatial analysis," Int. J. Disaster Risk Reduct., vol. 82, p. 103385, 2022, https://doi.org/10.1016/j.ijdrr.2022.103385
11. M. Alwi and B. W. Mutaqin, "Geospatial mapping of tsunami susceptibility in Parangtritis coastal area of Yogyakarta, Indonesia," Arab. J. Geosci., vol. 15, no. 15, p. 1332, 2022, https://doi.org/10.1007/s12517-022-10608-2
12. S. Biswas and A. Sil, "Tsunami vulnerability assessment and multi-criteria decision making analysis of Eastern Coast of India using GIS-based tools," KSCE J. Civ. Eng., vol. 27, no. 3, pp. 1270–1287, 2023, https://doi.org/10.1007/s12205-023-1493-y
13. S. Biswas and A. Sil, "Tsunami vulnerability assessment using GIS and AHP technique for Southern Coastal Region of India," Nat. Hazards Rev., vol. 25, no. 3, p. 04024019, 2024, https://doi.org/10.1061/NHREFO.NHENG-1594

14. F. Usman et al., "Assessing coastal population capacity in tsunami-prone areas: A grid-based approach," Jàmbá: J. Disaster Risk Stud., vol. 16, no. 1, pp. 1–10, 2024, https://doi.org/10.4102/jamba.v16i1.1685
15. Z. Hidayah, N. N. Rohmah, and M. K. Wardhani, "Coastal vulnerability study on potential impact of tsunami and community resilience in Pacitan Bay East Java," Forum Geogr., vol. 36, no. 1, 2022, https://doi.org/10.23917/forgeo.v36i1.17160

Numerical Design and Preliminary Evaluation of a U-Shaped Hydraulic Flume for Flood Early Warning Applications

Andi Patriadi[1*], *Esti* Wulandari[2], *Laksono Djoko* Nugroho[3], and *Ony Frengky* Rumihin[4]

[1,2,3,4]Universitas 17 Agustus 1945 Surabaya, Study Program of Civil Engineering, PO Box 60118, Indonesia

Abstract. The present study introduces the design and preliminary evaluation of a U-shaped hydraulic flume aimed at supporting flood early warning systems. Numerical simulations were employed to investigate flow characteristics under subcritical conditions, both with and without spur dike structures. The baseline scenario with an inlet velocity of 0.463 m/s demonstrated stable subcritical flow upstream and supercritical transition downstream, with velocity concentrated along the outer bank. The addition of two spur dikes successfully diverted the main current toward the channel centerline, reducing outer-bank velocity and mitigating erosion risk. However, high turbulence and concentrated velocity at the spur dike tips revealed the potential for localized scouring. Furthermore, the spur dikes caused a reduction in upstream water surface elevation, suggesting a secondary benefit in lowering flood levels entering the bend. These findings highlight the potential dual function of spur dikes in erosion control and flood regulation. Nonetheless, the study remains limited to numerical analysis, with physical validation required in future research to enhance the robustness of the results.

1 Introduction

Laboratory-scale physical model testing is a fundamental approach in hydraulic engineering studies, particularly when numerical modeling encounters limitations in capturing the complex interactions among flow, structures, and sediment. One of the key instruments in such testing is the flume, or experimental channel, which enables the simulation of actual flow conditions in a controlled scale environment. Straight flume channels have commonly been used for experiments involving uniform flow and sedimentation. However, to replicate the real conditions of natural rivers, which often feature bends, geometric modifications such as U-shaped or meandering channels are required. Experimental studies using curved flumes have proven essential in understanding bend flow dynamics, velocity distribution, and the interactions between flow and sediment transport [1,2]. Along with the increasing need for reliable disaster early warning systems and effective flood risk management, it is vital to ensure that every numerically-based predictive model is supported by accurate experimental data. In this context, flume channel designs with specific configurations can serve as ground

* Corresponding author: andipatriadi@untag-sby.ac.id

truth sources for the calibration and validation of predictive models that form part of early warning systems [3,4].

Furthermore, the adequacy of the pumping system greatly affects flow stability. The flow entering the flume through the pump system is at risk of generating excessive turbulence if not properly controlled, which can undermine the validity of experimental results [3,4]. To ensure the physical model accurately represents real-world conditions, the design scale must adhere to the principle of dominant force similarity (typically inertia and gravity forces) which in the context of open channel flow, is commonly governed by Froude similarity. The validity of physical testing results heavily depends on the accuracy of flow phenomenon replication and the system's ability to maintain steady discharge and consistent inflow characteristics [5,6]. In this regard, the present study designs a U-shaped flume with a water recirculation system using a pump. This preliminary design is intended to be used in future studies on bend flows and sedimentation processes. The primary focus lies in analyzing discharge requirements, pumping system design, and mitigating potential flow disturbances at the channel inlet. Additionally, an initial simulation with spur dikes along the flume's outer bank is included to investigate the flume's capability to model typical flood mitigation measures.

Previous research has extensively explored the use of flumes in understanding river flow dynamics from a physical perspective. A study by Blanckaert and Graf [7] demonstrated that bend geometry affects lateral pressure distribution and secondary forces, which in turn alter sedimentation patterns on the inner and outer banks of river bends. They utilized a curved flume with an adjusted curvature ratio to observe three-dimensional flow structures. Meanwhile, Ruquan et al. [8] employed a U-shaped flume to investigate the impact of weir structures on the formation of secondary flows and the distribution of longitudinal velocity. That study highlighted the importance of geometric smoothness and discharge stability in maintaining accurate representations of flow phenomena. Furthermore, research by Pan et al. [9] on laboratory meandering channels showed that minor variations in inflow discharge could significantly affect shear stress and turbulence distribution, particularly when inflow is not fully controlled and generates turbulence at the channel's entrance. Various approaches have also been undertaken to determine appropriate scaling in physical modeling. Liggett [10] emphasized the importance of maintaining similarity in dominant flow dynamics, such as the Froude number in open channel flows, to prevent distortions of physical phenomena at laboratory scale. One critical aspect that is often overlooked during the planning stage is the water circulation system and pump selection. Qian et al. [11] warned that mismatched pump capacity may lead to inlet disturbances, which can disrupt flow stability along the entire flume. Therefore, the pumping system design must be carefully matched to the head, discharge, and hydraulic requirements of the physical model (Table 1).

2 Methods

This study employs a fully numerical modeling approach to design and preliminarily evaluate a U-shaped open-channel flume intended to support flood early warning systems. The flume consists of two straight acrylic channel sections (each 2 m in length) connected by a semi-circular 180° bend, forming a U-shaped planform. The internal channel width is 0.3 m and wall height 0.35 m, with a bend centerline radius of 1.05 m. These geometric parameters were selected based on Froude similarity to ensure that hydraulic behavior at model-scale appropriately represents subcritical flow phenomena in natural rivers. The chosen flume dimensions fall within the range of previous curved flume studies [12–15], providing a basis for comparison and yielding data representative of field conditions at laboratory scale (Table 1). The bend configuration facilitates the development of secondary flow structures (helical currents and lateral velocity gradients) along the inner and outer banks, as documented in

https://doi.org/10.1051/e3sconf/202565205001

prior studies [2,7]. Such flow features are critical for observing river-like behavior and for generating empirical data to calibrate flood forecasting models in early warning systems.

Table 1. Geometric and flow parameters of curved flumes in previous studies compared to the present study

Previous research	Width × Height of flume (m)	Radius (m)	Length of Inlet and Outlet Channels (m)	Discharge (m^3/s)	Velocity (m/s)	Water depth (m)
[12,13]	0.8 × 0.8	2	16	0.05	0.52	0.12
[15]	0.47 × height not stated	2	8	0.0187	0.26	0.12
[14]	0.6 × 0.7	2.1	8	0.0187 0.0203 0.0219	0.260 0.282 0.304	0.12
Present study	0.3 × 0.35	1.05	4	0.0139	0.463	0.10

The flume is designed as a closed-circuit recirculation system with a submersible pump drawing water from a downstream reservoir and delivering it to the upstream inlet (Figure 1). At the downstream end, a stilling basin (energy absorber) dissipates the outflow energy to maintain stable conditions. A sediment trap is installed on the channel bed at the downstream section to collect any sediment transported by the flow. The channel walls are made of transparent acrylic to facilitate visual observation of flow patterns and water surface profiles during operation. For additional experimental scenarios, two model spur dikes (miniature groyne structures) are mounted along the outer bank of the bend, oriented perpendicular to the flow. These spur dikes are spaced approximately one spur-length apart and protrude into the flow without blocking the entire channel width. This configuration allows investigation of their influence on flow distribution and sediment dynamics as part of the flume's role in simulating flood mitigation measures.

All evaluations of the flume design were conducted via three-dimensional computational fluid dynamics (CFD) simulations, as no physical hydraulic model tests have been performed at this stage (a noted limitation of the study). A Volume of Fluid (VOF) multiphase modeling approach was employed to capture the free-surface flow, with water and air phases simulated simultaneously. Turbulent flow effects, including the formation of vortices in the curved section, were represented using an appropriate turbulence closure model. The simulation was configured to reflect a steady subcritical flow entering the flume and transitioning to supercritical conditions after the bend, in accordance with the design objectives. Preliminary trials at a higher inlet velocity (~1.0 m/s) indicated unstable supercritical behavior, so the baseline design focused on a subcritical inlet velocity of approximately 0.46 m/s. Specifically, the inlet boundary condition was defined by a uniform flow velocity, while the outlet boundary was controlled by imposing a fixed water surface elevation to achieve the target flow depth. The design discharge for the system (approximately 0.0139 m^3/s) was computed from the chosen cross-sectional area and target velocity (≈0.463 m/s) using the continuity relation (Eq. 1). This discharge corresponds to an initial flow depth on the order of 0.10 m, yielding a Froude number $Fr < 1$. The Froude number was calculated via the standard formula (Eq. 2) and confirmed to be well below unity, verifying that the flume operates in a subcritical flow regime as intended. The selected pump capacity and flow rate were iteratively adjusted (within practical limits) to ensure a stable circulation and minimize inlet turbulence.

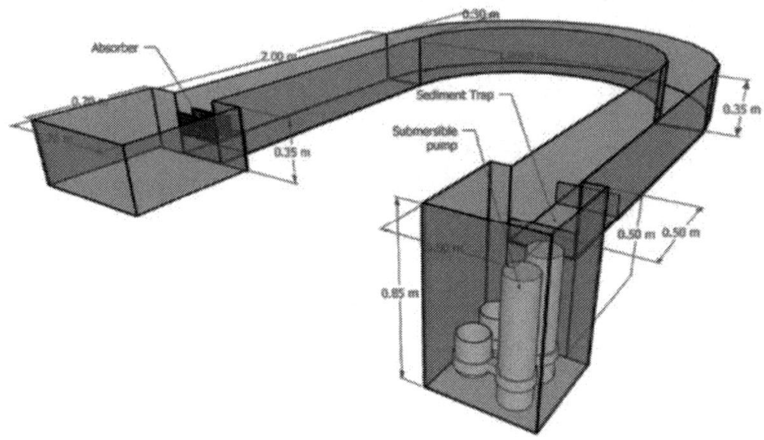

Fig. 1. Three-dimensional model of the U-shaped flume design

Discharge is then calculated using:

$$Q = v \times A \quad (1)$$

where Q is discharge (m^3/s), A is cross-sectional flow area (m^2), V is mean flow velocity (m/s). The discharge value can be adjusted based on initial testing and pump capacity to achieve optimal flow stability.

The Froude number is calculated by:

$$Fr = \frac{v}{\sqrt{g \cdot y}} \quad (2)$$

where Fr is Froude number (dimensionless), V is mean velocity (m/s), g is gravitational acceleration (9.81 m/s^2), and h is hydraulic depth, approximated by flow depth (m). Since $Fr < 1$, the flow is confirmed to be subcritical, aligning with the requirements for river behavior observation and supporting early warning system accuracy.

Before examining scenarios with spur dikes, a baseline simulation without any flow obstructions was performed to validate the numerical model. The resulting flow patterns (e.g. velocity distribution and free-surface superelevation in the bend) were compared qualitatively against well-established open-channel bend flow theory and previously published experimental results. Close agreement in these baseline characteristics gave confidence in the model's ability to replicate essential hydraulic phenomena. All findings and performance assessments of the flume thus far are derived from the CFD simulations; the physical flume has been designed but not yet constructed or tested, which remains a limitation of this preliminary study. This entirely numerical approach provides an initial proof-of-concept for the flume design and its expected flow behavior, while highlighting the need for future physical model validation.

3 Results and Discussion

3.1 Flow Characteristics without Spur Dikes

The initial simulation was performed for the baseline condition without spur dikes at an inlet velocity of 1.0 m/s. Figure 2 shows that in the bend section, a pronounced water surface rise occurred due to the relatively high inlet velocity compared to the flow depth. This rise impinged on the outer bank, significantly elevating the local water surface and creating a risk of overtopping the flume walls. Such unstable flow conditions are not recommended for experimental operation as they may damage the flume structure and compromise flow stability. Consequently, the 1.0 m/s scenario was considered an extreme case and was excluded from further analyses. Instead, a lower and more hydraulically stable subcritical velocity was applied.

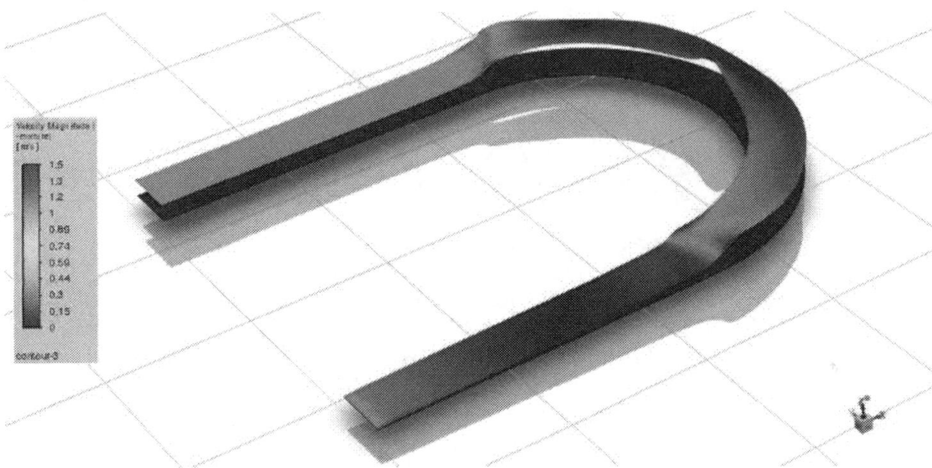

Fig. 2. Input flow 1 m/s

Figure 2. Velocity distribution at the free surface for the initial scenario (inlet velocity 1.0 m/s). A high-velocity zone (red region) is concentrated near the bend centerline, resulting in a hydraulic jump. The sudden transition from subcritical to supercritical flow generates turbulence and potential overtopping, making this scenario unsuitable for routine operation.

The subsequent simulation used an inlet velocity of 0.463 m/s. This configuration produced a stable subcritical flow upstream of the bend, which transitioned smoothly to supercritical conditions downstream. Figure 3a illustrates the surface velocity distribution, showing a gradual increase in velocity from upstream to downstream with no extreme surges. At cross-section A–A, the mean water depth was approximately 0.082 m (Figure 3b). The flow was relatively calm, with a nearly uniform depth profile, and the velocity distribution resembled the classical secondary flow pattern, where maximum velocities occurred near the surface and mid-channel. These results confirm that the 0.463 m/s scenario provides stable baseline conditions for subsequent spur dike analysis.

Figure 3a. Velocity distribution at the free surface (without spur dikes) for an inlet velocity of 0.463 m/s. Flow enters the bend under controlled subcritical conditions and accelerates progressively through the bend. High velocities are concentrated near the outer bank (orange regions), while lower velocities are found near the inner bank (green–yellow), consistent with the tendency of the main flow to press against the outer bend. No hydraulic jumps or overtopping are observed.

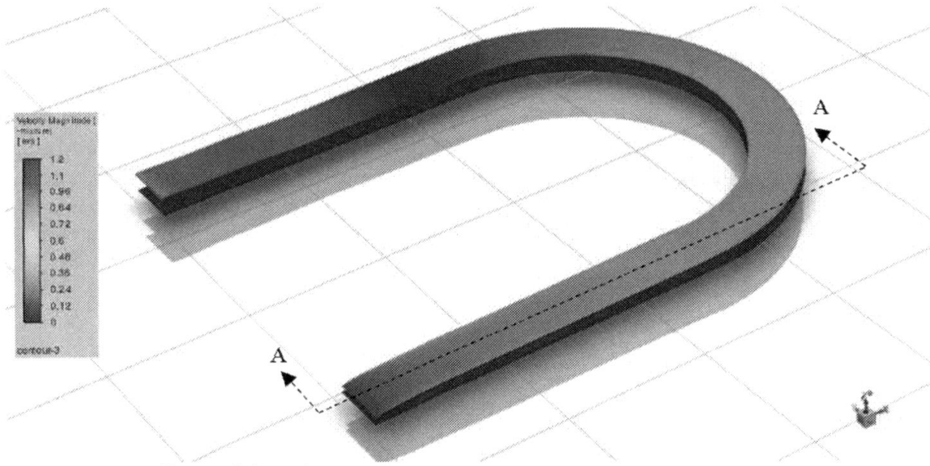

Fig. 3.a. Input flow veloicty 0.463 m/s

Figure 3b. Water surface profile at cross-section A–A (before the bend) for the scenario without spur dikes. The water depth was approximately 0.082 m, indicating a uniform and stable subcritical flow. The flat profile across the section reflects a relatively even velocity distribution unaffected by bend-induced acceleration. This serves as a reference for comparison with profiles downstream of spur dike installations.

Fig. 3.b. Section A – A

3.2 Influence of Spur Dikes

After establishing the baseline, two parallel spur dikes were introduced along the outer bank, and the simulation was repeated with the same inlet velocity of 0.463 m/s. Figure 4a shows that velocities decreased behind the spur dikes (blue–green regions), indicating flow deceleration and partial flow separation. The spur dikes successfully redirected the main current away from the outer bank, concentrating the high-velocity core (≈0.59 m/s) toward the channel centerline and inner bank. Compared to the baseline, the velocity distribution became narrower, and the outer bank was effectively shielded from direct high-velocity

https://doi.org/10.1051/e3sconf/202565205001

attack. However, the flow energy became concentrated at the spur dike tips, where turbulence intensified and local scouring was likely. This outcome aligns with prior studies, which reported that spur dikes reduce outer bank erosion but increase localized bed erosion near the structures.

Figure 4a. Velocity distribution at the free surface with two parallel spur dikes (inlet velocity 0.463 m/s). The structures (white blocks) create low-velocity shadow zones behind them. The high-velocity core (red region, ≈0.59 m/s) shifts inward, away from the outer bank. This demonstrates the effectiveness of spur dikes in reducing outer-bank velocity, but also indicates possible scouring at their tips due to concentrated turbulence.

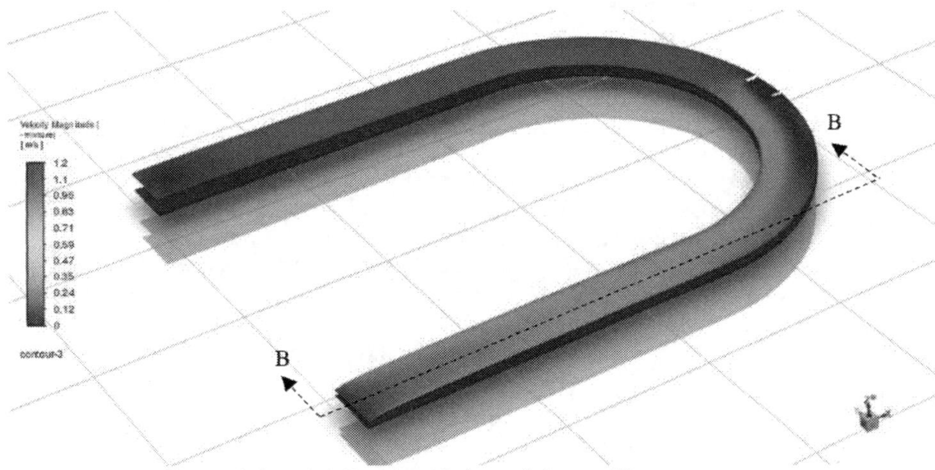

Fig. 4.a. Input flow veloicty 0.463 m/s with 2 paralel spur dike

The spur dikes also influenced the water surface elevation. Figure 4b presents the water surface profile at cross-section B–B, located downstream of the structures. Compared with cross-section A–A (Figure 3b), the depth decreased from 0.082 m to approximately 0.046 m. This reduction (~0.036 m) is attributed to the acceleration of flow through the narrowed effective width created by the spur dikes, consistent with the Bernoulli principle. The water surface slope across the section became more irregular, with slightly higher levels upstream of the spur dikes and lower levels downstream, reflecting turbulence and vortex formation around the structures. This effect must be considered in spur dike design, as excessive water surface depression may destabilize downstream banks.

Figure 4b. Water surface profile at cross-section B–B (downstream of the spur dikes). The depth decreased to ~0.046 m, compared with 0.082 m at A–A. This reduction reflects acceleration through the narrowed channel and conversion of potential to kinetic energy. Depth variation across the section indicates higher water near the spur dikes and lower levels toward the inner bank, suggesting flow redistribution and localized turbulence.

https://doi.org/10.1051/e3sconf/202565205001

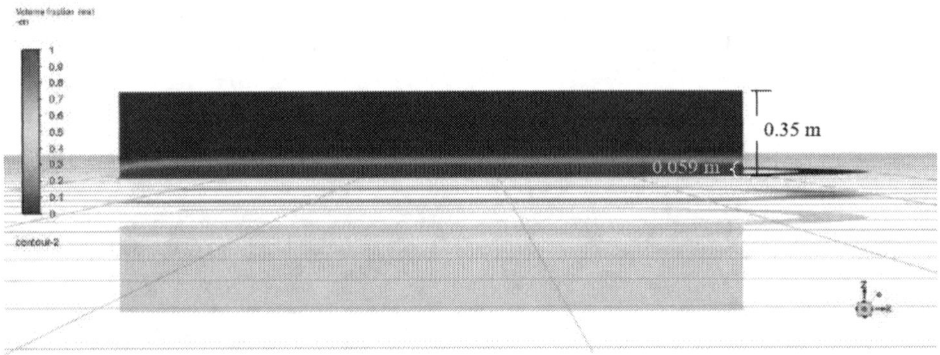

Fig. 4.b. Section B – B

Overall, the CFD simulations confirm the hydraulic role of spur dikes in modifying bend flow. Outer-bank velocities were successfully reduced, thereby lowering erosion risk in that region. However, the redirection of high-velocity flow toward the center and inner bank, combined with intensified turbulence near the spur dike tips, points to an increased risk of local scouring. Practically, this highlights the need for bed protection measures such as riprap at spur dike locations. Additionally, the observed water surface reduction upstream of the spur dikes suggests a potential secondary benefit: lowering flood wave elevations entering the bend. Thus, spur dikes may serve a dual function in river management—protecting banks from erosion and contributing to local flood regulation as part of early warning systems.

4 Conclusion

This study presented the design and preliminary evaluation of a U-shaped hydraulic flume using a numerical modeling approach to investigate flow behavior with and without spur dikes. The simulations confirmed that at an inlet velocity of 0.463 m/s, the flume produced stable subcritical flow conditions upstream of the bend, transitioning to supercritical conditions downstream. The introduction of spur dikes successfully reduced flow velocity along the outer bank, demonstrating their effectiveness in mitigating potential erosion, while simultaneously redirecting the high-velocity core toward the inner bend and concentrating turbulence around the spur dike tips. These outcomes indicate the dual role of spur dikes in providing outer bank protection and reducing local flood levels, but also highlight the risk of localized scouring near the structures. While the findings provide valuable insights into the hydraulic performance of the flume design, the absence of physical model validation remains a limitation, underscoring the need for future experimental work to strengthen the applicability of the results in flood early warning systems.

Acknowledgements

This research was financially supported by the Directorate of Research and Community Service Development (Direktorat Pengembangan Penelitian dan Pengabdian kepada Masyarakat – DPPM), Ministry of Higher Education, Science, and Technology, under the 2025 fiscal year funding scheme. The authors gratefully acknowledge this support.

References

1. D. Sisinggih, S. Wahyuni, and A. Rasyid, IOP Conf. Ser.: Earth Environ. Sci. **930**, 012033 (2021)
2. D. Termini, J. Hydraul. Eng. **135**, 575 (2009)
3. H. Chanson, *The Hydraulics of Open Channel Flow: An Introduction ; Basic Principles, Sediment Motion, Hydraulic Modelling, Design of Hydraulic Structures* (Arnold, London, 1999)
4. B. Chen, Y. Jin, X. Song, Y. Zhang, and D. Ni, J. Phys.: Conf. Ser. **2854**, 012035 (2024)
5. S. Erpicum, B. P. Tullis, M. Lodomez, P. Archambeau, B. J. Dewals, and M. Pirotton, Journal of Hydraulic Research **54**, 692 (2016)
6. V. Heller, Journal of Hydraulic Research **49**, 293 (2011)
7. K. Blanckaert and W. H. Graf, J. Hydraul. Eng. **127**, 835 (2001)
8. Y. Ruquan, L. Zexuan, L. Haiwang, and S. Jincheng, Physics of Fluids **35**, (2023)
9. Y. Pan, X. Liu, and K. Yang, Journal of Hydrology **607**, 127539 (2022)
10. J. A. Liggett, *Fluid Mechanics* (McGraw-Hill, New York, 1994)
11. B. Qian, J. Chen, P. Wu, B. Huang, and D. Wu, in *Volume 3B: Fluid Applications and Systems* (American Society of Mechanical Engineers, San Francisco, California, USA, 2019)
12. P. Xie, C. Li, S. Lv, F. Zhang, H. Jing, X. Li, and D. Liu, Sustainability **15**, 15862 (2023)
13. D. Liu, S. Lv, and C. Li, Water **16**, 2236 (2024)
14. M. Shokrian Hajibehzad, M. Shafai Bejestan, and V. Ferro, Water **12**, 3531 (2020)
15. A. Sardasteh, M. S. Bejestan, S. A. Ayyoubzadeh, and J. M. V. Samani, (n.d.)

Digital Disaster Education Park Vulnerability and Disaster Preparedness Measures (Study On Mount Semeru)

Anita Sulistyorini[1], *Tomy Rizky* Izzalqurny[2], *Muhammad Andry Prio* Utomo[3], *Emil* Azmanajaya[4], *Febi* Dwirahmadi[5], *Zahra* Klarisa[6], *Resti Novita* Sari[7], *Rifaldy Adinandra* Ferdiansyah[8], *Annisa* Hildaturrahma[9]

[1,6,7,9] *Department of Public Health Science, Faculty of Sports Science, State University of Malang, Indonesia*
[2] *Department of Accounting, Faculty of Economy and Business, State University of Malang, Indonesia*
[3] *Department of Biology, Faculty of Mathematics and Science, State University of Malang, Indonesia*
[4] *Department of Civil Engineering, Balikpapan State Polytechnic, Indonesia*
[5] *Global Health at School of Medicine, Griffith University, Queensland, Australia*
[8] *Department of Development Economics, Faculty of Economy and Business, State University of Malang, Indonesia*

Abstract. The 2021 eruption of Mount Semeru caused 61 deaths and the displacement of thousands people, exposed major shortcomings in community preparedness. One key factor contributing to these gaps is the limited availability of educational media that is accessible and relevant to the local context. This study addresses this issue by developing and validating a digital educational platform that was co-designed with local residents. The research offers two main novelties. Firstly, it combines comprehensive volcanic and geological information with practical preparedness guidelines that are presented on website. Secondly, the participatory co-design approach involving the local community contrasts with the usual top-down approach to disaster ICT initiatives. An educational website was produced and iteratively tested for feasibility with 34 individuals directly affected by the Semeru eruption, guided by a research and development (R&D) framework. Initial small-group testing (n = 10) yielded a 70% feasibility score, prompting significant revisions. Subsequent large-group testing (n = 24) demonstrated increased feasibility of over 92%, classifying the final product as 'highly feasible'. This research delivers a validated, context-specific digital tool that bridges the gap between government mitigation plans and grassroots needs, positioning the website as both a product and a model for interventions that strengthen community resilience.

* Corresponding author: anita.sulistyorini.fik@um.ac.id

1 Introduction

Indonesia, as an archipelagic country located at the intersection of the world's major tectonic plates, inherently lives with the reality of constant natural disaster risk. Located along the Pacific Ring of Fire, Indonesia is home to 127 active volcanoes. As a result, these geographical conditions place millions of its inhabitants under the shadow of a real volcanic threat. The disasters caused by volcanic activity are multifaceted. That is, their impact is not limited to primary eruptions such as pyroclastic flows and material fallout, but also triggers secondary hazards that cause damage. For example, these hazards include cold lava, clean water shortages, and damage to agricultural land due to ashfall [1]. Ultimately, this chain of impacts can systematically paralyze physical infrastructure and disrupt social order, health, and economic stability. It is in this context that the eruption of Mount Semeru on January 16, 2021, becomes a paradigmatic case study. This event clearly illustrates the complexity of post-disaster management challenges in Indonesia, which are not only geological events but also triggers for profound humanitarian crises.

The impact of the Semeru eruption actually goes beyond the statistics of 61 fatalities and 4,423 individuals forced to evacuate. More than that, this disaster has triggered a prolonged public health crisis, with 4,140 people recorded as requiring medical treatment [2]. Conditions in crowded evacuation centers with limited sanitation and unstable temperatures have exacerbated the spread of acute respiratory infections (ARI). Meanwhile, physical and psychological stress has also triggered other complaints such as myalgia and cephalgia, as reflected in the thousands of victims requiring medical treatment [3]. Even two years after the disaster, these health problems continue to haunt survivors. Furthermore, the situation is exacerbated by untreated burn cases and the emergence of long-term threats such as stunting in children. The prevalence of stunting itself is a tragic indicator of systemic failure. This is triggered by inadequate nutrition for pregnant and lactating mothers during emergencies, which ultimately has the potential to create intergenerational impacts that are difficult to reverse. Research also shows that the ability of communities around Semeru to adapt is greatly influenced by access to vital resources, social solidarity, and trust in early warning systems. However, conservative risk management policies and limited resources can hinder post-disaster recovery [4]. Therefore, the health crisis following the eruption of Mount Semeru is a multidimensional problem requiring long-term intervention in terms of physical and mental health, and basic nutrition.

In response to the disaster, the government implemented a series of mitigation measures as mandated by Law No. 24 of 2007. The construction of 1,951 permanent housing units was included in these plans, with the intention that they would serve as a permanent relocation solution [5]. Yet this large-scale reconstruction produced a paradox, with most residents rejecting the new housing and returning to areas that were prone to disaster. This outcome highlights a significant discrepancy with the global principle of 'Build Back Better' set out in the Sendai Framework for Disaster Risk Reduction, which emphasises that reconstruction efforts should also aim to reduce future vulnerability. This shortfall reflects the neglect of a key pillar of community resilience, namely economic resilience, which is essential for people's well-being and prosperity. A limited understanding of community resilience and the social dimensions of risk, an absence of effective methods to engage and empower communities, and a 'business as usual' mindset have constrained the implementation of sound disaster risk reduction and resilience-building strategies [6]. By locating permanent housing far from farmland and mining areas, the policy effectively cuts communities off from their livelihoods, forcing them to choose between physical safety and economic survival [7].

Therefore, the gap between the provision of safe housing and the socio-economic needs of the community is a crucial point that requires in-depth analysis. This condition shows that post-disaster recovery models that only focus on physical infrastructure are no longer adequate. Ongoing health issues also open up opportunities to explore innovative interventions such as digital health education. However, the successful adoption of such technology is highly dependent on the community's perception of its usefulness and ease of use. These aspects can be analyzed through the Technology Acceptance Model (TAM) framework. Therefore, this article aims to critically analyze the challenges in post-eruption recovery in Semeru through the theoretical lens of disaster risk reduction and community resilience. Ultimately, this study argues that only by integrating economic sustainability and the real needs of the community can recovery efforts shift from mere reconstruction to true and sustainable resilience building.

2 Experimental Details

2.1 Design and development procedures

This study adopts a research and development (R&D) approach, primarily aiming to design, produce and test the feasibility of a web-based educational medium. The development process involves systematic stages, from needs analysis and planning to prototyping, expert validation, and limited field testing, to ensure that the product is technically sound and pedagogically appropriate [8]. The end product is an interactive, web-based platform that provides disaster preparedness recommendations and general information about Mount Semeru. By integrating text, visuals, and interactive features, it is expected that this educational medium will enhance community knowledge and readiness for volcanic disasters, particularly among those living in disaster-prone areas.

The product development process is adapted from the ADDIE model, with the following procedures:

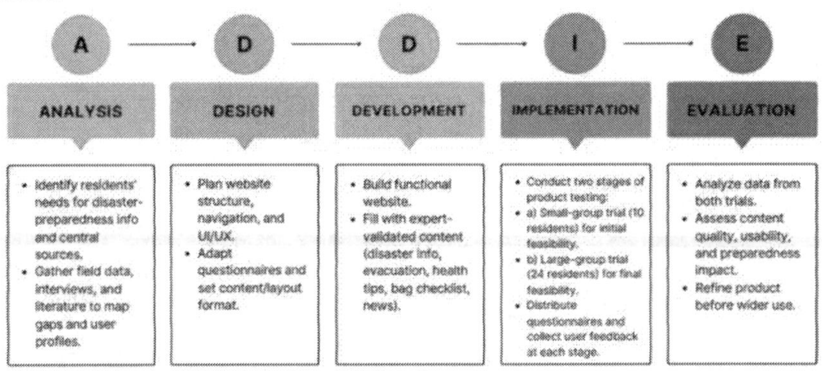

Fig. 1. ADDIE model stages applied in the development of the Digital Disaster Education Park on vulnerability and disaster preparedness measures.

a. Analysis

During this stage, researchers focus on identifying the specific disaster preparedness information needs of affected communities and the requirement for centralised information sources. This involves collecting field data, conducting interviews and studying literature to identify knowledge gaps, media preferences and obstacles that residents face when accessing disaster information. The results of this analysis inform the formulation of learning objectives and user characteristics.

b. Design

The design stage involves designing the structure and navigation flow of the site, the user interface (UI/UX) and adapting the questionnaire items to be used in the trial. The layout, menu, language style and information presentation format that best suit the users' characteristics are determined at this stage. The design results also include a blueprint for educational materials and product evaluation instruments.

c. Development
The development stage involves building a functional website according to the design and populating it with educational content validated by experts. This content includes general disaster information, volcanic eruption evacuation guidelines, media information, health promotion information, a list of items to include in a disaster preparedness kit and the latest news.

d. Implementation
During the implementation phase, the developed product was tested in two stages. First, a small-group trial involving 10 residents was conducted to measure initial feasibility and identify necessary improvements. Secondly, a larger group trial involving 24 residents was conducted to test final feasibility and broader community acceptance. Data was obtained by distributing questionnaires to respondents in both trials.

e. Evaluation
The evaluation stage involves analysing the quantitative data obtained from the two implementation stages. The aim of this analysis is to measure the final feasibility of the product in terms of its content quality, ease of use and the benefits it offers in terms of improving community preparedness. The results of the evaluation are then used to refine the web product before it is disseminated further.

2.2 Data Collection Method

This study involved 34 participants, all of whom were survivors of the Mount Semeru eruption and were selected using purposive sampling. This technique ensured that each participant met the following criteria: residing in the affected subdistricts (Candipuro or Pasrujambe); being of productive age (18–55 years); and owning and being able to operate digital devices. The participants were then divided into two groups: a small group of 10 for preliminary testing, and a large group of 24 for the main feasibility test. Data from these groups was collected using a rating scale questionnaire designed to measure the feasibility of the website, which served as the main instrument. The questionnaire used a 5-point Likert scale to evaluate three main dimensions, such as usability, content quality and visual design. To maintain data consistency and comparability, identical questionnaires were used in both test groups.

2.3 Data Analysis Method

All the collected data was quantitative and analysed using descriptive statistics. Questionnaire scores were processed to produce a percentage suitability value. The results of the small-group test were used to inform evaluation and product improvement before testing on a large group. The data from the large group then formed the basis for drawing final conclusions about the research product's suitability.

3 Result And Discussion

3.1 Result

The evaluation results for the respondents are presented below, including their scores and respective categories :

Table 1. Results of Small-Group Testing.

Responden	Value	Category
1	48	Moderately Feasible
2	54	Moderately Feasible
3	52	Moderately Feasible
4	51	Moderately Feasible
5	53	Moderately Feasible
6	51	Moderately Feasible
7	47	Moderately Feasible
8	52	Feasible
9	58	Feasible
10	57	Feasible

Based on the data collected from the respondents, 70% (7 out of 10) of the participants were classified as 'moderately feasible', with scores ranging from 47 to 54. This suggests that most respondents considered the subject to be moderately feasible. In contrast, 30% of respondents (three out of ten) were categorised as 'feasible', with scores ranging from 52 to 58. These results suggest that, although many participants viewed the subject as moderately feasible, fewer rated it as fully feasible. It is evident from this distribution that while a few respondents scored in the 'Feasible' category, the majority were classified as 'Moderately Feasible'. This suggests a general trend that most individuals have a moderate level of capability in terms of their preparedness for disaster scenarios.

The evaluation results of respondents concerning their scores and corresponding categories provide significant insights into their capabilities and areas for improvement, which can then be used to inform future training and development. This analysis focuses on interpreting the data in the context of disaster vulnerability and preparedness, particularly in relation to the Mount Semeru disaster area.

Respondents in the "Moderately Feasible" category have demonstrated acceptable performance levels, but require further development. Saiful Bahri, for instance, scored 48, suggesting he meets some basic disaster preparedness criteria but lacks the comprehensive knowledge and skills necessary for an effective disaster response. The other respondents in this category, with scores ranging from 47 to 54, demonstrate similar moderate performance. They may therefore benefit from targeted training or educational programmes focused on disaster risk reduction and management. Conversely, those categorised as 'Feasible', such as Mat Alek (52), Arifatul (58) and Anis (57), demonstrate a higher level of competence in disaster preparedness. Arifatul's score of 58 is particularly noteworthy, indicating strong capabilities in understanding and responding to potential disasters. The presence of these higher scores among a minority of respondents suggests that, while most are moderately feasible, there is a subset with more pronounced strengths in disaster management.

The evaluation results have several implications for enhancing disaster preparedness in vulnerable areas like Mount Semeru:

1. Targeted Training Programs: The findings highlight the necessity for targeted interventions aimed at those categorized as "Moderately Feasible." Educational programs focusing on disaster risk reduction strategies could significantly enhance their preparedness levels.
2. Utilization of Local Knowledge: Incorporating local wisdom and community engagement in disaster management strategies can improve overall effectiveness. As noted in related research, local communities possess valuable insights into disaster indicators and response mechanisms that can be leveraged to enhance preparedness efforts [9].

3. Strengthening Social Capital: Building social capital within communities is essential for effective disaster response. Trust between community members and local authorities can facilitate better communication and coordination during emergencies, as highlighted by studies on social capital's role in disaster management [10].

The evaluation results for the respondents are presented below, including their scores and respective categories:

Table 2. Results of Big-Group Testing.

Respondent	Value	Kategori Category
1	62	Feasible
2	70	Feasible
3	69	Feasible
4	62	Feasible
6	62	Feasible
8	52	Feasible
9	68	Feasible
10	63	Feasible
11	64	Feasible
12	61	Feasible
15	60	Feasible
17	64	Feasible
19	64	Feasible
20	69	Feasible
21	59	Feasible
22	61	Feasible
24	70	Feasible
25	68	Feasible
26	66	Feasible
27	55	Feasible
28	60	Feasible
29	60	Feasible
30	60	Feasible
33	59	Feasible

The data shows that all 24 respondents were categorised as 'feasible', with scores ranging from 52 to 70. Respondents 2 and 24 recorded the highest scores, both achieving 70, while respondent 8 recorded the lowest score of 52. The majority of respondents (approximately

https://doi.org/10.1051/e3sconf/202565205002

70.8%) scored between 60 and 70, with 60 being the most frequently reported score (4 respondents, or 16.7% of the sample). Respondents scoring 69 accounted for 12.5% of the sample (three respondents), while those with scores between 62 and 68 made up 50% of respondents. This distribution demonstrates a strong consensus on the feasibility of the subject, with no respondent scoring below 52.

The results of the feasibility questionnaire analysis for the Digital Disaster Education website, completed by 34 participants, are presented in Table 2. The findings indicate that the website's features, speed, stability and completeness of information, as well as the impact of its use in meeting the needs of post-disaster resettlement communities, were all evaluated. While requiring further refinement, the structured implementation of disaster mitigation is expected to be addressed with the introduction of the Digital Disaster Education website. The analysis shows that the Regional Disaster Management Agency (BPBD) is responsible for disseminating disaster-related information widely and in close proximity to the community. The launch of this website is anticipated to invigorate disaster mitigation efforts by providing easily accessible, understandable information to the public, thereby enhancing disaster preparedness.

3.2. Disscussion

3.2.1 Overview of the Semeru Volcano Area

Mount Semeru, located on the island of Java, Indonesia, is the highest peak on the island, with an elevation of approximately 3,676 meters above sea level (asl). In addition to being the highest peak in Java, this mountain is also an integral part of the Bromo Tengger Semeru National Park, a protected area renowned for its natural beauty. The diverse ecosystems found on Semeru make it especially unique, encompassing various types of forests such as Hill Dipterocarp Forest, Upper Dipterocarp Forest, Montane Forest, and Ericaceous Forest or mountain forest.

According to an informant from the East Java Regional Disaster Management Agency (BPBD), Semeru disasters are one of the types of disasters that frequently occur and have a national-scale impact. This is due to the volcanic activity of Mount Semeru, which is one of the most active volcanoes in Indonesia. Mount Semeru is located in East Java Province and experiences periodic eruptions. With its high level of activity, Semeru continues to be a focal point for volcanologists and authorities tasked with monitoring the potential volcanic disaster risks in Indonesia.

3.2.2 Impacts of the Semeru Volcano Disaster

The 2021 eruption of Mount Semeru resulted in 61 casualties, significant infrastructure damage and agricultural land destruction, forcing over 4,400 people to seek refuge in permanent shelters provided by the local government. This extensive impact was caused by a lack of community preparedness and difficulty in predicting the disaster. Displaced communities were unable to carry out their usual economic activities as a result of this impact, highlighting the need for disaster economic policy to maximise existing efforts and minimise injustice for vulnerable populations affected by the disaster.

The eruption also triggered issues such as hunger and disease outbreaks due to the limited food supply. Such impacts are often referred to as the indirect consequences of a volcanic eruption. It is estimated that the losses from the Semeru eruption, as calculated by PUSKAS (the Baznas Center for Strategic Studies), amounted to over IDR 310 billion. These losses include damage to public infrastructure and have affected various economic sectors, including plantations, agriculture, livestock, trade and mining. The losses were categorised into five areas: damage to residential areas, damage to infrastructure, negative economic impacts, social impacts and cross-sectoral impacts. The damage to residential areas was one of the most significant components of these losses, accounting for 65% of the total [2].

3.2.3 Semeru Recovery Strategy

As part of the economic recovery strategy following the Mount Semeru eruption in Lumajang Regency, East Java Province, a rehabilitation and reconstruction plan has been prepared for 2022-2024. This plan aims to foster collaboration between the government, regional authorities, businesses, communities, universities, academics and non-governmental organisations to rebuild all aspects of life for those affected by the disaster in Lumajang Regency. In the housing and settlement sector, the targets are to improve the disaster area's environment, assist residents with home repairs, and rebuild social facilities. In the infrastructure sector, meanwhile, the plan includes repairing public facilities, restoring government services and rebuilding vital infrastructure in the Semeru area. This plan, which focuses on the recovery of the community's physical assets or 'hardware', is a fundamental and crucial step. However, the findings of this study highlight a gap, namely the lack of focus on strengthening the community's knowledge capacity and self-preparedness, or community 'software' [11].

Efforts to address this with digital educational media face various contextual obstacles. Socio-economically, the community's priority is recovering their livelihoods, so finding the time and money to access digital content is challenging. Culturally, technology adoption must compete with traditional information channels. Technologically, the uneven internet infrastructure on mountain slopes poses a significant challenge [12]. These challenges are not isolated phenomena in Lumajang, but rather a reflection of broader patterns in various disaster-prone areas. Socio-economic barriers, such as limited access to devices and high internet costs in Semeru. However, even countries with advanced technological infrastructure, such as Japan, face challenges in different dimensions. Although ICT-based disaster education technology is popular with students, studies reveal the phenomenon of the 'black box of disaster education', whereby the involvement of the younger generation drops dramatically after leaving school as community ties weaken [13]. Furthermore, cross-country research confirms that the successful adoption of digital education tools depends heavily on 'cultural tuning', whereby materials and methods must align with local social values and norms to be accepted in the long term [14]. Experiences in various countries reinforce the same conclusion, which is that effective disaster recovery strategies cannot rely solely on technology. A hybrid approach is needed that consciously integrates digital solutions with strengthening existing, trusted social networks and community communication channels.

3.2.4 Website Development

In the prototype design process, observations are first made based on various research to provide several recommendations for disaster preparedness actions, ensuring that the recommendations are targeted and effective. Based on the results of the needs analysis phase, the next step is to design a website named "Mitigasi Bencana Gunung Berapi". This website will be developed with a user interface (UI) using the prototyping method. The following is a display of the website that has been created:

https://doi.org/10.1051/e3sconf/202565205002

Fig. 2. Mitigasi Bencana Gunung Berapi's Dashboard Website.

The research also examined the map distribution of volcanoes throughout Indonesia. The appearance is as follows:

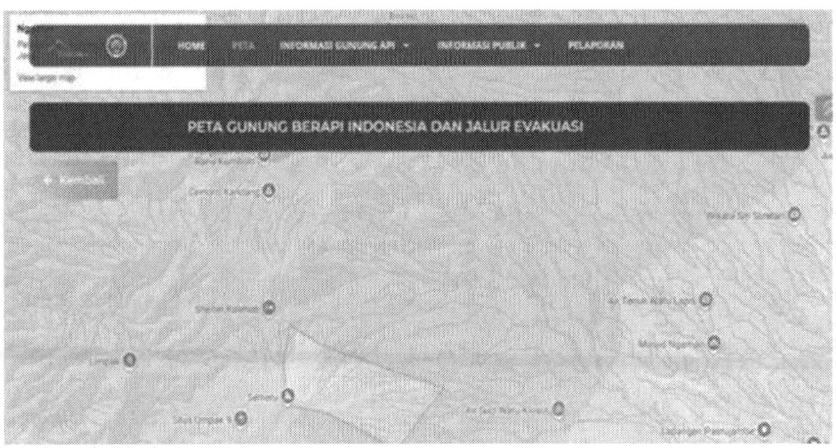

Fig. 3. Indonesian Volcanoes Map.

Apart from that, there is a feature that shows signs of an active volcano. The purpose of this feature is to increase public awareness of signs of a volcano that has the potential to erupt in the near future.

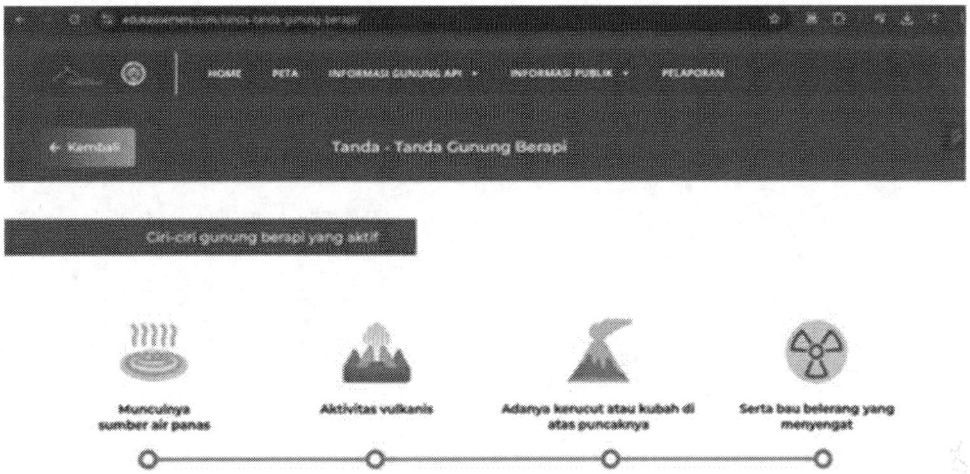

Fig. 4. Display of Volcano Signs Features.

In addition, the Semeru website features a public information section that can be used to raise awareness of physical vulnerabilities.

Fig. 5. Display of General Understanding of Disaster.

https://doi.org/10.1051/e3sconf/202565205002

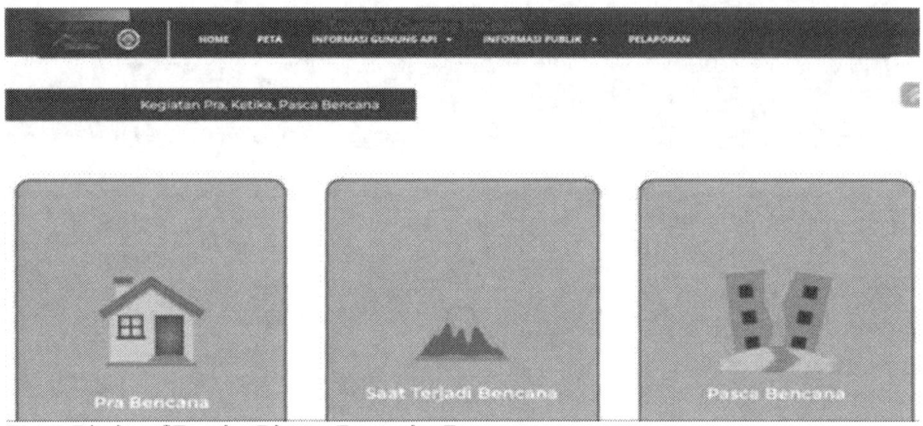

Fig. 6. Display of Eruption Disaster Evacuation Features

Fig. 7. Display of Information Media Features

In addition the system includes a reporting feature for volunteers and the surrounding community to help detect disaster events. These reports will be connected to the relevant agency's database. The aim is to ensure that disaster preparedness actions can take place appropriately and quickly by providing reports from volunteers and the public regarding signs of disaster events.

 https://doi.org/10.1051/e3sconf/202565205002

Fig. 8. Display of Reporting Features.

The final stage of this study involves publicly disseminating the developed website, thereby enabling stakeholders to access and utilise its content systematically. This initiative aims to raise community awareness and responsiveness, particularly with regard to volcanic eruptions, and to improve disaster preparedness. It is anticipated that these efforts will lead to a measurable reduction in adverse outcomes, including physical and material losses, in the event of a disaster.

4 Conclusion

This volcano disaster mitigation and preparedness innovation website is designed to measure disaster vulnerability and provide recommendations for community preparedness in facing eruptions, especially Mount Semeru. Keeping in mind the digital era, this website focuses on utilizing digital systems to support disaster preparedness. The feasibility test was carried out on media experts and two groups of respondents, namely a small group (10 people) and a large group (24 people) of residents affected by the Semeru eruption, with a total of 34 respondents. The results of the small group feasibility test showed 70% feasibility with several improvements, while the large group test stated the website was suitable for use (100%). This website is expected to be a health solution for Huntap Semeru residents.

Acknowledgments

The research team expresses gratitude to the State University of Malang, especially the Research and Community Service Institute (LPPM), for funding our activities. The research team also expresses gratitude to all relevant parties who have supported and assisted in providing data and information for this research.

References

1. A. Hidayat, M. A. Marfai, and D. S. Hadmoko, Eruption on Indonesia's volcanic islands: a review of potential hazards, fatalities, and management, IOP Conf. Ser. Earth Environ. Sci. **485**, 012061 (2020).

2. Volcanism Global Program, Report on Semeru (Indonesia), Bull. Glob. Volcanism Netw. **47**, (2022).
3. G. Yogadhita, M. Ariani, and B. Donna, Activation of the Health Cluster Coordination Post in Lumajang District Health Office During the Management of Mount Semeru Eruption Disaster, Prehosp. Disaster Med. **38**, s34 (2023).
4. J.-C. Thouret, E. Wavelet, M. Taillandier, B. Tjahjono, S. F. Jenkins, N. Azzaoui, and O. Santoni, Defining population socio-economic characteristics, hazard knowledge and risk perception: The adaptive capacity to persistent volcanic threats from Semeru, Indonesia, Int. J. Disaster Risk Reduct. **77**, 103064 (2022).
5. PUPR, Ditargetkan Rampung Akhir April 2022, Kementerian PUPR Selesaikan Pembangunan Huntap Korban Erupsi Semeru Secara Bertahap, (2022).
6. A. J. Imperiale and F. Vanclay, Conceptualizing community resilience and the social dimensions of risk to overcome barriers to disaster risk reduction and sustainable development, Sustain. Dev. **29**, 891 (2021).
7. A. J. Imperiale and F. Vanclay, Barriers to Enhancing Disaster Risk Reduction and Community Resilience: Evidence from the L'Aquila Disaster, Polit. Gov. **8**, 232 (2020).
8. L. Judijanto et al., *Metodologi Research and Development (Teori Dan Penerapan Metodologi RnD)* (2024).
9. M. Hirono and M. R. Nurdin, Local knowledge as the basis of disaster management and humanitarian assistance, Disasters **48**, (2024).
10. S. Carrasco, C. Ochiai, and L. M. Tang, Social Capital and Community Resilience in the wake of disasters, conflicts and displacements, Int. J. Disaster Risk Reduct. **115**, 105049 (2024).
11. D. Kusumawardani, Y. Rahmawati, M. N. Cahyadi, M. Rusli, and A. Martina, An analysis of the socio-economic impacts of the 2021 mountain Semeru Eruption on household level using PLS-SEM, Lett. Spat. Resour. Sci. **16**, 30 (2023).
12. S. Bachri, A. R. Rahman Hakiki, N. A. Wibowo, Sumarmi, R. Amini, Yosritzal, and E. Nursaribilah, Developing an education support system for disaster management through an ethnoscience-based digital disaster learning module, Int. J. Disaster Risk Reduct. **100**, 104214 (2024).
13. O. Uchida, S. Tajima, Y. Kajita, K. Utsu, Y. Murakami, and S. Yamada, Development and Implementation of an ICT-based Disaster Prevention and Mitigation Education Program for the Young Generation, Inf. Syst. Front. **23**, 1115 (2021).
14. G. Nakano and K. Yamori, Cultural tuning of a disaster education tool: A comparative study of Japan, El Salvador, and Mexico, Int. J. Disaster Risk Reduct. **105**, 104370 (2024).

E3S Web of Conferences **652**, 05003 (2025) https://doi.org/10.1051/e3sconf/202565205003
2nd SUSTAIN 2025

Multidimensional Assessment of Village Resilience to Landslide Disasters Using the RAP-DRR Method

Sapta Suhardono[1]*, *Oricya* Salma[2], *Rafika Kusuma* Hidayati[3], *Syahla Talitha* Khumaira[4], and *Silvania das Dores Tosa* Pereira[5]

[1,2,3,4,5]Department of Environmental Science, Faculty of Mathematics and Natural Sciences, Universitas Sebelas Maret, Surakarta, Indonesia 57126

Abstract. Landslides are one of the most frequent natural disasters in areas with steep slopes and high rainfall, such as Ngargoyoso Village, Karanganyar. This study aimed to assess the village's resilience level to landslide disasters from a multidimensional perspective using the RAP-DRR (Rapid Appraisal for Disaster Resilience in Rural areas) approach. The five resilience dimensions analyzed included social, economic, technological, institutional, and infrastructural aspects. Data were collected through questionnaire surveys and interviews with residents in landslide-prone areas, and were analyzed using the Multi-Dimensional Scaling (MDS) method in the Rapfish software. The results showed that the village's overall resilience index fell into the moderately sustainable category (score 55.64), with infrastructure being the strongest aspect (82.70) and technology the weakest (36.49). The social (63.35) and institutional (50.62) dimensions reflected relatively good community participation and institutional support, while the economic dimension (45.06) indicated a continued weakness in residents' financial resilience. These findings emphasized the importance of strategies to enhance community capacity, particularly in the technological and economic aspects, as a foundation for sustainable and adaptive disaster mitigation policies at the village level.

1 INTRODUCTION

Indonesia is one of the countries with a very high level of disaster vulnerability, particularly to hydrometeorological disasters such as floods, landslides, and strong winds [1]. Its location in the tropical region, right on the equator, and its topography dominated by mountains and hills make the country highly susceptible to extreme weather changes. Landslides are among the most frequent disasters, especially in areas with steep slopes and high rainfall. The risk of disaster increases when natural threats and vulnerabilities are not balanced by community mitigation capacity, as is the case in Central Java, where landslides have been the dominant disaster according to BNPB data since 2015. The natural characteristics of mountainous and hilly topography, combined with high rainfall,

[1] Corresponding author: sapta.suhardono@staff.uns.ac.id

make areas such as Karanganyar Regency in Central Java particularly disaster-prone. Data shows that Ngargoyoso Subdistrict is the highest-risk area with 101 landslide cases, followed by Jatiyoso (34 cases) and Jenawi (34 cases), where natural factors like the steep slopes of Mount Lawu interact with anthropogenic factors such as land-use change and vegetation degradation.

The impacts are quite severe, including damage to agricultural land, settlements, and disruption of inter-village road access. The impacts are multidimensional and interconnected. Physically, landslides destroy vital infrastructure such as housing, roads, and clean water networks. Data from the Regional Disaster Management Agency (BPBD) shows that landslides have caused serious damage to agricultural land and infrastructure and have disrupted access between villages, even resulting in fatalities and significant economic losses. Economically, this disaster causes substantial material losses and disrupts community livelihoods [2]. Socially, it leads to psychological trauma and displacement, further increasing victims' vulnerability [3]. The accompanying ecosystem damage also reduces the environment's carrying capacity in the face of future disasters.

Facing this reality, strengthening community capacity in disaster preparedness becomes crucial. One strategy currently promoted by the government is the development of Disaster Resilient Villages (Desa Tangguh Bencana), which are villages with the independent ability to recognize threats, organize resources, and respond to and recover from disaster impacts quickly and effectively [4]. This program emphasizes not only physical preparedness but also the development of social, economic, institutional, and educational capacities. Village disaster resilience must be assessed thoroughly and systematically, given the complex and multidimensional impacts of landslides [5].

Therefore, an approach that can describe village conditions holistically and participatively is required. One relevant method is RAP-DRR (Rapid Appraisal for Disaster Risk Reduction) using the Rapfish tool, which can evaluate village resilience to landslides from multiple dimensions, including social, economic, environmental, infrastructure, and institutional aspects. This study aims to assess the resilience level of Ngargoyoso Village to landslide disasters from a multidimensional perspective using the RAP-DRR approach. The assessment not only seeks to identify the level of vulnerability and existing capacity of the village but also to determine the most influential dimensions of resilience and uncover the weakest aspects that should be prioritized in resilience-enhancement strategies. By thoroughly understanding real field conditions based on data, this study is expected to provide a foundation for the formulation of adaptive, targeted, and sustainable disaster mitigation policies, and to promote community empowerment as the main actor in disaster risk reduction.

2 MATERIALS AND METHODS

Location

The research was conducted in Ngargoyoso Village, Ngargoyoso Sub-district, Karanganyar Regency, an area known for its landslide risk due to its location on the slopes of Mount Lawu, unstable soil structure, and high rainfall intensity. The site was selected purposively based on landslide incident reports from local residents and risk potential determined by topography and land use.

Materials

Data collection was conducted using a closed-ended questionnaire on an ordinal scale of 1 to 3. This simplified scale was selected to minimize respondent bias, ensure clarity, and make it easier for residents with diverse educational backgrounds to provide consistent answers.

The respondents consisted of 35 residents from landslide-prone areas in Ngargoyoso Village. The sample size was purposively determined to ensure representation of households directly exposed to landslide risks, while also maintaining a balance between representativeness and feasibility. The choice of 35 respondents was considered appropriate given the constraints of time, budget, and accessibility in conducting field surveys in disaster-prone terrain.

Data Analysis

This study employed a quantitative descriptive approach using the RAP-DRR method (Rapid Appraisal for Disaster Risk Reduction). It covers five main dimensions including technology, institutions, infrastructure, economy, and social aspects. Data were analyzed using Multidimensional Scaling (MDS) through the Rapfish software (Rapid Appraisal for Fisheries), which was specially modified into RAP-DRR to evaluate village resilience to landslides in a participatory manner.

Table 1. Sustainability Status Category [6]

Index Value	Sustainability Status
< 25	Not sustainable
25 – 50	Less sustainable
50 -75	Moderately sustainable
> 75	Sustainable

The MDS process involved the following steps: (1) ordination analysis to calculate the sustainability index for each dimension (2) leverage analysis to identify key attributes, and (3) Monte Carlo analysis (95 percent significance level) to examine errors [7], with the goodness of fit evaluation based on stress values less than 0.025 and a coefficient of determination (R^2) close to 1.

The scoring results were visualized in a triangular kite diagram, where the symmetry of the angles represented the proportion of the sustainability index, and the highest angle indicated the most sensitive attribute. The final output was a radar chart mapping the sustainability level of each dimension on a scale of 0 to 100, categorized from unsustainable to highly sustainable, providing a holistic overview of the village's capacity for landslide risk mitigation. The category of sustainability status can be seen in Table 1.

3 RESULT AND DISCUSSION

3.1 Multidimensional Scale

Based on the data processing results, a total index score of 55.644 was obtained, which was categorized as moderately sustainable (See Table 2). Based on the mapping results in the diagram, it was evident that infrastructure was the strongest dimension with the highest score among all others, indicating that the village's physical preparedness for disaster readiness was relatively good. On the other hand, the technology dimension received the lowest score, reflecting the weak implementation of disaster-related technology such as early warning information systems and technical mitigation tools. The social and institutional dimensions were in the moderately sustainable category, indicating a relatively good level of awareness and community participation in responding to landslide threats. In contrast, the economic dimension ranked low, suggesting the need to enhance household economic resilience against disaster risks.

Overall, Ngargoyoso Village demonstrated a fairly good level of resilience, particularly in terms of infrastructure. The high score in this dimension indicates that drainage systems, evacuation routes, and road accessibility are functioning relatively well, which provides a strong foundation for disaster preparedness. However, the low scores in the technological and economic dimensions reveal significant gaps. Limited knowledge and utilization of early warning systems, together with insufficient financial reserves, increase the community's vulnerability and may prolong recovery after landslides. These challenges suggest that efforts to improve technological literacy and to strengthen community-based economic institutions are essential. Enhancing these aspects would complement the existing infrastructure capacity and create a more comprehensive preparedness framework. In this way, resilience in Ngargoyoso can evolve beyond physical readiness toward a more integrated approach that combines infrastructure, technology, and economic empowerment.

Table 2. MDS Index Scores and Sustainability Level of Resilience

No.	Dimension	Number of Attributes	MDS Score	Sustainability Status
1	Social	5	63.35495684	Moderately Sustainable
2	Technology	5	36.48872868	Less Sustainable
3	Economy	5	45.05736489	Less Sustainable
4	Institutional	5	50.6222567	Moderately Sustainable
5	Infrastructure	5	82.70013449	Sustainable
	Total		55.6446883	Moderately Sustainable

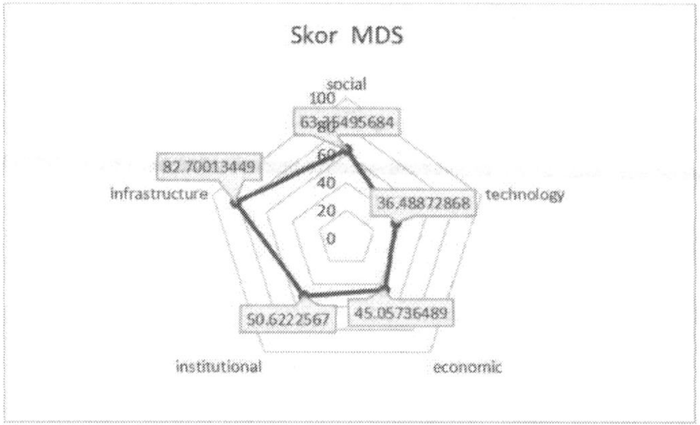

Fig. 1. Sustainability Index per Dimension in Ngargoyoso Village

3.2 Social Dimension

Based on the RAP-DRR analysis supported by the Monte Carlo test, it was revealed that the social dimension of the Ngargoyoso Village community demonstrated a sustainability index of 63.354, which fell into the moderately sustainable category. This result was not only supported by a reliability value of 0.705, which was significant at the

 https://doi.org/10.1051/e3sconf/202565205003

95 percent confidence level, but also reinforced by an optimal goodness of fit (stress value below 0.025 and R squared close to 1), confirming the validity of the research findings.

In the context of landslide disaster preparedness, which poses a real threat in this hilly region, three main pillars of social resilience emerged as key determinants. These included strong social solidarity with an RMS value of 8.727, adequate understanding of evacuation procedures with an RMS value of 8.311, and active participation in mitigation activities with an RMS value of 8.083. However, behind these strengths, significant challenges were identified in the low level of understanding of natural warning signs with an RMS value of 3.164. Only a few people in the study area were able to accurately recognize the early indicators of landslides.

This condition was further complicated by limited budgets for regular training, even though public enthusiasm for participation was relatively high. These findings aligned with the research of [8], which emphasized the crucial role of social capital in building disaster resilience in rural areas. To optimize existing social capacity, an integrated strategy is needed that combines the strengthening of local wisdom with technical interventions. The main recommendations include the development of educational modules based on local culture, the optimization of community-based early warning systems involving local leaders, and the increase of training frequency through strategic partnerships with universities and relevant institutions.

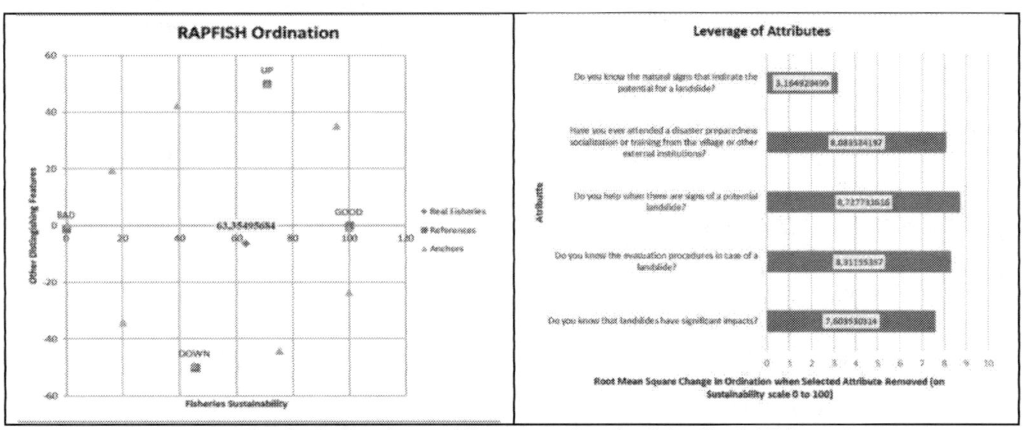

Fig. 3. Ordination and Leverage Graph of the Social Dimension

3.3 Economic Dimension

In the RAP-DRR analysis results, the economic dimension obtained a sustainability index score of 45.057, which was categorized as less sustainable. This score indicated that the economic resilience of Ngargoyoso Village residents in facing landslide disasters remained weak. Contributing factors included dependence on income sources located in landslide-prone areas, limited financial reserves, and minimal access to collective economic support. In general, the community did not yet have sufficient economic resilience to independently withstand and recover from the impacts of disasters.

The economic dimension covered aspects such as the exposure of livelihoods to disaster risks, the financial preparedness of the community in facing economic losses, and involvement in community-based economic support systems. The leverage analysis of attributes showed that among the five attributes examined, three had the most influence on the sustainability index. These were the impact of income loss due to landslides with an RMS value of 2.914, ownership of financial reserves such as savings or assets with an RMS value of 3.908, and participation in resilient economic groups such as cooperatives or microfinance organizations with an RMS value of 3.951. These three attributes showed the highest RMS values, indicating that they played a major role in shaping the economic vulnerability of the village community to disasters.

This was in line with the study by [9], which found that cooperatives as community-based economic institutions played an important role in strengthening economic resilience, particularly by increasing access to capital, providing entrepreneurship training, and offering business management assistance. Cooperatives could serve as strategic instruments for promoting local economic self-reliance in post-disaster recovery. The other two attributes were the location of primary livelihoods in landslide-prone areas with an RMS value of 1.801 and direct experience of economic losses caused by disasters with an RMS value of 2.104. Although these had less influence, they remained relevant for understanding the structure of economic resilience in the village.

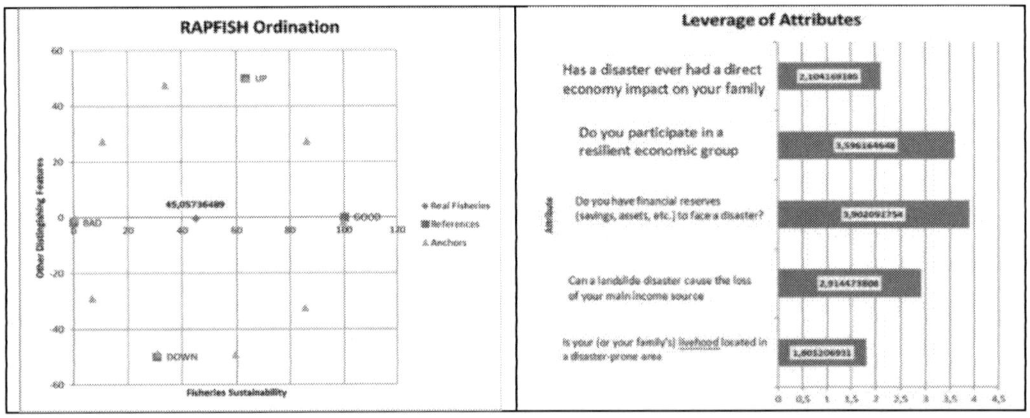

Fig. 4. Ordination and Leverage Graph of the Economic Dimension

3.4 Technological Dimension

Ngargoyoso Village faced serious challenges in the aspect of landslide mitigation technology. The RAP-DRR analysis results showed that the technology dimension in this village only achieved a sustainability score of 36.489, which fell into the less sustainable category. This low score was influenced by three main issues: limited mitigation training, low public understanding of early warning systems, and suboptimal soil conservation activities. The Monte Carlo statistical test (value 1.00487124) with an error rate below five percent confirmed that these findings were valid and required serious attention.

 https://doi.org/10.1051/e3sconf/202565205003

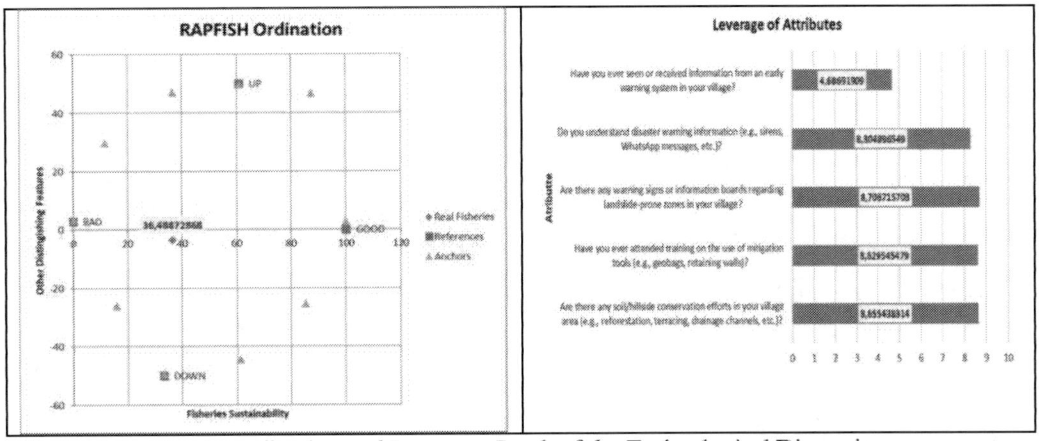

Fig. 5. Ordination and Leverage Graph of the Technological Dimension

The most critical factors hindering the progress of mitigation technology in Ngargoyoso were slope conservation with an RMS of 8.655, mitigation training with an RMS of 8.629, availability of warning signs with an RMS of 8.707, and understanding of disaster warning systems with an RMS of 8.305. The lack of training meant that residents were not familiar with using mitigation tools such as geobags. Meanwhile, the existing early warning systems, such as sirens or WhatsApp messages, were not effective because the community did not fully understand how to respond to them.

Reforestation and terracing activities as soil conservation efforts had also not been carried out optimally, even though research by [10] demonstrated that bioengineering approaches such as planting slope-stabilizing vegetation could significantly reduce landslide risk. Community participation in mitigation training remained very limited, involving only volunteer groups and a small portion of residents. In fact, community-based training, as recommended by [11], could be a solution to increase resident engagement. Experiences from other regions, such as Purbalingga which successfully implemented an IoT-based early warning system, could serve as inspiration for Ngargoyoso. However, given the limitations in technological literacy and infrastructure in this village, simpler approaches such as SMS-based applications or the use of traditional media like mosque loudspeakers and WhatsApp groups might be more effective.

3.5 Institutional Dimension

The institutional dimension in Ngargoyoso Village obtained a sustainability index of 50.622, which fell into the moderately sustainable category (Figure 6). This result was supported by the Monte Carlo test, which produced a value of 50.525 with a margin of only 0.09652, far below the five-point tolerance threshold. This small difference indicated that the analysis results were quite valid and could serve as a basis for assessing the institutional resilience of the village in facing landslide disasters.

https://doi.org/10.1051/e3sconf/202565205003

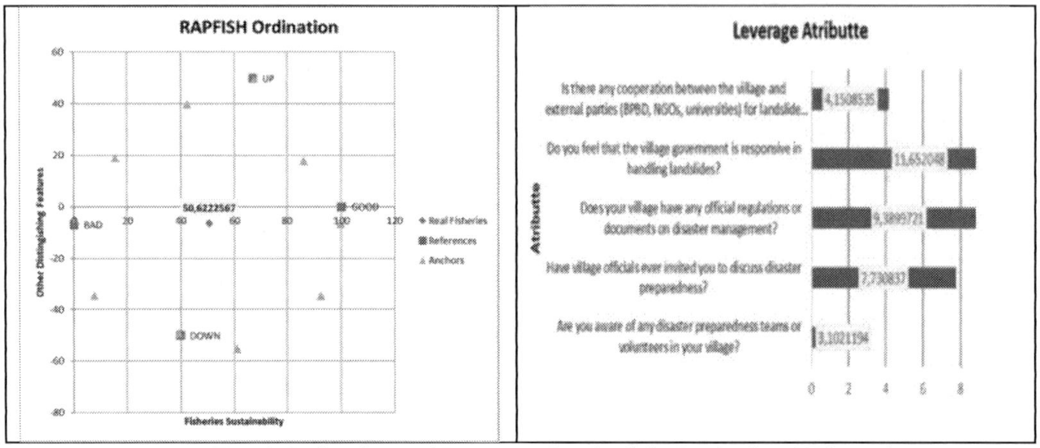

Fig. 6. Ordination and Leverage Graph of the Institutional Dimension

The institutional dimension in this study included the active roles of the village government, village officials, and community organizations in the formulation of disaster regulations, contingency planning, and regular mitigation training. These institutional roles were considered important because landslides posed a real threat in rural areas, including in Kendal Regency, Central Java, which had high landslide potential due to unstable morphology and land use practices [12].

This study focused on the involvement of village authorities and local institutions in mitigation processes, coordination, and decision-making related to disaster management. These findings were supported by research from [13], who stated that the ability of communities to deal with landslides was significantly influenced by institutional management at the village level, as well as the training and use of local wisdom among disaster volunteers. Furthermore, [14] emphasized that community resilience to disasters was strongly shaped by the strength of social capital and the role of local institutions, which encouraged more effective coordination and public participation.

The leverage analysis showed that out of the five assessed attributes, three had the greatest influence on the sustainability index. These included the responsiveness of the village government in handling landslides with an RMS of 11.652, the existence of official disaster-related documents or regulations with an RMS of 9.390, and the involvement of village officials in promoting disaster preparedness discussions with an RMS of 7.731. The other two attributes were cooperation with external parties such as the regional disaster management agency or NGOs, which had an RMS of 4.150, and the level of community knowledge regarding disaster response teams, which had an RMS of 3.102. Although these had less influence, they still contributed to the development of inclusive and adaptive institutions.

3.6 Infrastructure Dimension

The infrastructure dimension covered key aspects such as drainage conditions, road accessibility during rainfall, evacuation routes, housing location safety, and the availability of emergency evacuation facilities. The RAP-DRR analysis results showed that the infrastructure dimension achieved a sustainability score of 82.700, which fell into the sustainable category. The infrastructure dimension included crucial elements such as drainage conditions, road accessibility during heavy rain, evacuation routes, safety of residential locations, and the availability of emergency evacuation infrastructure.

Leverage analysis revealed that the three attributes with the highest influence on the sustainability index were the functionality of drainage during heavy rain (RMS = 10.237), the safety of main road access during the rainy season (RMS = 9.416), and the availability of clear and easily accessible evacuation routes (RMS = 8.765). These three attributes were the main determinants of the village's infrastructure resilience, as they directly affected evacuation speed, the smooth distribution of aid, and protection against infrastructure damage during disasters.

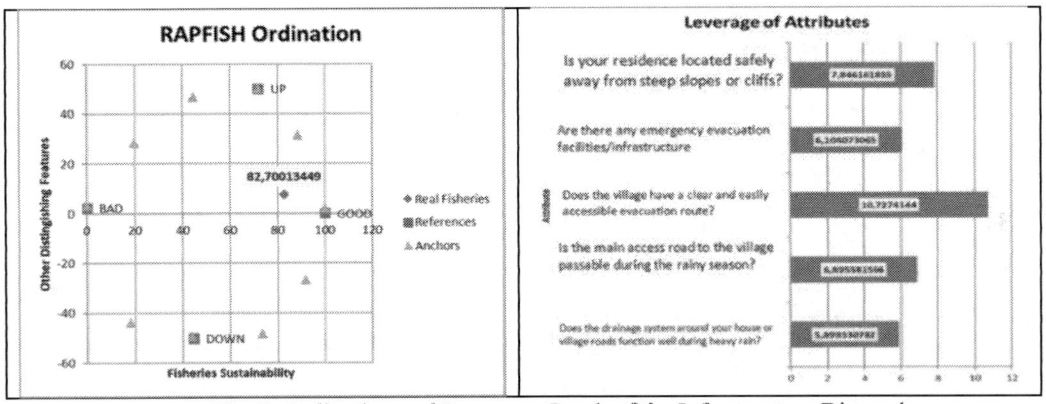

Fig. 7. Ordination and Leverage Graph of the Infrastructure Dimension

Two other attributes, namely the safety of residential locations from steep slopes or cliffs (RMS = 6.251) and the availability of evacuation infrastructure such as command posts, gathering points, or emergency tents (RMS = 5.604), although having lower influence, still contributed to community preparedness. The concept of resilient infrastructure was emphasized in the study by [15], which stated that infrastructure should be capable of absorbing disturbances, adapting, and recovering quickly from disasters through the principles of robustness, redundancy, rapidity, and resourcefulness.

Field findings showed that most areas of the village had drainage channels and evacuation route signs, although improvements were still needed in locations with steep terrain. However, infrastructure such as evacuation gathering points, directional signs, and evacuation routes were not yet fully connected or directed to clearly safe locations. Some gathering points were still located near hazard zones and lacked supporting emergency facilities. Although the infrastructure dimension had demonstrated good performance, efforts such as regular repair, relocation, and maintenance remained necessary to ensure the infrastructure functioned optimally in facing potential landslides.

4 CONCLUSION

This study shows that Ngargoyoso Village has a moderately sustainable level of resilience to landslide disasters, with a total index score of 55.64, based on a multidimensional assessment using the RAP-DRR method. Among the five dimensions analyzed, infrastructure scored the highest, indicating strong physical preparedness, while the technology and economic dimensions scored the lowest, reflecting limited early warning systems, disaster-related knowledge, and financial capacity. The social and institutional aspects are in a moderate condition but still require reinforcement through improved community participation, training, and cross-sector collaboration. To enhance overall resilience, targeted efforts are needed to strengthen the weaker dimensions through disaster education, capacity building, and economic empowerment at the community level.

ACKNOWLEDGMENTS

The authors gratefully acknowledge the Institute for Research and Community Service of Universitas Sebelas Maret for the research grant support provided through grant number 371/UN27.22/PT.01.03/2025.

REFERENCES

1. Sunimbar and I. S. Angin, "Analisis Geomorfologi Kejadian Longsor Di Kecamatan Wolotolo Kabupaten Ende," *J. Geogr.*, vol. 17, no. 2, pp. 14–22, 2021.
2. R. Haribulan *et al.*, "Jurnal Spasial Vol 6 . No . 3 , 2019 ISSN 2442-3262 KAJIAN KERENTANAN FISIK BENCANA LONGSOR DI KECAMATAN TOMOHON UTARA Jurnal Perencanaan Wilayah dan Kota Jurnal Perencanaan Wilayah dan Kota," vol. 6, no. 3, pp. 714–724, 2019.
3. R. A. Umaroh, "Strategi Penghidupan Masyarakat Korban Bencana Tanah Longsor," *J. Bumi Indones.*, vol. 5, no. 4, 2016.
4. R. S. Rini, P. Yuniarti, and W. Wianti, "Pemantauan Dan Evaluasi Kegiatan Pemberdayaan Dan Pengembangan Ketangguhan Masyarakat Desa Tangguh Bencana Di Provinsi Riau," *Sci. J. Reflect. Econ. Accounting, Manag. Bus.*, vol. 4, no. 3, pp. 588–597, 2021, doi: 10.37481/sjr.v4i3.339.
5. R. P. Robi and Y. Rahman, "Peningkatan Kesadaran Masyarakat Terhadap Kebencanaan Melalui Evaluasi Program Desa Tangguh Bencana (DESTANA) di Desa Sukaraja ," *J. Community Dev.*, vol. 5, no. 1, pp. 128–137, 2024.
6. N. S. R. KOESDARYANTO *et al.*, "Analysis of mangrove forest ecosystem sustainable management by coastal communities in Sampang District, Madura Island, Indonesia," *Indo Pacific J. Ocean Life*, vol. 8, no. 1, pp. 33–42, 2024, doi: 10.13057/oceanlife/o080104.
7. I. K. Sriwana, Y. Arkeman, D. Asma Assa, B. Besar Industri Hasil Perkebunan, J. Abdurahman Basalamah No, and S. Selatan, "ANALISIS KEBERLANJUTAN RANTAI PASOK AGROINDUSTRI KAKAO MENGGUNAKAN MULTI DIMENSIONAL SCALLING Analysis of Cacao Agroindustry Supply Chain Sustainability Using Multi Dimensional Scalling," *Indones. J. Ind. Res.*, vol. 16, no. 1, pp. 58–71, 2021.
8. A. Annisa and D. L. Setyowati, "Kapasitas Masyarakat dalam Upaya Pengurangan Risiko Bencana Tanah Longsor di Desa Tempur Kecamatan Keling Kabupaten Jepara Tahun 2018," *Edu Geogr. J.*, vol. 7, no. 1, pp. 83–94, 2019, [Online]. Available: http://journal.unnes.ac.id/sju/index.php/edugeo
9. I. S. Daher, A. S. Ramadlani, P. Masyarakat, U. S. Karawang, and J. Barat, "Pemberdayaan masyarakat melalui pendidikan berbasis program ekonomi koperasi," vol. 8, pp. 294–301, 2025.
10. D. Bosco, S. Prariz, M. Paul, A. Parangan, W. Wijaya, and L. Belakang, "Pemanfaatan Penahan Lereng Bioengineering Untuk Pembangunan Ibu Kota Negara Baru di Kalimantan Yang Ramah Lingkungan dan Berkelanjutan," *J. Reka Cipta*, vol. 1, pp. 1–7, 2024.
11. M. F. Hadian and R. Pancasilawan, "Implementasi Program Penanggulangan Bencana (Studi Pada Sub Kegiatan Pelatihan Pencegahan Dan Mitigasi Bencana Kabupaten/Kota Di Kelurahan Kalijaga Kota Cirebon)," *JANE-Jurnal Adm. Negara*, vol. 17, no. 1, pp. 10–19, 2025.
12. G. Yogiswara, T. T. Putranto, and D. Trisnawati, "Potensi Longsor di Kabupaten Kendal,

Provinsi Jawa Tengah Berdasarkan Penginderaan Jauh," *J. Geosains dan Teknol.*, vol. 3, no. 3, pp. 135–148, 2020, doi: 10.14710/jgt.3.3.2020.135-148.
13. R. A. Fitriyana and F. Hermanto, "Hidup Berdampingan Dengan Bencana, Studi Kasus Mitigasi Bencana Tanah Longsor (Kelurahan Susukan Kecamatan Ungaran Timur Kabupaten Semarang)," *Sosiolium J. Pembelajaran IPS*, vol. 4, no. 1, pp. 7–14, 2022, doi: 10.15294/sosiolium.v4i1.53174.
14. O. Shalih, H. Setiadi, T. Nurlambang, and W. Sumadio, "Toward a community resilience framework for disaster risk management. a case study: Landslide Cisolok in Sukabumi 2018 and Sunda strait tsunami in Pandeglang 2018," *E3S Web Conf.*, vol. 156, 2020, doi: 10.1051/e3sconf/202015601011.
15. W. Liu, M. Shan, S. Zhang, X. Zhao, and Z. Zhai, "Resilience in Infrastructure Systems: A Comprehensive Review," *Buildings*, vol. 12, no. 6, 2022, doi: 10.3390/buildings12060759.

https://doi.org/10.1051/e3sconf/202565206001

Disaster Management in Indonesian Schools: A Bibliometric Analysis

Ikhwan Amri[1], *Shefi* Jannati[2*], *Ihsan Weda* Prabaswara[3], and *Martanti Aji* Pangestu[4]

[1]Faculty of Teacher Training and Education, Universitas Samudra, Langsa 24416, Indonesia
[2,4]Department of Civil and Planning, Vocational College, Universitas Diponegoro, Semarang 50275, Indonesia
[3]Department of Risk and Disaster Reduction, University College London, London WC1E 6BT, United Kingdom

Abstract. Disaster management in school settings is essential for cultivating a culture of safety within school communities. This article presents a bibliometric review to investigate research trends on disaster management in Indonesian schools using the Google Scholar database. The metadata were filtered based on publication year (2010–2024), source type (journals and conference proceedings), and language (Bahasa Indonesia and English). A total of 659 publications were included, consisting of 78.8% journal articles and 21.2% conference papers. Most studies were concentrated on Java Island. The leading source of the included studies was *IOP Conference Series: Earth and Environmental Science*. The most productive contributors were S. Ayub and I.W. Gunada, publishing 10 documents. The keyword analysis generated seven thematic clusters: (1) disaster mitigation in elementary schools, (2) educational tools for disaster awareness, (3) knowledge and attitude toward disaster preparedness, (4) earthquake and tsunami preparedness in schools, (5) implementation of disaster-safe school programs, (6) disaster risk reduction education, and (7) flood preparedness in schools. Several topics, such as the involvement of students with special needs, structural safety assessments, and collaborative networks, have received limited attention. Strengthening collaboration and multidisciplinary approaches is essential to enhance the impact of research on disaster-safe school practices.

1 Introduction

As one of the five most populous countries in the world, Indonesia has one of the largest education systems. Simultaneously, the WorldRiskReport 2023 reported that Indonesia ranks as the second-highest country worldwide in terms of disaster risk index. Data from the Indonesian Ministry of Education, Culture, Research, and Technology (MoECRT) in 2024 indicated that around 57% of educational institutions are exposed to more than one natural hazard. These facts highlight the urgent need to integrate disaster management into school

* Corresponding author: shefijannati@lecturer.undip.ac.id.

management, particularly to protect the lives of the school-related community and ensure the continuity of education.

Children are the most at-risk group in school environments when disasters occur. They are more likely to suffer severe injuries, psychological distress, and educational disruptions [1]. Since children will eventually grow into adults, disaster education is essential as an investment in future resilience. This idea challenges the paternalistic paradigm that perceives children merely as a vulnerable group without agency. Disaster education extends beyond cognitive knowledge development to shape positive attitudes and behaviors toward preparedness. Therefore, schools provide an optimal environment for shaping mindsets and behaviors, enabling students to develop a culture of safety in their communities.

Disaster education has been implemented in Indonesia for a long time, but significant progress in its development began after the 2004 Aceh Tsunami. Since 2009, the Indonesian government has initiated pilot projects in several regions, initially known as "disaster-prepared schools" (*Sekolah Siaga Bencana* or SSB) [2]. In 2012, the National Disaster Management Authority issued guidelines for a program that was subsequently renamed "disaster-safe schools" (*Sekolah/Madrasah Aman Bencana* or SMAB). More recently, the program has been rebranded as "disaster-safe education units" (*Satuan Pendidikan Aman Bencana* or SPAB) to ensure its implementation across all educational institutions. The latest SPAB framework refers to the Comprehensive School Safety Framework (CSSF) 2022–2030, which consists of four key components: (1) enabling systems and policies as the foundation, (2) safer learning facilities as the first pillar, (3) school safety and educational continuity management as the second pillar, and (4) risk reduction and resilience education as the third pillar.

The integration of disaster management into school administration in Indonesia has made notable progress, such as the establishment of a five-year roadmap for disaster-safe schools (starting in 2015), the enactment of key ministerial regulations, the formation of Joint Secretariats for Disaster-Safe Schools at the national level, and the development of a monitoring and evaluation application called *Monev SPAB*. Despite these achievements, implementation at the regional level remains suboptimal, as central government policies are not perceived as mandatory obligations. Some studies have also identified several obstacles, including financial constraints, institutional challenges, human resource limitations, and collaboration issues [3].

Academics are one of the key actors who can support disaster management strategies in schools. The number of published academic works continues to grow, yet the scientific development, including research hotspots and gaps, has not been systematically mapped. To address this issue, a bibliometric review has the potential to fill this gap, as it can provide researchers with a broad overview of the research landscape [4]. Therefore, the current study aims to conduct a bibliometric review to investigate research trends on disaster management in Indonesian schools.

2 Research methods

2.1 Study design

This work utilized bibliometric analysis to map the research landscape development of disaster management in Indonesian schools. A bibliometric review differs from other literature review methodologies. For example, a narrative literature review provides an overview of a topic by synthesizing a few studies, typically conducted qualitatively and without standardized protocols or guidelines. Meanwhile, a systematic literature review is usually designed to answer specific research questions by applying inclusion and exclusion criteria when selecting scientific literature, sometimes followed by a meta-analysis if applicable.

Bibliometric analysis offers greater benefits to scholars by providing valuable insights into research trends, key influential entities, and collaboration patterns. It also helps uncover the evolution of a knowledge field and significant emerging themes. This study covers the following aspects: (1) general overview and publication growth, (2) the spatial distribution of research, (3) top sources, leading authors, and most cited documents, and (4) author keyword mapping.

2.2 Search strategy

The literature search was conducted using the Google Scholar database, the most popular academic search engine today. Google Scholar covered publications from various disciplines extensively. The selection of this database was also made due to its inclusivity in encompassing research published by local sources.

The document selection was executed using the Publish or Perish (PoP) software on November 3, 2024. This process is illustrated in detail in Figure 1. This study included documents written in both English and Bahasa Indonesia. The publication types were restricted to peer-reviewed articles from journals and conference proceedings.

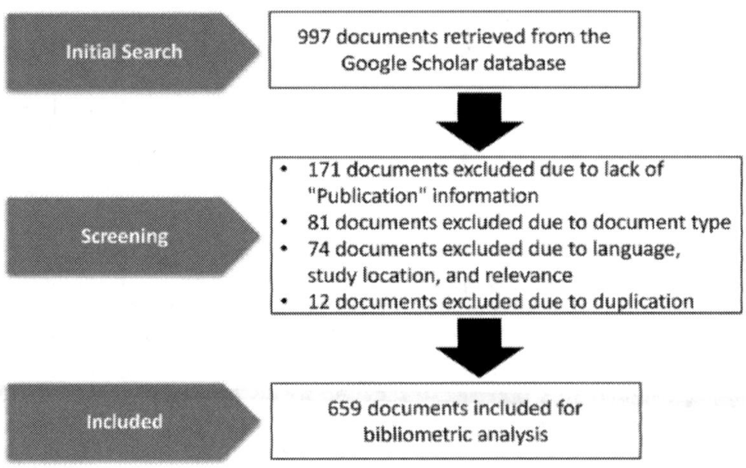

Fig. 1. Systematic workflow for document selection.

Technically, the "Title" column in the PoP software was filled with the following search terms: ("bencana" OR "kebencanaan" OR "disaster" OR "hazard") AND ("sekolah" OR "madrasah" OR "satuan pendidikan" OR "school"). Additionally, the "Keyword" column was assigned the term "Indonesia." The earliest publication year was set to 2010. The initial search retrieved a total of 997 documents.

The data obtained from the initial search was then filtered based on information completeness, document type, language, study location, relevance, and duplication. A total of 338 documents were excluded. The final dataset considered for analysis consisted of 659 documents.

The metadata exported from the PoP software (in .CSV file format) was thoroughly reviewed to ensure the completeness of essential information, particularly author names, publication years, and source titles. Since the exported metadata from PoP did not include author keywords and study locations, these details were manually added by verifying each document individually from the original sources. Additionally, the PoP-exported metadata was converted into the Scopus database export format to facilitate bibliometric analysis.

2.3 Data analysis

The data analysis techniques were supported by various software tools. Basic statistical summaries of metadata, particularly frequency analysis and graph construction, were performed using Microsoft Excel. The geographical distribution of research was visualized through the QGIS environment. Meanwhile, the analysis of author keywords was conducted using VOSviewer software, based on imported metadata in .CSV format that had been extracted and adjusted accordingly.

3 Results and discussion

3.1 General overview and publication growth

Based on a search of the Google Scholar database, this study identified at least 659 documents discussing disaster management in Indonesian schools, published between 2010 and 2024 (Table 1). In terms of document type, 78.8% were journal articles, while 21.2% were conference papers. These publications appeared in 499 outlets, consisting of 421 journals and 78 conference proceedings. A total of 1,757 individuals have contributed to research on this topic. Furthermore, the results revealed that the included studies have received a total of 5,902 citations, averaging 8–9 citations per document. The actual proportion of documents that were cited was 69.2%.

Table 1. Data Profile

Parameter	Value
Publication Years	2010–2024
Number of Documents	659
Sources	499
Authors	1,757
Total Citations	5,902

The annual publication growth rate began to rise significantly starting in 2018 (Figure 2). Compared to global trends on similar topics, the scientific output for disaster risk reduction education research started to escalate in 2014 [5]. From 2020 to the present, the number of published documents has consistently exceeded 80 per year, with a peak of 132 documents recorded in 2023. This trend appears to be associated with the increasing initiatives and commitments of relevant stakeholders in recent years to establish disaster-safe schools in Indonesia.

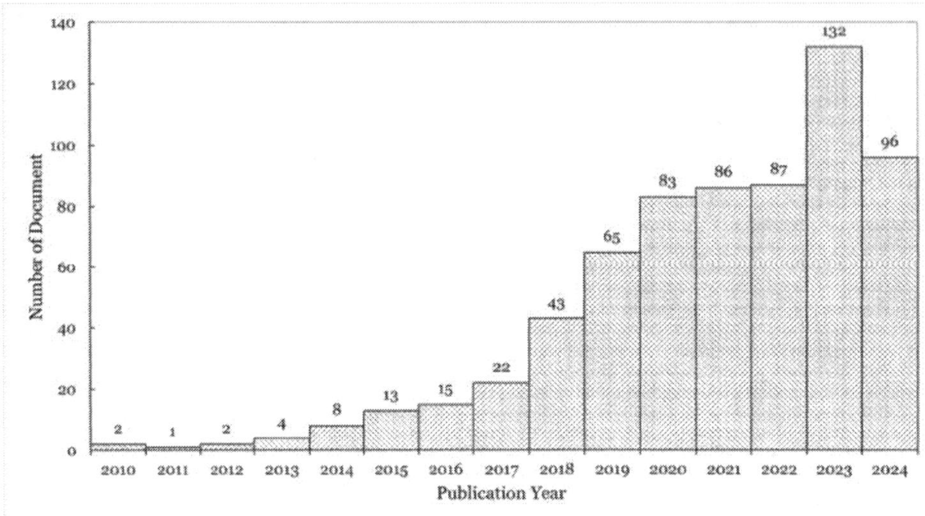

Fig. 2. Publication growth related to disaster management in Indonesian schools.

3.2 The spatial distribution of research

Although the annual publication growth trend increased, this study found that gaps still existed in the geographical distribution of research (Figure 3). In general, studies on disaster management in Indonesian schools were concentrated in Sumatra, Java, and the Nusa Tenggara Islands. The relatively high number of publications was likely closely related to the progress of the SPAB program implementation in these regions.

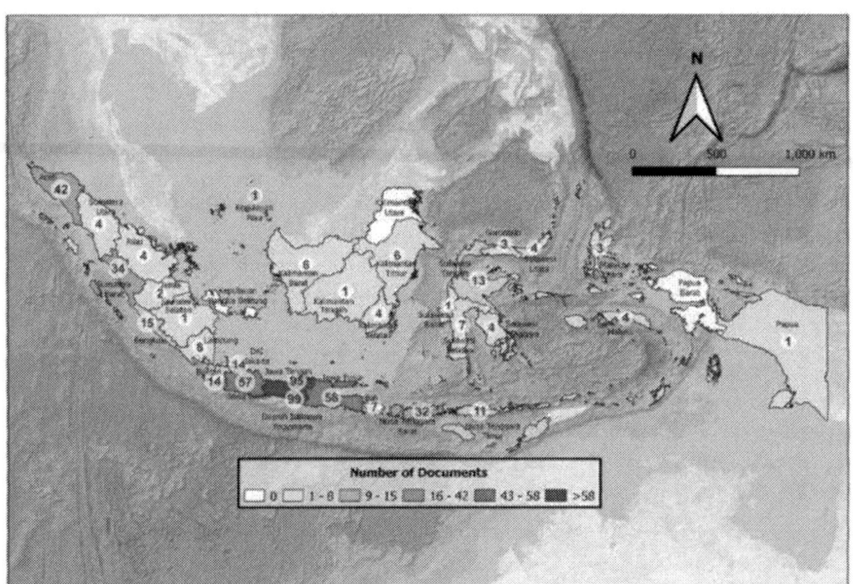

Fig. 3. Geographical distribution of the number of documents by province.

https://doi.org/10.1051/e3sconf/202565206001

The highest number of documents was identified in Daerah Istimewa Yogyakarta. It is worth noting that the implementation of the SPAB program has been intensive in this province in recent years, leading to its declaration as a "model province" in 2024. The three other provinces with the highest number of documents were Jawa Tengah, Jawa Barat, and Jawa Timur. These regions were among those with the highest disaster frequency in Indonesia [6].

Provinces in Kalimantan and most of the eastern Indonesian region had relatively fewer publications. Moreover, two provinces (i.e., Kalimantan Utara and Papua Barat) were not found to be the subject of any studies in this work. Such areas could be targeted for future research capacity development to enhance school resilience.

3.3 Top sources, leading authors, and most cited documents

This section examines the most influential sources, authors, and documents that have contributed to the research landscape of disaster management in Indonesian schools. The evaluation of sources and authors was primarily based on the number of published documents, while the most significant documents were determined by their total citation count.

Table 2 displays the top 10 sources with the highest number of publications, comprising three conference proceedings and seven journals. The first and second positions were occupied by conference proceedings published under IOP Publishing Ltd., namely *IOP Conference Series: Earth and Environmental Science* (33 documents) and *Journal of Physics: Conference Series* (14 documents). The third most productive source was *Edu Geography*, a journal published by Universitas Negeri Semarang, with a total of eight documents.

Table 2. Top 10 sources with the highest number of documents.

Source Title	Source Type	Publisher	Documents	Citations
IOP Conference Series: Earth and Environmental Science	Conference Proceeding	IOP Publishing Ltd.	33	158
Journal of Physics: Conference Series	Conference Proceeding	IOP Publishing Ltd.	14	92
Edu Geography	Journal	Universitas Negeri Semarang	8	41
International Journal of Disaster Risk Reduction	Journal	Elsevier Ltd.	6	377
AIP Conference Proceedings	Conference Proceeding	American Institute of Physics	6	8
Sumatra Journal of Disaster, Geography and Geography Education	Journal	Universitas Negeri Padang	5	3
Jurnal Ilmu Kebencanaan	Journal	Universitas Syiah Kuala	4	48
ORBITA: Jurnal Pendidikan dan Ilmu Fisika	Journal	Universitas Muhammadiyah Mataram	4	36
HIGEIA (Journal of Public Health Research and Development)	Journal	Universitas Negeri Semarang	4	23
Indonesian Journal of Environment and Disaster	Journal	Universitas Sebelas Maret	4	12

Most of the highly active journals were national publications linked to Indonesian universities, with much of the research appearing in journals specializing in education, geography, or disaster science. However, published works were also widely distributed across various disciplines, including environmental science, physics, health sciences, nursing, social sciences, engineering, psychology, and religious studies. Additionally, a significant portion of research had been published in community service journals, which tend to emphasize practical applications rather than theoretical or conceptual model development.

Among the top 10 sources, *International Journal of Disaster Risk Reduction* had the highest total citation count. Published by Elsevier Ltd., this journal is among the most reputable globally for disaster-related studies. Expanding networking and collaboration is crucial to increasing publications in high-impact journals, ensuring that research findings contribute more effectively to the global academic community.

The most productive researchers in disaster management studies in Indonesian schools were S. Ayub and I.W. Gunada, each with 10 documents (Table 3). E. Noviana followed closely with nine documents. Meanwhile, in terms of citation-to-document ratio, R.S. Oktari stood out as one of the most influential researchers, demonstrating a higher academic impact compared to others in the field.

Table 3. Top 10 authors with the highest number of documents.

Author Name	Affiliation	Documents	Citations
S. Ayub	Universitas Mataram	10	52
I.W. Gunada	Universitas Mataram	10	48
E. Noviana	Universitas Riau	9	105
R.S. Oktari	Universitas Syiah Kuala	7	226
E. Widowati	Universitas Negeri Semarang	7	75
K. Kosim	Universitas Mataram	7	40
O. Kurniaman	Universitas Riau	6	83
S.A. Sari	Universitas Negeri Medan	6	63
F.A. Kurniawan	Pujiono Centre	6	11
S.I.A. Dwiningrum	Universitas Negeri Yogyakarta	5	72

The affiliations of researchers included in this bibliometric dataset were gathered from the most recent Google Scholar profiles. Universitas Mataram, a higher education institution in Nusa Tenggara Barat, had the highest number of representatives among the top authors. Other leading authors were primarily affiliated with universities from various provinces, including Aceh, Sumatera Utara, Riau, Jawa Tengah, and Daerah Istimewa Yogyakarta.

The impact of scientific publications can be assessed by the number of times they are cited by other researchers. Table 4 lists the ten most-cited documents, with the majority focusing on disaster education and school preparedness. Four of these highly referenced studies were published in well-regarded international journals—three in *International Journal of Disaster Risk Reduction* and one in *Natural Hazards and Earth System Sciences*. The remaining studies appeared in various journals published by institutions based in Indonesia.

The study by Pramita et al. [7] recorded the highest number of citations (1,305), representing about 22% of the total citations in the dataset. This research highlighted the outcomes of a community service initiative that provided disaster-resilient school training at a vocational high school. The second most-cited document, authored by Sudrajat and Wibowo [8], accumulated 174 citations and examined students' comprehension of disaster management in a junior high school. Meanwhile, the third most-cited study, conducted by Hayudityas [9], received 152 citations and emphasized the significance of integrating disaster mitigation education into school curricula to assess student preparedness.

Table 4. Most cited documents.

Document Title	Author(s)	Source Title	Citations
Disaster-resilient school training at SMK Negeri 1 Bandar Lampung*	Pramita et al (2022)	*Journal of Social Sciences and Technology for Community Service*	1,305
Understanding disaster management for middle school students*	Sudrajat & Wibowo (2016)	*Jurnal Pendidikan Ilmu Pengetahuan Sosial Indonesia*	174
The importance of implementing disaster mitigation education in schools to assess student preparedness*	Hayudityas (2020)	*Jurnal Edukasi Nonformal*	152
Disaster mitigation education in elementary schools (an ethnopedagogical analysis study)*	Suarmika & Utama (2017)	*Jurnal Pendidikan Dasar Indonesia*	123
Improving disaster knowledge within high school students through geographic literacy	Kamil et al (2020)	*International Journal of Disaster Risk Reduction*	117
Enhancing community resilience towards disaster: The contributing factors of school-community collaborative network in the tsunami affected area in Aceh	Oktari et al (2018)	*International Journal of Disaster Risk Reduction*	104
From rapid visual survey to multi-hazard risk prioritisation and numerical fragility of school buildings	Gentile et al (2019)	*Natural Hazards and Earth System Sciences*	95
School community preparedness in facing disasters in Magelang Regency*	Lesmana & Purborini (2015)	*Jurnal Teknik Sipil*	83
Disaster mitigation learning model in social studies at middle schools*	Maryani (2010)	*Jurnal Geografi Gea*	82
Exploring minimum essentials for sustainable school disaster preparedness: A case of elementary schools in Banda Aceh City, Indonesia	Sakurai et al (2018)	*International Journal of Disaster Risk Reduction*	75

*Title translated from Bahasa Indonesia

The four most-cited studies published in prestigious international journals primarily focused on Aceh, with a particular emphasis on Banda Aceh City. This trend is likely associated with the lasting impact of the 2004 Aceh tsunami, which has made the affected region a key focus in disaster risk reduction studies within school settings. In essence, the 2004 tsunami played a pivotal role in driving increased awareness and implementation of disaster preparedness programs in Indonesian schools.

3.4 Author keyword mapping

The bibliometric network of author keywords was generated using the full counting method, considering only keywords in English for the analysis. The minimum keyword occurrence threshold was set at five. Out of 827 keywords, 43 met this criterion.

Network visualization is useful for identifying research hotspots based on keyword clustering (Figure 4). The size of each circle represents the frequency of publications using a particular keyword. The closer the circles are to each other, the stronger the relationship between the keywords. Keywords estimated to have strong connections are marked with the same color, forming thematic clusters.

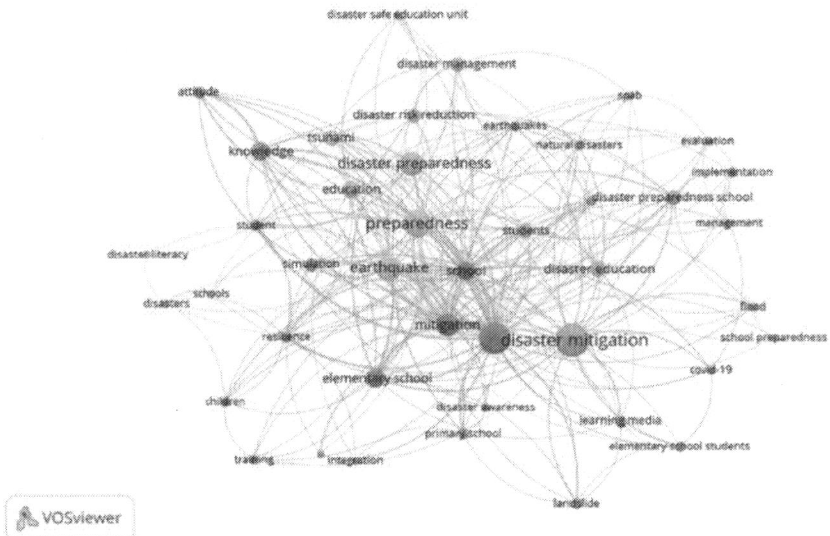

Fig. 4. Network visualization of author keywords.

Table 5 presents the interpretation of research hotspots based on the network visualization of author keywords. The bibliometric network analysis identified seven thematic clusters. These clusters represent the main themes explored by scholars in disaster management research within Indonesian schools.

Table 5. Clustering of author keywords.

Cluster	Number of Keywords	Representative Keywords	Label
I (red)	8	Elementary school, student, school, mitigation, disaster	Disaster mitigation in elementary schools
II (green)	7	Disaster mitigation, learning media, natural disasters	Educational tools for disaster awareness
III (dark blue)	7	Attitude, knowledge	Knowledge and attitude toward disaster preparedness
IV (yellow)	6	Earthquake, tsunami, preparedness	Earthquake and/or tsunami preparedness in schools
V (purple)	6	Disaster preparedness school, SPAB, implementation	Implementation of disaster-safe school programs
VI (light blue)	5	Education, disaster preparedness, disaster management, disaster risk reduction	Disaster risk reduction education
VII (orange)	4	Flood, school preparedness, disaster education, students	Flood preparedness in schools

Cluster I (red) pertains to the topic of disaster mitigation in elementary schools. Early education is a critical stage for building awareness and habits related to disaster risk reduction. While elementary school children have lower levels of independence, making them highly vulnerable to disasters, research suggested that they likely learn more efficiently than adults [10]. This advantage should be leveraged as a valuable asset in fostering a prepared community in the future.

Cluster II (green) includes studies focused on developing educational tools for disaster awareness. Traditional methods (e.g., lectures by external parties and teacher-centered approaches) tend to position students as passive recipients, limiting their role in disaster preparedness. Exploring technology and innovative, interactive teaching methods can make disaster education more engaging and effective [11]. Various media have been tested for their effectiveness in supporting disaster education in schools, including games, storytelling, comics, posters, booklets, pop-up books, videos, 3D miniatures, and augmented reality technology.

Cluster III (dark blue) groups studies that explore knowledge and attitudes toward disaster preparedness. Knowledge and attitudes play a crucial role in improving student preparedness and must be carefully considered in disaster risk reduction efforts. Several studies demonstrated a significant correlation between knowledge and attitudes toward preparedness [12]. Some research focused on assessing changes in knowledge and attitudes following educational interventions or evaluating the impact of disaster training programs.

Cluster IV (yellow) is associated with research on earthquake and/or tsunami preparedness in schools. These two natural hazards pose a high fatality risk, making it essential for exposed school communities to have adequate response capacities. Based on the MoECRT's database, 78% of schools in Indonesia are situated in regions vulnerable to earthquakes (the highest proportion among all hazards), which justifies why this issue warrants the greatest focus. Earthquake evacuation drills were widely implemented as an emergency response practice. In coastal areas at risk of megathrust earthquakes, tsunami preparedness was often prioritized. In particular, the 2004 Aceh tsunami served as a turning point for the increased implementation of disaster-safe school initiatives [13].

Cluster V (purple) primarily encompasses publications that explore the implementation of disaster-safe school programs. Among the most frequently identified formal models within this initiative were SSB and SPAB. Additionally, some strategies incorporated the concept of schools that emphasize both environmental sustainability and disaster mitigation, known as SWALIBA. Non-formal models also offer an alternative for independently establishing disaster-safe schools. Nonetheless, not all schools have teachers with the required expertise to initiate such programs or the capacity to consistently embed disaster risk knowledge into student learning. Initiatives introduced by external institutions (e.g., Green School, Child-Friendly School) should be aligned with disaster-safe school programs to enhance efficiency.

Regarding the SPAB program itself, which is currently the core practice of disaster-safe schools, most implementations so far have not accommodated all pillars. SPAB activities, as well as related research, tended to focus more on non-structural measures. Awareness campaigns, training sessions, and emergency drills were the most commonly conducted activities. Meanwhile, the SPAB pillar related to structural measures, such as building safety assessments, received relatively little attention.

Cluster VI (light blue) emphasizes research on disaster risk reduction education, aimed at equipping students with the knowledge, attitudes, and skills needed to effectively respond to disasters through both formal and non-formal methods. Essentially, disaster risk reduction education can be integrated into schools through seven key strategies: (1) intra-curricular activities, (2) local content subjects, (3) extracurricular programs, (4) formal schooling combined with community-based disaster risk reduction, (5) formal schooling combined with

https://doi.org/10.1051/e3sconf/202565206001

family-based disaster risk reduction, (6) pre-service training for civil servants, and (7) a hybrid model extending beyond formal education [14].

Cluster VIII (orange) highlights research hotspots on flood preparedness in schools. MoECRT's database indicated that 38% of schools in Indonesia are at risk of flooding, making it the second most significant hazard after earthquakes. However, many schools still lack comprehensive information on flood preparedness. To address this issue, preparedness education can enhance school communities' knowledge about floods and their risk reduction strategies [15].

The temporal development of research can be further examined through an overlay visualization of author keywords (Figure 5). Blue indicates terms with an earlier average occurrence time, while yellow represents terms that appeared more recently. Keywords with an older average annual publication date included "attitude", "evaluation", "implementation", "integration", and "disaster preparedness school". Meanwhile, keywords with a more recent average annual publication date included "flood", "disaster literacy", "disaster awareness", "disaster education", "SPAB", and "disaster-safe education unit".

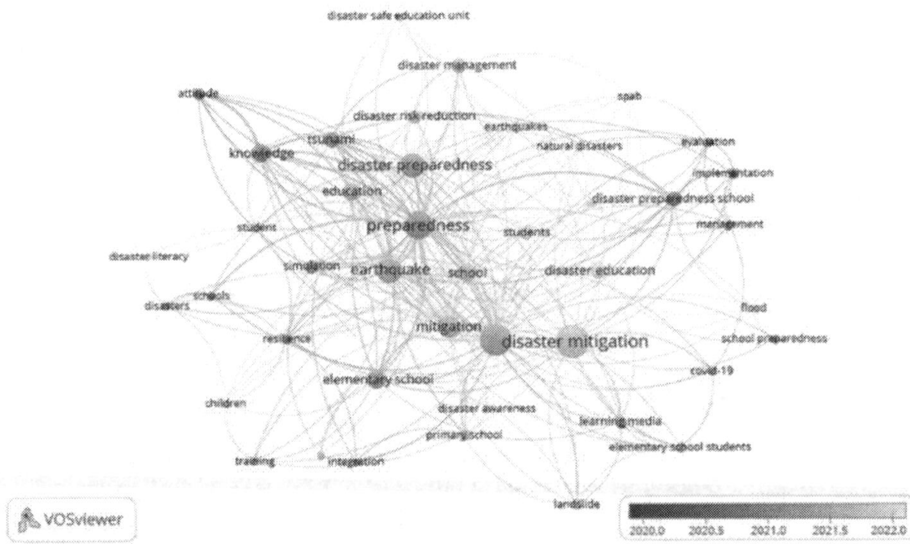

Fig. 5. Overlay visualization of author keywords.

Keyword mapping visualization can also reveal research gaps. Attention to students with special needs remained very limited, making this inclusion agenda a potential area for future studies. Other topics that could be explored include school building safety assessments, climate change adaptation, emergency education, and collaborative networks.

4 Conclusion

Amid the growing awareness of fostering a culture of disaster preparedness in schools, scholars can contribute by developing and recommending effective strategies with tangible impacts, as documented in scientific literature. This study has revealed a positive trend in the growth of publications related to disaster management in Indonesian schools. However, out of 659 documents, a geographical disparity was identified, as the research areas were primarily concentrated in several provinces, particularly in Sumatra, Java, and the Nusa Tenggara Islands.

The primary sources of publications on school-based disaster risk reduction programs in Indonesia were predominantly conference proceedings published by IOP Publishing Ltd., specifically through *IOP Conference Series: Earth and Environmental Science* and *Journal of Physics: Conference Series*. The most prolific contributors were S. Ayub and I.W. Gunada, both of whom are researchers at Universitas Mataram. The analysis of author keyword networks identified seven main clusters dominating research topics: (1) disaster mitigation in elementary schools, (2) educational tools for disaster awareness, (3) knowledge and attitude toward disaster preparedness, (4) earthquake and tsunami preparedness in schools, (5) implementation of disaster-safe school programs, (6) disaster risk reduction education, and (7) flood preparedness in schools.

Future studies may incorporate other relevant keywords (e.g., disaster education) that were not covered in this review. In addition to journals and conference proceedings, other sources such as books and grey literature can also be utilized to synthesize knowledge. Furthermore, a more comprehensive review of research on school safety should not be limited to natural-induced hazards, as existing risks also include technological hazards, biological and health-related hazards, conflict and violence, as well as everyday dangers and threats.

This research was funded by the vocational college, Diponegoro University, under the research and development program 2025. The views expressed in this document are those of the authors and do not necessarily reflect the views of the funding agency.

References

1. C. Kousky, Impact of natural disasters on children. The Future of Children **26**, 73–92 (2016). https://doi.org/10.1353/foc.2016.0004
2. R.S. Oktari, K. Shiwaku, K. Munadi, Syamsidik, R. Shaw, Enhancing community resilience towards disaster: The contributing factors of school-community collaborative network in the tsunami affected area in Aceh. Int. J. Disaster Risk Reduct. **29**, 3–12 (2018). https://doi.org/10.1016/j.ijdrr.2017.07.009
3. S. Hasbi, Z. Hanim, S. Bin Husain, The implementation optimization of school development plan in flood disaster mitigation policy in tropical rainforest (Case study at state junior high school 5 Samarinda). Soc. Sci. Humanit. Open **7**, 100440 (2023). https://doi.org/10.1016/j.ssaho.2023.100440
4. A.H. Achadi, I. Amri, D. Ruslanjari, R. Tanaka, Two decades of bibliometric exploration on leadership in disaster management. Disaster Adv. **17**, 24–32 (2024). https://doi.org/10.25303/176da024032
5. I.P. Canlas, Three decades of disaster risk reduction education: A bibliometric study. Nat. Hazards Res. **3**, 326–335 (2023). https://doi.org/10.1016/j.nhres.2023.02.007
6. A. Wibowo, N. Rohman, Rusdah, A.H. Achadi, I. Amri, Clustering Indonesian Provinces by disaster intensity using K-Means algorithm: a data mining approach. Disaster Adv. **17**, 1–8 (2024). https://doi.org/10.25303/1712da0108
7. G. Pramita, S. Saniati, M.A. Assuja, M.P. Kharisma, F.A. Hasbi, C.F. Daiyah, S.P. Tambunan, Pelatihan sekolah tangguh bencana di SMK Negeri 1 Bandar Lampung. J. Soc. Sci. Technol. Community Serv. **3**, 264–271 (2022). https://doi.org/10.33365/jsstcs.v3i2.2177

8. S. Sudrajat, S. Wibowo, Pemahaman tentang menejemen bencana alam siswa sekolah menengah pertama. Jurnal Pendidikan Ilmu Pengetahuan Sosial Indonesia **3**, 168–189 (2016). https://doi.org/10.21831/jipsindo.v3i2.11697
9. B. Hayudityas, Pentingnya penerapan pendidikan mitigasi bencana di sekolah untuk mengetahui kesiapsiagaan peserta didik. Jurnal Edukasi Nonformal **1**, 94–102 (2020).
10. S.M. Frank, M. Becker, A. Qi, P. Geiger, U.I. Frank, L.A. Rosedahl, W.M. Malloni, Y. Sasaki, M.W. Greenlee, T. Watanabe, Efficient learning in children with rapid GABA boosting during and after training. Curr. Biol. **32**, 5022–5030 (2022). https://doi.org/10.1016/j.cub.2022.10.021
11. N.R. Desilia, J. Lassa, R.S. Oktari, integrating disaster education into school curriculum in Indonesia: a scoping review. Int. J. Disaster Manag. **6**, 263–274 (2023). http://dx.doi.org/10.24815/ijdm.v6i2.34867
12. A.I. Benardi, Sumarmi, Budijanto, S. Bachri, Atta-Ur-Rahman, F. Wulandari, Disaster preparedness in proximity of Merapi Volcano, Indonesia: Is there any relationship in knowledge and attitude of senior high school students?. Int. J. Saf. Secur. Eng. **13**, 245–254 (2023). https://doi.org/10.18280/ijsse.130207
13. A. Sakurai, M.B.F. Bisri, T. Oda, R.S. Oktari, Y. Murayama, Nizammudin, M. Affan, Exploring minimum essentials for sustainable school disaster preparedness: A case of elementary schools in Banda Aceh City, Indonesia. Int. J. Disaster Risk Reduct. **29**, 73–83 (2018). https://doi.org/10.1016/j.ijdrr.2017.08.005
14. A. Amri, J.A. Lassa, Y. Tebe, N.R. Hanifa, J. Kumar, S. Sagala, Pathways to disaster risk reduction education integration in schools: Insights from SPAB evaluation in Indonesia. Int. J. Disaster Risk Reduct. **73**, 102860 (2022). https://doi.org/10.1016/j.ijdrr.2022.102860
15. M. Intaramuean, A. Nonomura, T. Boonrod, Empowering flood preparedness: Enhancing flood knowledge, risk perception, and preparedness among primary school learners in flood-affected southern Thailand. Prog. Disaster Sci. **26**, 100410 (2025). https://doi.org/10.1016/j.pdisas.2025.100410

https://doi.org/10.1051/e3sconf/202565206002

Enhancing Disaster Resilience through Sustainable Farming in Flood-Prone Areas of the Bengawan Solo Delta Areas

Raden Danang Aryo Putro Satriyono[1*], Eko Setiawan[2], Yenny Nur Chasanah[3]
Septin Puji Astuti[4]

[1,2] Industrial Engineering, Universitas Muhammadiyah Surakarta, Surakarta, Indonesia
[3] Civil Engineering, Universitas Muhammadiyah Surakarta, Surakarta, Indonesia
[4] Enviromental Study, Universitas Islam Negeri Raden Mas Said, Surakarta, Indonesia

Abstract. High-intensity flooding in Indonesia's Bengawan Solo Delta, especially within the Ngawi region, threatens the sustainability of agricultural communities due to climate change impacts and unstable precipitation patterns. This research explores the implementation of sustainable agriculture as a disaster mitigation strategy to build resilience in flood-prone rural areas. Using a mixed-methods approach and the House of Risk (HoR) framework, the study identifies critical vulnerabilities and prioritizes necessary interventions. It was found that adopting sustainable practices such as intercropping, conservation tillage, and integrated pest management effectively improves soil health and mitigates flood damage. Moreover, the study emphasizes the vital role of community-based approaches that leverage local knowledge to support early warning systems and post-disaster recovery. The findings recommend integrating sustainable agriculture into disaster risk management policies as a dual solution to reduce vulnerability while ensuring food security and environmental preservation. This study offers an actionable framework for stakeholders to build community resilience against climate-related disasters.

1 Introduction

Ensuring the resilience of food systems is a cornerstone of sustainable development, requiring that individuals and institutions can effectively absorb and recover from disruptions [1, 2]. This is particularly challenging in agricultural regions exposed to severe climate-related hazards. In Indonesia's central Bengawan Solo basin, farmers in Sragen and Ngawi Regencies confront a persistent threat from riverine floods, a phenomenon that has shaped the region's history and continues to undermine its food security [11, 12, 13]. While the immediate impacts of floods on crop production are well-documented [6, 7], the underlying risk factors—ranging from land-use change to waste management [8]—are complex and interconnected. Traditional top-down approaches to disaster management [5] are often insufficient, creating a need for more integrated strategies that leverage local knowledge and enhance community resilience from the ground up [9, 10, 14, 15].

* Corresponding author: rda715@ums.ac.id

This research responds to this need by proposing an innovative framework to develop and prioritize flood risk mitigation strategies specifically for the farming communities of this vulnerable region. Our study moves beyond conventional risk assessment by addressing three interconnected research objectives: the high-resolution mapping of flood risk zones, the identification of both biophysical and socio-economic drivers of this risk, and the formulation of a prioritized action plan focused squarely on enhancing farmer food security. The primary contribution of this work lies in its unique methodological synthesis. We integrate the spatial analytical capabilities of Geographic Information Systems (GIS) with qualitative data from stakeholder engagement (surveys and interviews) and a comprehensive literature review. These inputs are then channeled into the House of Risk (HoR) model, a powerful framework for translating complex risk data into clear, actionable, and prioritized mitigation measures. This approach ensures that the resulting strategies are not only scientifically robust but also contextually relevant and directly aligned with the goal of building a more resilient and secure food system in the Bengawan Solo basin.

By operationalizing this integrated framework, this study aims to move beyond a purely academic diagnosis of the problem toward the co-creation of practical solutions. The complexity of the Bengawan Solo socio-ecological system, with its interacting environmental pressures and community dynamics, necessitates a structured yet flexible approach. The House of Risk model is particularly well-suited for this challenge, as it provides a transparent mechanism for weighing diverse risk factors and systematically linking them to targeted mitigation actions, thereby facilitating a more participatory and evidence-based planning process. The goal is to produce a tangible roadmap for local government agencies, non-governmental organizations, and farmer groups, offering clear guidance on where to invest resources for maximum impact. In doing so, this research seeks to provide a replicable model for other river basins in Indonesia and across Southeast Asia facing similar challenges at the intersection of climate change, agriculture, and disaster risk.

2 Research methods

This study employed a multi-stage, mixed-methods research design to systematically investigate flood risk and develop targeted mitigation strategies for agricultural communities in the Bengawan Solo basin. The methodological framework was designed to be sequential, with the outputs of each stage providing critical inputs for the next, ensuring a logical progression from problem identification to actionable recommendations.

The research commenced with **Stage 1: Problem Formulation**, an exploratory phase grounded in a comprehensive literature review, preliminary field observations, and initial discussions with key local stakeholders. This foundational work allowed for the precise definition of the research problem and the establishment of clear objectives.

Following this, **Stage 2: Research Design** involved the meticulous planning of the study's operational parameters. This stage defined the scope and timeline, spanning a nine-month period from April to December 2024, focusing on farmers affected by Bengawan Solo flood risk in Sragen and Ngawi Regencies. A robust data collection strategy was developed, specifying the need for both spatial data (for GIS mapping) and non-spatial data, including ten years of historical flood records and socio-economic indicators. Data sources were triangulated from official publications by the Central Bureau of

https://doi.org/10.1051/e3sconf/202565206002

Statistics (BPS), regional Agriculture Offices (*Dinas Pertanian*), the Bengawan Solo River Basin Agency (BBWS), and supplemented with media reports and prior academic studies. The research instruments included semi-open-ended interview questionnaires, voice recorders, and video recorders to ensure thorough data capture. This stage also formalized the integrated approach, combining qualitative methods (in-depth interviews, Focus Group Discussions) with quantitative tools (GIS, House of Risk).

Stage 3: Risk and Impact Identification was the core data-gathering phase. Here, a deep dive into identifying flood risk factors and their specific impacts on farmer food security was conducted through a combination of literature synthesis, surveys, and in-depth interviews. This process culminated in a series of Focus Group Discussions (FGDs), which were crucial for validating and enriching the data. These FGDs were strategically composed to ensure a multi-stakeholder perspective, bringing together representatives from affected farmer groups, the Agriculture Offices, the BBWS, the Regional Disaster Management Agencies (BPBD) of both regencies, and the Environmental Agency (DLH) of Sragen. The primary output of this stage was a validated, comprehensive list of context-specific flood risk factors and their corresponding impacts.

In **Stage 4: Risk Evaluation**, the research transitioned from identification to analysis. This stage operationalized the first phase of the House of Risk (HoR-1) model. Through participatory FGDs, the identified risks and impacts were systematically evaluated based on criteria such as severity and frequency. This collaborative process ensured that the quantitative risk evaluation was grounded in the lived experiences and expert opinions of the stakeholders, resulting in a prioritized list of risks.

Finally, **Stage 5: Formulation of Mitigation Priorities** utilized the outputs from the preceding stages to develop strategic recommendations. The second phase of the House of Risk (HoR-2) model was applied in further FGDs to systematically link the highest-priority risks to potential mitigation actions. To enhance the robustness of these recommendations, the results were cross-referenced and validated against existing literature and best practices through a process of literature triangulation. This final stage yielded the study's primary outcome: a prioritized, evidence-based formulation of flood risk mitigation strategies specifically tailored to enhance the food security of farmers in Sragen and Ngawi Regencies.

3 Results and discussion

Table 1. Impacts of Flood Risk for Farmers in Sragen Regency (2025–2030)

No.	Impact	Impact Value
1	Displacement and evacuation of residents	4.285
2	Loss of assets and infrastructure	4.714
3	Loss of life	1.000
4	Damage to public social facilities (schools, places of worship, village offices)	4.142
5	Disruption of community social activities	5.000
6	Economic losses	4.000
7	Disruption of trade and industrial activities	4.000
8	Ecosystem degradation	4.142

No.	Impact	Impact Value
9	Water pollution	4.285
10	Infectious diseases	4.000

3.1. Analysis of Flood Impacts on Farming Communities

The initial phase of the House of Risk (HoR) framework was dedicated to identifying and quantifying the multifaceted impacts of flooding on the agricultural communities in Sragen Regency. As detailed in Table 1, the study cataloged ten distinct impacts, revealing that the consequences of flooding extend far beyond immediate agricultural losses. The findings indicate that the most severe impact, with the highest value (5.000), is the disruption of community social activities. This underscores the profound effect that floods have on the social fabric, impeding collective activities, damaging social cohesion, and isolating households, which in turn weakens the community's capacity for collective response and recovery.

Closely following this are the tangible losses of assets and infrastructure (4.714), which represent the direct economic devastation experienced by farmers, including damage to homes, farm equipment, and vital infrastructure like roads and irrigation channels. Other significant impacts include the displacement and evacuation of residents (4.285) and water pollution (4.285), highlighting the dual threat to both physical safety and environmental health, as floodwaters often carry contaminants that degrade water sources and soil quality. The damage to public social facilities (4.142) and ecosystem degradation (4.142) further illustrate the systemic nature of the damage, affecting everything from education and community governance to the long-term health of the local environment.

Interestingly, while still a critical concern, the loss of life received the lowest impact value (1.000). This may suggest that while floods are highly disruptive, the presence of informal community-based warning systems or local knowledge in evacuation allows residents to protect themselves from the ultimate harm, even as their livelihoods and social structures suffer immensely. Overall, the analysis of impacts paints a clear picture of a complex disaster profile where social disruption and economic losses are the most pervasive and damaging consequences for the farmers in the region.

Table 2. Flood Risk Mitigation Strategies in Sragen Regency (2025–2030)

No.	Mitigation Strategies - Sragen	Effectiveness Value	Difficulty Level	Effectiveness/ Difficulty
1	Building infrastructure for bamboo waste and post-harvest trash treatment	288.778	3	96.259
			2	144.389
2	Raising roads	183.814	3	61.271
3	Strengthening social capital	182.376	3	60.792
4	Building reservoirs	288.762	4	72.191
			4	72.191
5	Building community-based early warning system	153.352	3	51.117
			1	153.352

3.2. Prioritization of Flood Risk Mitigation Strategies

Following the evaluation of flood impacts, the second phase of the House of Risk (HoR) model was employed to identify, evaluate, and prioritize potential mitigation strategies. This process, detailed in Table 2, assessed each strategy based on its potential effectiveness against the identified risks and the practical difficulty of its implementation. The ratio of effectiveness to difficulty was then used to generate a clear hierarchy of strategic priorities.

The analysis reveals that the most efficient and highly recommended strategy is building infrastructure for bamboo waste and post-harvest trash treatment, which achieved the highest effectiveness-to-difficulty ratio (96.259). This finding is significant, as it points to a practical, high-impact intervention. The accumulation of agricultural and natural debris is a primary cause of blockages in waterways, which exacerbates flooding. By addressing waste management directly, the community can achieve a substantial reduction in flood risk with a relatively moderate level of difficulty, making it a strategic first step.

The next priority is building reservoirs (72.191). While this is a highly effective measure for water retention and flood control, its high difficulty level (4) reflects the significant capital investment, technical expertise, and political will required for such a large-scale infrastructure project. Following this are two strategies with similar priority scores: raising roads (61.271) and strengthening social capital (60.792). Raising roads is a critical structural adaptation for maintaining connectivity and access during flood events, while strengthening social capital represents a vital non-structural approach aimed at enhancing community organization, trust, and collective action. Lastly, building a community-based early warning system (51.117), while ranked lower on this specific metric, remains an essential component of a comprehensive mitigation plan, directly contributing to the protection of lives and assets.

3.3. Discussion: Pathways Toward an Integrated Resilience Strategy

The findings from the HoR analysis provide a clear roadmap for developing a resilient agricultural system in the Bengawan Solo basin. The systemic nature of the identified impacts—spanning social, economic, and environmental domains—necessitates a multi-pronged mitigation approach that moves beyond single-solution thinking. The prioritized strategies in Table 2 collectively form the basis of a hybrid strategy that intelligently blends structural ("hard") and non-structural ("soft") interventions.
This dual approach is critical. For instance, the highest-priority impact, disruption of social activities, cannot be solved by infrastructure alone; it requires soft interventions like strengthening social capital, which is ranked as a key mitigation strategy. Conversely, mitigating the severe impact of loss of assets and infrastructure necessitates hard solutions like building reservoirs and raising roads. The HoR model effectively validates this hybrid model, demonstrating that the most effective overall strategy is not a choice between hard or soft measures, but a strategic integration of both. The top-ranked strategy—waste management infrastructure—is itself a perfect example of a socio-technical solution that combines physical infrastructure with essential community behavior change.

3.4. The Central Role of Governance and Farmer-Centric Adaptation

Translating these prioritized strategies into action requires robust governance and cross-sectoral collaboration. The implementation of large-scale projects like reservoirs and road elevation falls under the purview of government agencies such as the Bengawan Solo River Basin Agency (BBWS) and public works departments, and necessitates aligning regional spatial planning with flood risk assessments. However, the success of community-level strategies, including the early warning system, social capital initiatives, and localized waste management, depends on the active participation of farmer groups, village governments, and local agencies like the BPBD and DLH. This underscores the need for a co-governance model where top-down support and resources empower bottom-up, farmer-centric initiatives.

Ultimately, the resilience of the local food system hinges on the adaptive capacity of the farmers themselves. The prioritized strategies provide a framework for empowerment. For example, investing in a community-based early warning system and strengthening social capital equips farmers with the tools and networks needed to prepare for and respond to floods effectively. Furthermore, linking waste management infrastructure to behavioral and technological interventions such as promoting composting or bioenergy conversion through educational outreach can create new value streams and transform a risk factor into an economic opportunity. By placing farmers at the center of the solution supported by appropriate infrastructure and collaborative governance it is possible to transition from a reactive cycle of disaster and recovery to a proactive state of sustainable and resilient agriculture.

4 Conclusion

This study successfully provides a comprehensive and actionable framework for enhancing food system resilience among the flood-prone farming communities of the Bengawan Solo basin. By synergizing stakeholder-driven data with the analytical rigor of the House of Risk (HoR) model, this research moves beyond a conventional risk assessment to deliver a prioritized roadmap for targeted intervention in Sragen and Ngawi Regencies. The findings reveal that the true burden of flooding extends far beyond agricultural damage, with the most severe consequences being the profound disruption of community social activities and the devastating loss of assets and infrastructure. This nuanced understanding underscores that effective resilience-building must address not only the physical landscape but also the socio-economic fabric that underpins community well-being and adaptive capacity.

In response to these complex challenges, this research has identified and prioritized a set of pragmatic, high-impact mitigation strategies. Crucially, the analysis points to the development of infrastructure for agricultural waste management as the most efficient and strategic starting point, directly addressing a key driver of waterway blockage with a tangible, community-level solution. This is complemented by a clear mandate for a hybrid strategy that balances large-scale structural interventions, such as the construction of reservoirs and the elevation of critical road infrastructure, with essential non-structural measures like strengthening social capital and implementing community-based early warning systems. This dual approach ensures that physical defenses are reinforced by empowered, informed, and cohesive communities capable of collective action.

The successful implementation of this prioritized agenda hinges on a robust framework of collaborative governance. This study highlights that achieving meaningful flood resilience is not the responsibility of a single entity but requires the integrated and sustained effort of a network of actors. It necessitates coordinated action between technical agencies like the Bengawan Solo River Basin Agency (BBWS), regional disaster management bodies (BPBD), and local government departments, working in close partnership with the farmer groups who possess invaluable local knowledge. Harmonizing regional spatial planning with on-the-ground realities and fostering participatory processes are therefore critical for ensuring that interventions are effective, equitable, and sustainable.

In conclusion, strengthening agricultural resilience in the Bengawan Solo basin demands a decisive shift from a reactive posture to an integrated, adaptive, and inclusive strategy. The evidence-based approach presented here, grounded in ecological restoration, strategic infrastructural reinforcement, and profound socio-economic empowerment, offers a clear path forward. Through dedicated cross-sectoral cooperation and an unwavering focus on farmer-centered innovation, it is not only possible to mitigate the recurring threat of floods but also to fundamentally transform these vulnerable agricultural systems into resilient, sustainable, and productive landscapes capable of withstanding the climate and environmental shocks of the future. This research, therefore, serves as both a specific guide for the region and a replicable model for other vulnerable river basins facing similar challenges.

ACKNOWLEDGMENTS

Researchers would like to acknowledge Muhammadiyah Surakarta University for the opportunities and facilities provided

REFERENCES

1. Bene, C., Frankenberger, T.R., Nelson, S., Constas, M.A., Collins, G., Langworthy, M., dan Fox, Karyn, **2023**, "Food system resilience measurement: principles, framework and caveats", *Food Security*, **15**:1437–1458.
2. Fanzo dkk., **2021**, "Viewpoint: Rigorous monitoring is necessary to guide food system transformation in the countdown to the 2030 global goals". *Food Policy,* **104**: 102163.
3. Toth, A., Rendall, S., dan Reitsma, F., **2015**, "Resilient food systems: a qualitative tool for measuring food resilience", *Urban Ecosystems*, **19**: 19–43.
4. Forbord, M. dan Vik, J., **2017**, "Food, farmers, and the future: Investigating prospects of increased food production within a national context", *Land Use Policy*, **67**: 546–557.
5. Coppola, D.P., **2015**, *Introduction to International Disaster Management*, 3rd edition, Butterworth-Heinemann, Oxford.
6. Kaur, G., Singh, G., Motavalli, P.P., Nelson, K.A., Orlowski, J.M., dan Golden, B.R., **2020**, "Impacts and management strategies for crop production in waterlogged or flooded soils: A review", *Agronomy Journal*, **112**:1475–1501.
7. MacMahon, A., Smith, K., dan Lawrence, G., **2015**, "Connecting resilience, food security and climate change: lessons from flooding in Queensland, Australia", *Journal of Environmental Studies and Sciences,* **5**:378–391, DOI 10.1007/s13412-015-0278-0
8. Iloka, N.G., **2016**, "Indigenous knowledge for disaster risk reduction: An African perspective", *Jàmbá: Journal of Disaster Risk Studies*, **8**(1): a272. http://dx.doi. org/10.4102/jamba.v8i1.272.
9. Farid, M., Gunawan, B., Kusuma, M.S.B., Habibi, S.A., dan Yahya, A., **2020**, "Assessment of flood risk reduction in Bengawan Solo river: a case study of Sragen Regency", *International Journal of GEOMATE,* **18**(70): 229–234.
10. Rustinsyah, R. Prasetyo, R.A., dan Adib, M., **2021**, "Social capital for flood disaster management: Case study of flooding in a village of Bengawan Solo Riverbank, Tuban, East Java Province", *International Journal of Disaster Risk Reduction*, **52:** 101963-1–101963-10.
11. Mawandha, H.G., Wignyosukarto, B.S., dan Jayadi, R., **2018**, "Mini polders as alternative flood management in the lower Bengawan Solo River, Indonesia", *Irrigation and Drainage*, **67**(S1): 72–80.
12. Ait Zamzami, H.,Elaanzouli, M., Zahrani, A., Boumeaza, O., Aachrine, B.,Saidi, J., Musiyam, M., & Boumeaza, T.(2025). Remote Sensing Assessment of Tourism's Groundwater Consumption: Evapotranspi-ration Analysis in the Northern Piedmont of the Western High Atlas, Morocco. Forum Geografi. 39(1), 112-124

13. Koliou, M., dan de Lindt, J.W., McAllister, T.P., Ellingwodd, B.R., Dillard, M., dan Cutler, H., 2020, "State of the research in community resilience: progress and challenges", *Sustainable and Resilient Infrastructure*, **5**(3): 131–151.
14. Alausa, O. A., Adaradoun, O. S., Adekunle, G. T., Priyono, K. D. (2023). Leveraging Geospatial Technology for Enhanced Utility Management: A Case Study in Electrical Distribution Power Systems. Forum Geografi. 37(2), 164-177.
15. Setiawan, E. Pramana, G.A., 2023, "Improving the effectiveness of disaster mitigation in Wonogiri Regency, Indonesia using House of Risk method", *Indonesian Journal of Geography*, 55(3): 421–432.

https://doi.org/10.1051/e3sconf/202565206003

Enhancing Deep Early Eco-Knowledge for Students About Green Campus Program Through Student Centered Video Assignment

Robby Hilmi Rachmadian[1*], *Sumarmi*[2], *Sugeng* Utaya[3], *Hety Dwi* Lestari[4] and *Muhammad Ariandhi* Saputro[5]

[1,2,3,4,5]Department of Geography, State University of Malang, 65145 Malang, Indonesia

Abstract. Increasing ecological knowledge among students can have an impact on their awareness and intention to protect their surroundings. Universities play a key role in shaping environmentally conscious students. Assigning students a project to create a video about "A Day in my life at Cakra Green Campus" aims to provide an understanding of the green campus program, involve students in campus environmental conservation, and enhance creativity in video production. Therefore, the purpose of this study is to decide the perceptions of new UM students about the video assignment "A Day in My Life at Cakra Green Campus" and to decide the influence of the video assignment on the eco-knowledge of new UM students. This study is an ex post facto study with a quantitative descriptive approach. 435 First year students at the State University of Malang (UM) who had completed the video assignment participated in this study using a questionnaire adapted based on the indicators of the UI GreenMetric World University Ranking. Data analysis used descriptive statistics with the aid of SPSS version 22 to describe the level of eco-knowledge among students. The results of the study show that UM first-year students have a particularly proficient level of eco-knowledge, at 72.2% after completing the video assignment. Additionally, the video assignment "A Day in My Life at Cakra Green Campus" had a 51.5% influence on students' eco-knowledge, which can serve as a basis for UM to evaluate and provide environmental education tailored to the characteristics of its students.

1 Introduction

Environmental issues such as climate change, air pollution, waste management, and lack of environmental awareness have prompted higher education institutions to play an active role in supporting sustainable development [1]. Higher education institutions have a moral and strategic responsibility to instill environmental awareness values in their students. As the younger generation and agents of change, students play a crucial role in shaping the future of a nation, including the development of environmentally conscious character [2]. Therefore,

* Corresponding author: robby.hilmi.2407219@students.um.ac.id

the development of Green Campus programs is a tangible manifestation of an institution's commitment to accelerating the principles of sustainable environmental practices.

Green Campus does not only focus on the application of environmentally friendly technology and the management of environmentally friendly infrastructure. Green Campus also emphasizes the importance of student involvement in understanding, responding to, familiarizing themselves with, and applying ecological values in their daily lives. One of the challenges in involving students in environmental awareness efforts is the cycle of students entering and graduating each year. First year students, transitioning learners from secondary to higher education, have different understandings and habits regarding sustainability values. Therefore, enhancing environmental awareness among First year students is a crucial effort to strengthen literacy and early eco-knowledge regarding the Green Campus concept, which should be implemented from the very beginning of their studies [3].

Early eco-knowledge plays a fundamental role in fostering sustainable pro-environmental attitudes and behaviors among First year students. However, in practice, learning activities and campus orientation programs for First year students are often presented in a theoretical manner and lack contextual relevance, thereby reducing the effectiveness of students' understanding and engagement in recognizing their campus environment. A purely theoretical learning approach tends to lack sufficient participatory space for students to explore and reflect on environmental issues in a personal manner. Based on previous research, a participatory and experience-based learning approach has proven effective in fostering a deeper understanding of the environment among students.

To address existing challenges, a student-centered video assignment learning approach offers the potential to improve students' eco-knowledge. This learning approach has an indirect impact on the intention to strengthen the environment to overcome existing challenges. Through video assignments about life on the Green Campus, students are not only encouraged to observe and describe environmentally friendly behavior on campus but also involved in reflecting on their own behavior in support of the Green Campus program. Video as a medium of expression provides First year students with the opportunity to convey environmental narratives from a personal perspective in an authentic and relevant manner. Additionally, video assignments enhance students' creativity in improving their communication skills and critical thinking about environmental issues on their campus. This concept approaches individual commitment or willingness to engage in environmental activities. It is closely related to various pro-environmental behaviors, such as recycling, water conservation, waste separation, and other environmentally friendly practices. These behaviors are directly integrated with the UI GreenMetric indicators for waste, water, energy, and climate change.

Contextually designed video assignments based on students' real-life experiences are believed to enhance emotional and cognitive engagement in the learning process. The internalization of ecological values does not stop at the knowledge level but also touches on deeper aspects of attitude and awareness. The use of digital technology through video creation also strengthens multimodal literacy, which is an important component of 21st-century competencies. Therefore, it is important to build environmental understanding and creativity among First year students through innovative learning strategies that can explore students' understanding, attitudes, and skills toward the Green Campus program [4].

The importance of building environmental awareness and creativity among First year students is one of the integrated efforts. Universities must innovate in conducting assessments to measure students' level of environmental awareness in a manner appropriate to their characteristics [5]. Based on existing issues and potential, the objective of this study is to determine the perceptions of new UM students regarding the video assignment "A Day in My Life at Cakra Green Campus" (VA-CGC) and to assess the impact of the video assignment on the eco-knowledge of new UM students. Thus, this research is expected to contribute to the development of a learning model that supports the transformation of students' ecological behavior through a participatory, reflective, and experience-based approach.

2 Literature Review

2.1 Theoretical framework

The theory and practice related to this research aim to inform and create a framework for universities and researchers. This is to be considered when implementing project-based learning, namely video production for First year students. This research is characterized as an active learning approach to increase student engagement and creativity in higher-order cognitive abilities based on social cognitive theory and constructivism theory.

Student engagement can be enhanced through mastery experiences, verbal persuasion, social assignments, and physiological conditions [6]. The level of student engagement can improve students' understanding and self-confidence, the higher the level of student engagement, the higher the level of student understanding. Students with a high level of understanding can demonstrate motivation, self-correction, and build experiences and challenges for students. In this study, the level of student understanding is related to environmental issues and green campus programs, or what is referred to as eco-knowledge [7]. The level of eco-knowledge is analyzed through the perceptions of first year students after completing a learning experience designing and creating videos.

Students, as a unique social group, have a high ability to adapt in terms of values and behavioral intentions, and they play an important role in environmental protection and sustainable development. Therefore, it is theoretically and practically very important to prioritize and increase environmental awareness of university students. Through student-centered video-based learning, it is possible to intensively build emotional experiences, a basic psychological approach that contains information and changes a person's perception of the environment. Research shows that when individuals are immersed in a video, they mentally construct a virtual world triggered by the narrative, then form beliefs and attitudes consistent with the story, thereby influencing pro-environmental behavioral intentions.

The constructivist theory theoretically underpins the video-making activities of students in this study. Students act as teachers and impart knowledge in video assignments. This study follows the theory of learning through the creation of real and meaningful products about learning as reconstruction. In this study, the videos created by students are products of the reconstruction of understanding, showing what is known and expanding into what is unknown. Thus, cognitive theory, student engagement, and constructivist theory support the research objective of providing a learning experience that fosters students' eco-knowledge.

2.2 Student-centered video assignment

Improving students' understanding and involvement in creating a sustainable campus environment is very important. Students' understanding can be shaped through lectures and extracurricular activities related to environmental issues [8]. One way to improve students' understanding is through video assignments that involve student creativity. Videos made by students are audio-visual works that demonstrate the learning and knowledge gained by students.

Student-centered video assignments facilitate students in exploring concepts, expressing understanding, and presenting findings in creative ways [9]. Video assignments enable students to develop critical thinking skills, digital literacy, and visual communication

https://doi.org/10.1051/e3sconf/202565206003

skills. Several studies have shown that video-based assignments can increase students' emotional and cognitive engagement in the learning process. Student-created videos foster a sense of ownership over the learning process and increase intrinsic motivation. Additionally, video creation promotes active learning as students must research, plan, create a script, record, and edit, all of which involve critical thinking skills [10].

The implementation of video assignments in environmental education has great potential to increase eco-knowledge among students [11]. Creating environment-themed videos can strengthen students' connection with environmental issues around them. Students who create environmental campaign videos show a significant increase in pro-environmental literacy and behavior compared to students who only take part in conventional learning. Therefore, authentic experiences in observing and documenting environmentally friendly practices on campus through video assignments enable students to develop essential contextual understanding in environmental education.

2.3 Eco-knowledge

A sustainable campus environment can be achieved through green campus initiatives. Green campus initiatives aim to integrate environmental sustainability issues into institutions to address global sustainability issues. The role of the campus is to promote sustainable development internally and externally through teaching, research, community service, governance, and the values of sustainable development for future generations [12]. The implementation of green campus initiatives can encourage environmental education that aims to integrate academic discipline with environmental issues and enhance students' eco-knowledge. Therefore, incorporating sustainable environmental concepts into campus activities and learning processes can enhance students' understanding and engagement [13].

Learning processes that teach about the environment, sustainability, and the impact of human activities on the environment contribute significantly to improving students' ecological knowledge [14]. Project-based education, environmental simulations, and problem-based learning approaches have been proven effective in improving understanding of the relationship between humans and the environment. Therefore, environmental education must be transformative, not just informative, so that the knowledge acquired by students can be internalized into concrete actions.

There are several factors that influence students' eco-knowledge levels, including educational background, exposure to environmental issues, involvement in environment-based activities, and use of digital media. Students who are actively involved in conservation activities and environmental campaigns demonstrate higher levels of eco-knowledge compared to students who only receive passive learning. Eco-knowledge not only serves as declarative knowledge but also as an important predictor of pro-environmental behavior. High levels of eco-knowledge are positively correlated with intentions and actual actions for environmental conservation, such as reducing waste, walking, conserving water, and choosing environmentally friendly products [15].

3 Methodology

3.1 Research Design

This research is a survey-type study with a quantitative descriptive approach, which can be interpreted as research conducted after implementing treatment on the research subjects. Survey research is a quantitative research procedure conducted to identify perceptions that reflect attitudes, behaviors, understanding, characteristics, or phenomena obtained through a sample within a population. The data sources are derived from the perceptions of First year students at the State University of Malang (UM) using a questionnaire administered after the Campus Life Orientation Program for First year Students (PKKMB), which took place in August and September 2024. The data obtained were then analyzed using statistical tests to determine the level and average results of First year students' perceptions, identify the relationships between variables, and provide recommendations for future research.

3.2 Population and sample

The population in this study was all new UM students at the undergraduate and diploma levels in the 2024 cohort. UM students are divided into ten faculties with a total of 11,567 students. The sample determination in this study is probabilistic in nature, using proportionate stratified random sampling to obtain representative respondents and reduce bias in sample determination. The researcher used the Slovin formula in determining the sample in this study, as follows.

$$n = \frac{N}{N.e^2+1} \quad (1)$$

$$n = \frac{11567}{11567.(0,05)^2+1}$$

$$n = \frac{11567}{29,9}$$

$$n = 387$$

Based on sample calculations using the Slovin formula, the sample in this study consisted of 387 respondents. However, during the data collection process, there were 435 respondents who met the criteria. Therefore, the sample size for this study was set at 435, resulting in a diversification of characteristics and responses from the respondents. The diversity of responses from the respondents can enrich the perceptions of First year students in the video assignment "A Day in My Life at Cakra Green Campus" (VA-CGC) as well as the level of eco-knowledge among First year students at UM.

3.3 Instrument and data collection

This study used an online questionnaire to survey new UM students' perceptions of the VA-CGC and their level of eco-knowledge regarding green campus initiatives conducted in August–September 2024. The online questionnaire data collection was carried out with the assistance of social media and the UM PKKMB committee. The questionnaire instrument was adapted from the UI GreenMetric (UIGM) questionnaire standards to assess the results of measurements of commitment, first year students' understanding of UIGM indicators and eco-knowledge, and perceptions of the VA-CGC. The questionnaire was validated by seeking input from experts, who assessed it for content validity and conducted a pilot test. The pilot testing aimed to evaluate the questionnaire's ability to accurately measure the intended variables. During the pilot test, experts were asked to examine the questionnaire and offer feedback on questions that were relevant and appropriate for capturing the research objectives. The questionnaire's suitability was evaluated using Cronbach's Alpha with a value of 0.859, indicating that this questionnaire can be used to analyze the relationship between video assignments and eco-knowledge.

The questionnaire consists of three sections with a total of 40 questions. The first section relates to the demographic information of the respondents. The second section investigates students' knowledge of various criteria in the implementation of the UM Green

Campus. The third section explores the perceptions of First year students in completing the VA-CGC regarding their understanding and commitment to preserving the campus environment. These three sections of questions aim to identify several indicators, including student involvement, student understanding, and the exploration of eco-knowledge among students. This research questionnaire is closed-ended, using a four-point Likert scale, which consists of 1 (representing strongly disagree), 2 (representing disagree), 3 (representing agree), and 4 (representing strongly agree). Additionally, the researcher conducted a preliminary survey to check the validity and reliability of the questionnaire.

3.4 Data analysis

Determining the level of eco-knowledge among new UM students using data analysis based on predetermined assessment criteria. At this stage of analysis, the maximum score is determined based on the maximum score found in the research instrument. The next stage is to determine the total score of participants by adding up the scores obtained from respondents' answers to each question item. Determining the level of eco-knowledge among students in the green campus program, based on the following formula.

$$Eco\text{-}knowledge = \frac{Students'\ score}{Maximum\ score}\ x\ 100 \qquad (2)$$

Measurement of students' perception levels toward the VA-CGC based on the following formula.

$$Video\ Assignment = \frac{Students'\ score}{Maximum\ score}\ x\ 100 \qquad (3)$$

The next step is to determine the assessment criteria for eco-knowledge levels and students' perceptions toward the VA-CGC based on the calculated scores obtained within the assessment range in Table 1 below.

Table 1. Range of eco-knowledge scores among UM's students.

Score Interval	Criteria
$X > 81$	Very good
$65{,}5 < X \leq 81$	Good
$51{,}5\ X \leq 65{,}5$	Low
$X \leq 51$	Very low

The data on the average eco-knowledge scores and video assignment scores, which have been normally distributed, were then tested using simple linear regression with the assistance of SPSS version 22. This aims to determine the effect of the video assignment on the eco-knowledge level of new UM students. Assuming that if the significance value is < 0.05, it means that the video assignment influences students' eco-knowledge. However, if the significance value is > 0.05, it means that the video assignment does not influence students' eco-knowledge.

4 Results

4.1 Students' perception about video assignment

The results of this study aim to determine the perceptions of First year students at Malang State University (UM) in 2024 regarding the assignment of Campus Life Orientation Program for First year Students (PKKMB) in the form of a video entitled "A Day in My Life at Cakra Green Campus," using descriptive quantitative analysis. The following are the results of the data analysis of student perceptions, which were analyzed using SPSS 22 for Windows and Microsoft Excel.

Table 2. Descriptive statistics of students' perceptions of video assignments.

Variable	N	Mean	Median	Mode	Std. Dev	Min	Max
Video Assignment	435	88,25	89,3	100	10,71	36	100

Source: Author's statistics

Based on the descriptive statistical calculations, the results of UM first year students' perceptions of the VA-CGC show an average score of 88.25, a median score of 89.3, a standard deviation of 10.71, a minimum score of 36, and a maximum score of 100. These results need to be reviewed in more detail based on the categorization of students' perceptions of the assignment VA-CGC. The categorization results can be seen in Table 3.

Table 3. The score value of students' perceptions of video assignments..

Interval Score	Criteria	F	%
$X > 81$	Very good	320	73,6
$65{,}5 < X \leq 81$	Good	106	24,4
$51{,}5\ X \leq 65{,}5$	Low	7	1,6
$X \leq 51$	Very low	2	0,4
Total		**435**	**100**

Source: Author's statistics

Based on the results of the categorization of First year students' perceptions of the VA-CGC in Table 4, it shows that First year students feel that the assignment is very good for providing an understanding of the green campus concept. This is evidenced by the results of the perception measurement among students, with 320 respondents (73.6%) in the "very good" category, 106 respondents (24.4%) in the "good" category, 7 respondents (1.6%) in the 'low' category, and 2 respondents (0.4%) in the "very low" category.

https://doi.org/10.1051/e3sconf/202565206003

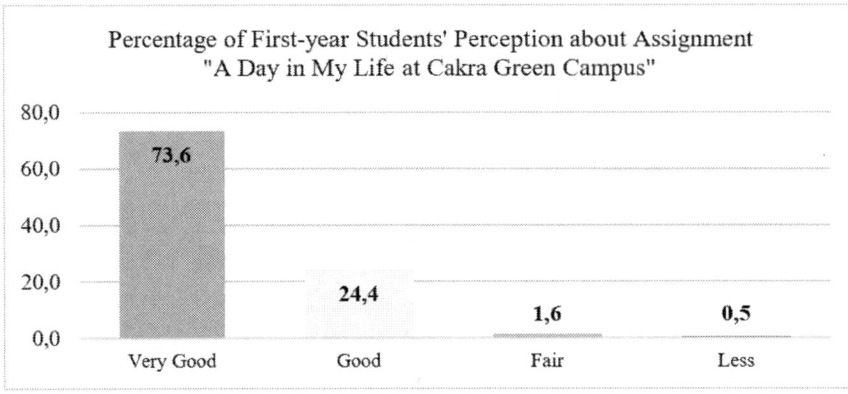

Fig. 1. Percentage of students' perceptions toward the video assignment "A Day in My Life at Cakra Green Campus"

4.1.1 Students' perception about video assignment based on gender

Measuring first year students' perceptions of the video assignment based on gender is one of the things that need to be known. This aims to determine the level of perception between male and female students. Based on the measurement of First year students' perceptions of the VA-CGC female students have higher scores than male students, as presented in the following Table 4.

Table 4. Descriptive statistics of students' perceptions of the VA-CGC based on gender.

Variable	N	Mean	Median	Mode	Std. Dev	Min	Max
Male	159	87,04	89,3	130,458	11,422	50	100
Female	276	88,95	92,9	104,717	10,233	36	100

Source: Author's statistics

Based on the descriptive statistical calculations, the results of perceptions toward the video assignment about first year students' understanding of the green campus program show differences in average scores based on respondents' gender. The average score for male students is 87.04, and the average score for female students is 88.95.

4.1.2 Students' perception about video assignment based on faculties

Measuring First year students' perceptions of the VA-CGC based on their faculty of origin in relation to their eco-knowledge in the green campus program is one of the things that needs to be known. This aims to determine the level of environmental awareness among students across faculties at UM. Based on the measurement of First year students' perceptions of the video assignment, first year students from the Faculty of Mathematics and Natural Sciences (FMIPA) had the highest average score compared to other faculties. The measurement results are presented in the following Table 5.

Table 5. Descriptive statistics of students' perceptions of the VA-CGC based on faculty.

Variable	N	Mean	Median	Mode	Std. Dev	Min	Max
Faculty of Mathematics and	68	90,44	92,9	103,404	10,169	61	100

 https://doi.org/10.1051/e3sconf/202565206003

Variable	N	Mean	Median	Mode	Std. Dev	Min	Max
Natural Sciences (FMIPA)							
Faculty of Economics and Business (FEB)	48	89,66	92,9	90,735	9,526	71	100
Faculty of Educations (FIP)	53	88,55	89,3	114,99	10,732	61	100
Faculty of Vocational Studies (FV)	34	87,82	89,3	153,157	12,376	50	100
Faculty of Social Sciences (FIS)	55	87,59	89,3	157,367	12,545	36	100
Faculty of Letters (FS)	57	87,47	85,7	92,005	9,592	64	100
Faculty of Psychology (FPsi)	15	87,13	89,3	87,085	9,332	71	100
Faculty of Sport Sciences (FIK)	38	87,12	85,7	93,409	9,665	71	100
Faculty of Engineering (FT)	60	85,89	85,7	127,021	11,27	57	100
Faculty of Medicine (FK)	7	80,10	75	224,553	14,986	64	100

Source: Author's statistics

Based on the descriptive statistical calculations, it shows that the perception of First year students toward the VA-CGC indicates a significant difference in average scores based on faculty affiliation. The highest average score was obtained by FMIPA students at 90.44, followed by FEB students at 89.66, FIP students at 88.55, FV students at 87.82, FIS students at 87.59, FS students at 87.47, FPsi students at 87.13, FIK students at 87.12, FT students scored 85.89, and FK students scored 80.10.

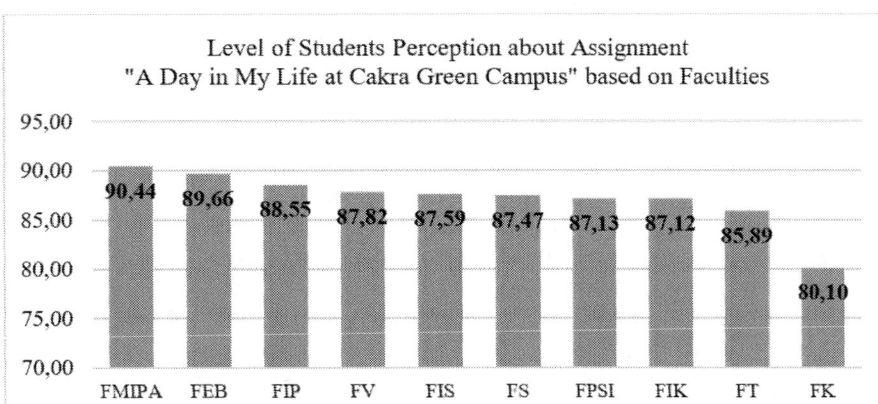

Fig. 2. Percentage of students' perceptions toward the VA-CGC based on faculty.

4.2 Students' eco-knowledge about green campus

The purpose of this analysis is to determine the level of understanding of First year students at Malang State University (UM) in 2024 regarding the implementation of the green campus program, particularly in terms of eco-knowledge. The following are the results of the analysis of student perception data, which were analyzed using SPSS 22 for Windows and Microsoft Excel. Based on the measurement of First year students' eco-knowledge regarding the green campus program, the results are considered good, as explained in Table 6 below.

Table 6. Descriptive statistics on the level of eco-knowledge among students at UM.

Variable	N	Mean	Median	Mode	Std. Dev	Min	Max
Eco-Knowledge	435	84,22	85,6	86	8,029	27	97

Source: Author's statistics

Based on these descriptive statistical calculations, it shows that the results of new UM students' perceptions of eco-knowledge in the green campus program are an average score of 84.22, a median score of 85.6, a standard deviation of 8.029, a minimum score of 27, and a maximum score of 97. These results need to be reviewed in more detail based on the categorization of students' perceptions of eco-knowledge in the green campus program. The categorization results can be seen in Table 7.

Table 7. Measurement results of first year students' eco-knowledge levels.

Interval Score	Criteria	F	%
X > 81	Very good	314	72,2
65,5 < X ≤ 81	Good	116	26,7
51,5 X ≤ 65,5	Low	2	0,5
X ≤ 51	Very low	3	0,7
Total		**435**	**100**

Source: Author's statistics

Based on the results of the eco-knowledge level categorization in Table 8, it shows that First year students have a very good level of eco-knowledge regarding their understanding of the green campus concept. This is demonstrated by the results of the measurement of students' perceptions, with 314 respondents (72.2%) in the very good category, 116 respondents (26.7%) in the good category, 2 respondents (0.5%) in the low category, and 3 respondents (0.7%) in the very low category.

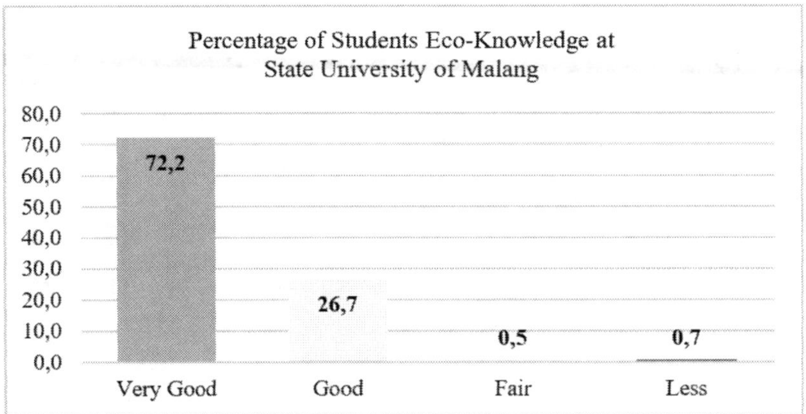

Fig. 3. Percentage of students' Eco-Knowledge at State University of Malang.

4.2.1 Students' eco-knowledge about green campus based on gender

Measuring the level of eco-knowledge regarding the green campus program based on the gender of respondents is one of the things that needs to be known. This aims to determine the level of perception between male and female students. Based on the measurement of the level of eco-knowledge regarding the green campus program, female students have higher scores than male students, as presented in Table 8 below.

Table 8. Descriptive statistics of eco-knowledge levels based on gender.

Variable	N	Mean	Median	Mode	Std. Dev	Min	Max
Male	159	83,89	85,6	65,813	8,113	39	97
Female	276	84,41	85,6	63,827	7,989	27	97

Source: Author's statistics

Based on these descriptive statistical calculations, the results of eco-knowledge levels regarding First year students' understanding of the green campus program show differences in average scores based on the gender of the respondents. The average score for male students was 83.89, while the average score for female students was 84.41.

4.2.2 Students' eco-knowledge about green campus based on faculties

Measuring the level of eco-knowledge based on faculty of origin in relation to First year students' understanding of the green campus program is one of the things that needs to be known. This aims to determine the level of eco-knowledge among first year students across faculties at UM. Based on the measurement of First year students' perceptions of eco-knowledge, first year students from the Faculty of Sports Sciences (FIK) have the highest average score compared to other faculties. The measurement results are presented in the following Table 9.

Table 9. Descriptive statistics of eco-knowledge levels based on faculties.

Variable	N	Mean	Median	Mode	Std. Dev	Min	Max
Faculty of Sport Sciences (FIK)	38	86,36	87,9	46,233	6,8	71	96
Faculty of Mathematics and Natural Sciences (FMIPA)	68	85,38	86,35	55,498	7,45	69	97
Faculty of Letters (FS)	57	84,67	84,8	25,023	5,002	71	94
Faculty of Economics and Business (FEB)	48	84,65	85,6	51,543	7,179	63	97
Faculty of Engineering (FT)	60	84,03	84,1	65,876	8,116	69	97
Faculty of Vocational Studies (FV)	34	83,58	86,75	117,515	10,84	39	95
Faculty of Psychology (FPsi)	15	83,29	84,1	84,763	9,207	68	96
Faculty of Educations (FIP)	53	82,88	83,3	71,24	8,44	46	97
Faculty of Social Sciences (FIS)	55	82,62	84,1	103,035	10,151	27	96
Faculty of Medicine (FK)	7	78,46	78	67,226	8,199	69	92

Source: Author's statistics

Based on the descriptive statistical calculations, it shows that the results of the eco-knowledge measurement of the green campus program indicate that there is a significant difference in average scores based on faculty origin. The highest average score was obtained by FIK students at 86.36, followed by FMIPA students at 85.38, FS students at 84.67, FEB students at 84.65, FT students at 84.03, FV students at 83.58, FPsi students at 83.29, and FIP students at 82.88, the score for FIS students was 82.62, and the score for FK students was 78.46.

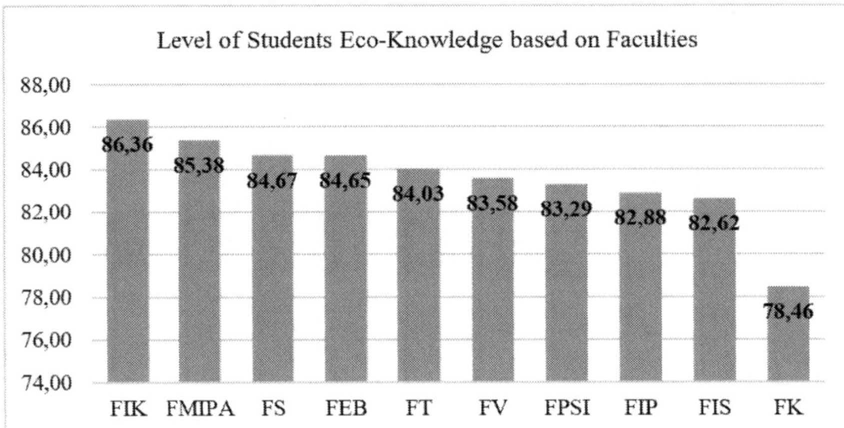

Fig. 4. Percentage of students' eco-knowledge levels based on faculties.

4.3 The effect of video assignment on students' eco-knowledge

The results of this analysis aim to determine the effect of video assignments on the eco-knowledge of first year students at UM. The following are the results of data analysis showing that video assignments have a significant effect on the eco-knowledge of first year students, as explained in Table 10 below.

Table 10. Summary Regression Analysis.

Model	R	R Square	Adjusted R Square	Std. Error of the Estimate	T	Sig.
1	,716	,512	,510	5,591	-	-
Video Assignment	-	-	-	-	16,959	,000

Predictors: (Constant), Video Assignment
Dependent Variable: Eco-Knowledge
Source: Author's statistics

Based on statistical analysis, it shows that the significance value is < 0.05, which means that video assignments influence the eco-knowledge level of new UM students. In addition, the adjusted R Square value is 0.510, which means that video assignments have a 51% effect on increasing the eco-knowledge of new UM students. Therefore, the VA-CGC has a positive impact on enhancing new UM students' understanding of the green campus program.

5 Discussion

Providing first year students with an understanding of the green campus program and sustainable campus environmental conservation practices during the Campus Life Orientation Program for First year Students (PKKMB) is an effort to deepen eco-knowledge

https://doi.org/10.1051/e3sconf/202565206003

from an early stage. Efforts to deepen First year students' eco-knowledge of the green campus program are also a step toward addressing environmental issues on campus. Enhancing first year students' understanding of the green campus program can be achieved through assignments that introduce indicators of green campus implementation. One innovative strategy has been implemented by the State University of Malang (UM) through an assignment requiring first year students to create a VA-CGC.

The VA-CGC is one of the innovative strategies in introducing UM as a campus that promotes the green campus concept. First year students design an individual video project that can enhance understanding of the green campus program. As the first generation of students, they need to be given an understanding and practice related to accelerating sustainable campus environments. This video assignment aims to provide a personal perspective on the importance of environmental conservation efforts and inspire fellow academic community members in campus environmental conservation initiatives. Another objective of this video assignment is to enhance eco-knowledge and ecological awareness among UM first year students and support the Sustainable Development Goals (SDGs) within the campus environment.

(a) (b)

Fig. 5. **(a)** Results of the video assignment "A Day in My Life at Cakra Green Campus"; **(b)** Students took concrete action by disposing of trash in trash bins in their video content.

UM students can choose several topics related to green campus indicators, including waste and garbage, water, energy, transportation, carbon emissions, and education. Video recordings must be done on campus, and students are encouraged to be creative in conveying the information. Videos must be in portrait format with a 9:16 aspect ratio and a duration of

2 minutes. The results of the VA-CGC must be published on Instagram Reels and TikTok. Involving first year students in this video assignment are expected to enhance engagement and understanding of the green campus program. The results of the VA-CGC can be accessed via the following link: https://www.instagram.com/umgreencampus/tagged/.

An innovative strategy in the form of a VA-CGC was implemented as an effort to provide understanding and increase the involvement of First year students in the green campus program. The evaluation process for this activity involved measuring first year students' perceptions of the VA-CGC and their level of eco-knowledge regarding the green campus program. Based on the measurement of students' perceptions of the VA-CGC in Table 5, it shows that students' perceptions fall into the "very good" category. The very good perception level indicates that students' perceptions of the video assignment are highly beneficial for their understanding and engagement in accelerating the green campus program. Efforts to provide environmental understanding are not limited to the video assignment during the PKKMB activity but are implemented continuously.

Fig. 6. The collection of video assignments for "A Day in My Life at Cakra Green Campus" is integrated on UM Green Campus' Instagram.

Students have experienced several positive impacts through the completion of an assignment involving the creation of a VA-CGC. The positive impacts felt by students include gaining an understanding and experience of exploring their campus environment. Providing first year students with the opportunity to explore the campus environment helps them understand both the physical conditions and the learning culture present at UM. Students can also enhance their creativity and collaborate through video filming and editing activities. Additionally, students gain knowledge about the UM Green Campus concept and sustainability issues within the campus environment. The UM Green Campus concept needs to be introduced to students so that each generation of students studying at UM can adapt and apply environmentally friendly behavior within the campus area.

Biological influences shape certain behavior patterns through different socialization processes. In practice, female students tend to show a higher level of environmental awareness than male students. This is influenced by the more prominent affective tendencies and empathetic attitudes of female students, making them more responsive to environmental issues such as cleanliness, nature conservation, and waste management. Male students take a more practical approach and are less involved in activities that require attention to environmental details. Although not absolute, this tendency reflects the influence of gender on the formation of environmental awareness.

The results of the study show that the natural sciences faculty has higher knowledge. This is because there are more female students than male students. This is in line with existing research that women are more concerned about environmental issues and tend to be quicker to understand renewable energy that has been designed. The relationship between attention and behavior appears to depend on the context. Female students have a deeper understanding of the process of linking attitudes and behavior. This can help in developing better models of pro-environmental behavior, designing more effective experiments, and improving policy interventions. The new environment has led to a special focus on information relevant to pro-environmental decisions.

The UM Green Campus concept emphasizes the creation of a sustainable campus environment through various environmental actions involving the entire UM academic community. UM Green Campus measures various innovative efforts made by universities in the fields of green infrastructure, energy conservation, climate change mitigation actions, waste management, water conservation, use of emission-free transportation, and environmental education. First year students can create video content that incorporates several aspects of the Green Campus concept. Therefore, in addition to impacting new UM students' understanding of the Green Campus concept, the VA-CGC also impacts students' eco-knowledge.

The green campus program will provide individualized and flexible educational opportunities for students. This demonstrates the potential to make environmental education more useful and appealing to today's students. Environmental education provides a more in-depth and interactive learning experience. Students will deepen their understanding of environmental issues and inspire them to move towards a more sustainable future. This study can encourage more effective environmental education and contribute to a sustainable future by preventing further isolated global environmental degradation. The use of media and direct experience will maximize environmental awareness among students and provide more personalized and customized environmental education measures and sustainable policies.

The implementation of the VA-CGC is related to efforts to increase eco-knowledge among first year students from an early stage. Table 9 shows that the eco-knowledge level of students has a very good average score. This indicates that through the VA-CGC, first year students can explore the campus environment and learn about the concept of a green campus. Enhancing eco-knowledge among first year students can serve as a foundation for increasing their awareness and involvement in environmental conservation. Several studies have shown that eco-knowledge can influence the improvement of pro-environmental behavior and environmental attitudes. Through the enhancement of eco-knowledge among students, it is hoped that their involvement in the Green Campus concept will increase from the very beginning.

Enhancing eco-knowledge among students is crucial because educated students can make informed decisions regarding sustainable environmental issues. Additionally, students with high eco-knowledge can share best practices with their peers and the academic community.

Previous research has shown that project-based assignments can encourage more active learning and engage students through an inquiry-based approach in changing their perceptions of sustainable environments. In addition to assigning project-based tasks, there are several ways to implement sustainable environmental education in higher education institutions, such as hosting guest lectures and fostering interdisciplinary academic relationships.

5.1 Research limitations

This study is limited to the student population of a single university, so more respondents are needed to avoid bias in the research. The study was conducted within a single university environment, limiting the generalizability of the findings to other institutions with different environmental policies and campus cultures. In addition, the subjects of this study were first year students who did not yet fully understand how to fill out the research questionnaire, so further guidance was needed to ensure the validity of the data obtained. This study focuses on students' experiences and perceptions through video assignments, which may not comprehensively capture all aspects of environmental knowledge development. Since students create video content independently, the quality and depth of the information presented may vary based on creativity, engagement, and prior knowledge. Personal bias and selective presentation of experiences may influence how the Green Campus program is described. Technical difficulties and varying levels of digital literacy among students may affect their ability to express their understanding of the Green Campus program. Despite these limitations, this study provides valuable insights into the potential of student-centered video assignments as a creative medium for fostering environmental knowledge early on and promoting environmental awareness on campus.

6 Conclusion

This study shows that assigning videos based on a student-centered approach is an effective strategy for improving students' early eco-knowledge related to green campus programs. Through active student involvement in designing, recording, and delivering educational messages about campus environmental issues, students' ecological knowledge not only develops conceptually but also reflectively and practically. The analysis results indicate that students' perceptions of the video assignment are positive, suggesting that learning that involves students in technology and creativity can enhance understanding of environmental issues.

Video assignments also have a significant effect on the level of eco-knowledge among first year students. This shows that video assignments can engage students in identifying campus environmental issues in greater depth, understanding sustainability principles, and linking local issues to the global context. Video assignments encourage students to access a wider range of learning resources, work collaboratively, and use digital literacy to creatively and persuasively communicate environmental messages. Therefore, it can be concluded that the implementation of student-centered video assignments is a relevant learning activity for developing meaningful and contextual eco-knowledge in higher education focused on sustainability.

References

1. H. Li, J. Du, and D. Chow, Perceived environmental factors and students' mental wellbeing in outdoor public spaces of university campuses: A systematic scoping

review. Building and Environment. **265(1)**, 112023 (2024). https://doi.org/10.1016/j.buildenv.2024.112023.

2. L. Bertolini, D. Gerritsen, and K. Sol, Supporting students to become agents of change: Introducing and evaluating the Transition Cycle approach to teaching transformative skills. Futures **164(4)**, 1-14 (2024). https://doi.org/10.1016/j.futures.2024.103459.
3. Z. Feng, H. (Cynthia) Hou, and H. Lan, Understanding university students' perceptions of classroom environment: A synergistic approach integrating grounded theory (GT) and analytic hierarchy process (AHP). Journal of Building Engineering. **3(5)**, 1-20 (2024). https://doi.org/10.1016/j.jobe.2024.108446.
4. H. Kashefiyeh, H. Beyti, L. Medghalchi, and F. Haghparast, The impact of environmental knowledge on the creativity of architectural designs: A study on master of "Energy in Architecture" students in Iran. Thinking Skills and Creativity. **5(4)**, 101693 (2024). https://doi.org/10.1016/j.tsc.2024.101693.
5. R. H. Rachmadian, S. Sumarmi, H. Masruroh, S. Utaya, and Y. Suharto, Membentuk Kesadaran dan Keterlibatan Mahasiswa sebagai Aktor Penggunaan Transportasi dan Energi Berkelanjutan di Perguruan Tinggi. Journal of Education Action Research. **8(1)**, 169 (2024). https://doi.org/10.23887/jear.v8i1.76919.
6. P. Dorożyński and K. Dorożyńska, The role-playing game to increase students' activity and engagement in the teaching process – A pilot study of research & development campaign. Currents in Pharmacy Teaching and Learning. **14(8)**, 1046-1052 (2022). https://doi.org/10.1016/j.cptl.2022.07.015.
7. N. Humaida, An overview of key concepts in environmental knowledge: From ecology to sustainable development, OMNICODE Journal (Omnicompetence ... **1(2)**, 90-96 (2022).
8. C. Stuckrath, J. Rosales-Carreón, and E. Worrell, Conceptualisation of Campus Living Labs for the sustainability transition: An integrative literature review. Environmental Development. **5(4)**, 1-15 (2025). https://doi.org/10.1016/j.envdev.2025.101143.
9. H. Wallace and J. VanderMolen, Teaching Health Education Through the Development of Student Centered Video Assignment. Frontiers in Public Health. **7(3)**, 1-7 (2019). https://doi.org/10.3389/fpubh.2019.00312.
10. S. L. Kuhlmann, R. Plumley, Z. Evans, M. L. Bernacki, J. A. Greene, K. A. Hogan, M. Berro, K. Gates, and A. Panter, Students' active cognitive engagement with instructional videos predicts STEM learning. Computers and Education. **2(1)**, 1-20 (2024). https://doi.org/10.1016/j.compedu.2024.105050.
11. A. Khamhaengpol, T. Nokaew, and P. Chuamchaitrakool, Development of STEAM activity "Eco-Friendly Straw" based science learning kit to examine students' basic science process skills. Thinking Skills and Creativity. **5(3)**, 10-16 (2024). https://doi.org/10.1016/j.tsc.2024.101618.
12. M. Haggag, A. Oulefki, A. Amira, F. Kurugollu, E. S. Mushtaha, B. Soudan, K. Hamad, and S. Foufou, Development of STEAM activity "Eco-Friendly Straw" based science learning kit to examine students' basic science process skills. Thinking Skills and Creativity **5(3)**, 10-16 (2024). https://doi.org/10.1016/j.tsc.2024.101618.
13. R. Hilmi Rachmadian and dan Heni Masruroh, Implementasi Strategi Inovatif Perguruan Tinggi dalam Menanamkan Sustainability Awareness pada Sivitas Akademika melalui Program Green Campus. Jurnal Ilmu Lingkungan **23(1)**, 10-22 (2025). https://doi.org/10.14710/jil.23.1.10-22.
14. B. Hu, C. Wang, Z. Xia, C. Lu, and X. Ju, Environmental education in low-income and middle-income countries: A systematic review and meta-analysis. Journal of

Environmental Psychology. **1(5)**, (2025). https://doi.org/10.1016/j.jenvp.2025.102613.

15. A. Fubani, M. Diheim, N. Makhya, and W. Velasufah, Pengetahuan dan perilaku ramah lingkungan mahasiswa, Journal of Character and Environment. **1(2)**, 138-160 (2024). https://doi.org/10.61511/jocae.v1i1.2024.478.

 https://doi.org/10.1051/e3sconf/202565206004

Assessing Village Disaster Resilience Through Gap Analysis: Insights from the Implementation of SNI 8357:2017 in Kemuning Village

Sunarto[1*], *Marcella Paramitha Susilo* Dewi[2], *Reinata Aulia* Fahimah[3], *Zalda Aniqoh* Nursafithri[4], *Sapta* Suhardono[5], and *Malfa Amorita* Iftitani[6]

[1,2,3,4,5,6]Department of Environmental Science, Faculty of Mathematics and Natural Sciences, Universitas Sebelas Maret, Surakarta, Indonesia 57126

Abstract. Disaster risk is a major challenge for rural communities in Indonesia, particularly those located in high-risk areas such as Kemuning Village, which is prone to landslides. This study aimed to assess the village's disaster resilience using gap analysis based on the Indonesian National Standard (SNI) 8357:2017 on Disaster Resilient Villages. A descriptive quantitative approach was employed through structured questionnaires and in-depth interviews with key stakeholders. The assessment focused on five dimensions: basic services, disaster management systems, risk management, emergency preparedness, and recovery preparedness. The results showed that Kemuning Village had strong performance in risk management and community participation, with a zero gap score. However, critical gaps were found in emergency preparedness and post-disaster recovery, with the largest gap score reaching -2.0. These gaps reflected limited early warning systems, inadequate evacuation routes, and insufficient disaster documentation. This study concludes that while the community has demonstrated strong social cohesion and local initiatives, further efforts are needed to improve technical infrastructure, local capacity, and cross-sectoral collaboration. The findings offer practical insights for integrating disaster risk reduction into village development planning and for replicating effective practices in other disaster-prone areas.

*Corresponding author: rm.sunarto@staff.uns.ac.id

1 INTRODUCTION

Indonesia is an archipelagic country located in Southeast Asia and partly in the Oceania region. Geographically, it lies between two continents, Asia and Australia, and between two oceans, the Indian Ocean and the Pacific Ocean. Its astronomical and geological location contributes significantly to the vulnerability to natural disasters across various regions in Indonesia [1]. In addition, Indonesia has a tropical climate with two seasons, which causes extreme changes in weather, temperature, and wind direction. Combined with the country's diverse surface topography and rock formations, this increases the likelihood of hydrometeorological disasters such as floods, landslides, forest fires, and droughts. Given the high potential for natural disasters, Indonesia requires a sustainable and integrated disaster management strategy. This strategy is essential to minimize risks and impacts on people, the environment, and infrastructure.

Disaster management involves prevention, hazard risk reduction, community preparedness, emergency response, recovery through rehabilitation and reconstruction, and sustainable development efforts aimed at reducing disaster risk [2]. Disaster risk reduction (DRR) is a series of efforts to reduce the negative impact of disasters, with the goal of protecting lives, assets, and the environment from both natural and human-made threats [3]. The implementation of DRR cannot be solely the responsibility of the government; it requires active involvement from all elements of society. The community plays a strategic role, as it is the first to experience and respond to disasters in its surroundings. One important approach in DRR is community-based resilience, which emphasizes the ability of communities to independently face, manage, and recover from disasters. This approach directly involves communities in planning, implementation, and evaluation of DRR efforts, thereby enhancing collective capacity and awareness to build more resilient communities.

An active DRR program in Indonesia is the Disaster Resilient Village Program (Destana), which is a national community-based approach to reducing disaster risks at the village level [4]. The National Disaster Management Agency (BNPB) introduced this program to protect all Indonesian citizens, particularly the poor, who are often the most vulnerable to disasters. Representatives from Destana villages are also involved in the Disaster Risk Reduction Forum (FPRB), which aims to ensure disaster-resilient development in vulnerable areas with a strong emphasis on risk reduction [5]. This underlines the importance of local community involvement in enhancing capacity and resilience to various disaster threats in their respective areas. According to [6], Destana is designed to improve community preparedness, response, and recovery through active participation in disaster planning, training, and simulations.

Parts of Karanganyar Regency, particularly Ngargoyoso District, have steep topography and high rainfall, making the area prone to landslides [7]. Around 4,797.25 hectares of the total 6,533.942 hectares of land are categorized as vulnerable. Kemuning Village is classified as a landslide-prone area due to its location in a mountainous region with an elevation of 800 to 1500 meters above sea level, slope gradients of 30 to 40 percent, and annual rainfall of 3000 to 4000 mm [8]. Landslides in this area are triggered by geophysical factors, land use practices, and high rainfall and it lies within a medium to high soil vulnerability zone [9]. To improve community preparedness and resilience, the Destana program plays a crucial role. Currently, Ngargoyoso District has established Destana programs in five villages: Kemuning, Nglegok, Berjo, Ngargoyoso, and Girimulyo.

This study aimed to evaluate the implementation of the Destana Program in Kemuning Village by comprehensively examining the type of evaluation that was needed, the program's achievements up to that point, and by identifying its strengths and

https://doi.org/10.1051/e3sconf/202565206004

weaknesses. In addition, the analysis focused on supporting and hindering factors that influenced the program's effectiveness, both from within the community and from external support such as policies, resources, and inter-agency coordination.

2 MATERIALS AND METHODS

Location

This study was conducted in Kemuning Village, Ngargoyoso District, Karanganyar Regency, Central Java Province. Kemuning Village was selected as the research location because it is a highland area that was vulnerable to disasters, particularly landslides. In addition, the village was also known as a tourist destination with various natural attractions such as tea plantations, the Mount Lawu hiking trail, and waterfalls, which made it potentially susceptible to significant losses in the event of a disaster.

Materials

This study utilized a questionnaire as the primary tool for data collection. The questionnaire was developed based on the Indonesian National Standard (SNI) 8357:2017 concerning Disaster Resilient Villages and Urban Communities. It consisted of 28 indicators representing five main dimensions: basic services, disaster management systems, risk management, emergency preparedness, and recovery preparedness. Each indicator was assessed using a scale from 1 to 5, based on the level of implementation in the village or urban community.

Each indicator was translated into a series of clear and measurable questions adapted to the local context of Kemuning Village. To ensure content validity, the questionnaire was reviewed by experts in disaster management and pre-tested with a small group of respondents before formal distribution. Respondents were asked to provide answers using a Likert-type scale from 1 to 5, where 1 = very low, 2 = low, 3 = moderate, 4 = high, and 5 = very high. This scale was chosen to capture the degree of implementation of each indicator in a way that is both practical for respondents and analytically useful for gap analysis. The questionnaire was also used as a guide during in-depth interviews with the Head of the Kelompok Peduli Kemuning (Kemuning Community Care Group).

Data Analysis

The data analysis technique in this study used a descriptive quantitative approach based on questionnaires and structured interviews. Each indicator was analyzed using the Gap Analysis method to identify the level of disparity between the actual condition and the ideal condition of village disaster resilience indicators based on SNI 8357:2017 [10].

The assessment scale consisted of five levels of achievement: 1 = Very Low, 2 = Low, 3 = Moderate, 4 = High, and 5 = Very High. To measure the gap between actual and ideal conditions, the formula used was Gap = Ideal Score - Actual Score, with interpretations as shown in Table 1.

Table 1. Gap interpretation

Gap Value	Sustainability Status
Gap = 0	The indicator has been ideally met
Gap > 0	There are shortcomings in the implementation
Gap < 0	Exceeds expectations (rare and needs further investigation)

 https://doi.org/10.1051/e3sconf/202565206004

3 RESULT AND DISCUSSION

3.1 Quality and Access to Basic Services

Basic services included various aspects such as education, healthcare, transportation, public services, and community safety. These services needed to be available, accessible, and of adequate quality so that the community could live with dignity and resilience in the face of risks [11]. It showed that most indicators had not yet reached the ideal level. Several indicators, such as educational facilities, healthcare facilities, and environmental management, had purple bars nearly as high as the ideal ones, indicating that services in these areas had been functioning well. This could be explained by the fact that Kemuning Village already had sufficient access to education and healthcare, supported by adequate infrastructure and active involvement of the village government in managing public services in a formal and organized manner.

However, there were indicators with significant discrepancies between actual and ideal scores, such as transportation, cultural activities, and security protection. This was likely due to the low prioritization and limited resource allocation for social and cultural aspects, which were non-structural in nature. This was likely due to the low prioritization and limited resource allocation for social and cultural aspects, which were non-structural in nature, highlighting the need for improvements in disaster management systems, risk management, and basic service access [8].

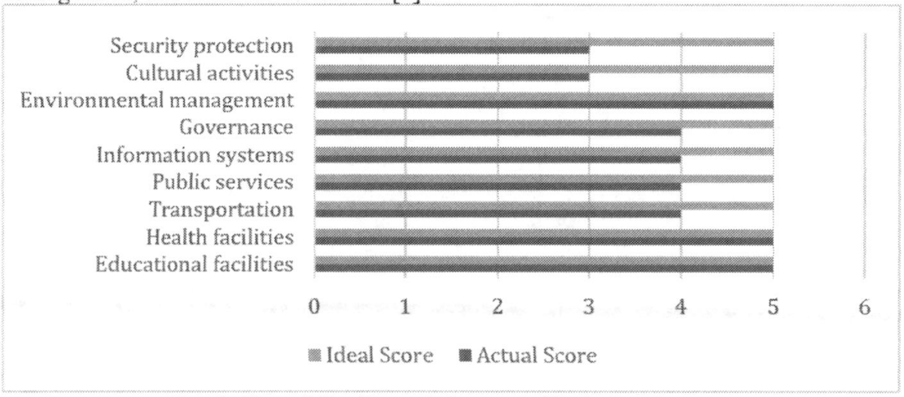

Fig. 1. Gap of Quality and Access to Basic Services

3.2 Disaster Management System Foundations

The foundation of the disaster management system referred to structural and institutional aspects such as regulations, local policies, risk assessments, adaptive development planning, and the integration of climate change issues. This pillar formed the basis for creating a targeted and sustainable system. Law No. 24 of 2007 on Disaster Management emphasized the importance of coordination, integration, partnership, and empowerment in every stage of disaster management. These principles aimed to ensure that disaster handling was carried out using a holistic and inclusive approach. The role of agencies such as BPBD and support from institutions like BMKG was also crucial [12].

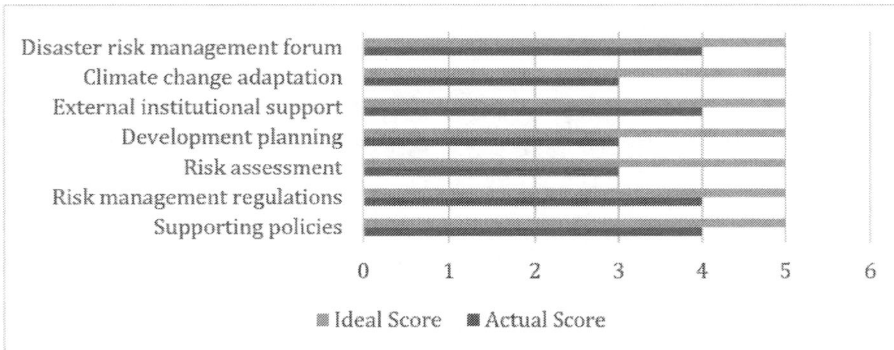

Fig. 2. Gap of Disaster Management System Foundations

Figure 2 showed a significant disparity between ideal and actual scores in most of the indicators evaluated in Kemuning Village. Overall, the disaster management system was available and functioning at a minimum level, but had not yet reached the optimal standard expected under the village resilience framework. The gaps indicated the need for reinforcement in both strategic and technical aspects. Supporting policies and risk management regulations received relatively high actual scores, reflecting the presence of institutional structures and basic norms supporting disaster management activities. This was evident from the establishment of the Kemuning Community Care Group (KPK), which had been active since 2020 and continued to carry out various disaster mitigation activities.

In addition, collaboration between the village government and external institutions such as the Karanganyar Regency BPBD and PT Telkom for the installation of an Early Warning System (EWS) further strengthened the institutional role in the disaster management system. However, lower actual scores on indicators such as risk assessment, development planning, and climate change adaptation alignment showed that the existing system was still sectoral and not fully integrated. Monitoring systems like the EWS, although available, were not yet functioning optimally in detecting potential disasters or disseminating real-time information to the public. Similarly, not all village development planning documents systematically accommodated disaster-related aspects, which weakened the integration between development and risk mitigation. Climate adaptation programs, such as reforestation and river cleanup, remained sporadic and lacked structured long-term planning. Another indicator, namely the optimization of disaster risk management forums, showed limitations in cross-sectoral participation and engagement. Although the KPK had played an active role, it had not fully involved vulnerable groups, youth, or local businesses in the planning and implementation of mitigation activities.

3.3 Disaster Risk Management

Disaster risk management referred to a series of efforts to prevent and reduce the impact of disasters through education, training, and interregional cooperation. To reduce the risks of a disaster, increasing understanding through knowledge was crucial. Awareness of disaster preparedness could be promoted through outreach aimed at educating the public. This aligned with BNPB's initiatives, which emphasized that raising disaster awareness through outreach was vital in reducing disaster impacts[13]. Research by [14] in Palu showed that disaster management education significantly improved community preparedness. This indicated that active community involvement in local

https://doi.org/10.1051/e3sconf/202565206004

disaster preparedness groups and pre-disaster readiness stages could substantially increase community participation.

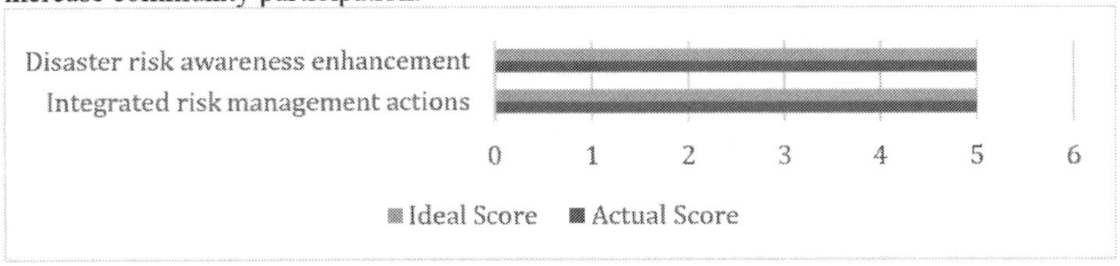

Fig. 3. Gap of Disaster Risk Management

Figure 3 showed excellent performance on the two indicators analyzed: interregional coordination and community capacity building in disaster risk reduction. The actual scores for both indicators reached the maximum value (5), equal to the ideal score, indicating no gap. This achievement reflected that the disaster risk management strategy in Kemuning Village had been implemented comprehensively and demonstrated a high level of success. Interregional coordination was evident in the collaborative relationships between KPK as a local actor, the village government, and the regency-level BPBD. This cooperation enabled proportional distribution of roles between formal and informal institutions during mitigation activities and disaster response.

Additionally, the availability of standby vehicles to respond to disasters such as landslides and preparedness for tourist-related accidents in the Kemuning area showed that the village had built a planned and anticipatory response capacity. Community capacity building was also carried out intensively through regular disaster training, education, and outreach. Mitigation information was disseminated through various media, including a repeater radio communication system used to expand information coverage during emergencies. The EWS also functioned as a risk awareness tool, although its use had not yet been fully optimized. The direct involvement of residents in various mitigation activities indicated that risk management had moved beyond institutional responsibility and become part of the community's collective culture. With no gap in the scores shown in this graph, it could be concluded that Kemuning Village had successfully implemented inclusive, community-based disaster risk management.

3.4 Emergency Preparedness

Preparedness referred to a variety of efforts organized to face the possibility of disasters. Since disaster situations were difficult to predict, emergency response actions needed to be implemented quickly once a disaster occurred. In the context of disaster management, preparedness not only included anticipatory measures such as the preparation of evacuation plans and the provision of supporting infrastructure but also involved education, training, and active community participation in disaster simulations. According to the National Disaster Management Agency (BNPB), preparedness was a series of activities carried out to anticipate disasters through proper and effective organization and action.

https://doi.org/10.1051/e3sconf/202565206004

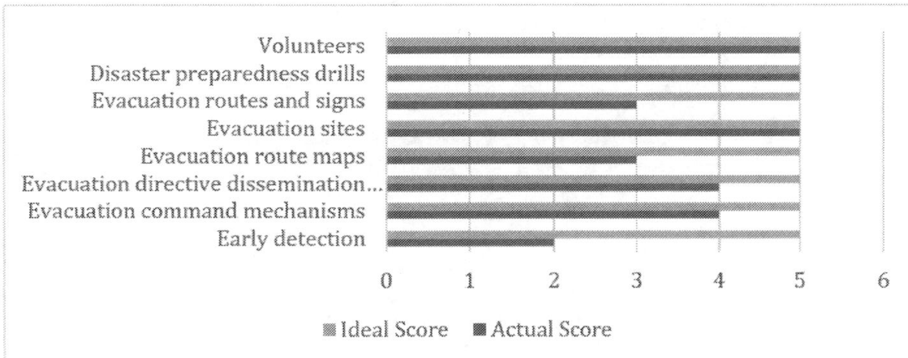

Fig. 4. Gap of Emergency Preparedness

Figure 4 showed that the level of community preparedness in Kemuning Village in facing disasters had not yet reached an optimal level. Eight main indicators were measured, ranging from detection and early warning mechanisms to evacuation systems and the presence of volunteers, all of which revealed significant gaps between actual and ideal conditions. This indicated that the infrastructure and social capacity of the community to deal with disasters were not yet fully ready. The availability of early detection and early warning mechanisms was a crucial initial step in risk reduction. Technology played an important role in strengthening community preparedness. Innovations such as early warning systems, evacuation routes, and disaster response facilities and infrastructure could help communities prepare for emergencies. However, in reality, these mechanisms were still largely unavailable in many villages, including Kemuning Village.

One of the indicators with the lowest achievement was the availability of early detection mechanisms, which showed that warning systems or devices such as ground motion or crack sensors had not been adequately installed in vulnerable areas. This posed a major challenge, as without an early detection system, communities would be too late in recognizing landslide threats, especially during the rainy season. Additionally, the mechanisms for receiving early warnings and disseminating evacuation instructions had not yet functioned optimally. Many residents had not received timely and accurate information during emergencies. Nevertheless, community groups such as KPK (Kemuning Community Care Group), which had been equipped with disaster knowledge and access to relevant information, were able to respond more quickly and effectively to disaster threats [15].

KPK also played an active role in building awareness and enhancing community capacity through outreach, volunteer training, and the development of community-based emergency response plans. KPK encouraged the village government to become more proactive in providing mitigation infrastructure. The availability of disaster response volunteers indicated that the community had begun to get involved, although further technical training and equipment support were still needed.

3.5 Recovery Preparedness

Recovery preparedness was one of the crucial aspects of disaster risk management, focusing on a region's ability to carry out post-disaster recovery processes quickly, in a structured, and sustainable manner. The main goals of recovery preparedness were to minimize long-term impacts, accelerate the recovery of community life, and ensure the continuity of infrastructure and essential services. In this study, recovery preparedness

https://doi.org/10.1051/e3sconf/202565206004

was assessed using two main indicators: the availability of early recovery mechanisms at the village/urban ward level, and the existence of sustainable recovery planning for various strategic assets at high risk of disaster damage.

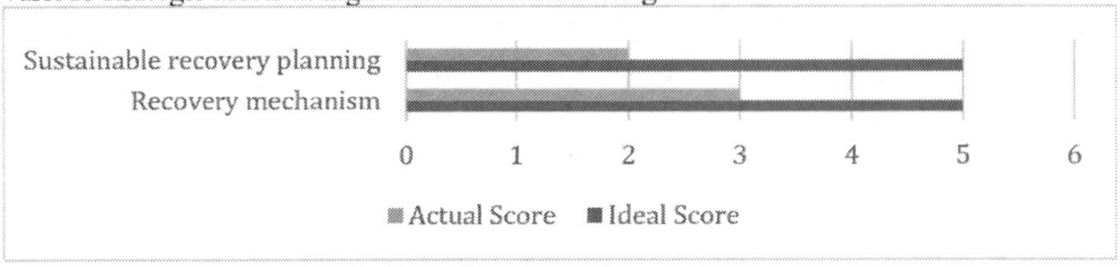

Fig. 5. Gap of Recovery Preparedness

Figure 5 indicated that for the first indicator, which involved the availability of various early recovery mechanisms at the village/urban ward level, the ideal score was set at 90%, while the actual score achieved was only 60%. Thus, there was a gap of -30% (60% – 90%), suggesting that early recovery efforts at the local level had not yet been optimized. This situation reflected the need to enhance the capacity and readiness of village officials in designing and implementing post-disaster recovery mechanisms.

Meanwhile, for the second indicator, which assessed sustainable recovery planning for strategic assets and properties, the ideal score was also set at 90%, while the actual score stood at 70%. The resulting -20% gap indicated that although planning efforts had been made, they still fell short of the expected standards. This highlighted the need for updating and strengthening recovery plans, especially for vital assets that played a key role in maintaining public services.

3.6 Gap Analysis of Community Resilience Indicators in Kemuning Village

The Disaster Resilient Village Program (Destana) in Kemuning Village had several key strengths, one of which was the high level of community awareness regarding disaster risks. This was reflected in the Disaster Risk Management indicator, which showed a GAP score of 0, indicating that the actual condition aligned with the ideal standard. The community actively participated in activities such as communal work (gotong royong), emergency response training, and disaster forums, which demonstrated that disaster education had been well-received and internalized. The active involvement of residents in disaster mitigation efforts also strengthened social cohesion and village preparedness.

In addition, Kemuning Village had established a volunteer structure and a Disaster Risk Reduction (DRR) Forum, which supported local institutional disaster management efforts. Although there was still a GAP of 1 for this indicator, the presence of active volunteers showed a collective awareness in building an organized system. The commitment of village officials was also a significant strength, as seen in the responsiveness of governance to disaster issues, the allocation of village funds for Destana activities, and active involvement in disaster-related training and planning—even though the disaster management system was still in the process of being strengthened.

https://doi.org/10.1051/e3sconf/202565206004

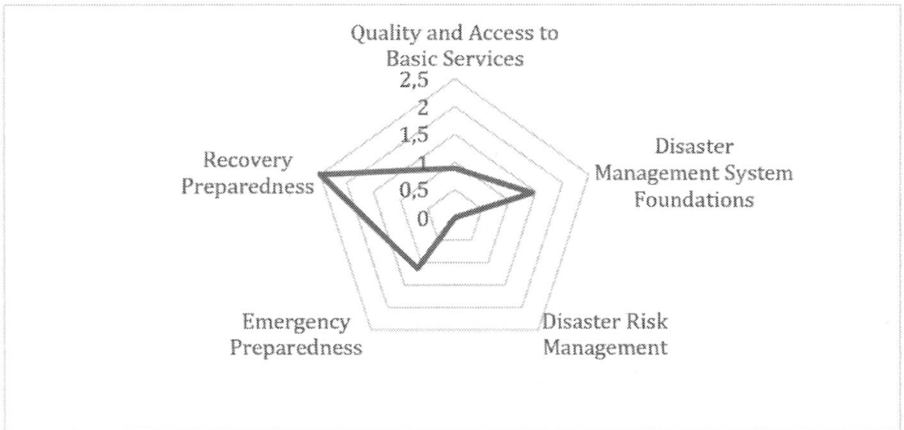

Fig. 6. Gap Analysis Diagram

One of the main weaknesses in the implementation of the Destana Program in Kemuning Village was the suboptimal documentation and evaluation system. A GAP score of 1 in the Basic Disaster Management System category reflected the lack of a fully integrated reporting and monitoring process. The limited documentation and absence of routine evaluations made it difficult to assess the program's effectiveness and to precisely identify its successes and shortcomings, which could hinder continuous improvement. Furthermore, emergency preparedness in Kemuning Village still faced major challenges, with the highest GAP score of 2.0. This indicated a high dependency on external assistance due to limited internal capacity, both in terms of human resources, equipment, infrastructure, and funding. Evacuation routes, gathering points, and early warning systems were still inadequate, and evacuation drills were rarely conducted. This situation posed serious risks during a disaster, as it hampered quick response and effective self-evacuation.

Based on the evaluation of the Destana Program implementation in Kemuning Village, several strategic recommendations were proposed to strengthen community resilience in facing disaster risks. First, it was necessary to strengthen the emergency preparedness system by improving the availability and functionality of early warning systems in landslide-prone areas, as well as arranging evacuation routes equipped with clear signage and good accessibility. Disaster evacuation simulations also needed to be carried out regularly to enhance community responsiveness. Second, the internal documentation and evaluation system had to be optimized by establishing a disaster activity reporting team and implementing periodic evaluations, so that the program's progress could be monitored and developed systematically.

Moreover, the capabilities of volunteers and village institutions must be augmented through technical training and increased community engagement, encompassing vulnerable groups, in disaster risk reduction forums. The village was also expected to reduce reliance on external aid by developing emergency logistics self-sufficiency and establishing village-owned enterprises, with part of the revenue allocated to mitigation activities. The integration of disaster risk management into village development planning documents, such as the Village Mid-Term Development Plan (RPJMDes) and the Village Government Work Plan (RKPDes), was also a crucial step to ensure program sustainability. In addition, expanding disaster education and literacy through local media and fostering cross-sector partnerships (including government, private sector, academia, communities, and media) was strongly recommended to enhance collaboration and broaden the reach of mitigation information.

4 CONCLUSION

The Disaster Resilient Village Program (Destana) in Kemuning Village had shown significant progress in building community resilience to disasters, particularly in the areas of disaster risk management and community participation. The main strengths of the program lay in the high level of public awareness, the establishment of a volunteer structure and risk reduction forum, as well as the commitment of village authorities. However, the evaluation based on the Indonesian National Standard (SNI 8357:2017) also revealed several weaknesses, such as dependence on external assistance, suboptimal activity documentation, and limited preparedness infrastructure, including evacuation routes and early warning systems. The ideal targets had not yet been fully achieved, particularly in terms of emergency preparedness and post-disaster recovery. Therefore, it was necessary to strengthen internal systems, integrate risk planning into village development, enhance the technical capacity of volunteers, and foster cross-sector collaboration so that Kemuning Village could become a sustainable and self-reliant model of disaster resilience.

ACKNOWLEDGMENTS

The authors gratefully acknowledge the Institute for Research and Community Service of Universitas Sebelas Maret for the research grant support provided through grant number 371/UN27.22/PT.01.03/2025.

REFERENCES

1. E. Normelani, A. Putri, M. Efendi, and W. P. Danarto, "Letak Strategi dan Pengembangan Potensi Sumber Daya Alam di Kalimantan Selatan," *J. Geogr. (Geografi Lingkung. Lahan Basah)*, vol. 3, no. 2, p. 67, 2022, doi: 10.20527/jgp.v3i2.6778.
2. I. Trisnawati, "Partisipasi Masyarakat Dalam Penanggulangan Bencana Untuk Mewujudkan Kampung Siaga Bencana (KSB) di Desa Sindangjaya," *J. Geogr. Educ. Univ. Siliwangi*, vol. 4, no. 1, pp. 44–51, 2023, [Online]. Available: http://jurnal.unsil.ac.id/index.php/geoducation
3. J. P. Pratama, L. P. Dewo, and H. K. Rahmat, "Model Sinergitas Pentahelix dalam Rangka Pengurangan Risiko Bencana di Indonesia: Sebuah Tinjauan Pustaka," *J. Curr. Res. Disaster Response Emerg. Manag. e-ISSN xxxx-xxx*, vol. 1, no. 1, pp. 1–6, 2024.
4. F. F. Salman, K. Kusnandar, and S. Sugihardjo, "Peningkatan Kapasitas Kegempaan Masyarakat Melalui Program Desa Tangguh Bencana," *JMM (Jurnal Masy. Mandiri)*, vol. 7, no. 5, p. 4322, 2023, doi: 10.31764/jmm.v7i5.16955.
5. R. F. Muzib, B. L. Handayani, and J. Mulyono, "Solidaritas Pokja Masyarakat Hadapi Potensi Tsunami Melalui Wadah Destana," *J. Urban Sociol.*, vol. 4, no. 2, p. 103, 2021, doi: 10.30742/jus.v4i2.1774.
6. Rahmi Fadiah Nasution, Evi Bunga Lestari, and Usiono Usiono, "Peran Pendidikan Kesiapsiagaan Bencana dalam Meningkatkan Kesadaran pada Remaja," *J. Bintang Pendidik. Indones.*, vol. 3, no. 1, pp. 119–128, 2024, doi: 10.55606/jubpi.v3i1.3491.
7. A. M. Prameswari, "Analisis Kerawanan Longsor Berbasis Sistem Informasi Geografis Di Kecamatan Ngargoyoso Kabupaten Karanganyar," 2021, [Online]. Available: http://eprints.ums.ac.id/id/eprint/94700%0Ahttp://eprints.ums.ac.id/94700/1/Naskah

Publikasi.pdf
8. P. Wijayanti *et al.*, "Strategy for Mitigating Landslide Disaster Risk through Improving the Status of Disaster Resilient Villages (DESTANA) Case Study in Kemuning Village, Ngargoyoso District, Karanganyar Regency," *Soc. Humanit. Educ. Stud. Conf. Ser.*, vol. 3, no. 1, pp. 307–313, 2020, doi: 10.20961/shes.v3i1.45075.
9. I. Ammelia, D. S. Lestari, G. T. Al Ghazy, and Y. A. Wibowo, "Integrasi Materi Kebencanaan Pada Jenjang Sekolah Dasar Di Kecamatan Ngargoyoso, Jawa Tengah," *Indones. J. Environ. Disaster*, vol. 1, no. 1, pp. 60–72, 2022, doi: 10.20961/ijed.v1i1.66.
10. P. E. Phyu, N. Effendi, and P. Purnawan, "Assessment of Residential Satisfaction using Gap Analysis: A Case Study in Naypyitaw City, Myanmar," *J. Tek. Sipil*, vol. 20, no. 1, pp. 176–190, 2024, doi: 10.28932/jts.v20i1.7428.
11. L. Natika, "Transformasi Pelayanan Publik Di Era Digital: Menuju Pelayanan Masa Depan Yang Lebih Baik," *World Public Adm. J.*, vol. 6, no. 1, pp. 1–11, 2024, doi: 10.37950/wpaj.v6i1.2040.
12. M. F. Adnan, "Wafda 2023," vol. 7, pp. 18583–18592, 2023.
13. M. R. Pahleviannur, "Edukasi Sadar Bencana Melalui Sosialisasi Kebencanaan Sebagai Upaya Peningkatan Pengetahuan Siswa Terhadap Mitigasi Bencana," *J. Pendidik. Ilmu Sos.*, vol. 29, no. 1, pp. 49–55, 2019, doi: 10.23917/jpis.v29i1.8203.
14. D. Mangemba and S. A. Mangundap, "Pengaruh Edukasi Manajemen Bencana Terhadap Kesiapsiagaan Masyarakat Pesisir Pantai Menghadapi Bencana," *J. Ber. Kesehat.*, vol. 17, no. 1, pp. 142–146, 2024.
15. M. Meili and Z. Sembiring, "Analisis Kesiapsiagaan Masyarakat dalam Menanggulangi Bencana Alam di Indonesia (Studi pada Bencana Alam di Sumatera Utara)," *Al-DYAS*, vol. 4, no. 2, pp. 1006–1019, 2025, doi: 10.58578/aldyas.v4i2.5593.

Changes in Lifestyle and Economic Behavior of Communities After the Semeru Volcanic Eruption: Promoting Economic Resilience and Environmental Sustainability in Permanent Settlements

Tomy Rizky Izzalqurny[1*], *Anita* Sulistyorini[2], *Muhammad Andry Prio* Utomo[3], *Febi* Dwirahmadi[4], *Rifaldy Adinandra* Ferdiansyah[5], *Resti Novita* Sari[6], *Zahra* Klarisa[7] and *Elman* Prastyo[8]

[1,8]Department of Accounting, Faculty of Economics and Business, Universitas Negeri Malang, Indonesia
[2,6,7]Department of Public Health Science, Faculty of Sports Science, Universitas Negeri Malang, Indonesia
[3]Department of Biology, Faculty of Mathematics and Science, Universitas Negeri Malang, Indonesia
[4]Global Health at School of Medicine, Griffith University, Queensland, Australia
[5]Department of Development Economics, Faculty of Economics and Business, Universitas Negeri Malang, Indonesia

Abstract. This study aims to explore the changes in lifestyle and economic behavior of communities affected by the Semeru volcanic eruption, following their relocation from disaster-prone areas to Permanent Settlements (Huntap). The primary focus of the research is to examine how financial literacy, retirement planning, financial management, risk tolerance, and social support contribute to fostering behavioral changes that promote economic resilience and environmental sustainability. Using a qualitative approach, the study involves in-depth interviews with the Lumajang and East Java Regional Disaster Management Agencies (BPBD), the head of the Huntap neighborhood, local community leaders, and the distribution of questionnaires to gather descriptive statistical data on economic and environmental behavioral changes. The findings indicate that the transition to permanent settlements has led to significant changes in financial management practices and lifestyle. Improved financial literacy, a better understanding of retirement planning, and enhanced social support have encouraged the adoption of more sustainable economic practices. Additionally, there is an observable shift in resource management and the application of sustainability principles in the new settlement environment. The study recommends strengthening financial literacy programs based on sustainability and developing social support models that promote environmentally friendly lifestyle changes in Huntap areas.

* Corresponding author: tomyrizky.Izzalqurny.fe@um.ac.id

1 Introduction

Natural disasters are phenomena that often occur without warning, causing significant disruptions to the social and economic lives of affected communities. Indonesia, located in the Pacific Ring of Fire, regularly faces natural disaster threats, including volcanic eruptions, earthquakes, and tsunamis [1]. One major disaster was the eruption of Mount Semeru in December 2021, which displaced over 10,000 people and severely damaged or destroyed approximately 2,000 homes. The eruption had significant human rights implications, disrupting access to education, healthcare, and clean water, underscoring the need for enhanced legal protection and social responsibility. Post-disaster recovery often faces considerable challenges, particularly in restoring psychological stability, where social support plays a critical role in fostering resilience and post-traumatic growth. Additionally, the extensive damage to educational infrastructure necessitates swift reconstruction efforts to ensure children's access to quality education [2]. The broad economic impact of such disasters demands adaptive social protection policies, including investments in early warning systems to mitigate future risks [3]. The Mount Semeru eruption and other disasters also highlight the importance of collaboration between governments and communities to enhance disaster mitigation and recovery efforts.

In the context of recovery efforts led by the government, the construction of permanent housing (Hunian Tetap or Huntap) is a crucial step in providing adequate shelter for disaster victims. However, post-disaster recovery extends beyond physical rebuilding to encompass broader social and economic dimensions [4]. A significant post-disaster change involves shifts in economic, consumption, and production patterns, particularly among those previously reliant on agriculture and livestock sectors [5]. Many people lose their livelihoods due to disasters, necessitating adaptation to new models of living, such as micro-enterprises, creative industries, or other occupations aligned with their skills [6]. This process demands innovation and adaptability to enhance long-term economic resilience [7]. Furthermore, social support, such as community participation, has proven to play a vital role in economic and social recovery after disasters [8]

Economic recovery after disasters requires an integrated, community-based approach [9] Such an approach not only focuses on physical reconstruction but also strengthens the community's economic capacity through skill training, local economic empowerment, and microfinance initiatives. These programs are critical for diversifying income sources and improving economic resilience amidst uncertainties caused by disasters [9]. Community-based socio-economic strengthening initiatives have been successful in several areas, such as Gili Trawangan, where social capital played a key role in recovery.

In addition to addressing financial challenges, rapid economic recovery must ensure that the process does not compromise environmental sustainability. Environmental sustainability principles must be integral to the construction of permanent housing, including considerations such as energy efficiency and proper waste management [10]. The United Nations Environment Programme emphasizes that post-disaster recovery must integrate sustainability principles by prioritizing the use of environmentally friendly building materials, efficient waste management, and renewable energy adoption.

Neglecting environmental considerations in recovery efforts risks exacerbating ecosystem damage, as seen in some cases in Aceh, where reconstruction in disaster-prone areas created new vulnerabilities for impoverished communities [11]. Conversely, rehabilitation programs that incorporate sustainability concepts, such as mangrove planting as a natural barrier against sea waves, can enhance community resilience while preserving the environment [12]. Thus, recovery strategies that prioritize environmental sustainability not only protect ecosystems but also support the long-term livelihoods of affected communities [8].

This study explores the extent to which environmental sustainability can be integrated into the construction of permanent housing in areas affected by the Mount Semeru disaster and how impacted communities can build economic resilience by adapting to lifestyle changes and consumption patterns. Sustainable housing development requires an integrative approach, such as the selection of environmentally friendly building materials and efficient waste management, which have proven effective in reducing environmental damage after disasters (10). Moreover, supporting local economic diversification, such as through the development of community-based tourism, can strengthen community resilience and promote sustainable economic recovery (14). In this context, it is essential to consider sustainability indicators encompassing social, economic, and ecological factors to create a holistic recovery approach.

This research aims to explore two key dimensions in the post-disaster recovery context of Mount Semeru. The first dimension is lifestyle changes, including consumption patterns, resource management, and adaptation to new ways of living (15). The second dimension is the community's economic behavior, including household expenditure management, new income generation strategies, and leveraging emerging economic opportunities post-disaster (14). Community-based approaches emphasizing local economic empowerment have been proven to accelerate sustainable social and economic recovery (16). This study provides policy recommendations for governments and other institutions to enhance the effectiveness of post-disaster recovery efforts while promoting inclusive and sustainable recovery. Strategies such as revitalizing microfinance systems, community-based development, and strengthening community adaptation capacities have demonstrated positive outcomes in the context of long-term recovery in Indonesia [8].

2 Literature Review

2.1 Social and Economic Impacts of Disasters

Natural disasters often cause extensive damage, both physical and social, affecting all aspects of community life. Research indicates that natural disasters, such as the eruption of Mount Semeru, have significant impacts on the social and economic structures of affected communities, including the loss of primary livelihoods and increased dependence on external assistance . The importance of developing flexible and responsive social protection policies to support disaster-affected communities, particularly in the context of inclusive economic recovery [3]. Post-disaster social and economic recovery frequently requires adaptation to new ways of life, highlighting how disasters compel communities to restructure consumption and production patterns, particularly in agriculture and livestock sectors. Similarly, community-based economic rehabilitation programs have been identified as key to rebuilding local economic resilience, especially for individuals who have lost their jobs due to disasters [13].

Beyond economic impacts, the psychosocial effects of disasters are often broader and can influence long-term socio-economic recovery. Psychological trauma from disasters can slow recovery processes, making social support from both governments and communities a crucial element in rehabilitation efforts. Support enhances communities' capacity to recover and face future challenges[7].

2.2 Community-Based Approaches to Economic Recovery

Community-based approaches have proven to be effective methods for post-disaster recovery. Strengthening social capital and empowering local economies play crucial roles in restoring community resilience after disasters [14]. Programs involving communities in decision-making and local resource management can accelerate economic and social recovery. Local economic empowerment, such as skill training and access to microfinance, is essential to enhance communities' capacities to create sustainable economic opportunities[9] . Approaches centred on community economic empowerment and locally relevant skill training can bolster long-term economic resilience[7]. Liu-Lastres et al. (2020) found that micro-enterprises and tourism-based economies are highly effective in diversifying income sources for affected communities, thereby accelerating socio-economic recovery [15]. Studies in disaster-affected areas show that well-organized communities are better equipped to tackle economic challenges posed by disasters.

2.3 Environmental Sustainability in Post-Disaster Recovery

Environmental sustainability plays a critical role in post-disaster recovery strategies. The construction of permanent housing (Hunian Tetap or Huntap) after disasters must integrate environmental sustainability principles to avoid further damage to ecosystems already threatened by disasters [10]. For instance, using environmentally friendly building materials and renewable energy technologies in Huntap construction can mitigate negative environmental impacts while accelerating the economic recovery of communities.
post-disaster environmental rehabilitation, such as mangrove planting to prevent coastal erosion and serve as natural tsunami barriers, can enhance community resilience to future natural disasters [12]. This research suggests that sustainable natural resource management can be a vital component of holistic and integrated recovery strategies. Neglecting environmental sustainability in post-disaster recovery can lead to further ecosystem damage, ultimately exacerbating vulnerability to future disaster, it is essential to ensure that recovery efforts focus not only on physical reconstruction but also on measures to protect the environment, supporting the long-term resilience of affected communities.

2.4 Economic Diversification and Social Resilience

Economic diversification is a key element in strengthening post-disaster social resilience. To address this, economic sectors must be diversified through initiatives such as skill training, enhancement of micro-enterprises, and the development of community-based tourism sectors, all of which contribute to strengthening economic and social resilience. Economic diversification based on creative industries and independent small businesses can increase communities' resilience to disasters [8]. Programs supporting local economic sectors, such as handicraft production or technology-based businesses, help communities develop economic capacities that are more resistant to disaster-related shocks.
Furthermore, Lassa, Petal, and Surjan (2023) argue that social resilience must also be a priority in the recovery process [2]. Active community participation in managing recovery efforts can strengthen social bonds and expedite economic recovery. In studies of disaster recovery in Aceh, Partelow (2021) found that communities that effectively collaborated in recovery management were able to mitigate the negative impacts of disasters and accelerate social and economic recovery processes [14].

2.5 Evaluation of Sustainability and Economic Resilience

The evaluation of sustainability in the context of post-disaster recovery must encompass three primary dimensions: social, economic, and ecological [13]. The social dimension involves the role of communities in managing lifestyle changes and consumption patterns following a disaster. The economic dimension focuses on strengthening local economic capacities and

https://doi.org/10.1051/e3sconf/202565206005

diversifying income sources to reduce dependency on sectors vulnerable to disasters. The ecological dimension emphasizes the application of sustainability principles in physical reconstruction and efficient environmental management.

3 Methodology

This study examines lifestyle changes and economic behavior among communities following the eruption of Mount Semeru, with a focus on economic resilience and environmental sustainability in permanent housing settlements (Huntap). The approach adopted is qualitative, employing a case study method on Mount Semeru. This site was chosen due to the significant impact of the 2021 eruption, which was declared a national disaster and continues to affect the community as of 2024. The severity of the disaster resulted in the construction of Indonesia's largest Huntap, consisting of 1,951 houses built within a short period. The establishment of this new settlement has prompted significant lifestyle changes among its residents.

The data used in this study include both primary and secondary sources. Primary data were obtained through in-depth interviews with residents of Huntap to gather information on lifestyle changes, economic behavior, and the challenges faced after the disaster. Key informants for this research include community leaders (RW heads) and prominent local figures. Direct observations were also conducted within Huntap to monitor daily activities such as consumption patterns, resource management, and economic practices. Secondary data were collected from official documents issued by the Regional Disaster Management Agency (BPBD), reports and policies from local governments, and statistical data from the Central Statistics Agency (BPS). Internet-based references were used to complement and expand the scope of information, including current trends, public opinions, and updates relevant to post-disaster lifestyle and economic behaviors. Literature on disaster mitigation, economic adaptation, and environmental sustainability was also reviewed to provide theoretical context for the research.

The data analysis process follows the Miles and Huberman (1984) model, modified to fit the needs of the study. The process begins with data reduction, wherein relevant information from interview transcripts, observational notes, and supporting documents is selected. The filtered data are then presented in the form of narratives and visualizations, such as diagrams or tables, to facilitate interpretation. Finally, conclusions are drawn based on data triangulation to validate findings, enhance credibility, and minimize potential bias in interpretation.

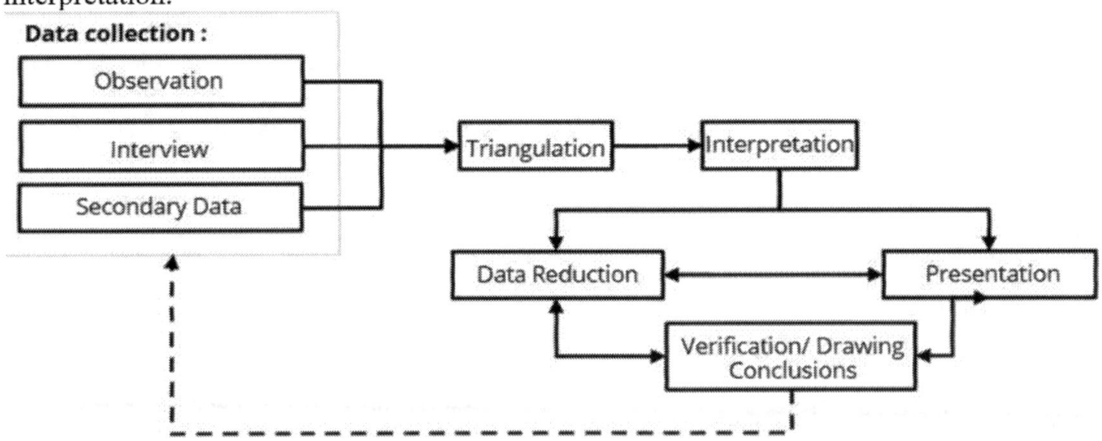

Fig. 1. Flowchart of Data Analysis Model (Analysis Model Source: Miles & Huberman (1984) with Modifications).

This study employs data triangulation to ensure the validity and reliability of the data by integrating findings from interviews, observations, and document analysis, thereby providing a more comprehensive perspective. The triangulation process involves cross-verification with multiple informants. Through this approach, the research aims to holistically uncover the socio-economic changes experienced by communities post-disaster and offer relevant recommendations for public policy and sustainable disaster management.

4 Discussion

This study highlights the socio-economic dynamics of communities affected by the eruption of Mount Semeru and relocated to permanent housing (Huntap) in Sumbermujur Village. Through field observations, in-depth interviews, and secondary data analysis sourced from BPBD, BPS, and various government and NGO documents, it was found that the recovery process occurs multidimensionally with the involvement of multiple stakeholders. Three key themes underpinning this adaptive process include economic transformation, strengthening social capital, and environmental sustainability, which are integrally linked in supporting sustainable recovery.

Fig. 2. Observation and Interviews with BPBD and Permanent Housing Community.

4.1 Changes in Lifestyle and Adaptation to a New Way of Life

Before the disaster, the majority of the residents in the Permanent Housing (Huntap) of Sumbermujur Village relied on agriculture and sand mining as their primary sources of livelihood. However, after their land was buried due to the eruption of Mount Semeru, the community was forced to shift to microenterprises and new forms of employment in order to survive. Many residents started small businesses, such as handicrafts, processed food products, and some even sought jobs outside the village, hoping to find better opportunities in other areas.

According to BPBD Lumajang (2023), 65% of households in the Huntap are now involved in small enterprises based on local handicrafts, such as bamboo weaving, crafts made from waste materials, and food products. The government and NGOs have conducted various skill training programs, which have been attended by more than 75% of the heads of households in the Huntap. One head of household shared:

"Before, we were farmers and miners. Now, many of us have switched to making handicrafts or working outside the village. The training we've received has been quite helpful in starting new businesses, although the income is unpredictable."

However, even though many residents have shifted to small businesses, a significant number of them feel that their income has actually decreased compared to before. One resident, who used to earn IDR 3 million per month from farming and mining, now struggles to reach IDR 2 million per month.

"Before, our income was stable, around IDR 3 million per month, but now it's often unpredictable. Sometimes it's less than IDR 2 million. We find it harder to survive now."

This unstable income is partly due to market fluctuations in the craft products and the dependency on the informal sector, which often does not provide a steady income. Most of the community members are now reliant on handicrafts and local products, which, despite gaining access to markets through social media and online platforms, do not always guarantee stable sales.

Table 1. Comparison of Household Income Before and After Relocation to Huntap.

Aspect	Before the Disaster	Post-Disaster
Household Income	IDR 3,000,000/month	> IDR 2,000,000/month (unstable)
Type of Work	Sand Mining Worker & Farmer	Small Business (Grocery Store), external jobs, continued work at former job

In addition, at the time of relocation, electricity was provided free of charge by the government to help alleviate the economic burden of the residents. However, after a few months, the electricity subsidy was terminated, which further increased the financial strain on the community. One resident shared:

"At the beginning, electricity was free, but now we have to pay for it ourselves."

The recovery process not only depends on economic aspects but also on strengthening social capital that connects residents through various collective activities. Gotong royong (mutual cooperation) has become a central form of social capital that strengthens solidarity among the community, particularly in the construction of public facilities such as mosques, village halls, and road infrastructure. According to an interview with the Sumbermujur Village government, gotong royong plays a significant role in accelerating physical development and rebuilding trust among the residents.

The Head of Sumbermujur Village stated:

"Mutual cooperation here is not just for building physical structures like the mosque and roads, but also for restoring the trust between residents that was shaken after the disaster."

Cooperatives have become an essential tool for strengthening social capital. 60% of households in the Huntap area are registered as cooperative members. The cooperative not only provides access to microfinance but also serves as a space for discussions to create new business opportunities. One cooperative member shared:

"Through the cooperative, we can purchase raw materials at a lower price, which helps us produce crafts more efficiently."

4.2 Behavior of the Community's Change in Huntap Based on Environmental Sustainability

Environmental sustainability plays a crucial role in rebuilding the ecosystems affected by the disaster. The government-facilitated reforestation program in the Huntap area has successfully planted over 1,000 trees to prevent soil erosion and enhance environmental

quality. This initiative not only benefits the ecosystem but also creates green spaces that provide advantages for the community.
An interview with a recipient of the reforestation program revealed:
"We are not just planting trees, but we also take care of them. This gives us a sense of ownership of our environment."

Additionally, community-based waste management practices that have been implemented in Huntap have resulted in compost made from organic waste and crafts from inorganic waste, which are sold by residents to supplement household incomes. This program not only supports environmental sustainability but also creates additional economic opportunities. The government has also established a waste management center and communal livestock shelters, which are managed by the community. This program not only reduces waste but also creates additional economic opportunities. The active participation of the community in waste management and the maintenance of communal shelters helps strengthen social independence within the community.

The findings of this research show that the success of post-disaster recovery is highly influenced by the synergy between economic transformation, social capital strengthening, and environmental sustainability. Although many have successfully transitioned to small businesses and accessed markets through online platforms, most of the community has experienced a significant decline in income. Factors such as dependency on the informal sector, market instability, and the location of the settlement being far from previous workplaces have become the main barriers to more sustainable economic recovery.

Moreover, although the electricity subsidy initially helped residents survive, the termination of this subsidy has increased the financial burden on residents, whose incomes are now lower than before. Therefore, recovery programs that are more focused on economic empowerment, skills training, and market access are more needed than direct aid, which often proves unsustainable. Recovery strategies based on local potential and sustainability approaches must be prioritized in the future. This includes strengthening cooperatives, skills education, and providing access to renewable energy that can improve long-term quality of life.

5 Conclusion

The recovery of communities impacted by the eruption of Mount Semeru requires an integrated, multidimensional approach that encompasses economic transformation through livelihood diversification and local business empowerment, strengthening social capital by fostering community solidarity through cooperatives and cooperation (gotong royong), and environmental sustainability through efficient natural resource management and the application of environmentally friendly technologies. These three dimensions must synergize to create an inclusive and sustainable recovery while building the community's resilience to future disaster risks.

6 Acknowledgments

The research team expresses gratitude to the State University of Malang, especially the Research and Community Service Institute (LPPM), for funding our activities. The research team also expresses gratitude to all relevant parties who have supported and assisted in providing data and information for this research.

References

1. Djalante R (2018) Review article: A systematic literature review of research trends and authorships on natural hazards, disasters, risk reduction and climate change in Indonesia. Natural Hazards and Earth System Sciences 18:1785–1810. https://doi.org/10.5194/nhess-18-1785-2018
2. Lassa J, Petal M, Surjan A (2023) Understanding the impacts of floods on learning quality, school facilities, and educational recovery in Indonesia. Disasters 47:412–436. https://doi.org/10.1111/DISA.12543
3. Dartanto T (2022) Natural disasters, mitigation and household welfare in Indonesia: Evidence from a large-scale longitudinal survey. Cogent Economics & Finance 10:. https://doi.org/10.1080/23322039.2022.2037250
4. Mushtaha AW, Alaloul WS (2025) Sustainability factors influencing post-disaster reconstruction projects: Critical review and bibliometric analysis. Sustainable Futures 10:100930. https://doi.org/10.1016/J.SFTR.2025.100930
5. Padhan N, Madheswaran S (2023) Effectiveness of post-disaster coping strategies among the farming households in the coastal districts of Odisha, India. Natural Hazards Research 3:66–75. https://doi.org/10.1016/J.NHRES.2022.12.006
6. Godde CM, Mason-D'Croz D, Mayberry DE, Thornton PK, Herrero M (2021) Impacts of climate change on the livestock food supply chain; a review of the evidence. Glob Food Sec 28:100488. https://doi.org/10.1016/J.GFS.2020.100488
7. Zhang Z, Cui P, Hao J, Li N, Zeng Z, Liu Y, Zou Q, Huang C, Wu S (2021) Analysis of the impact of dynamic economic resilience on post-disaster recovery "secondary shock" and sustainable improvement of system performance. Saf Sci 144:105443. https://doi.org/10.1016/J.SSCI.2021.105443
8. Tuhkanen H (2023) Shifting power through participation in post-disaster recovery: A scoping review. International Journal of Disaster Risk Reduction 97:104041. https://doi.org/10.1016/J.IJDRR.2023.104041
9. Rifai, Hakim A, Sarwono, Rozikin M, Riyadi BS (2021) Post-Disaster Policies And Socio-Economic Recovery Based On Local Community Empowerment (A Case Study In Lombok, Indonesia). Journal of Southwest Jiaotong University 56:809–819. https://doi.org/10.35741/ISSN.0258-2724.56.6.71
10. Ismail FZ, Halog A, Smith C (2017) How sustainable is disaster resilience?An overview of sustainable construction approach in post-disaster housing reconstruction. Int J Disaster Resil Built Environ 8:555–572. https://doi.org/10.1108/IJDRBE-07-2016-0028
11. McCaughey JW, Daly P, Mundir I, Mahdi S, Patt A (2018) Socio-economic consequences of post-disaster reconstruction in hazard-exposed areas. Nat Sustain 1:38–43. https://doi.org/10.1038/S41893-017-0002-Z;SUBJMETA
12. Sunkur R, Kantamaneni K, Bokhoree C, Ravan S (2023) Mangroves' role in supporting ecosystem-based techniques to reduce disaster risk and adapt to climate change: A review. J Sea Res 196:102449. https://doi.org/10.1016/J.SEARES.2023.102449
13. Nugroho F, Alifa S (2025) Empowerment and Post-disaster Poverty Reduction by a Local Community Organisation in Indonesia. International Journal of Community and Social Development 7:76–94. https://doi.org/10.1177/25166026251319886
14. Partelow S (2021) Social capital and community disaster resilience: post-earthquake tourism recovery on Gili Trawangan, Indonesia. Sustain Sci 16:203–220. https://doi.org/10.1007/S11625-020-00854-2/TABLES/6
15. Liu-Lastres B, Mariska D, Tan X, Ying T (2020) Can post-disaster tourism development improve destination livelihoods? A case study of Aceh, Indonesia.

Journal of Destination Marketing & Management 18:100510. https://doi.org/10.1016/J.JDMM.2020.100510

 https://doi.org/10.1051/e3sconf/202565207001

Revolutionizing Plastic Waste Management: A Bibliometric Analysis of Superworm and Mealworm Biodegradation Potential

Muhammad Al-Irsyad[1*], *Nurnaningsih Herya* Ulfah[2], *Anita* Sulistyorini[3], *Melati Nastiti Ningrum* Abidin[4], *Shohifatul* Ilma[5], and *Anu* Surach[6]

[1,2,3,4,5] Department of Public Health, Faculty of Sport Science, Universitas Negeri Malang, Malang 65141, Indonesia
[6] Faculty of Public Health, Ramkhamhaeng University, Bangkok 10240, Thailand

Abstract. The rapid increase in plastic use has contributed to a 12% accumulation of plastic waste, ranking it as the fourth largest global waste category. Conventional degradation methods remain largely ineffective, prompting interest in biological alternatives. Recent studies highlight superworms (Zophobas morio) and mealworms (Tenebrio molitor) as potential agents for plastic biodegradation. This study applies bibliometric analysis using ProQuest data from 2019–2024. The co-occurrence keyword analysis identified 53 keywords, 940 links, and a total link strength of 2521. Publications rose from 2 in 2019 to a peak of 8 in 2023, with 6 already in 2024. Polymers emerged as the leading journal (9 papers). China was the most active country, followed by Poland, Japan, Malaysia, Spain, and Italy. Common keywords associated with T. molitor and Z. morio were "insect," "larva," and "protein," reflecting the focus on biological and nutritional aspects. Findings underscore the potential of superworms and mealworms in reducing plastic waste through biodegradation. They also open avenues for further research into enzymes, nutrients, and consumption, emphasizing the link between nutrition and enzymatic efficiency. Continued study is essential to deepen understanding and maximize the role of these insects in addressing global plastic waste.

1 Introduction

Rapid economic growth has led to a high market demand, particularly for plastic materials used for packaging and finished goods. Plastic is more commonly used compared to other packaging materials due to its advantages: ease of packaging, airtight sealing, portability, practicality, cost-effectiveness, strength, light weight, durability, and ability to be shaped as needed [1,2]. The extensive use of plastic results in a significant amount of plastic waste. Globally, plastic waste reached 242 million tons, making up 12% of all solid waste worldwide and ranking fourth in 2016 [3].

* Corresponding author: muhammad.irsyad.fik@um.ac.id

The increase in plastic waste has a detrimental impact on the environment because plastic waste is durable and difficult to degrade, leading to waste accumulation and environmental pollution [2]. Generally, plastic can be decomposed through several methods, such as landfilling, thermal oxidation, incineration, and chemical recycling. These methods are not effective in solving the global plastic problem because they are costly and can produce toxic compounds [4].

Recent research indicates that there are biological agents from the Tenebrionidae family that can degrade plastic, specifically superworms (*Zophobas morio*) and mealworms (*Tenebrio molitor*) [5]. Based on DNA barcode analysis of the COI gene, the average genetic difference between superworms and mealworms is 20.9% [6]. Superworms range in length from 3 to 6 cm, which is four times longer than mealworms. According to previous studies, superworms can also chew and consume various types of plastic as their sole food source [7,8].

Superworms and mealworms are believed to have the ability to degrade plastic synergistically with the bacteria in their gut. The gut bacteria of superworms and mealworms play a crucial role in depolymerization, insecticide resistance, and biodegradation of xenobiotic components in the insect digestive system [8,9]. This plastic degradation capability of superworms and mealworms does not interfere with their larval life cycle and continues until they become adult beetles, thus supporting the plastic degradation process through their life cycle [7].

This study aims to reveal the development of research on superworms and mealworms in their role as plastic biodegradation agents through a bibliometric analysis. The bibliometric analysis is conducted to visualize research trends and uncover potential future research opportunities related to superworms and mealworms as plastic biodegradation agents. The novelty of this study lies in its specific focus: to date, no bibliometric study has specifically examined superworms and mealworms in this context. Previous bibliometric works have remained broad, either discussing plastic biodegradation in general without specifying plastic type or degradation agent, as in Akinpelu & Nchu [10] and Yang et al. [11], or considering insect agents in a general sense, as in Boctor et al. [12]. By narrowing the lens to superworms and mealworms, this study addresses a critical gap and provides a more detailed understanding of species-specific trends in plastic biodegradation research.

2 Methods

2.1 Design Study

The research data were sourced from the ProQuest database, with a total of 49 published articles ranging from 2019 to 2024 (last five years). This ensures that the analyzed research falls under the category of recent studies. The search terms used for the articles included keywords relevant to the topic, namely "superworm," "mealworm," "Zophobas morio," "Tenebrio molitor," "biodegradation," and "plastic." The bibliometric analysis was conducted using VOSviewer software, which is closely related to the visualization of research trends mapped based on scientific data. This analysis can present the knowledge structure of each research field based on timeline analysis, co-occurrence analysis, and density analysis.

This study used only the ProQuest database to keep the analysis consistent and manageable. ProQuest provides wide coverage of peer-reviewed journals, ensures reliable data quality, and offers standardized metadata suitable for bibliometric tools like VOSviewer. Although some studies from other databases may not be included, this choice gives a clear and focused dataset, while the limitation is acknowledged for future research.

2.2 Inclusions and Exclusions Criteria

The study applied the following criteria to ensure the relevance and reliability of the bibliometric dataset:

Inclusion criteria:

a. Articles published between 2019–2024, to capture the most recent and relevant developments in superworm and mealworm research.
b. Peer-reviewed publications available in the ProQuest database.
c. Articles containing specific keywords related to the research scope, namely "superworm," "mealworm," "Zophobas morio," "Tenebrio molitor," "biodegradation," and "plastic."
d. Publications directly addressing the role of superworms and/or mealworms in plastic biodegradation.

Exclusion criteria:

a. Articles outside the 2019–2024 time frame.
b. Non-scholarly sources (news articles, blogs, opinion pieces, conference posters without full papers).
c. Studies focusing on other insect species or plastic biodegradation methods without involving Zophobas morio or Tenebrio molitor.
d. Duplicated entries within the database.

3 Result

3.1 Descriptive Analysis of Bibliometric Indicators

Bibliometric research involves bibliographic data and publication analysis using VOSviewer software to obtain data on trends, relationships, and patterns of specific topics. The co-occurrence keyword analysis resulted in 53 keywords, 940 connections (Links), and a Total Link Strength of 2521.

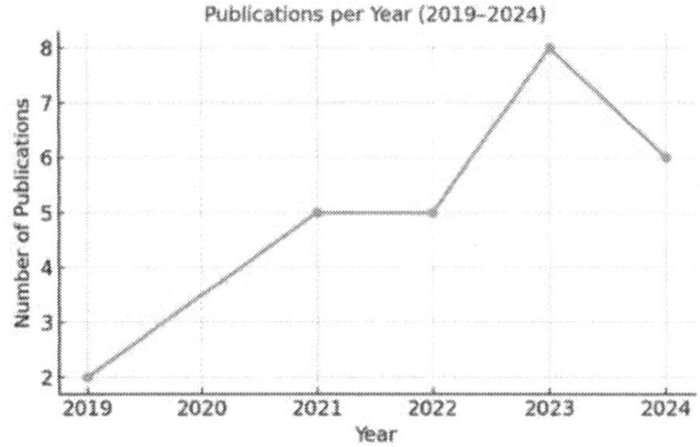

Fig. 1. Publications by Year (2019–2024)

Figure 1 shows the yearly trend of publications on the use of *Tenebrio molitor* and *Zophobas morio* as plastic biodegradation agents. Research activity began modestly in 2019 with only two papers. It increased gradually, reaching stability in 2021 and 2022 with five publications each, before peaking in 2023 with eight publications. The six papers recorded for 2024 are likely underreported as the year is not yet complete. This indicates a clear upward trajectory in scholarly interest over the last five years.

Table 1. Main Journals Publishing on the Topic

Journal	Articles	% of Total
Polymers	9	42.9%
Animals	2	9.5%
Sustainability	2	9.5%
Energies	2	9.5%
Communications Earth & Environment	1	4.8%
The European Zoological Journal	1	4.8%
Scientific Reports	1	4.8%
PLoS One	1	4.8%
PLoS Biology	1	4.8%
Corrosion and Materials Degradation	1	4.8%

Table 1 summarizes the journals most active in this field. Polymers stand out as the dominant outlet with 42.9% of the total publications, far exceeding other journals. Secondary contributions come from Animals, Sustainability, and Energies, each accounting for around 9.5% of the output. The remaining journals contribute single publications each, suggesting that while the research area is gaining traction, it is still concentrated within a limited number of journals.

Table 2. Main Contributing Countries

Country	Articles	% of Total
China	2	28.6%
Poland	1	14.3%
Japan	1	14.3%
Malaysia	1	14.3%
Spain	1	14.3%
Italy	1	14.3%

Table 2 highlights the geographical distribution of contributions. China leads with 28.6% of the publications, followed by Poland, Japan, Malaysia, Spain, and Italy, each contributing 14.3%. This reflects a diverse but still relatively small global network of research groups.

3.2 Keyword Analysis

The bibliometric analysis related to superworms and mealworms as plastic biodegradation agents was conducted over the past five years. The results of the co-occurrence keyword analysis provide a comprehensive interpretation of the visualization of research mapping and can be used as a reference for identifying and analyzing frequently used keywords, hence aiding in understanding related research.

Three cluster groups were identified among the 53 keywords in this study, indicating three main topics of focus in this field. A total of 940 links connected each keyword with a total link strength of 2521, demonstrating the interrelationship among the research topics.

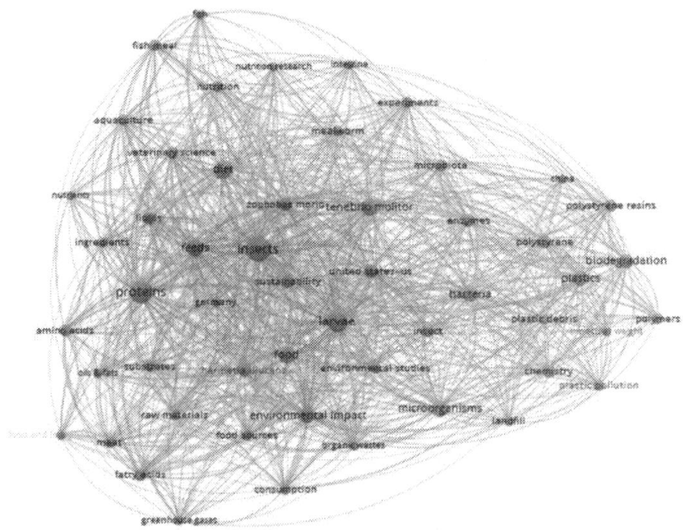

Fig. 2. Network Visualization of Keyword Occurrence Related to Superworms and Mealworms in Their Role as Plastic Biodegradation Agents

Figure 2 shows a broad overview of future research opportunities related to superworms and mealworms in their role as plastic biodegradation agents. The figure shows three main clusters distinguished by color: red related to insects and food sustainability, green related to plastic degradation and microorganisms, and blue related to diet and nutritional studies. The topic "superworm" does not appear in the data and is only found under the scientific term "Zophobas morio." The topics "mealworm," "Tenebrio molitor," and "Zophobas morio" have primary connections to the topics "insect," "larva," and "protein." The figure shows that the topics "Tenebrio molitor" and "Zophobas morio" with "biodegradation" and "plastic" are not as frequently studied compared to the topics "Tenebrio molitor" and "Zophobas morio" with "insect," "larva," and "protein.".

 https://doi.org/10.1051/e3sconf/202565207001

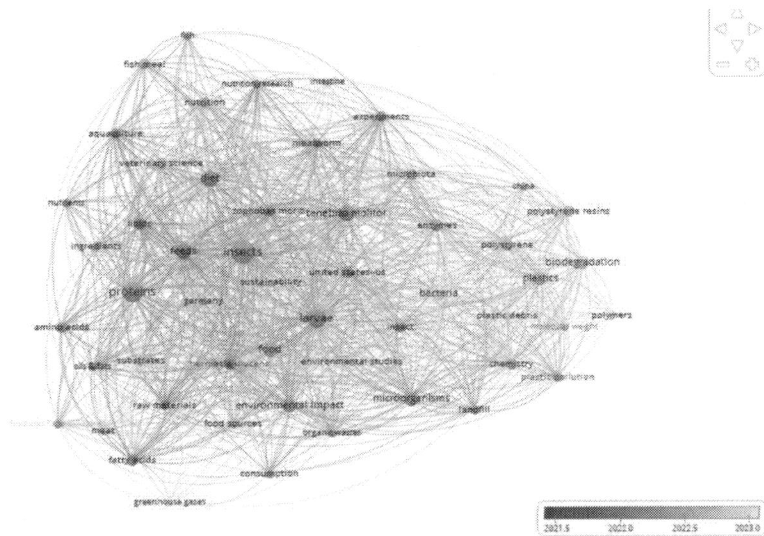

Fig. 3. Overlay Visualization of Keyword Occurrence Related to Superworms and Mealworms in Their Role as Plastic Biodegradation Agents

In Figure 3, various colors represent the publication years of research over the past five years. The topic "superworm" is not found in the data and only appears under the scientific term "Zophobas morio." The figure shows that the topics "mealworm," "Tenebrio molitor," and "Zophobas morio" have been studied since mid-2021. In contrast, the topics "biodegradation" and "plastic" began to be studied from mid-2022. Thus, research on superworms and mealworms as plastic biodegradation agents is still relatively new.

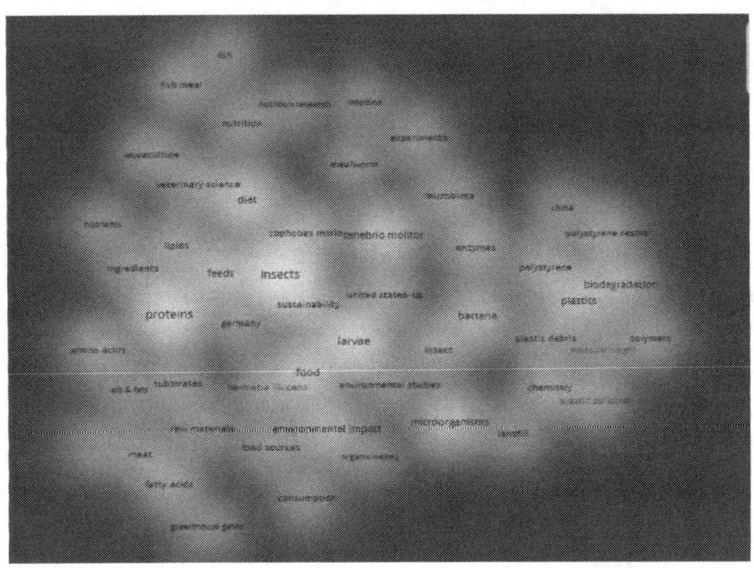

Fig. 4. Density Visualization of Keyword Occurrence Related to Superworms and Mealworms in Their Role as Plastic Biodegradation Agents

Figure 4 shows a visualization of the depth of a research topic, indicated by differences in color contrast. The topic "superworm" is not found in the data and only appears under the scientific term "Zophobas morio." The figure shows that the topics "mealworm," "Tenebrio molitor," and "Zophobas morio" with "biodegradation" and "plastic" have a subtle color contrast, indicating that these topics are still rarely studied or have not been explored in depth. In contrast, the topics "insect" and "larvae" have a prominent color contrast, indicating that these topics have been more extensively studied.

3.3 Research Opportunities

Based on the presented figures, several dominant topics are evident within the studied fields, categorized by cluster groups. The red cluster is dominated by the topics "insects," "proteins," "feeds," "food," "sustainability," and "environmental impact." The green cluster features dominant topics such as "plastics," "polymers," "polystyrene," "bacteria," "microorganisms," and "enzymes." The blue cluster includes dominant topics like "nutrition," "diet," "fish," "fish meal," and "aquaculture."

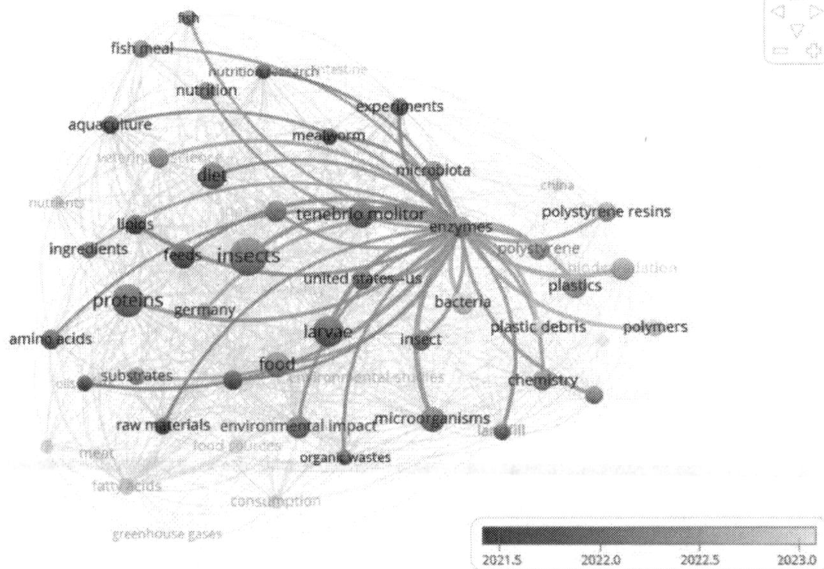

Figure 5. Future Research Opportunities Related to Superworms and Mealworms in Their Role as Plastic Biodegradation Agents

Future research opportunities in the field of superworms and mealworms as plastic biodegradation agents can be seen in Figure 5, which presents the interrelation of the topic "enzymes" with other topics. Topics that are not connected can represent new opportunities for future research planning. For example, the topic "enzymes" is not connected with "nutrients" and "consumption," which could lead to new findings related to the influence of nutrition on the enzymatic efficiency of superworms and mealworms in plastic biodegradation.

4 Discussion

Three main topic groups on superworms and mealworms as plastic biodegradation agents were found in the research visualization using VOSviewer. These topics are closely related to each other, indicating the broad scope of knowledge in this area. Bibliometric analysis plays a role in revealing the relationships between research topics, providing insights into the trends and patterns of a study's topics [13].

The three main groups or clusters in this research are color-coded: red for insect and food sustainability, green for plastic degradation and microorganisms, and blue for diet and nutritional studies. The red cluster shows a research focus on using insects as an alternative protein source and sustainable feed, as well as their environmental impact. The green cluster focuses on the ability of superworms and mealworms as plastic biodegradation agents, involving the role of microorganisms and enzymes in the process. The blue cluster discusses the nutritional aspects of mealworms and superworms as animal feed, particularly in the context of aquaculture.

These three clusters are interconnected, demonstrating that the strong relationship between the topics of superworms, mealworms, and plastic degradation represents a multidisciplinary approach that can be utilized by the scientific community. This is evidenced by a significant increase in the number of studies focusing on the ability of superworms and mealworms as plastic biodegradation agents, particularly regarding the reduction of environmental impacts compared to other methods, such as landfill, incineration, thermal oxidation, and chemical recycling [4]. Even with the assistance of other microorganisms like bacteria, fungi, and algae, which are known to have potential as plastic biodegradation agents, microbes still require a long time to degrade plastic [14]. Previous research has also revealed that superworms and mealworms can chew and eat various types of plastic as their only food source [4]. The larvae of superworms and mealworms are believed to have the ability to degrade plastic due to the bacteria in their guts, which contribute synergistically. Gut bacteria in larvae are known and believed to play a crucial role in insecticide resistance, depolymerization, and biodegradation of xenobiotic components in the insect digestive system [8].

Beyond academic mapping, these clusters also align with global sustainability agendas. The red cluster on insect-based protein and sustainable feed supports SDG 12 (Responsible Consumption and Production) by promoting alternative resources that reduce reliance on conventional, resource-intensive inputs. The green cluster on plastic degradation and microorganisms contributes to SDG 13 (Climate Action), as insect-mediated biodegradation offers lower-emission alternatives to incineration or chemical recycling, while also fitting within circular economy principles where plastic waste is converted into valuable byproducts. The blue cluster on nutrition and aquaculture reinforces these links by showing pathways for integrating insect biomass into sustainable food systems. Together, these connections highlight that advancing research on Tenebrionidae larvae has not only scientific but also policy relevance for building low-carbon and circular solutions to the global plastic crisis.

In addition to the cluster interconnections, the study results indicate that the role of superworms and mealworms as plastic biodegradation agents has not been extensively and deeply explored. Based on publication years, research on superworms and mealworms as plastic biodegradation agents is still relatively new. Therefore, insights into superworms and mealworms as plastic biodegradation agents remain relevant for further investigation with various modifications or more specific studies. This contrasts with studies on superworms and mealworms as alternative protein feed sources, which have been extensively studied.

This bibliometric analysis acknowledges certain considerations related to the database, research focus, and keyword selection. The study exclusively utilized research from the ProQuest database, which may not encompass all relevant studies from other sources. The research focus, determined through co-occurrence, timeline analysis, and density analysis, highlights the need for further clarification on the context and quality variation within each publication. The keywords analyzed were selected carefully, though there is potential for the inadvertent omission of emerging terms in the field. Furthermore, this analysis prioritizes a quantitative approach, leaving room for more in-depth examination of the methodologies employed in the studies. The selection of studies was also influenced by the time frame, acknowledging the evolving nature of the research landscape. Bias may also occur in data collection, authorship trends, and publication patterns in the ProQuest database, which can affect data validity.

Based on the results of this study, there is promising potential for using superworms and mealworms to reduce plastic waste accumulation through biodegradation, offering an alternative solution to the global plastic problem. Additionally, future studies using bibliometric analysis should address the limitations identified in this study, expand the scope of topics, and deepen the analysis of results

Compared to broad-scope bibliometric studies of plastic biodegradation, our species-focused analysis highlights different research priorities. Yang et al. [11] report a long-term, field-level expansion of biodegradation studies across decades, identifying major country-level contributors and thematic hotspots, while Akinpelu & Nchu [10] provide a 2000–2021 bibliometric baseline demonstrating steady growth and key institutional players. Boctor et al. [12] synthesize insect-level mechanistic literature and emphasize enzyme exaptation and species diversity in plastivory. Our analysis complements these works by narrowing the lens to Tenebrionidae (superworms and mealworms), revealing early signals for nutrition–enzyme interactions as an emerging micro-theme. However, because our study used ProQuest (2019–2024) and a limited sample (n=49), our findings should be viewed as hypothesis-generating; integrating WoS and Scopus in future analyses would enable direct quantitative alignment with the broader trends reported by Yang et al. and Akinpelu & Nchu.

5 Conclusion

This study uses bibliometric analysis to illustrate the evolutionary mapping of research related to superworms and mealworms in their role as agents of plastic biodegradation, highlighting the publication landscape and scientific mapping. The study emphasizes co-occurrence, timeline analysis, and density analysis, revealing trends and relationships between research topics. The analysis revealed 53 keywords, 940 connections, and a Total Link Strength of 2521, providing comprehensive insights and promising potential regarding superworms and mealworms in their role as plastic biodegradation agents. Future research should extend beyond bibliometric mapping to experimental and applied studies. Key priorities include: (1) identifying and characterizing gut-derived enzymes with high catalytic efficiency; (2) testing the influence of diet and co-feeding strategies on larval degradation capacity; (3) scaling pilot systems that combine insect rearing with plastic waste processing to evaluate feasibility under real-world conditions; and (4) conducting comparative studies across databases (e.g., WoS, Scopus, ProQuest) to capture a more comprehensive evidence base. Advancing these directions would not only strengthen scientific understanding but also translate the potential of *Tenebrionidae* larvae into practical, scalable interventions for global plastic waste management.

Acknowledgements

Financial support for this research was provided by Universitas Negeri Malang. The authors declare that there is no conflict of interest in this research.
Specify the contribution to the work and write-up of the manuscript for each person listed as author

1. Muhammad Al-Irsyad: Conceptualization, methodology, supervision, writing—review & editing.
2. Nurnaningsih Herya Ulfah: Data curation, formal analysis, writing—original draft, writing—review & editing.
3. Anita Sulistyorini: Validation, formal analysis, writing—review & editing.
4. Melati Nastiti Ningrum Abidin: Investigation, data curation, visualization, writing—original draft.
5. Shohifatul Ilma: Investigation, resources, data curation, writing—review & editing.
6. Anu Surach: Conceptualization, project administration, writing—review & editing, international collaboration.

References

1. Nugroho AS. Pengolahan Limbah Plastik LDPE dan PP untuk Bahan Bakar dengan Cara Pirolisis. Jurnal Litbang Sukowati. 2020 Oct 16;4(1):10. https://doi.org/10.32630/sukowati.v4i1.166
2. Luqmana I, Putra I, Ma'rufah N. Degradation Rate of Several Plastics Type Using Mealworm (Tenebrio molitor L.) and Germany Superworm (Zophobas atratus F.). Journal of Enviromental Tech. 2022;23(1). https://www.researchgate.net/publication/359127736_Laju_Degradasi_Beberapa_Jenis_Plastik_Menggunakan_Ulat_Hongkong_Tenebrio_molitor_L_dan_Ulat_Jerman_Zophobas_atratus_F_Degradation_Rate_of_Several_Plastics_Type_Using_Mealworm_Tenebrio_molitor_L_and_Ge
3. Kaza S, Yao L, Bhada-Tata P, Woerden F Van. What a Waste 2.0: A Global Snapshot of Solid Waste Management to 2050. The World Bank, editor. Washington DC: World Bank Group; 2018. https://openknowledge.worldbank.org/entities/publication/d3f9d45e-115f-559b-b14f-28552410e90a
4. Jiang S, Su T, Zhao J, Wang Z. Biodegradation of polystyrene by tenebrio molitor, galleria mellonella, and zophobas atratus larvae and comparison of their degradation effects. Polymers (Basel). 2021 Oct 1;13(20). https://doi.org/10.3390/polym13203539
5. Peng BY, Li Y, Fan R, Chen Z, Chen J, Brandon AM, et al. Biodegradation of Low-Density Polyethylene and Polystyrene in Superworms, Larvae of Zophobas Atratus (Coleoptera: Tenebrionidae): Broad and Limited Extent Depolymerization. Environmental Pollution. 2020 Nov 1;266. https://doi.org/10.1016/j.envpol.2020.115206
6. Park HC, Jung BH, Han T, Lee YB, Kim SH, Kim NJ. Taxonomy of introduced commercial insect, Zophobas atratus (Coleoptera: Tenebrionidae) and a comparison of DNA barcoding with similar tenebrionids, Promethis valgipes and Tenebrio molitor in Korea. Korean Journal of Sericultural Science. 2013 Nov 30;51(2):185–90. https://doi.org/10.7852/jses.2013.51.2.185
7. Yang Y, Wang J, Xia M. Biodegradation and mineralization of polystyrene by plastic-eating superworms Zophobas atratus. Science of the Total Environment. 2020 Mar 15;708. https://doi.org/10.1016/j.scitotenv.2019.135233
8. Peng BY, Sun Y, Wu Z, Chen J, Shen Z, Zhou X, et al. Biodegradation of polystyrene and low-density polyethylene by Zophobas atratus larvae: Fragmentation into microplastics, gut microbiota shift, and microbial functional enzymes. J Clean Prod. 2022 Sep 20;367(132987). https://doi.org/10.1016/j.jclepro.2022.132987

9. Yang SS, Ding MQ, He L, Zhang CH, Li QX, Xing DF, et al. Biodegradation of polypropylene by yellow mealworms (Tenebrio molitor) and superworms (Zophobas atratus) via gut-microbe-dependent depolymerization. Science of the Total Environment. 2021 Feb 20;756. https://doi.org/10.1016/j.scitotenv.2020.144087
10. Akinpelu EA, Nchu F. A Bibliometric Analysis of Research Trends in Biodegradation of Plastics. Vol. 14, Polymers. MDPI; 2022. https://doi.org/10.3390/polym14132642
11. Yang SS, Wu WM, Pang JW, He L, Ding MQ, Li MX, et al. Bibliometric analysis of publications on biodegradation of plastics: Explosively emerging research over 70 years. J Clean Prod. 2023 Nov 20;428:139423. https://doi.org/10.1016/j.jclepro.2023.139423
12. Boctor J, Pandey G, Xu W, Murphy D V., Hoyle FC. Nature's Plastic Predators: A Comprehensive and Bibliometric Review of Plastivore Insects. Polymers (Basel). 2024 Jun 12;16(12):1671. https://doi.org/10.3390/polym16121671
13. Puri V, Mondal S, Das S, Vrana VG. Blockchain Propels Tourism Industry—An Attempt to Explore Topics and Information in Smart Tourism Management through Text Mining and Machine Learning. Informatics. 2023 Mar 1;10(1). https://doi.org/10.3390/informatics10010009
14. Kundungal H, Gangarapu M, Sarangapani S, Patchaiyappan A, Devipriya SP. Efficient biodegradation of polyethylene (HDPE) waste by the plastic-eating lesser waxworm (Achroia grisella). Environmental Science and Pollution Research. 2019 Jun 1;26(18):18509–19. https://doi.org/10.1007/s11356-019-05038-9

GLÈH for Ecology Intelligence: A Sustainable Lifestyle Practice at School

Daska Azis[1*], *Suhendro*[2], *Ade* Nailan Sadida[3], *Ade* Asyifa SP[4], *Sugeng* Widodo[5]

[1]Departement of Geography Education, Universitas Syiah Kuala, 23111 Banda Aceh, Indonesia
[2]Departement of Geography Education, Universitas Pendidikan Indonesia, 40154 Bandung, Indonesia
[3]Darul Quran Aceh Senior High School, 23361 Banda Aceh, Indonesia
[4]Departement of Medical Education, Universitas Syiah Kuala, 23111 Banda Aceh, Indonesia
[5]Departement of Geography Education, Universitas Lampung, 35141 Bandar Lampung, Indonesia

Abstract. The objective of this study is to measure the effectiveness of GLÈH in enhancing ecological intelligence across eight indicators: environmental awareness, ecological knowledge, environmental concern, eco-friendly behavior, critical thinking about environmental impacts, ecological responsibility, sustainable lifestyle, and collaboration in environmental actions. The research employed an experimental method with a quantitative approach. All data were obtained through testing and analyzed using inferential statistical techniques, specifically R square, to determine the contribution of GLÈH to each individual indicator. Our general findings show a significant improvement in ecological intelligence following the GLÈH intervention. Moreover, we identified the specific contribution of GLÈH to each indicator: environmental awareness (48.3%), ecological knowledge (52.1%), environmental concern (70.5%), eco-friendly behavior (60.6%), critical thinking about environmental impacts (53.2%), ecological responsibility (69.1%), sustainable lifestyle (61%), and collaboration in environmental actions (65.4%). These results indicate that GLÈH contributes substantially to shaping students' environmental awareness and attitudes in the school context.

1 Introduction

Twenty-first century education plays a strategic role in ensuring the realization of sustainable development across various sectors of human life [1, 2]. Education for sustainable development represents one of the earliest educational trends to adopt an environmental protection approach [3]. Several scholars have argued that education which integrates collaborative aspects in environmental action also shows a positive correlation with ecological behavior [4, 5, 6]. Cultivating an ecological culture, accompanied by critical thinking skills, is seen as a key foundation for building a sustainable future [7].

In recent years, the trend of ecological education for sustainability has grown significantly across Southeast Asia, including Indonesia [8]. The conceptual framework of

* Corresponding author: daska@usk.ac.id

https://doi.org/10.1051/e3sconf/202565207002

ecological intelligence continues to be explored, given the persistent anthropogenic pressures that place increasing strain on the planet. Ecological intelligence is recognized as a new form of awareness that addresses globally shared environmental issues [9]. But how important is this concept, and why does it continue to warrant investigation? Our review of previous studies suggests that strengthening the dimensions of ecological intelligence in primary education can contribute to shaping environmentally conscious thinking and fostering a consistent sustainable lifestyle [10, 11].

In Germany, formal education is expected to consistently promote sustainable lifestyle habits in daily life [12, 13]. What does this imply? Attitudes and behaviors toward sustainable practices will become increasingly crucial in the future [14, 15]. According to Mróz et al, fostering a positive attitude toward nature must be grounded in the acquisition of sound knowledge [12]. There is an urgent need to educate and enhance teachers' knowledge, awareness, skills, and attitudes within the teaching and learning process for sustainable development [10].

The pressing question currently shaping our thinking is: how can educators meaningfully engage students in sustainability actions at school. Through increased global awareness of the challenges facing the planet that are disrupting ecosystems and therefore life as we know it, it is hoped that more individuals will choose to change their consumption habits, voluntarily adopt a more sustainable lifestyle, or align their professional activities with efforts to restore the planet [14]. So, what should we explicitly do as living beings inhabiting this Earth. In this article, we would like to introduce the concept and application of GLEH in schools.

In Aceh, there is a locally rooted term commonly used in everyday communication to represent a sustainable lifestyle: GLÉH. This concept translates to "clean" and embodies the ethos of ecological responsibility. It consists of 20 learning syntaxes in several stages such as the introduction stage, contextual stage, and action stage. We previously introduced this concept in detail in the article "Green Life Education Habit (GLEH) for Sustainability Lifestyles based Locally in Schools" [1]. In a previous investigation, GLÉH was implemented as an instructional framework within the Merdeka Curriculum, specifically under the theme of sustainable living [2]. The concept promotes a set of syntaxes or learning steps that encourage collaborative, critical, and creative engagement. This aligns with the views of several scholars who argue that integrating sustainability principles into educational processes is essential for fostering a holistic and inclusive approach to sustainable development action in schools [4]. Through the GLÉH concept, this sustainability-focused research aims to measure the improvement in ecological intelligence resulting from the implementation of the GLÉH module.

Ecological intelligence itself is defined as a new form of awareness that addresses global environmental issues with a different ecological approach [6]. Ecological intelligence depends not only on knowledge or cognitive components of environmental issues but also on perceptions and sensitivity to their impacts [7]. This awareness can lead to more sustainable choices, as individuals will be able to recognize ecological threats and respond appropriately [14]. We summarized from several sources to measure ecological intelligence based on several indicators such as environmental awareness [15], ecological knowledge [13], environmental concern [7], eco-friendly behavior [11], critical thinking about environmental impacts [9], ecological responsibility [8], sustainable lifestyle [3] and collaboration in environmental action [5].

These indicators form our measurement framework for identifying each individual's level of ecological intelligence. Specifically, we also examine the individual contribution of each ecological intelligence indicator. This approach allows us to determine whether GLÉH is effective in enhancing students' ecological intelligence or, conversely, whether its impact is limited or less effective.

2 Method

2.1 Overview Method

This study employed a quantitative approach using an experimental method. The research design applied was a pretest-posttest only group design. This technique was chosen to ensure optimal sample representation, thereby enhancing the reliability and generalizability of the results obtained.

2.2 Research site

This research was conducted in Sekolah Penggerak (curriculum reform pioneer schools) implementing the Merdeka Curriculum in Banda Aceh, Indonesia. The study involved two schools: JHS 2 Banda Aceh and JHS 18 Banda Aceh, with a total of 360 students participating as research subjects.

2.3 Data Collection

The data collection technique used in this study was a test consisting of 40 validated items. These items were designed to measure eight indicators of ecological intelligence: environmental awareness, ecological knowledge, environmental concern, eco-friendly behavior, critical thinking about environmental impacts, ecological responsibility, sustainable lifestyle, and collaboration in environmental action [9], and each indicator was assessed through five validated items.

2.4 Data Analysis

After collecting the empirical data, a quantitative analysis was conducted using inferential statistics. Specifically, the coefficient of determination (R^2) was applied to measure the contribution of the GLÉH module to each indicator of students' ecological intelligence.

3 Result and Discussion

Educational institutions play a crucial role in shaping the attitudes, behaviors, and mindsets of future decision-makers while equipping them with essential knowledge, skills, and critical thinking abilities [1]. Ecological intelligence is among the competencies needed to realize national development visions aligned with the principles of sustainable development. The use of the GLÉH module as a teacher's guide for implementing environmentally themed learning centered on sustainable lifestyles offers a unique appeal by integrating sustainability concepts with local wisdom from Aceh [8]. This concept has demonstrated a substantial contribution to enhancing students' ecological intelligence in schools (Figure 1).

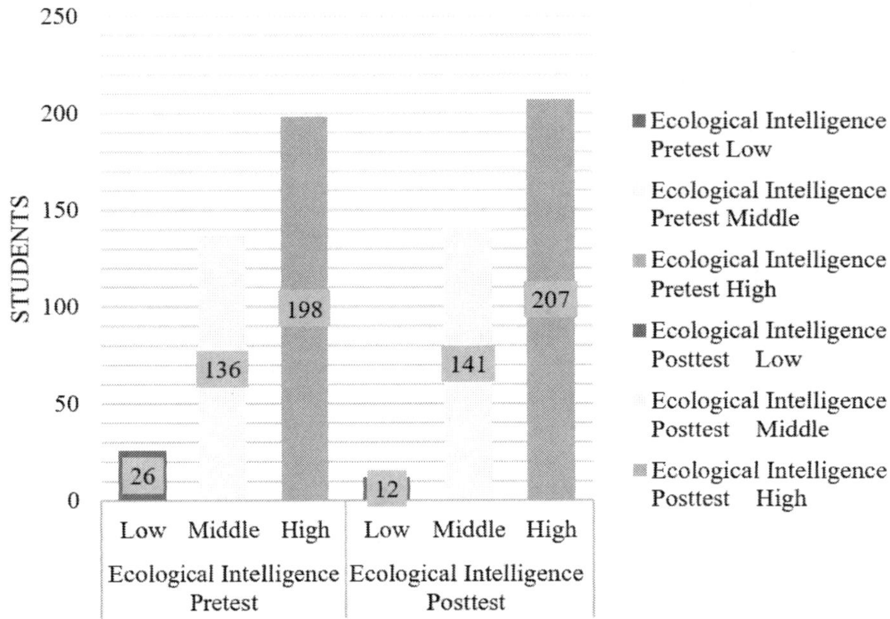

Fig 1. Development of Environmental Awareness

Figure 1 illustrates a significant improvement in students' ecological intelligence from pretest to posttest. If we consider the improvement in students' intermediate and advanced capabilities, there has been an increase from the previous pretest. In addition, if we look at the data for students in the low category, there has been a decline after undergoing treatment through GLEH learning. This shift indicates a positive development in students' ecological intelligence and, overall, these findings show that most students have successfully improved their understanding and awareness of ecological issues after participating in the GLÉH module.

The GLÉH module has proven effective in enhancing ecological intelligence. Previous theories support these findings, such as the Real-World Learning framework proposed by Brundiers et al. published in Sustainability in Higher Education, which emphasizes that sustainability education through modules like GLÉH requires students to be actively engaged with real-world issues involving authentic, complex, and often uncertain contexts [3]. This module includes contextual and action-based stages such as environmental campaign projects using posters and the 4R (Reduce, Reuse, Recycle, Replace) framework. In a similar vein, Goleman in his book Ecological Intelligence: The Hidden Impacts of What We Buy, underscores the importance of "ecological footprint reflection" as a core component of ecological awareness [2].

The ecological footprint reflection is one of the key components embedded within the learning syntax of the GLÉH module, particularly during the introductory phase. At this stage, students are encouraged to reflect on their ecological footprints before progressing to contextual and action-based phases. The idea behind this reflection is to cultivate students' ecological intelligence by prompting them to critically evaluate their past behaviors and make wise, environmentally responsible decisions moving forward. These several stages are the advantages and novelty of the GLEH module.

Therefore, it can be synthesized that this finding has significant implications for the effectiveness of modules such as GLÉH in supporting the development of ecological intelligence among students in schools. Beyond assessing the overall development of ecological intelligence, we also examined the specific contribution of each individual indicator in fostering this growth. In this study, ecological intelligence was measured using eight distinct indicators, including:

Statistical Testing Test

Table 1. GLÈH's contribution to environmental awareness

Test Description	Significance Value	Decision
GLÈH on Ecological Intelligence	Sig. (2-tailed) < 0,05	Significantly influential
GLÈH on Environmental Awareness	Sig. (2-tailed) < 0,05	Significantly influential
GLÈH on Ecological Knowledge	Sig. (2-tailed) < 0,05	Significantly influential
GLÈH on Environmental Concern	Sig. (2-tailed) < 0,05	Significantly influential
GLÈH on Environmentally Friendly Behavior	Sig. (2-tailed) < 0,05	Significantly influential
GLÈH on Critical Thinking About Environmental Impact	Sig. (2-tailed) < 0,05	Significantly influential
GLÈH on Ecological Responsibility	Sig. (2-tailed) < 0,05	Significantly influential
GLÈH on Sustainable Lifestyle	Sig. (2-tailed) < 0,05	Significantly influential
GLÈH on Collaborating in Environmental Action	Sig. (2-tailed) < 0,05	Significantly influential

Source: Data Processing Results (2025)

We explain each indicator specifically as follows:

GLÈH on Environmental Awareness

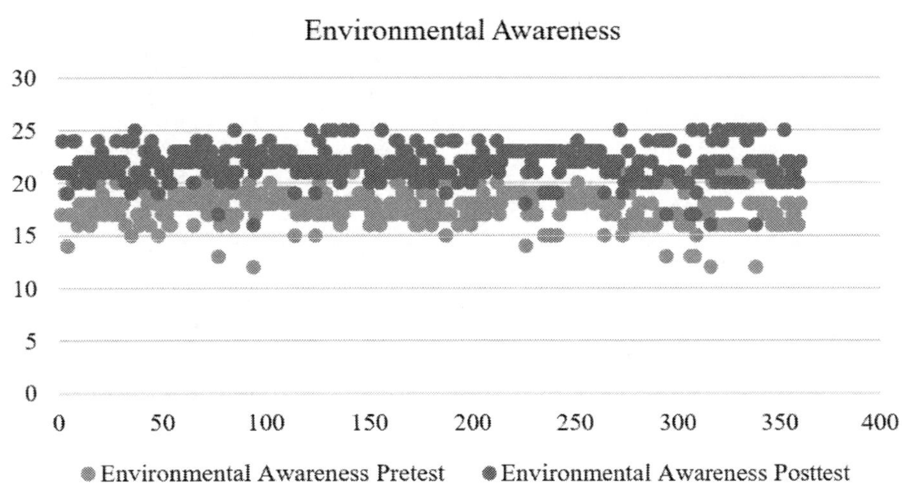

Fig. 2. Development of Environmental Awareness

Significance R^2:

Table 2. GLÈH's contribution to environmental awareness

Model Summary[b]				
Model	R	R Square	Adjusted R Square	Std. Error of the Estimate
1	.483[a]	.233	.231	1.419
a. Predictors: (Constant), Pretest environmental awareness				
b. Dependent Variable: Posttest environmental awareness				

Source: Data Processing Results (2025)

The results of the analysis indicated that the GLÉH module had a statistically significant effect on students' environmental awareness, with a Sig. (2-tailed) value of less than 0.05. This result supports the acceptance of the alternative hypothesis (Ha) and the rejection of the null hypothesis (Ho). Specifically, the contribution value of 0.483 (48.3%) reflects a moderate yet positive impact of the GLÉH module on enhancing students' environmental awareness. Several previous studies have mentioned that the integration of sustainable practices significantly increases students' environmental awareness, which is included in this GLEH module [3].

GLÈH on Ecological Knowledge

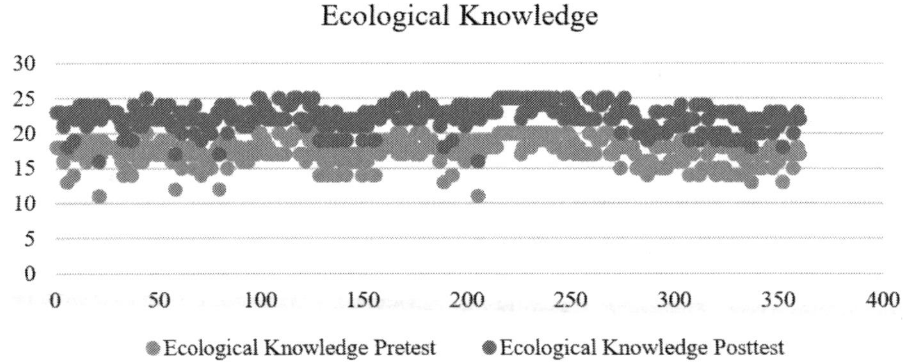

Fig. 3. Development of Ecological Knowledge
Significance R^2:

Table 3. GLÈH's contribution to ecological knowledge

Model Summary[b]				
Model	R	R Square	Adjusted R Square	Std. Error of the Estimate
1	.521[a]	.272	.270	1.520
a. Predictors: (Constant), Pretest ecological knowledge				
b. Dependent Variable: Posttest ecological knowledge				

Source: Data Processing Results (2025)

The test results found that GLÈH had a significant effect on strengthening students' ecological knowledge with a Sig. (2-tailed) value < 0.05. This means that Ha is accepted and

Ho is rejected. Specifically, the influence or contribution of the value 0.521 (52.1%) of the use of the GLÈH module on strengthening the ecological knowledge of students is classified as moderate and has a positive effect. Several studies have shown that applying these concepts in sustainable practices can improve ecological knowledge [6].

GLÈH on Environmental Concern

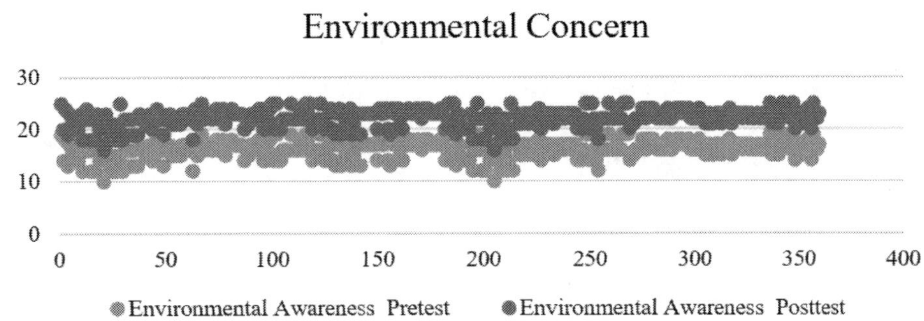

Fig. 4. Development of Environmental Concern

Significance R^2:

Table 4. Contribution of GLÈH to Environmental Concern

Model Summary[b]				
Model	R	R Square	Adjusted R Square	Std. Error of the Estimate
1	.705[a]	.497	.495	1.242
a. Predictors: (Constant), Pretest Environmental Concern				
b. Dependent Variable: Posttest Environmental Concern				

Source: Data Processing Results (2025)

The analysis revealed that the GLÉH module had a statistically significant effect on strengthening students' environmental concern, with a Sig. (2-tailed) value of less than 0.05. This indicates that the alternative hypothesis (Ha) is accepted, while the null hypothesis (Ho) is rejected. Specifically, the contribution value of 0.705 (71%) suggests a high and positive influence of the GLÉH module on reinforcing students' environmental concern. We trace the correlation between this study and these findings. We trace studies in Sweden showing that these sustainability practices can effectively increase students' environmental concern [4]. This is further reinforced by studies on practical approaches to advancing education for sustainability and sustainable development goals [1].

GLÈH on Environmentally Friendly Behavior
Significance R^2:

Table 5. GLÈH's contribution to environmentally friendly behavior

Model Summary[b]				
Model	R	R Square	Adjusted R Square	Std. Error of the Estimate
1	.606[a]	.367	.365	1.583

 https://doi.org/10.1051/e3sconf/202565207002

Model Summary[b]				
Model	R	R Square	Adjusted R Square	Std. Error of the Estimate
a. Predictors: (Constant), Pretest environmentally friendly behavior				
b. Dependent Variable: Posttest environmentally friendly behavior				

Source: Data Processing Results (2025)

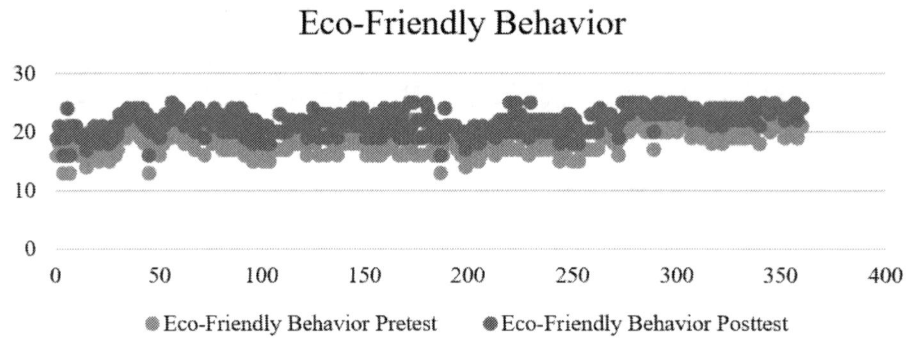

Fig. 5. Environmentally Friendly Development

The analysis indicated that the GLÉH module had a statistically significant effect on strengthening students' eco-friendly behavior, with a Sig. (2-tailed) value less than 0.05. This result supports the acceptance of the alternative hypothesis (Ha) and the rejection of the null hypothesis (Ho). Specifically, the contribution value of 0.606 (61%) reflects a moderate yet positive influence of the GLÉH module on enhancing students' environmentally friendly behavior. Several studies we tracked support the findings of this study, namely that GLEH learning is effective in promoting Environmentally Friendly Development [4].

GLÈH on Critical Thinking about Environmental Impact

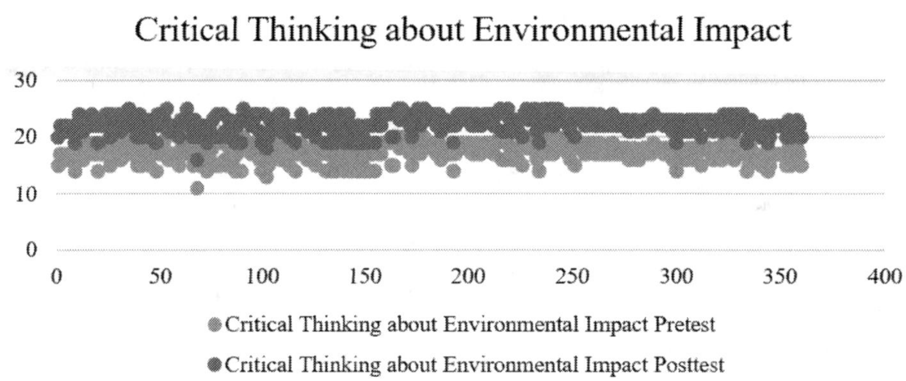

Fig. 6. Development of Critical Thinking on Environmental Impact

Significance R^2:

Table 6. GLÈH's contribution to critical thinking about environmental impact

Model Summary[b]				
Model	R	R Square	Adjusted R Square	Std. Error of the Estimate
1	.532[a]	.283	.281	1.345
a. Predictors: (Constant), Pretest Critical Thinking about Environmental Impact				
b. Dependent Variable: Posttest Critical Thinking about Environmental Impact				

Source: Data Processing Results (2025)

The analysis showed that the GLÉH module had a statistically significant impact on strengthening students' critical thinking regarding environmental impacts, with a Sig. (2-tailed) value of less than 0.05. This finding confirms the acceptance of the alternative hypothesis (Ha) and the rejection of the null hypothesis (Ho). Specifically, the contribution value of 0.532 (53.2%) indicates a moderate yet positive influence of the GLÉH module on enhancing students' critical thinking about environmental issues. We explored whether sustainability practices packaged in a GLEH module are effective in improving critical thinking about environmental impact [7].

GLÈH on Ecological Responsibility

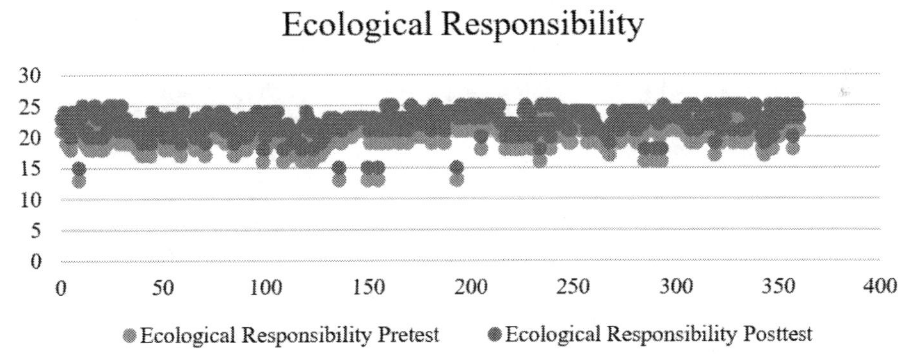

Fig. 7. Development of Ecological Responsibility

Significance R^2:

Table 7. GLÈH's contribution to ecological responsibility

Model Summary[b]				
Model	R	R Square	Adjusted R Square	Std. Error of the Estimate
1	.691[a]	.478	.476	1.417
a. Predictors: (Constant), Pretest Ecological Responsibility				
b. Dependent Variable: Posttest Ecological Responsibility				

Source: Data Processing Results (2025)

https://doi.org/10.1051/e3sconf/202565207002

The inferential statistical analysis revealed that the GLÉH module had a statistically significant effect on strengthening students' ecological responsibility, with a Sig. (2-tailed) value less than 0.05. This result supports the acceptance of the alternative hypothesis (Ha) and the rejection of the null hypothesis (Ho). Specifically, the contribution value of 0.691 (69.1%) indicates a moderate yet positive influence of the GLÉH module on enhancing students' sense of ecological responsibility. In India, schools certified as green schools that require the implementation of sustainable concepts can contribute to the development of ecological responsibility in students [4].

GLÈH on Sustainable Lifestyle

Significance R^2:

Table 8. GLÈH's contribution to sustainable lifestyles

Model Summary[b]				
Model	R	R Square	Adjusted R Square	Std. Error of the Estimate
1	.610[a]	.372	.371	1.575
a. Predictors: (Constant), Pretest Sustainable Lifestyle				
b. Dependent Variable: Posttest Sustainable Lifestyle				

Source: Data Processing Results (2025)

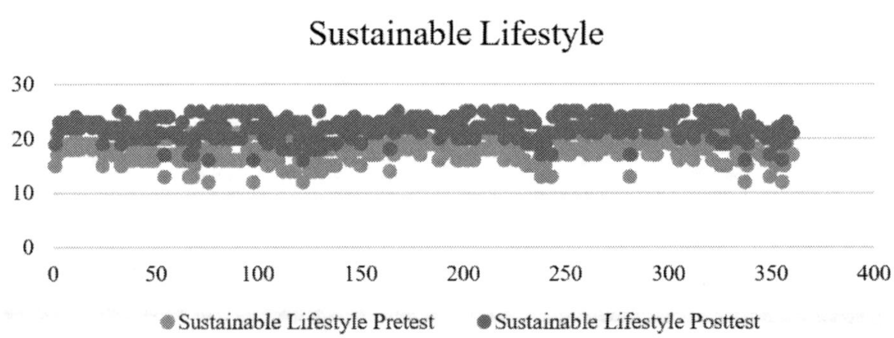

Fig. 8. Development of Sustainable Lifestyles

The inferential statistical analysis revealed that the GLÉH module had a statistically significant effect on strengthening students' sustainable lifestyles, with a Sig. (2-tailed) value of less than 0.05. This finding supports the acceptance of the alternative hypothesis (Ha) and the rejection of the null hypothesis (Ho). Specifically, the contribution value of 0.610 (61%) indicates a relatively low yet still positive influence of the GLÉH module on promoting sustainable lifestyle practices among students. It is no surprise to us that GLEH can make a positive contribution, because the actions carried out in the GLEH module are based on sustainable concepts such as waste recycling and environmental campaigns. We trace this concept back to previous studies that have been conducted on sustainability elements such as solar energy, behavioral change, and recycling practices in educational institutions, which can improve the sustainable lifestyle of students in terms of both understanding and application [5].

https://doi.org/10.1051/e3sconf/202565207002

GLÈH on Collaborating in Environmental Action
Significance R^2:

Table 9. GLÈH's contribution to collaboration in environmental action

Model Summary[b]				
Model	R	R Square	Adjusted R Square	Std. Error of the Estimate
1	.654[a]	.428	.426	1.439
a. Predictors: (Constant), Pretest Collaborating in Environmental Action				
b. Dependent Variable: Posttest Collaborating in Environmental Action				

Source: Data Processing Results (2025)

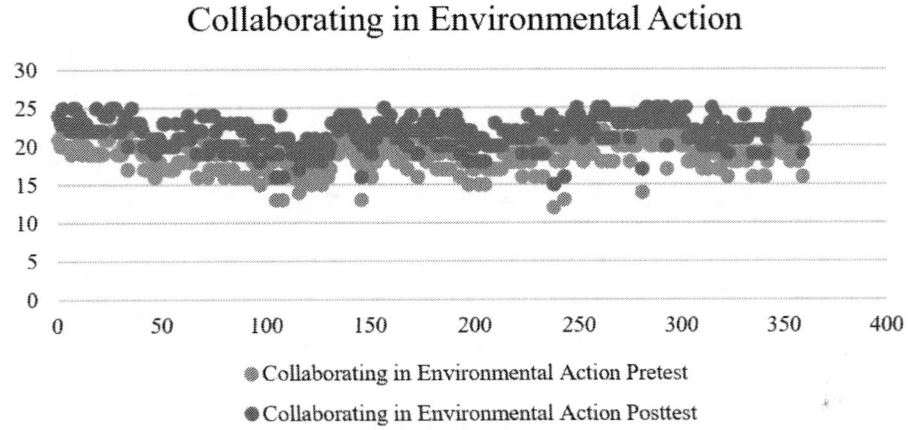

Fig. 9. Development of Collaborating in Environmental Action

The inferential statistical analysis demonstrated that the GLÉH module had a statistically significant effect on strengthening students' collaboration in environmental action, with a Sig. (2-tailed) value less than 0.05. This supports the acceptance of the alternative hypothesis (Ha) and the rejection of the null hypothesis (Ho). Specifically, the contribution value of 0.654 (66%) indicates a moderate yet positive influence of the GLÉH module on enhancing students' ability to collaborate in environmental initiatives. We found several studies supporting the idea that this sustainability-based GLEH concept can encourage collaboration or action among students in environmental campaigns at school [6].

https://doi.org/10.1051/e3sconf/202565207002

Fig. 10. The Impact Trajectory of GLÉH on Ecological Intelligence

Furthermore, we conclude that the GLÉH module contributed variably across each indicator of ecological intelligence: environmental awareness (48.3%), ecological knowledge (52.1%), environmental concern (70.5%), eco-friendly behavior (60.6%), critical thinking about environmental impacts (53.2%), ecological responsibility (69.1%), sustainable lifestyle (61%), and collaboration in environmental action (65.4%), visualized in Figure 10.

It is worth noting that indicators related to cognitive domains, such as ecological knowledge and critical thinking, showed relatively lower contributions even within the moderate category compared to indicators associated with the psychomotor domain, such as eco-friendly behavior, ecological responsibility, sustainable lifestyle, and environmental collaboration. This synthesis indicates that the GLÉH module is particularly effective in fostering students' psychomotor development in environmental contexts rather than their cognitive abilities. This insight offers valuable direction for refining and enhancing the instructional design of the GLÉH learning syntax in future applications.

Several studies that we have tracked previously support the findings in this study. GLEH has been proven effective in addressing eight indicators of ecological intelligence because it integrates cognitive, affective, and psychomotor dimensions in sustainability-based learning. Improvements in ecological knowledge (52.1%) and environmental awareness (48.3%) indicate the role of the module in strengthening basic understanding and critical awareness of environmental issues [7]. Significant contributions to environmental concern (70.5%) and ecological responsibility (69.1%) reflect the module's success in fostering ecological concern and responsibility through direct experience and self-reflection [7].

Furthermore, the increase in eco-friendly behavior (60.6%) and sustainable lifestyle (61%) shows that contextual learning encourages attitudinal changes to become real behavior [52, 53]. The factor of active participation also explains why the indicator of collaboration in environmental action (65.4%) was achieved, as the module emphasizes cooperation in environmental projects at school [5]. Meanwhile, the contribution to critical thinking about environmental impacts (53.2%) demonstrates the module's effectiveness in training participants to reflectively analyze ecological impacts [7]. Overall, the GLEH module works holistically through real activities and this concept strongly promotes learning that provides direct experience through activities that have a positive impact on their development [6], the highest level in Dale's cone of experience theory as described in previous studies [7], connecting knowledge, attitudes, skills, and collective actions to promote a sustainable lifestyle.

4 Conclusion

The effectiveness of GLÉH on ecological intelligence demonstrated a positive and statistically significant impact, evidenced by consistent improvement between the pre-treatment and post-treatment phases. Specifically, GLÉH contributed substantially across all indicators of ecological intelligence. We found that the module contributed to environmental awareness (48.3%), ecological knowledge (52.1%), environmental concern (70.5%), eco-friendly behavior (60.6%), critical thinking on environmental impacts (53.2%), ecological responsibility (69.1%), sustainable lifestyle (61%), and collaboration in environmental action (65.4%). These findings provide encouraging evidence of an innovative educational breakthrough that can be implemented in schools throughout the Aceh Province. Notably, the GLÉH module integrates local Acehnese cultural values with core principles of sustainability, offering a promising approach to advancing ecological intelligence in educational settings. A limitation of this study is that the GLEH module only focuses on sustainable lifestyles, while the Pancasila profile covers many themes. This leaves room for further research to develop similar modules that elaborate on other themes.

The Acknowledgements, We would like to express our gratitude to the Institute for Research and Community Service (LPPM), Universitas Syiah Kuala, Indonesia, for their financial support in the publication of this research.

Fundings, This research was funded by the Institute for Research and Community Service (LPPM), Universitas Syiah Kuala.

Data availability statement, All data in this study are available upon request to the author.

Author contribution statement, All authors have contributed substantially to the conception, design, analysis, and interpretation of data for this study. All authors have read and approved the final version of the manuscript submitted for publication.

References

1. A. Rasulov, Use of foreign experiences in ensuring effectiveness of geography education. *Conf. Zone* (2022). https://conferencezone.org/index.php/cz/article/view/848/822
2. Suhendro, D.S. Sugandi, M. Ruhimat, The urgency of HOTS-oriented learning and assessment towards quality of education in facing Indonesia Sustainable Development Goals (SDGs) 2030. Proc. 5th Asian Educ. Symp. (AES 2020), 237–250 (2020). https://doi.org/10.2991/assehr.k.210715.052
3. P.M. Acosta Castellanos, A. Queiruga-Dios, From environmental education to education for sustainable development in higher education: a systematic review. *Int. J. Sustain. High. Educ.* 23, 622–644 (2022). https://doi.org/10.1108/IJSHE-04-2021-0167
4. R. Rozenszayn, O.B.Z. Assaraf, When collaborative learning meets nature: Collaborative learning as a meaningful learning tool in the ecology inquiry-based project. *Res. Sci. Educ.* 41, 123–146 (2011). https://doi.org/10.1007/s11165-009-9149-6

5. M. Kim, H.T. Tan, A collaborative problem-solving process through environmental field studies. *Int. J. Sci. Educ.* 35, 357–387 (2013). https://doi.org/10.1080/09500693.2012.752116
6. D. Gan, I. Alkaher, T. Segal, Incorporating collaborative learning in teacher education to foster self-efficacy to implement environmental citizenship: an action research. *Int. J. Sustain. High. Educ.* 24, 700–718 (2023). https://doi.org/10.1108/IJSHE-07-2021-0258
7. A. Shutaleva, Ecological culture and critical thinking: building of a sustainable future. *Sustainability* 15, 13492 (2023). https://doi.org/10.3390/su151813492
8. V.R. Savage, Ecology matters: sustainable development in Southeast Asia. *Sustain. Sci.* 1, 37–63 (2006). https://doi.org/10.1007/s11625-006-0002-9
9. N.A. Güven, M.A. Uyulgan, An active learning framework for ecological intelligence: Using activities of multiple intelligences to achieve ecological awareness. *Sci. Educ. Int.* 32, 358–367 (2021). https://doi.org/10.33828/sei.v32.i4.11
10. E. Arbuzova, N. Alexandrova, Ecological mindset in education: integrating sustainable practices into educational programs. *Reliab. Theory Appl.* 19, 1194–1201 (2024). https://doi.org/10.24412/1932-2321-2024-681-1194-1201
11. F. Kılıçarslan, T.A. Durhan, Environmental behaviors and ecological intelligence of university students. *Int. J. Environ. Geoinformatics* 10, 93–105 (2023). https://doi.org/10.30897/ijegeo.1159071
12. S. Venghaus, M. Henseleit, M. Belka, The impact of climate change awareness on behavioral changes in Germany: changing minds or changing behavior?. *Energy Sustain. Soc.* 12, 8 (2022). https://doi.org/10.1186/s13705-022-00334-8
13. P. Lindemann-Matthies, E. Hoyer, M. Remmele, Collective public commitment: young people on the path to a more sustainable lifestyle. *Sustainability* 13, 11349 (2021). https://doi.org/10.3390/su132011349
14. G. George, S.J.D. Schillebeeckx, T.L. Liak, The management of natural resources: an overview and research agenda. In: *Managing Natural Resources*, 1–32 (2018). https://doi.org/10.4337/9781786435729.00009
15. N.S. Rianti, et al., Uncovering the effectiveness of the project-based learning model and ecological intelligence: impact on environmental problem-solving ability in senior high school. *Jambura Geo Educ. J.* 5, 33–43 (2024). https://doi.org/10.37905/jgej.v5i1.24461

Green Education in Addressing Climate Change: An Empirical Study of Ecological Literacy through the Adiwiyata Program

Novika Adi Wibowo[1*], *Sumarmi* Sumarmi[2], *Alfyananda Kurnia* Putra[3], *Waode Yunia* Silviariza[4], *Batchuluun* Yembuu[5], *Chyta* Anindhyta[6], *Singgih* Prihadi[7], *Dedy Zain* Mochtar Asri[8], *Aura Julita* Andiansyah[9], *Annisa Anna* Firdausi[10], *and A Riyan Rahman* Hakiki[11]

[1,2,3,4,8,9,10]Department Geography, Faculty of Social Science, Universitas Negeri Malang, Malang, Indonesia

[5]Department Geography, Faculty of Earth and Environmental Sciences, Mongolian National University of Education, Ulaanbaatar, Mongolia

[6]Education Management, Postgraduate School, Universitas Negeri Jakarta, Jakarta, Indonesia

[7]Department of Geography Education, Faculty of Teacher Training and Education, Sebelas Maret University, Surakarta, Indonesia

[11]Geography Education Study Program, Faculty of Teacher Training and Education, Lambung Mangkurat University, Banjarmasin, Indonesia

Abstract. Environmental education plays an important role in students' ecological literacy on climate change issues and is a challenge in Adiwiyata Schools. This research aims: (1) to examine the implementation of the Adiwiyata school program in addressing climate change; and (2) to analyze students' ecological literacy in Adiwiyata schools. There is a difference with previous research which focused on the effectiveness of the green school program, this research focuses on the study of ecological literacy with a case study of the Adiwiyata school. The study employs a mixed-methods approach with an exploratory sequential design. The research subjects included four senior high schools in Pati Regency. Data were collected through questionnaires and interviews and then analyzed using Ridder's interactive analysis model. The findings reveal that the integration of environment-based actualization learning, including project-based learning in waste management activities, tree planting activities known as the "Green Planet Project," school environmental cleaning activities known as the "Clean Campus Project", and river cleaning programs in surrounding areas known as the "Clean River Project", significantly enhanced students' caring aspects, practical competencies, and ecological knowledge. The Adiwiyata School Program proved effective in developing and enhancing ecological literacy to address climate change through collaborative and sustainable curricula and activities.

* Corresponding author: novika.wibowo.fis@um.ac.id

1 Introduction

Global developments in the 21st century are characterized by the urgent need to address environmental issues. Accelerated technological development and unprecedented population expansion have driven Earth in the 21st century to reach a critical environmental crisis point, generating multiple interconnected ecological challenges. Increasing environmental issues such as climate change, sea pollution, and environmental degradation have become global challenges. One contributing factor to these issues is the lack of awareness and concern for the environment among individuals, including students. Many students still have a limited understanding of environmental issues, reflected in their minimal participation in environmental conservation activities and lack of awareness regarding the impact of human activities on ecosystems. There are findings in previous research that student ecoliteracy in Pati Regency is still low [1]. Based on the New Ecological Paradigm (NEP), the environmental care attitudes of students in Adiwiyata schools in Pati were only in the moderate category with an index of 69.67%. These data indicate that the younger generation has not yet understood their role in addressing environmental problems.

The younger generation, including students, has a strategic role in addressing environmental problems. Their strategic role lies in their superior adaptive attitudes, technological literacy, and critical thinking in seeking innovative solutions to complex environmental problems. The younger generation is not only capable of advocating sustainable practices but also initiating various leadership programs that support environmental conservation, such as environmental cleanup campaigns, reforestation projects, sustainable tourism development, organic farming promotion, waste reduction strategies, and water conservation programs. Therefore, enhancing ecological understanding from school age becomes crucial, as in-depth knowledge of ecology can expand environmentally friendly behavior among students. The low ecological literacy condition in Pati demands the implementation of more innovative and effective educational strategies to improve students' ecological literacy and environmental awareness. Efforts to improve ecological literacy Ecological literacy is in line with the concept of Education for Sustainable Development (ESD) initiated by UNESCO in the early 1990s and proclaimed by the United Nations through the Decade of Education for Sustainable Development, which focuses on caring attitudes, active participation of the younger generation, and critical thinking skills in facing climate change.

The concept of Education for Sustainable Development (ESD) has been implemented in Germany, integrated with Planetary Health, which aims to shape students who are aware of the impact of climate change on human health. Students are trained to think critically about the relationship between the environment, society, and health. In addition, students are also trained to present their simple research findings on climate change to the community [2]. The concept of Education for Sustainable Development (ESD) is also implemented in the US, UK, and Nordic countries in the form of Outdoor Education for Sustainable Development (OESD). Its implementation focuses on changing students' mindsets by combining geography, biology, chemistry, and social science learning. Learning is conducted based on nature, place, community, and technology to train students' systematic competencies.

A significant effort that has been implemented in schools to increase environmental awareness and students' ecological literacy is through the Adiwiyata School program. This program aims to encourage schools to implement environmentally friendly principles and develop students' character to care about environmental sustainability. To achieve the program's objectives, four main components are implemented to realize Adiwiyata schools: (1) developing environmentally friendly principles, (2) implementing environment-based curricula, (3) conducting participatory environmental activities, and (4) managing eco-friendly support facilities. Environmental care attitude can be assessed based on three criteria: (1) recognizing that current environmental issues are crucial problems that must be immediately addressed with effective solutions; (2) consistently supporting every policy or regulation related to environmental protection; (3) actively participating in efforts to solve environmental problems. Based on this, environmental actualization in the Adiwiyata School program becomes an important aspect to examine. Learning about ecology should not be limited to theory but must also be manifested in concrete actions or active participation from students to address environmental problems.

In the ecopedagogy approach, it is emphasized that environmental education is not only cognitive but also transformative because it can motivate students to change their perspectives. The ecopedagogy approach emphasizes that environmental education is not only cognitive, but also transformative, which can encourage students to change their perspectives and behaviors towards the environment. Therefore, the Adiwiyata program in Indonesia can be a local practice that supports the global discourse on climate education and sustainability. Students' environmental care character can be improved through habituation and also an active school culture in addressing environmental problems. This is in line with the Theory of Planned Behavior which states that a person's behavior depends on their awareness. Factors that influence this are attitudes towards behavior, subjective norms, and perceived behavioral control. The application of this theory can be applied with the Adiwiyata program in the school environment to instill habits and awareness in high school students to realize Education for Sustainable Development (ESD).

Environmental actualization can be implemented through Project-Based Learning (PBL) models, which have been proven to enhance ecological literacy knowledge and students' learning outcomes in environmental conservation materials through project creation involving waste and used goods utilization into more valuable items. Environment-based actualization approaches enable students to directly apply the ecological knowledge they acquire through various real activities that impact the surrounding environment, ultimately enhancing ecological literacy. This is in line with previous research findings that school participation in the eco-school program in Flanders can increase students' environmental knowledge [3]. The implementation of green schools can not only influence behavior but can also have an impact on climate change because it uses an environmental education curriculum, such as that implemented in Afghan schools that includes climate change issues. This curriculum provides basic knowledge about climate, instills an attitude of environmental concern, and encourages students to adapt and mitigate the negative impacts of climate change.

Ecological literacy is defined as the ability to understand basic ecological concepts underlying environmental dynamics and problems. Through ecological literacy, individuals can make appropriate decisions and implement innovative solutions in facing complex environmental challenges. Ecological literacy skills should ideally be introduced from elementary school age so that children's cognitive, affective, and psychomotor development can develop optimally. However, based on the results of observations, the level of ecological literacy among students in Pati is still relatively low. This condition emphasizes the need for more effective educational strategies to improve ecological literacy. Environmental actualization strategies become relevant for enhancing deeper ecological literacy because these approaches not only present environmental materials theoretically but also foster students' active participation in waste management and environmental problem-solving. Therefore, this research investigates the effectiveness of the Adiwiyata school program in addressing climate change, focusing on SMA Negeri 1, SMA Negeri 2, SMA Negeri 3, and SMA Negeri 1 Kayen in Pati Regency, to examine how the program enhances ecological literacy and promotes active student participation in environmental sustainability efforts related to climate change mitigation.

2 Methods

2.1 Research Design

This research aims to investigate the implementation conditions of the Adiwiyata School program in Pati Regency and analyze students' ecological literacy abilities in Adiwiyata Schools. This study employs a mixed-methods approach combining qualitative and quantitative approaches. Mixed methods are popular because they can provide more complete explanations with data presentation accompanied by descriptions of research findings. Data validity and reliability will be ensured through the integration of qualitative and quantitative methods. The research design implements an exploratory sequential design. The exploratory sequential design phases are divided into the first phase with a qualitative approach for data interpretation, while the second phase applies a quantitative approach for data confirmation and verification.

2.2 Research Subjects

Ecological literacy data were collected from various senior high schools in Pati Regency. The selection of research locations was determined based on the frequency of environmental issues in these areas and the study's focus on Adiwiyata Schools, specifically SMAN 1 Pati, SMAN 2 Pati, SMAN 3 Pati, and SMAN 1 Kayen. In line with the findings of previous research which prove that Pati Regency often experiences environmental problems such as flooding, drought and garbage piles. Within these selected schools, purposive sampling was employed to determine research subjects based on students' average learning outcomes, with criteria focusing on both the highest and lowest performing students to ensure comprehensive representation

2.3 Research Instruments

Ecological literacy was assessed using instruments employing four-point Likert scales ranging from very low (score 1), low (score 2), high (score 3), and very high (score 4). The research instrument was developed collaboratively by a research team consisting of two Professors with more than 30 years of experience, one Senior Lecturer with more than 10 years of experience, and two Geography Education Doctoral Lecturers with research fields in environmental education and learning. Validation was carried out with a limited trial of 40 students, including 10 students of SMA Negeri 1 Pati, 10 students of SMA Negeri 2 Pati, 10 students of SMA Negeri 3 Pati and 10 students of SMA Negeri 1 Kayen. Validity and reliability tests used SPSS version 25 with consideration of the level of accuracy and effectiveness in proving validity and reliability data. The results of the analysis proved that the ecological literacy data is included in the valid category with a sig value > r table = 0.312. Furthermore, the reliability of the data in the ecological literacy questionnaire is included in the reliable category with a cronbach's alpha value > r table showing a cronbach's alpha reliability coefficient > 0.60.

2.4 Data Collection and Analysis Techniques

Comprehensive data were collected in 7 days through questionnaires distributed using Google Forms to Adiwiyata School students. Data analysis employed Ridder's interactive analysis model, encompassing four sequential stages: data collection, data reduction, data presentation, and conclusion drawing. The study utilized both quantitative and qualitative data sources. Quantitative data were gathered through a 35-item questionnaire designed to measure ecological literacy levels. Qualitative data were obtained through in-depth interviews to contextualize and explain quantitative findings, thereby providing a comprehensive understanding of students' ecological literacy in Adiwiyata Schools. The results of quantitative and qualitative data were then analyzed to obtain conclusions regarding the study of ecological literacy on the Adiwiyata school program.

3 Results and Discussion

3.1 Implementation of the Adiwiyata School Program

Ecological literacy research provides valuable insights when examined within the context of Adiwiyata Schools. The Adiwiyata program operates on three foundational principles: educative, collaborative, and sustainable approaches. Education serves as a critical driver for enhancing ecological literacy. Previous research demonstrates that learning experiences create lasting impacts on individual knowledge acquisition. The integration of environmental education has emerged as an effective strategy for strengthening ecological literacy. This aligns with previous findings that identify environmental education as the primary mechanism for improving ecological understanding. Furthermore, environmental education encompasses key areas including natural ecosystem functions, human environmental impacts, and sustainable resource management practices.

The Adiwiyata school program aims to foster environmental protection commitment while achieving sustainable development goals. This study's findings confirm that the program's core indicators center on integrating environment-based curricula into educational frameworks. Such curriculum implementation represents a systematic approach to environmental management within school settings. Adiwiyata Schools in Pati Regency operationalize these environment-based curricula through three key components: developing comprehensive lesson plans, creating environment-focused learning materials, and implementing hands-on environmental activities.

This approach reflects a global trend in environmental education integration. Similar initiatives have been successfully implemented across various countries: Portugal has incorporated environmental curricula to enhance ecological literacy and engage students in conservation activities [4]; Romania utilizes environmental education as a strategic tool for conservation, environmental protection, and knowledge enhancement [5]; while Japan emphasizes formal environmental education to support sustainable development objectives [6]. In Pati Regency, the planning and implementation of environment-based curricula are carefully aligned with existing subject matter content across all Adiwiyata Schools.

Environment-based curricula are characterized by the development of learning activities through environmental actualization approaches. This comprehensive implementation requires active participation from all school community stakeholders, including teachers, academic staff, school maintenance personnel, and most importantly, students in environmental actualization learning processes. The program's effectiveness is further enhanced through the involvement of external stakeholders, particularly parents and surrounding communities, who provide crucial support for environmental actualization learning implementation [7]. Furthermore, the curriculum design also emphasizes the integration of theoretical learning materials with practical environmental actualization activities. Environmental actualization learning encompasses three primary focus areas: school-based environmental management practices, the development of environmental value-oriented policies, and the generation of innovative environmental conservation strategies. The implementation of these environmental actualization learning activities is illustrated in Figure 1.

Fig. 1. Environmental Actualization Learning

Figure 1 demonstrates how environmental actualization learning is embedded within the daily educational activities of students and teachers across Adiwiyata Schools. The implementation varies across four different school programs:

a) SMA Negeri 1 Pati implements environmental education through project-based learning focused on plastic waste recycling. Research indicates that project-based learning effectively enhances students' critical environmental awareness [8]. Through this approach, students gain practical experience in transforming waste materials into valuable and economically viable products. Students also develop an understanding of plastic waste hazards, particularly from repeated use. This environmental knowledge empowers students to prevent environmental degradation and respond proactively to climate change challenges [9].
b) SMA Negeri 2 Pati operates the "Green Planet Project" through a tree planting initiative. This program engages not only students as primary participants but also includes broader school community involvement from teachers and staff. This enhances students' environmental literacy through learning experiences about plant cultivation. This practical learning approach aligns with research demonstrating that environmental experiences improve environmental literacy. This also aligns with Edgar Dale's theory that learning experiences are effective when integrated into learning [10]. Tree planting serves as a crucial intervention for mitigating global warming and climate change impacts.

c) SMA Negeri 3 Pati features the distinctive "Clean Campus Project". This daily morning initiative requires student participation in community service activities to maintain school environmental cleanliness. Maintaining waste-free environments, both organic and inorganic, contributes significantly to temperature regulation and climate stability within the school environment [11].
d) SMA Negeri 1 Kayen conducts the "Clean River Project", targeting river systems surrounding the school through monthly cleaning activities. This initiative specifically addresses flood prevention, a recurring environmental challenge at this location. The collaborative efforts of teachers, students, and staff in implementing river cleaning programs effectively reduce waste accumulation and mitigate environmental problems. Research supports that direct environmental interaction through action-based approaches enhances environmental literacy while reducing aquatic ecosystem damage, contributing to SDG 14: Life Below Water [12].

3.2 Students' Ecological Literacy in Adiwiyata Schools

Ecological literacy, also referred to as ecological intelligence, encompasses an individual's capacity to comprehend the complex interconnections within Earth's living systems. This concept extends beyond theoretical understanding to include an individual's conscious engagement with environmental issues, enabling them to implement concrete conservation actions. The primary goal of ecological literacy within Adiwiyata Schools is to develop sustainable thinking approaches across all subjects and student age levels. Furthermore, ecological literacy is evaluated through three core dimensions: caring aspects, practical competency aspects, and knowledge aspects. The comprehensive assessment results of students' ecological literacy across Senior High Schools in Pati Regency are presented in Table 1.

Table 1. Students' Ecological Literacy in Adiwiyata Schools

Indicator	Statistics	Adiwiyata Schools			
		SMAN 1 Pati	SMAN 2 Pati	SMAN 3 Pati	SMAN 1 Kayen
Caring	Mean	33,83	30,58	32,93	38,70
	Percentage (%)	70,47%	63,70%	68,59%	66,93%
	Category	High	High	High	High
Practical Competency	Mean	33,28	34,18	32,00	31,83
	Percentage (%)	69,32%	71,20%	66,67%	66,30%
	Category	High	High	High	High
Knowledge	Mean	31,43	31,70	32,73	41,08
	Percentage (%)	65,47%	66,09%	68,18%	63,75%
	Category	High	High	High	High

The caring indicator reveals that SMA Negeri 1 Pati achieved the highest score of 70.47%, placing it in the high category for students' ecological literacy. This demonstrates strong environmental concern among students, evidenced by their consistent practice of separating organic and inorganic waste. These findings support previous research indicating that environmental concern develops through active participation in environmental management and control activities. Furthermore, student environmental concern is enhanced when students actively engage with their environment through conservation practices. Research in China shows that high school students have a positive attitude towards environmental issues, but its application in real action is still limited because learning facilities and strategies are less than optimal [13]. Meanwhile, in Germany itself, it shows higher achievements where high school students not only have concern (caring) but are also involved in climate-friendly schools, for example energy and emission management, this is also supported by the integration of the Education for Sustainable Development (ESD) curriculum [2]. Therefore, literacy in Indonesia is relatively high, but there is a need for sustainable practices and structural support.

The practical competency indicator demonstrates that SMA Negeri 2 Pati achieved the highest score of 71.20%. This high performance is evidenced through students' active participation in environmental activities, including waste collection and reforestation projects within the school environments. These results support previous research showing that environmental practices effectively enhance ecological literacy. The approach also reflects Kolb's theory, which emphasizes that students best internalize knowledge through direct, practical engagement with their environment. Similar findings from research in China further validate this approach, where students demonstrated improved environmental understanding through practical waste processing activities. Programs in China, through the Green School Program policy, involve students in recycling, energy management, and waste management, raising awareness of sustainable living [14]. This context emphasizes similar applications that focus on waste management as part of the school curriculum. Meanwhile, in Germany, ecological literacy education is integrated into the program through outdoor education activities such as reforestation, energy conservation, and climate management in the school environment [2]. This aspect shows that student involvement increases practical competence in ecological skills. In Indonesia, through the context of environmental education, the focus is still on school-based practices, compared to Germany, which receives stronger structural support.

The knowledge indicator reveals that SMA Negeri 3 achieved the highest score of 68.18%. This demonstrates that students possess a substantial understanding of environmental issues and challenges. The integration of environmental education through the Adiwiyata program serves as the foundation for enhancing knowledge components within ecological literacy. Environmental education effectively builds both competencies and knowledge, equipping students with the ability to identify environmental problems and develop appropriate solutions. This is in line with previous findings that students' attitudes toward the environment depend on their ecological knowledge and literacy acquired through environmental education. Moreover, incorporating environmental education into learning processes helps students develop critical thinking skills, enabling them to approach environmental challenges with more significant analytical depth and selectivity. Indonesia's knowledge achievement can be compared with that of China and Germany. China demonstrated that the Green School program strengthens students' ecological literacy through the integration of national policies [14]. This approach emphasizes cognitive mastery through curriculum and government regulations. Meanwhile, Germany states that Education for Sustainable Development (ESD) is oriented towards development based on direct experience, for example, the Schools 4 Future program, through student activities in examining energy use and calculating carbon footprints, emphasizes direct experience and a participatory approach [15].

Differences in school achievement across indicators of caring, practical competency, and knowledge can be seen through the environmental culture developed in each country's schools. Schools with an environmental culture have the potential to foster caring awareness in students through collective norms, values, and habits that encourage pro-environmental behavior. Practical competency indicators are more prominent in schools with infrastructure such as recycling facilities, green spaces, integrated waste management facilities, laboratories, and optimal resources. Meanwhile, knowledge indicators are generally higher in schools that integrate environmental education into the formal curriculum, embracing visionary leadership and encouraging learning innovation. Ecological literacy is not solely a student achievement but is closely related to school culture, the availability of adequate resources, and leadership contributions to the implementation of sustainability strategies. This study has several limitations related to the location and subjects used, making the findings not fully generalizable to schools with different characteristics. The variables are also limited to measuring caring, practical competency, and knowledge, while contextual variables including student motivation, family participation, school infrastructure, and sociocultural norms have not been studied in depth. Further research is recommended to expand the subject, add variations in variables, and explore contextual variables that impact the achievement of ecological literacy to obtain comprehensive and applicable results.

4 Conclusion

Research findings demonstrate that the Adiwiyata program effectively integrates environmental education across participating schools. Adiwiyata Schools in Pati Regency implement diverse activities to operationalize the program, including: (a) incorporating environmental education into academic subjects through project-based learning focused on plastic waste recycling, (b) conducting tree planting initiatives through the "Green Planet Project", (c) maintaining school environmental cleanliness via the "Clean Campus Project", and (d) implementing community river cleaning through the "Clean River Project". The study reveals that students' ecological literacy achieved high performance across all three dimensions: caring, practical competency, and knowledge aspects. These results show that the various activities in the Adiwiyata program effectively build strong environmental awareness among students. The program's success demonstrates that it can serve as a practical model for other schools seeking to enhance environmental education and address climate change issues.

Acknowledgements

The author would like to thank the Institute for Research and Community Service (LPPM) of Malang State University for funding this research under contract number 24.2.64/UN32.7.2/LT/2025. This research has no intention or conflict of interest toward any individuals or groups.

References

1. N. A. Wibowo, S. Sumarmi, S. Utaya, S. Bachri, and Y. Kodama, "Students' Environmental Care Attitude: A Study at Adiwiyata Public High School Based on the New Ecological Paradigm (NEP)," *Sustain.*, vol. 15, no. 11, 2023, doi: 10.3390/su15118651.
2. C. Wittlich, L. MöNter, and H. Lathan, "Planetary Health and Education for Sustainable Development: An Integrative Approach with Focus on Climate Change-related Human Health Risks and Their Thematization in School Contexts," *J. Educ. Sustain. Dev.*, vol. 17, no. 2, pp. 200–217, 2023, doi: 10.1177/09734082241238087.
3. D. Petkou, A. Kounani, M. Tsiouni, and E. Afedoulidou, "Can Sustainable Schools Influence Environmental Consciousness and Behavior in Early Childhood? The Perspectives of Early Childhood Educators," *Educ. Sci.*, vol. 15, no. 7, pp. 1–20, 2025, doi: 10.3390/educsci15070916.
4. P. João, P. Sá, M. H. Henriques, and A. V. Rodrigues, "Sustainable Development in Basic Education Sciences in Portugal—Perspective of Official Curriculum Documents,"

Sustain., vol. 14, no. 9, 2022, doi: 10.3390/su14095651.
5. M. Cernicova-Buca, G. M. Dragomir, V. Gherheş, and A. Palea, "Students' Awareness Regarding Environment Protection in Campus Life: Evidence from Romania," *Sustain.*, vol. 15, no. 23, pp. 1–18, 2023, doi: 10.3390/su152316444.
6. K. Velempini, "Assessing the Role of Environmental Education Practices Towards the Attainment of the 2030 Sustainable Development Goals," *Sustain.*, vol. 17, no. 5, pp. 1–20, 2025, doi: 10.3390/su17052043.
7. A. A. Akinsemolu and H. Onyeaka, "The role of green education in achieving the sustainable development goals: A review," *Renew. Sustain. Energy Rev.*, vol. 210, no. August 2024, p. 115239, 2025, doi: 10.1016/j.rser.2024.115239.
8. R. Sánchez-García and S. Reyes-de-Cózar, "Enhancing Project-Based Learning: A Framework for Optimizing Structural Design and Implementation—A Systematic Review with a Sustainable Focus," *Sustain.*, vol. 17, no. 11, pp. 1–24, 2025, doi: 10.3390/su17114978.
9. S. L. Colombo, S. G. Chiarella, C. Lefrançois, J. Fradin, A. Raffone, and L. Simione, "Why Knowing about Climate Change Is Not Enough to Change: A Perspective Paper on the Factors Explaining the Environmental Knowledge-Action Gap," *Sustainability*, vol. 15, no. 20, p. 14859, 2023, doi: 10.3390/su152014859.
10. Y. S. Lee, A. Rashidi, A. Talei, H. J. Beh, and S. Rashidi, "A Comparison Study on the Learning Effectiveness of Construction Training Scenarios in a Virtual Reality Environment," *Virtual Worlds*, vol. 2, no. 1, pp. 36–52, 2023, doi: 10.3390/virtualworlds2010003.
11. A. Djirong, K. Jayadi, A. Abduh, A. Mutolib, R. F. Mustofa, and A. Rahmat, "Assessment of Student Awareness and Application of Eco-Friendly Curriculum and Technologies in Indonesian Higher Education for Supporting Sustainable Development Goals (SDGs): A Case Study on Environmental Challenges," *Indones. J. Sci. Technol.*, vol. 9, no. 3, pp. 657–678, 2024, doi: 10.17509/ijost.v9i3.74385.
12. Sumarmi *et al.*, "Implementing the OBE Model in Plastic Waste Management Using the 4R EPR Pattern for Green Campus," *Int. J. Environ. Impacts*, vol. 7, no. 3, pp. 455–473, 2024, doi: 10.18280/ijei.070308.
13. C. Chen, Q. An, L. Zheng, and C. Guan, "Sustainability Literacy: Assessment of Knowingness, Attitude and Behavior Regarding Sustainable Development among Students in China," *Sustain.*, vol. 14, no. 9, 2022, doi: 10.3390/su14094886.
14. X. Yuan *et al.*, "Promoting education for sustainable development through the green school program to achieve SDGs: Insights from a case study of Beijing's primary and secondary schools," *Int. J. Chinese Educ.*, vol. 13, no. 2, pp. 1–30, 2024, doi: 10.1177/2212585X241259192.
15. O. Wagner, L. Tholen, S. Albert-Seifried, and J. Swagemakers, "Empowering Students to Create Climate-Friendly Schools," *Energies*, vol. 17, no. 9, 2024, doi: 10.3390/en17092199.

Integrating Huma Betang values into conflict management education: development of a flip book chart in Sampit, Central Kalimantan

Mifdal Zusron Alfaqi [1*], *Siti* Awaliyah[2], *Andhika* Yudha Pratama[3], *Abd. Mu'id* Aris Shofa[4], *Ali* Sunarno[5], *Rika* Safitri Nur Azizah[6], *Risa* Khusnul Khotimah[7], *Anisa* Amalia Maisaroh[8] *Lakasmana* Prammana Agung [9], *Muhammad* Adel Nurfirmansyah[10], and *Satya* Robbi Pratama[11]

[1,2,3,4,6,7,8,9,10,11] Universitas Negeri Malang, East Java
[5] Universitas Palangkaraya Central Kalimantan Indonesia

Abstract Huma Betang is a philosophy of life of the Dayak people in Central Kalimantan that upholds deliberation, mutual cooperation, togetherness and tolerance. These cultural values are relevant to fostering awareness of peaceful conflict resolution in a pluralistic society. This research aims to develop an interactive Flip Book Chart based on Huma Betang values as an effort to integrate local wisdom into conflict management courses. The method used is Research and Development (R&D) with the ADDIE model. Data collection techniques include observation, interviews, questionnaires, using a mixed methods approach. Qualitative data describes the development process, while quantitative data from a four-point Likert scale. The validation results show that Flip Book Chart is very feasible, with a score of 95% from material experts and 96.6% from media experts. This media increases students' understanding of conflict, resolution strategies, and appreciation of Huma Betang values. Effectiveness reached 85.2% in small groups (8 students) and 89.6% in large groups (35 students). Material validators emphasized the depth of philosophical values, while media validators focused on visual design, interactivity, and format suitability. Media evaluation included student assessment of educational standards, integration of Huma Betang cultural values, and media relevance to context conflict resolution.

1 Introduction

Indonesia is known as a country with extraordinary cultural diversity, encompassing diverse ethnicities, races, religions, and cultures. This diversity is a national asset that enriches the nation's identity, but it also carries the potential for complex social conflict. Social conflict, such as tensions between ethnic groups, often arises from the failure of understanding differences and implementing effective resolution strategies [1]. One real-life example is the

* Corresponding author: mifdal.zusron.fis@um.ac.id

Sampit conflict in Central Kalimantan in 2001, which involved the Dayak and Madurese tribes. This conflict caused material losses, loss of life, and long-term impacts in the form of trauma and social inequality between ethnic groups . This tragedy reflects the urgency of instilling conflict resolution skills among the community, including students, as a preventive measure to prevent future conflict escalation. Education is a strategic tool for instilling the values of harmony and tolerance in a pluralistic society, in line with the global need to build inclusive and peaceful communities [2].

The people of Central Kalimantan possess a local wisdom known as Huma Betang, a Dayak philosophy of life reflected in their traditional longhouses. Literally, Huma Betang is a rectangular traditional house inhabited by various families from diverse backgrounds [3]. More than just a physical structure, Huma Betang represents the values of togetherness, deliberation, mutual cooperation, tolerance, and equality, emphasizing a humanist approach to minimizing social friction [4]. This philosophy has proven effective in post-conflict reconciliation in Sampit, as it is rooted in values that have long grown within the local community. Huma Betang offers a preventive and solution-oriented approach to conflict management, which is relevant not only in the local context, but also in the global discourse on cultural preservation and sustainable development. In a modern context, Huma Betang has the potential to be integrated into community designs, such as community-based cultural centers, and smart tourism destinations (*smart tourism*), such as a virtual tour of the Betang house, to promote Dayak cultural heritage [5].

This study adopts the theoretical framework of "living space" (*lived space*), which views space as a social construct laden with cultural meaning and collective experience. Huma Betang reflects*lived space* through communal practices that strengthen the social bonds and cultural identity of the Dayak people. This study uses the concept of socio-ecological resilience that emphasizes the community's ability to adapt to social and environmental changes through local values [6]. Huma Betang, with values such as deliberation and tolerance, supports social resilience by facilitating intergroup dialogue and preventing conflict escalation. In the context of higher education, the integration of local wisdom such as Huma Betang into the curriculum, particularly Conflict Management courses, is important to shape individuals who are able to manage conflict contextually and culturally based. There is a gap in the current conflict management course material, which still focuses on universal conflict theory. The curriculum lacks content that integrates local wisdom with in-depth knowledge of conflict management. Consequently, conflict management materials are deemed less relevant to students when faced with the realities of social conflict in their lives. Another problem is that learning media tends to be conventional, with lectures or presentations that are less able to visualize complex case studies of social conflict. Therefore, student engagement and understanding of complex topics such as the Sampit conflict and its resolution are less than optimal.

To overcome this challenge, learning media are needed that are interactive, relevant, and capable of connecting theoretical concepts with cultural realities.*Flip book chart* was chosen as a learning medium due to its ability to present material visually and interactively, which increases student engagement and understanding of complex topics such as conflict management [7]. The use of flip book charts has been proven effective in improving student understanding because it allows them to learn the material in a visual, engaging, and interactive manner, thus helping to strengthen memory and engagement in the learning process . Supported by research by Alfarizi (2021), it can increase student understanding by up to 30% compared to traditional lecture methods [8]. This medium allows students to understand conflict scenarios and their resolutions based on Huma Betang values, such as togetherness and deliberation, in a concrete context that is relevant to local culture.

This research aims to develop*flip book chart* Huma Betang-based interactive learning media for Conflict Management courses in higher education. This media is designed to translate Huma Betang values such as togetherness, deliberation, mutual cooperation, and tolerance into relevant learning scenarios. The goal is for students to understand and apply the principles of culture-based conflict resolution. This research also explores Huma Betang learning media to strengthen social cohesion and support the preservation of Dayak cultural heritage and promote sustainable tourism in Central Kalimantan. Thus, this research contributes not only to higher education but also to cultural preservation and inclusive community development.

2 Method

Purpose of this research method is to obtain an in-depth understanding of local wisdom values Huma Betang and their practical application in Central Kalimantan society, which have potential to be developed as teaching materials for conflict management courses. The results from data analysis of Huma Betang values will used for developing learning media, a Flip Book Chart. This research from beginning to end lasted for approximately three months and starting in May 2025. This research uses a Research and Development/R&D framework with ADDIE model. Problem and potential identification regarding Huma Betang was conducted prior to field research. In subsequent stages, from data collection to product trials, a mixed methods approach combining qualitative and quantitative research will be used [9]. This approach was chosen to develop flip book chart based on the values of Huma Betang as an interactive learning media for the Conflict Management course, while ensuring that the development of the media is relevant to the cultural context of the Dayak community in Sampit, Central Kalimantan.

Data collection was carried out through an approach*mixed methods* to gain a deep understanding of the values of Huma Betang and develop valid learning media. Data collection techniques include: (1) qualitative, namely by conducting: a) in-depth interviews

: Conducted with key informants, namely Dayak traditional elders, Dayak cultural experts, and members of the indigenous community who live around the Betang house. The interviews aim to explore the philosophical values of Huma Betang (togetherness, deliberation, mutual cooperation, tolerance) and their practices in daily conflict resolution, including post-conflict experiences in Sampit.

b) Participatory observation : conducted by visiting traditional Huma Betang houses in Sampit to observe the life of the indigenous community, social interactions, and deliberation practices. This observation includes observing social dynamics in a heterogeneous environment. c) Documentation study: analyzing documents related to Huma Betang, such as historical records, Dayak cultural literature, and reports on the Sampit conflict, to understand the historical and cultural context of Huma Betang philosophy.

(2) quantitative, namely by conducting questionnaires : used to collect data at the stage of validating materials and media.*flip book chart*A Likert-scale questionnaire was developed to assess the effectiveness, relevance, and appeal of the learning media from the perspectives of students, lecturers, and expert validators (media and cultural experts). This questionnaire was also used to measure students' understanding of Huma Betang-based conflict resolution after using the media.

Data analysis was conducted using different approaches for qualitative and quantitative data: (1) qualitative: using the Miles & Huberman model, which consists of three stages: (a) data reduction to select and simplify relevant information from interviews, observations, and workshops; (b) presentation of data in the form of narratives and main themes, such as the values of Huma Betang in conflict resolution; and (c) drawing conclusions to identify patterns and findings that support the development of*flip book chart*The results of the qualitative analysis form the basis for compiling media content, including conflict scenarios and solutions based on Huma Betang values.

(2) quantitative: using a Likert scale (1-4) to analyze the results of the validation questionnaire. The material validator emphasizes the depth of philosophical values and the relevance of the content, while the media validator focuses on visual design, interactivity, and the suitability of the media format to user needs and educational goals;

https://doi.org/10.1051/e3sconf/202565207004

on the other hand, student assessment towards media includes meeting educational standards, integrating Huma Betang cultural values, and the relevance of media in the context of conflict resolution.

This study uses the ADDIE model to develop*flip book chart* based on Huma Betang, with the following five stages:

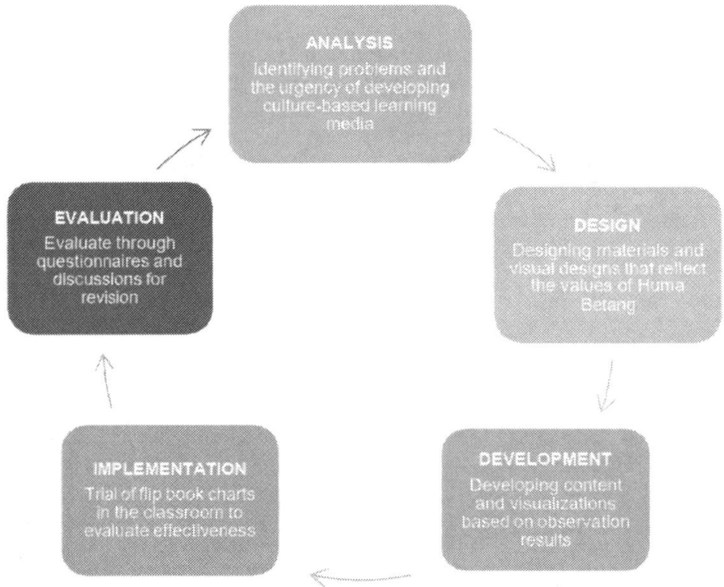

Fig. 1. ADDIE Model Based on Huma Betang
Source: Researcher, 2025

The ADDIE model has five stages: Analysis identified the need for culture-based learning media for the Conflict Management course through a study of Huma Betang literature and interviews with traditional elders. Design comprised materials reflecting Huma Betang values (togetherness, deliberation, mutual cooperation, tolerance) with visual designs themed on Dayak culture. Development resulted in a prototype based on interviews, observations, and workshops, validated by cultural and media experts. Implementation involved small group (8 students) and large group (35 students) trials in class, followed by FGDs to evaluate student experiences. Students are given a case study of a real-life conflict and are directed to analyze it using a flipbook guide. Implementations process involves group discussions, presentations of analysis results, and collaborative reflection under the guidance of a lecturer. Evaluation used a Likert scale questionnaire (1-4) and FGDs to assess aspects of material, visual design, interactivity, and format suitability, with revisions based on feedback. Qualitative analysis used the Miles & Huberman model (data reduction, narrative/theme presentation, conclusion drawing) to compile content, while quantitative analysis used a Likert scale to assess the validation of the material (depth of philosophical values, relevance) and media (visual design, interactivity, format suitability) as well as student assessments of educational standards, integration of Huma Betang values, and relevance in conflict resolution.

3 Results and Discussions

3.1 Huma Betang Values as a Foundation for Conflict Resolution Learning in Sampit, Central Kalimantan

Conflict resolution learning rooted in the values of Huma Betang aligns with the substance of conflict management learning. Huma Betang values are closely aligned with the socio- cultural realities of the Dayak community in Sampit, Central Kalimantan. Using local wisdom as a pedagogical foundation, this approach strengthens the link between higher education and the real needs of the community, while equipping students with the cultural sensitivity essential for constructive social engagement [10]. Philosophy *Huma Betang*is a representation of the values of the Dayak people's lives, which uphold the principle of harmony between humans, nature, and the creator. Symbols such as*Crispy Sticks*and principles *Not Yet True*not only a cultural heritage, but also a source of ethical values that are applicable in modern social dynamics, including in conflict resolution [11]. These values instill awareness of the importance of togetherness, deliberation, mutual cooperation, and tolerance as social capital in facing the complexity of contemporary conflicts, ranging from land disputes to conflicts between community groups.

Draft *There is* as a center of social harmony, as well as the Betang settlement system with its collective model that relies on mutual cooperation, provides important lessons about the value of social cohesion. Social cohesion refers to the level of solidarity, attachment, and sense of belonging among members of a community. In the context of Dayak society, social cohesion is reflected through the active involvement of residents in maintaining harmony together, helping each other in daily activities, and resolving conflicts through deliberation [12]. These values are an important foundation in strengthening an inclusive, tolerant, and just community life. In learning, this can be adapted through an approach*problem-based learning* and case studies that position students as facilitators in designing solutions to real- life conflicts, such as agrarian conflicts resulting from plantation investments. By placing value*togetherness* As a basis, students are trained to understand complex social dynamics and design fair and sustainable conflict resolution strategies.

Draft *There is* as a center of social harmony and settlement system *Betang* Mutual cooperation-based communities reflect the values of social cohesion, namely solidarity, attachment, and a sense of belonging among community members. In Dayak society, this is evident through the active participation of residents in maintaining harmony, helping each other, and resolving conflicts through deliberation. These values form the foundation of an inclusive, tolerant, and just community life. In learning, this principle can be applied through a problem-based learning approach and case studies, where students act as facilitators in formulating solutions to real-life conflicts, such as agrarian disputes resulting from plantation investments. This approach trains students to understand complex social dynamics and design fair and sustainable resolution strategies.

Deliberation as a pillar in the Belom Bahadat philosophy of life. Reflected in the Mamapas Lewu tradition, a Dayak tradition to restore social harmony after conflict. Through deliberation involving the entire community and traditional and religious leaders, collective decisions are made based on cultural values, resulting in resolutions accepted by all parties. Mamapas Lewu aims to cleanse nature (petak danum) from danger and create inner and outer peace, while also serving as a mechanism for conflict resolution based on deliberation and consensus to maintain socio-cultural balance.

Integration of deliberation values into learning through*Focus Group Discussion* (FGD) and negotiation simulations are educational strategies that foster communication skills, empathy, and a culture of dialogue in students. This approach reflects the spirit of deliberation among the Ngaju Dayak ancestors, which was embodied in the Tumbang Anoi Agreement in 1894, when 152 Dayak sub-tribes gathered at Huma Betang Damang Batu to resolve a major conflict peacefully and justly. This history is proof that even the most complex problems can be resolved through inclusive participation and consensus.

The spirit of mutual cooperation contained in *Surviving Petak Danum* And *Rebutting*Reflecting collective solidarity to maintain environmental and social sustainability. In learning, this approach can be realized through collaborative projects involving the community, for example through ecologically based conflict resolution activities.
such as water source protection campaigns or participatory land conflict mitigation. Customary practices such as*Mesanggar Memapas Lewu*It can also be an educational medium to align traditional values with modern approaches in maintaining collective welfare.

The spirit of mutual cooperation contained in traditional practices*Surviving Petak Danum* And *Rebutting*reflects the collective solidarity of the Dayak people in maintaining environmental and social sustainability. This tradition emphasizes the importance of community cooperation in protecting natural resources such as water and forests, which are culturally considered sacred and have high spiritual value. In learning, this approach can be realized through collaborative community-based projects, for example through water source protection campaigns, forest rehabilitation, or participatory land conflict mitigation. Customary practices such as *Mesanggar Memapas Lewu*It also serves as an educational medium to align traditional values with modern approaches to maintaining collective well- being. Students not only learn ecological and social concepts but also develop cultural sensitivity and responsibility for environmental sustainability.

Tolerance in principle*Hasunda*is the key to reconciliation in Sampit after the ethnic conflict, building harmony amidst diversity. By incorporating customary law such as the customary oath into education, one can learn to appreciate local pluralism and justice. Educational supervision that uses regional languages and culture*Huma Betang*creating equal and inclusive dialogue. Central Kalimantan Provincial Regulation No. 16 of 2008 concerning Dayak customary institutions recognizes customary institutions, strengthening culture-based conflict resolution as a complement to national law to maintain togetherness.

By integrating values such as*Garing Trunk, Belom Bahadat, Pali, Mamapas Lewu, Greeting Petak Danum,*And *Rebutting*In the learning process, conflict management education not only produces graduates who are theoretically competent, but also sensitive to the local socio-cultural context. This approach reinforces the idea*inclusive development from a Dayak perspective*, which places local communities as the main actors in the social transformation process.

Conflict resolution learning is based on Huma Betang values using a relevant, contextual approach. The values of togetherness, deliberation, mutual cooperation, and tolerance form a strong foundation for developing inclusive and sustainable conflict resolution strategies. Through methods such as focus group discussions (FGDs), case studies, and collaborative projects, students not only understand conflict management theory but also internalize local Dayak wisdom in building social harmony and respecting differences. This approach fosters cultural sensitivity and prepares students to become agents of peace in a pluralistic society.

Table 1. Classification of Huma Betang Values as a Foundation for Conflict Resolution Learning

The Value of Huma Betang	Description	Application in Conflict Resolution
Crispy Sticks	Symbol of harmony between humans, nature, and the creator, reflecting the balance of life.	Instilling awareness of the ecological and social interrelationships in conflict resolution, such as land disputes.
Not Yet True	The principle of living in society that emphasizes deliberation, togetherness, and justice.	Encourage conflict resolution through deliberation and consensus to achieve a solution that is acceptable to all parties.
There is	The concept of a center of social harmony that maintains balance within the community.	Strengthening social cohesion through active participation in maintaining harmony and resolve conflicts.
Mompas Lewu	Customary traditions for restoring social harmony after conflict through community deliberation.	Culture-based conflict resolution mechanisms for maintaining social and spiritual balance.
Surviving Petak Danum	Mutual cooperation practices to maintain sustainability environment, such as water sources and forests.	Encourage cooperation communities in mitigating environmental conflicts, such as plantation land conflicts.
Rebutting	Collective solidarity in protecting the environment and social welfare.	Building participatory conflict resolution strategies for social and ecological sustainability.
Hasunda	The principle of tolerance is the key to reconciliation in multicultural society.	Building harmony in amidst diversity, support the inclusive resolution of ethnic conflicts.

Source: researcher, 2025

3.2 Development of a Flip Book Chart integrating the values of Huma Betang in Sampit, Central Kalimantan

This research aims to develop learning media*Conflict Resolution Flip Book Chart*Based on the philosophical values of Huma Betang of the Dayak people in Central Kalimantan for the Conflict Management course. This media is designed to integrate local values such as deliberation, mutual cooperation, togetherness, and tolerance in the context of conflict resolution. The development process uses the Addie model, which consists of the following stages:*analysis, design, development, implementation, dan evaluation*.

Level of analysis (*analysis*), based on the results of observations and interviews, it shows a gap between the delivery of conflict resolution theory material and the contextual reality of students. Students tend not to understand how local cultural values such as Huma Betang can be used as a basis for conflict resolution. This is the basis for the need to develop informative media while grounding the values of local wisdom that live in society, such as in the context of the Sampit conflict. The values of Huma Betang were chosen because this philosophy has been historically proven to be able to reconcile ethnic conflicts, especially between the Dayak and Madurese communities in the early 2000s in Sampit [13]. Based on interviews, Huma Betang is defined as a typical Dayak longhouse that reflects togetherness, tolerance, deliberation, mutual cooperation, and mutual respect, with the principle of living together as one big family under one roof.

Design stage (*design*) on the visual *Flip Book Chart* designed according to the Dayak cultural context and interactive learning needs.*Flip Book* made in three sizes (A3, A4, and portable), with typical Kalimantan colors (brown, green, cream), and contains illustrations of Betang houses and the social life of the Dayak people. The content structure includes: (1) Chapter 1 introduction, which explains the introduction to the huma betang philosophy and the Sampit conflict which is a valuable lesson about the importance of peace, tolerance, and wise conflict management; (2) Chapter 2 conflict resolution and management, a local culture-based approach, such as facilitating cross-ethnic dialogue conducted in the traditional hall or Betang House; (3) Chapter 3 huma betang philosophy, as a cultural value that is a guideline for the life of the Dayak people in Central Kalimantan; (4) Chapter 4 closing, explaining the role of youth as agents of change by actively involving them in cultural preservation activities and inter-ethnic dialogue.

Validation by material experts resulted in a feasibility score of 95%. The indicators assessed included the accuracy of concepts, depth of content, relevance to the learning objectives, and integration of cultural and philosophical values. The experts provided specific input to deepen the historical narrative of the Sampit conflict, emphasizing the importance of presenting events not only factually but also contextually, so that students can understand the root causes and dynamics of conflict. Additionally, they highlighted the need to strengthen the role of Huma Betang philosophy, particularly in the context of deliberation and collective decision-making, as a fundamental principle of conflict resolution.

Meanwhile, validation by media experts obtained a feasibility score of 96.6%. The aspects evaluated included visual design, illustration quality, interactivity, readability of text and images, and the suitability of the format with learning standards. Media validators emphasized that the Flip Book Chart successfully combined aesthetic appeal and pedagogical clarity, making it both engaging and easy to follow for students. They also recommended strengthening visual consistency and ensuring that illustrations reflect local cultural symbols to enhance contextual authenticity.

The combined validation results indicate that the Flip Book Chart is feasible, contextual, and capable of providing a meaningful learning experience. Students not only gain knowledge about the Sampit conflict but also develop an understanding of conflict resolution strategies and appreciation of the Huma Betang philosophy. Effectiveness testing further reinforced this feasibility. In small group trials (8 students), the effectiveness score reached 85.2%, while in large group trials (35 students), it reached 89.6%. These results demonstrate that the media is effective in enhancing students' cognitive understanding, affective appreciation, and value-based reflection.

In summary, material validators placed emphasis on philosophical depth and historical accuracy, while media validators focused on design quality, interactivity, and cultural representation. Additionally, students' evaluations included indicators such as alignment with educational standards, integration of Huma Betang cultural values, and relevance of the media to real conflict resolution contexts. Altogether, these validation indicators confirm that the Flip Book Chart is a valid, effective, and culturally rooted learning medium.

Fig 2. Flip Book Chart Design
Source: researcher, 2025

Implementation stage *(implementation),* The Flip Book Chart media was piloted on two test groups. A small group of eight students used this media to analyze interpersonal conflicts in everyday life through simulation and reflection activities. Meanwhile, a large group of 35 students utilized the media to analyze group-based social conflicts through case studies and conflict mapping. The implementation results showed that students were able to map the roots of the conflict, identify resolution strategies based on deliberation and tolerance, and critically link the findings to conflict management theory. In addition, understanding of the Huma Betang philosophy as a symbol of inclusivity and identity of the Dayak community also increased [14]. The effectiveness of the media showed a very good category, with scores of 85.2% in the small group and 89.6% in the large group.

At the implementation stage (implementation), the Flip Book Chart media was piloted on two groups of students to test its practicality and effectiveness in real classroom learning. In the small group trial consisting of eight students, the media was used to analyze interpersonal conflicts in everyday life. The learning process was facilitated by the teacher, who acted as a learning guide and facilitator. The teacher explained the learning objectives, provided instructions on how to use the Flip Book Chart, and gave initial examples of interpersonal conflict scenarios. Students then worked in pairs and small teams to conduct simulation and reflection activities, where they identified the root causes of interpersonal conflicts, mapped possible resolution paths, and reflected on the role of tolerance and deliberation. The teacher's role was to scaffold the learning process, provide feedback during discussions, and ensure that students connected their reflections to the broader framework of conflict management theory.

In the large group trial involving 35 students, the media was implemented through the analysis of group-based social conflicts using case studies and conflict mapping activities. The teacher began by introducing a real-life social conflict case relevant to the students' context, then guided the students in dividing into smaller groups for collaborative work. Each group used the Flip Book Chart to map the root causes of conflict, design alternative resolution strategies, and present their findings to the class. The students' role was more active and collaborative, as they were required to critically analyze, debate, and connect conflict cases to the principles of the Huma Betang philosophy. The teacher's role in this stage was to moderate group presentations, ensure balanced participation, and highlight key theoretical insights so that students could link their practical findings to established concepts in conflict management.

Research such as "Development of Digital Flipbook Media Based on Local Wisdom to Improve Students' Critical Thinking Skills" by Rini Febriani, Hairunisa, and Syafruddin (2025) proves that flipbook media that integrates local values can improve students' achievement in critical thinking [15]. Likewise, research "Development of Digital Flipbook Based on Local Wisdom" by Syarifah Muthohharoh, Arum Ratnaningsih, and Suyoto (2024) on the material of gotong royong (mutual cooperation) for grade IV, using the ADDIE model, produced a very feasible media with high practicality (percentage 87.5%) and effectiveness 73.67%. Furthermore, the development of flippingbooks at SMP Negeri 3 Kabila also showed that the media met the aspects of validity, practicality, and effectiveness in learning [15].

The results of implementation showed that students were able to map conflict roots, identify resolution strategies based on deliberation and tolerance, and critically connect their findings to conflict management theory. Moreover, their understanding of the Huma Betang philosophy as a symbol of inclusivity, collectivism, and Dayak identity also increased significantly. The effectiveness of the media showed a very good category, with scores of 85.2% in the small group and 89.6% in the large group. These findings indicate that the Flip Book Chart not only supports cognitive and analytical skills but also facilitates value internalization and collaborative learning, with teachers playing a guiding role and students actively engaging as learners, analysts, and problem-solvers.

Evaluation stage (evaluation) was conducted through qualitative data analysis using the Miles and Huberman model. The evaluation results showed that the use of Flip Book Chart media encouraged students' conceptual and reflective understanding, as well as increased appreciation of local values in conflict management practices. This evaluation reinforced the finding that the media was not only effective in conveying material but also contextually relevant in building cultural awareness among students. The Flip Book Chart media's high cultural integration is one of its key benefits. The media not only imparts theoretical knowledge about conflict management but also fosters cultural awareness and appreciation in students by incorporating the principles of the Huma Betang philosophy. As part of their reflective learning, students are able to cultivate inclusion, debate, and tolerance by connecting academic concepts with their local wisdom, which gives the learning process greater value.

Conversely, a significant drawback of the Flip Book Chart media is its dependence on teacher involvement. The success of learning tasks like simulations, case studies, and reflections greatly relies on the teacher's ability to facilitate discussions and help students connect their experiences to theoretical concepts critically. In the absence of proactive facilitation, there's a chance that students might concentrate solely on superficial observations, diminishing the depth of conceptual and reflective comprehension that the media aims to achieve

4 Conclusion

The findings of this research indicate that the Flip Book Chart media grounded in Huma Betang values is successful in enhancing students' comprehension of conflict, bolstering their reflective abilities, and promoting cultural awareness in learning about conflict resolution. Incorporating values like deliberation, mutual cooperation, and tolerance establishes a solid basis for students to critically assess conflicts and create inclusive solutions. By means of case studies and simulations, students could relate theoretical ideas to real world situations, thus achieving the research aim of creating a culturally grounded and pedagogically sound educational tool.

It is recommended for future studies to broaden the application of this media to various contexts and topics, including multicultural education, civic education, or peace studies, to evaluate its flexibility and potential for wider use. Future research might additionally concentrate on creating a digital format of the Flip Book Chart to enhance accessibility and engagement. Moreover, engaging a broader sample of students from different areas would offer richer perspectives on how local knowledge can be incorporated into conflict resolution education across diverse cultural contexts.

References

1. D. Wulandari, "Implementasi Program Pemajuan Kebudayaan Desa: Tinjauan Pemberdayaan Masyarakat Berbasis Budaya," *Jurnal Pendidikan dan Kebudayaan*, vol. 9, no. 1, pp. 20–34, Jun. 2024, doi: 10.24832/jpnk.v9i1.4489.
2. P. Sumantri, Y. Tanjung, and S. Aymanda Nababan, "Pendidikan perdamaian terintegrasi dalam ilmu pengetahuan sosial berbasis kearifan lokal Pela Gandong," *Jurnal Pendidikan Ilmu Pengetahuan Sosial Indonesia*, vol. 7, no. 3, pp. 195–205, Sep. 2022.
3. N. P. Intani, S. Nadzifah, A. L. Hakim, and M. H. Asy'ari, "Perang Sampit (konflik Suku Dayak dengan Suku Madura) pada tahun 2001," *SOSIAL : Jurnal Peneitian Ilmu-Ilmu Sosial*, vol. 23, no. 2, pp. 14–8, Oct. 2022, [Online]. Available: http://sosial.unmermadiun.ac.id/index.php/sosial
4. I. E. A. Pelu and J. Tarantang, "Interkoneksi nilai-nilai Huma Betang Kalimantan Tengah dengan Pancasila," *Jurnal Studi Agama dan Masyarakat*, vol. 14, no. 2, p. 119, Sep. 2018, doi: 10.23971/jsam.v14i2.928.
5. R. Nasa and G. Nuwa, "Resolusi konflik berbasis kearifan lokal dalam kehidupan etnis Sikka Krowe," *Edukatif : Jurnal Ilmu Pendidikan*, vol. 4, no. 1, pp. 1–6, Nov. 2022, doi: 10.31004/edukatif.v4i1.1720.
6. A. Loi, D. Kwirinus, and H. Peri, "The Concept of 'Huma Betang' as a Model of the Values of the Philosophy of Life of the Dayak Community in Central Kalimantan," *Jurnal Pendidikan Sosiologi dan Humaniora*, vol. 14, no. 2, p. 219, Oct. 2023, doi: 10.26418/j-psh.v14i2.64914.
7. M. E. Daryati, D. Suryadi, Wembrayarli, and M. Hatta, "Integration of local wisdom literacy based on flipbook worksheets to enhance students' creative problem-solving in the foundation phase of the independent curriculum," vol. 07, no. 02, pp. 255–268, 2024, [Online]. Available: http://ejournal.radenintan.ac.id/index.php/al-athfaal
8. M. S. Alfarizi, "Efektivitas penggunaan media flipbook dalam pembelajaran PAI untuk meningkatkan minat belajar siswa pada mata pelajaran PAI kelas VII di UPT SMP Negeri 9 Gresik," UIN Sunan Ampel Surabaya, Surabaya, 2023.
9. Ade Rahayu, "Metode Penelitian dan Pengembangan (R&D) : Pengertian, Jenis dan Tahapan," *DIAJAR: Jurnal Pendidikan dan Pembelajaran*, vol. 4, no. 3, pp. 459–470, Jul. 2025, doi: 10.54259/diajar.v4i3.5092.
10. Dakir, "Pengelolaan Budaya Inklusif Berbasis Nilai Belom Bahadat pada Huma Betang dan Transformasi Sosial Masyarakat Dayak Kalimantan Tengah," *Religió: Jurnal Studi Agama-Agama*, vol. 7, no. 1, pp. 28–54, 2017.
11. E. Chilwanto, G. Rahmad, and A. Saefulloh, "Upacara Adat Mamapas Lewu (Studi Kasus di Kota Kasongan Kalimantan Tengah)," *Jurnal Kewarganegaraan*, vol. 5, no. 2, 2021.
12. S. R. Usop and I. Rajiani, "Indigenous Indonesian Dayak Traditional Wisdom in Reducing," *Indonesian Journal of Geography*, vol. 53, no. 3, pp. 310–317, 2021.
13. Muthohharoh, S., & Ratnaningsih, A. (2024). Pengembangan Flipbook Digital Berbasis

Kearifan Lokal Pada Pembelajaran Pendidikan Pancasila Dan Kewarganegaraan Materi Gotong Royong Kelas Iv Sd Negeri Secang. *Didaktik: Jurnal Ilmiah PGSD STKIP Subang*, *10*(1), 1930-1937.

14. P. Sumantri, Y. Tanjung, S. A. Nababan, Integrated Peace Education in Social Sciences Based on Local Wisdom of Pela Gandong. PIPSI Journal (Indonesian Social Studies Education Journal). 7, 195-205 (2022).
15. Iman, S. H., Yusuf, M., Abdjul, T., & Odja, A. H. (2023). Pengembangan E-Modul Kearifan Lokal Berbantuan Aplikasi Flipping Book pada Materi Getaran, Gelombang dan Bunyi di SMP Negeri 3 Kabila. *ORBITA: Jurnal Pendidikan dan Ilmu Fisika*, *9*(2), 294-304.

Mangrove Awareness and Coastal Resilience Education through Geo-game Design: Bridging Ecological Knowledge and Student Experience

Alfyananda Kurnia Putra[1*], *Lailatul* Umami[2], *Yusuf* Suharto[3] and *Diky* Al Khalidy[4]

[1,2,3,4]Departement of Geography, Universitas Negeri Malang, Malang, Indonesia

Abstract. In this study, we present the Lusi Eco-Quest, a place-based geo-game designed to enhance environmental literacy and spatial thinking in geography education. Situated on Pulau Lusi, a coastal island formed by sedimentation following the 2006 Lapindo mudflow, the game incorporates localized themes of landform transformation, coastal resilience, and mangrove conservation in an interactive role-playing format. Developed using RPG Maker MZ to inquiry-driven missions rooted in real-world ecological contexts. The evaluation involved 36 participants who utilized the User Experience Questionnaire – Short Version (UEQ-S) and the System Usability Scale (SUS), supplemented by an expert review. The results indicated an excellent user experience (mean UEQ-S score: 1.98) and acceptable usability (SUS score: 69), with recommendations for enhanced instructional support for future use. The game facilitates experiential and problem-based learning by enabling players to analyze spatial risks, simulate conservation practices and engage with environmental narratives. The findings suggest that Lusi Eco-Quest is an effective educational tool that integrates geospatial technology with local environmental content, offering a replicable model for enhancing student engagement and ecological understanding in disaster-affected coastal settings.
Keywords. Environmental Geography, Geo-game, Coastal Resilience, Mangrove Conservation, Pulau Lusi.

1 Introduction

Despite the widespread recognition of the ecological significance of mangrove ecosystems, the rapid pace of their degradation continues to outstrip the evolution of effective educational interventions. This degradation is driven by intensifying anthropogenic pressures and climate-induced transformations of coastal environments [1]. This research is anchored in the urgent necessity to reconceptualize geography education as a vehicle for cultivating ecological literacy and resilience-oriented thinking in teacher education [2]. This study aims to achieve this through pedagogical models that are contextually rooted, digitally facilitated, and experientially meaningful for learners.

* Corresponding author: alfyananda.fis@um.ac.id

This challenge is particularly acute within the Indonesian educational landscape, where the national geography curriculum mandates the study of coastal ecosystems and disaster mitigation [3]. However, preliminary investigation revealed that conventional pedagogical approaches, such as fieldwork, face significant logistical and pedagogical constraints, including limitations related to cost, time, accessibility, and safety [4]. These constraints often result in a reliance on textbook-based, decontextualized instruction, which fails to foster a deep, experiential understanding of complex socio-ecological systems. Consequently, our needs analysis identified a clear pedagogical gap: the demand for an innovative educational tool that can simulate place-based environmental inquiry, offering an accessible and engaging alternative to traditional fieldwork while aligning with curricular goals.

Within the global discourse on innovative pedagogy, serious games and digitally mediated geo-games have emerged as promising tools for fostering spatial cognition, systems-level ecological understanding, and emotional engagement with environmental complexity [5]. These interactive environments are increasingly recognized for their potential to simulate socio-environmental processes, scaffold inquiry-driven learning, and provide students with opportunities to explore the implications of environmental decisions in low-risk virtual spaces. Previous empirical studies have substantiated these affordances, particularly in environmental and geographical education contexts [6].

However, several critical gaps persist. A substantial proportion of existing educational games are disconnected from locally grounded ecological contexts [7]. They frequently rely on generalized or Western-centric ecological narratives that lack cultural and geographical specificity. Moreover, these games are rarely aligned with the national curricula or designed to engage students in authentic, place-based environmental problem-solving. These limitations are especially pronounced in underrepresented educational regions, such as those in the Global South. While promising, the application of this technology reveals critical gaps that our study seeks to address. A substantial proportion of existing educational games remain disconnected from locally grounded ecological contexts, often relying on generalized or Western-centric ecological narratives that lack cultural and geographical specificity. Moreover, few are explicitly designed to align with national curricula or to engage students in authentic, place-based environmental problem-solving, a limitation especially pronounced in regions like the Global South.

This dual disconnect, which is both epistemological and pedagogical, calls for a reorientation of the design of game-based learning tools. There is a clear need for geo-games that integrate localized environmental knowledge while responding to curricular mandates and learner contexts. In response, this study introduces the Lusi Eco-Quest, a role-playing geo-game designed for the Pulau Lusi mangrove environment in East Java, Indonesia. The landscape of Pulau Lusi is the result of sediment accumulation from large-scale mudflow disasters (MFDs). This hybrid coastal site represents a unique intersection of anthropogenic transformation, ecological recovery, and community-based conservation. As such, it offers a compelling case for embedding environmental reasoning, resilience discourse, and spatial learning into immersive digital experiences.

The principal innovation of this research lies in the integration of place-responsive environmental content, geospatial thinking, and interactive digital media. The development of Lusi Eco-Quest followed a user-centered design methodology and was evaluated using the System Usability Scale (SUS) and the User Experience Questionnaire Short Version (UEQ-S). Through this dual evaluation approach, this study contributes in two significant ways. First, it presents a replicable model for embedding real-world environmental scenarios into curriculum-aligned digital learning tools. Second, it provides empirical insights into how students construct ecological understanding and spatial awareness through digital interactions and immersive engagement. Ultimately, this research responds to the broader imperative of grounding environmental education in lived ecological realities. It advances geography education not only as a means of delivering content but also as a transformative praxis that integrates technological innovation with ecological sensitivity, critical spatial reasoning, and pedagogical authenticity.

2 Method

2.1 Research Design

This study adopted a Research and Development (R&D) approach with the primary objective of producing a functional and contextually relevant digital educational game branded as "Lusi Eco-Quest." [8]. The development process was systematically structured using the PLOMP model, with three main phases: (1) preliminary investigation, (2) Design and Prototyping, and (3) Assessment and Evaluation, as presented in Figure 1. The model was selected for its ability to facilitate a structured and systematic approach to the development ofeducational tools, allowing for continuous improvement based on feedback and iterative enhancement of the geo-game product.

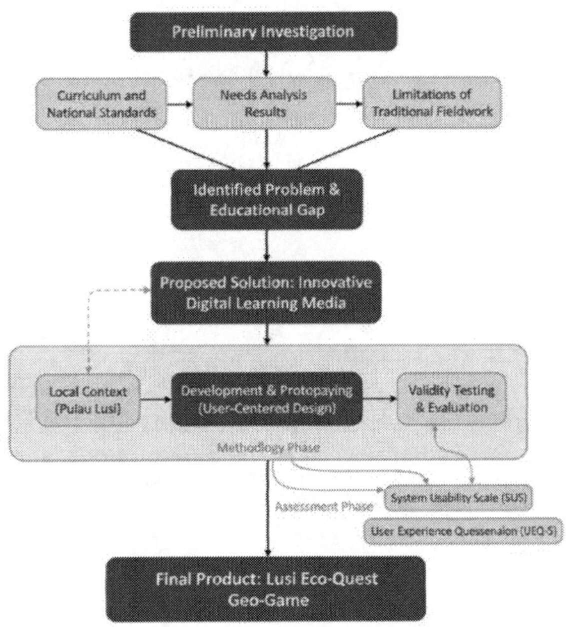

Fig 1. Methodological Flowchart for the Design and Validation of the Lusi Eco-Quest Geo-Game

2.1.1 Preliminary Investigation

The preliminary investigation focused on identifying the design criteria and functional requirements of the geo-game. This phase included a review of existing geo-games, environmental simulation software, and interactive map-based learning tools [9]. Additionally, needs assessments were conducted through user analysis to capture target user expectations related to content accuracy, interface intuitiveness and narrative relevance. Technical constraints, platform compatibility, and user experience expectations were also documented as design constraints.

2.1.2 Design and Prototyping

The Design and Prototyping Phase emphasized the development of the initial geo-game build using RPG Maker MZ as the authoring platform. Core game mechanics were programmed to simulate ecological interactions, role-based tasks, and interactive environmental challenges. The design activities involved were as follows:

1. Mapping the digital landscape of Pulau Lusi to ensure that the game accurately represents the island's ecological features and real-world dynamics (*Environment Modeling*).
2. Creating game characters, storylines, and item databases relevant to the mangrove ecosystem and designed to engage players in meaningful learning about coastal resilience (*Asset Generation*).
3. This involves defining the player-game-environment logic using event triggers, variable control, and conditional branches to simulate dynamic interactions between the player and the game environment (*Interaction Scripting*).
4. User interfaces that balance functionality and aesthetic minimalism were constructed to ensure that the game was intuitive, visually appealing, and easy to navigate (*Interface Design*).

Expert Judgement was integral to evaluating the game at various stages of development to ensure its feasibility and educational value. Experts in game design, ecology, and education were consulted to assess the quality, usability, and educational alignment of this game.

2.1.3 Assessment and Evaluation

The assessment and evaluation phase involved structured usability testing to assess the prototype's technical reliability and experiential quality. This stage includes controlled play sessions, bug identification, performance logging, and interface audits. Two standardized instruments were deployed: the System Usability Scale (SUS) to evaluate functional usability, and the User Experience Questionnaire – Short Version (UEQ-S) to assess hedonic aspects, such as engagement, enjoyment, and perceived innovation.

2.2 Research Object

The research object in this study is the Lusi Eco-Quest, designed to foster students' understanding of biodiversity and coastal resilience through immersive, localized digital learning. The game content is centered around Pulau Lusi, a human-induced island in Sidoarjo Regency, East Java, Indonesia. Pulau Lusi was selected due to its geographical significance, environmental sensitivity, and its potential to serve as a place-based learning platform in geography education. The geographical location of the study area is shown in Figure 2.

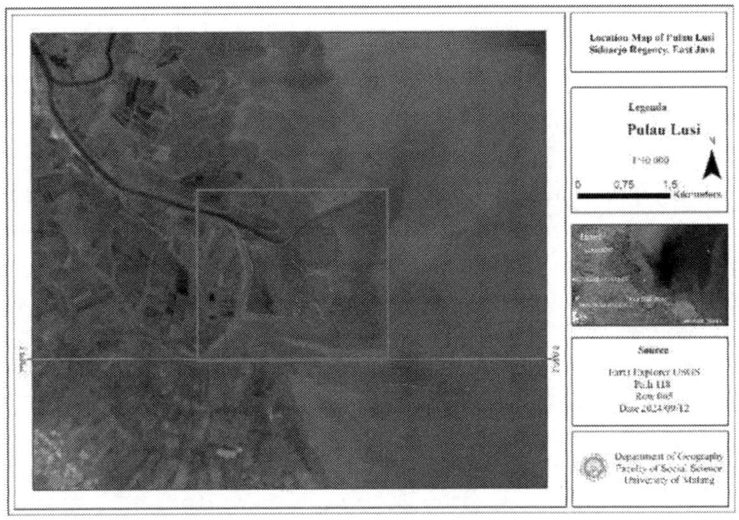

Fig 2. Location Map of Pulau Lusi, Sidoarjo Regency, East Java, Indonesia

2.3 Data Collection and Analysis Procedure

Data were collected using three main instruments: (1) expert judgement, (2) the UEQ-S, and (3) the SUS. The data gathered from these instruments were analyzed quantitatively and qualitatively to evaluate the usability, experience, and content validity of the Lusi Eco-Quest. The feedback obtained from the Expert Review was analyzed qualitatively. Experts suggested areas for improvement in game design, content accuracy, and educational effectiveness. Furthermore, feedback from students gathered during the game's evaluation was categorized to understand their perceptions of the game's educational value, its usability, and their overall engagement.

The SUS scores were calculated to assess the usability of the game. The responses were analyzed using a standard conversion method, where odd-numbered items were converted using the following formula: $Adjusted\ value = x - 1$, and even-numbered items with $Adjusted\ value = 5 - x$. The adjusted scores were then summed, and the total score was multiplied by 2.5 to obtain the final SUS score. This score was compared with industry benchmarks to determine the overall usability of the game. Additionally, the UEQ-S scores were analyzed to assess both the pragmatic and hedonic qualities of the user experience. The mean, variance, and standard deviation for each item were calculated to provide a comprehensive understanding of the game's strengths and weaknesses regarding usability and engagement. The results were then categorized into different levels of user experience, such as excellent, good, or poor, based on the international UEQ benchmarks.

3 Result and Discussion

3.1 Development Process, Design, and Characteristics of the Lusi Eco-Quest Geo-Game

The development of the Lusi Eco-Quest geo-game adhered to a user-centered design (UCD) methodology, emphasizing iterative stages of analysis, design, development, and evaluation to ensure its relevance, usability, and pedagogical effectiveness [10]. This approach was critical to integrating the specific needs identified from the preliminary investigation and needs analysis (as elaborated in the Introduction section) into the final product [11]. The entire development process unfolded over a six-month period, involving a dedicated team of geography educators, game developers, and environmental experts.

The initial phase involved comprehensive content analysis of the Indonesian national geography curriculum relevant to coastal ecosystems and disaster mitigation, coupled with in-depth studies of Pulau Lusi's unique geo-environmental characteristics and socio-ecological dynamics. This informed the foundational learning objectives for the geo-game, focusing on enhancing ecological literacy, spatial thinking, and fostering resilience-oriented decision-making.

Subsequently, the design phase commenced with the creation of detailed storyboards and game mechanics blueprints. These blueprints outlined the narrative flow, mission structures, character roles, and interactive elements. The geo-game was specifically developed as a role-playing geo-game to facilitate immersive learning within the context of environmental geography. RPG Maker MV was chosen as the primary development platform due to its flexibility in creating compelling virtual environments suitable for educational narratives, and its robust event system which allowed for complex story and mission design. For the precise integration and interpretation of geospatial data, ArcGIS was utilized for spatial data integration, particularly in the interpretation of satellite imagery of the Pulau Lusi ecosystem. This ensured the accurate reflection of real geographical conditions and allowed for the precise embedding of local content into authentic, place-based educational narratives [12].

Lusi Eco-Quest incorporates a carefully selected set of gamification elements designed to enhance engagement and motivation: (1) Plan (strategic decision-making), (2) Level (progression and skill acquisition), (3) Ranking (social comparison and achievement), (4) Progress bar (visual feedback on task completion), (5) Virtual currency (reward system and resource management), (6) Consequences (impact of choices), and (7) Rewards or penalties (reinforcement of desired behaviors). These elements were chosen based on established principles of game-based learning to promote active participation and sustained interest.

Pedagogically, the Lusi Eco-Quest is designed to improve ecological literacy regarding the functions and dynamics of mangrove ecosystems. It endeavors to enhance critical awareness of coastal resilience issues through a problem-based learning methodology that reflects real-world challenges in coastal regions. Users engage in virtual environmental exploration, which is intended to enhance their spatial thinking skills in a contextualized manner and foster emotional connections with the surrounding environment [13]. The integration of local thematic content, geographic visualization, and digital interaction positions Lusi Eco-Quest as a strategic tool for elucidating coastal areas in alignment with technological advancements. The game is structured around a series of missions that users must complete, incorporating narrative elements into gameplay. This exploration of information about coastal areas and mangrove ecosystems, along with challenges in the form of questions, provides users with an authentic, simulated experience of Pulau Lusi. Furthermore, Lusi Eco-Quest simulates a reflective and action-based process, thereby internalizing the values of sustainability and ecological responsibility.

3.2 The Formation of Pulau Lusi

The emergence of Pulau Lusi constitutes a unique instance of anthropogenically influenced coastal geomorphological transformation within the Porong Estuary in East Java. Located in the Sidoarjo Regency, the island began to form after the Lapindo mud eruption in 2006, which released substantial volumes of sediment-laden water into the Porong River system. This persistent sediment influx expedited fluvial deposition at the river mouth, culminating in the gradual development of a depositional landform, eventually designated as Pulau Lusi [12]. These processes exemplify rapid deltaic progradation, an uncommon occurrence in contemporary coastal geomorphology, particularly one indirectly instigated by an industrial disaster [13].

Fig 3. Temporal Satellite Imagery of Pulau Lusi Formation (2005–2025)

Temporal analysis using satellite imagery (Figure 3) elucidates the spatial development of the island from 2005 to 2020. In 2005, the site exhibited no discernible land masses. By 2010, sediment accretion initiated the formation of a small mudflat, which experienced significant expansion by 2015 and was further stabilized by the growth of natural vegetation, particularly mangrove colonization, by 2020. The spatiotemporal progression of Pulau Lusi exemplifies the transformation of estuarine sediment plumes into stable landforms through continuous deposition and ecological succession. This transition from a depositional substrate to a biologically productive island illustrates a hybrid system influenced by both natural and anthropogenic disturbances.

3.3 Integration of Coastal Resilience in the Lusi Eco-Quest Geo-game

Coastal resilience refers to the ability of coastal landforms and ecosystems to endure, absorb, and recover from natural and anthropogenic environmental stressors [12]. This concept encompasses the interaction of geomorphological stability, ecological integrity, and hydrodynamic processes, which collectively determine a coastline's capacity to persist and adapt under changing conditions. In the unique context of Pulau Lusi, resilience is not a static attribute but rather a dynamic result of the interactions among sediment dynamics, vegetation succession, and human intervention within the post-disaster coastal environment of the Porong Estuary.

Pulau Lusi serves as an exemplar of a newly formed geomorphic unit arising from substantial sediment discharge attributable to the Lapindo mudflow disaster. Sediment deposition at the estuarine outlet facilitated the formation of an accretional landform, which subsequently experienced rapid ecological colonization, predominantly by mangroves. This process, marked by sediment stabilization, root-binding by pioneer mangrove species, and organic matter accumulation, illustrates a natural trajectory of ecosystem resilience [13]. Vegetation presence is pivotal in attenuating wave energy, enhancing shoreline stabilization, and promoting sediment retention, thereby contributing to the island's physical resilience.

These feedback mechanisms correspond to resilience indicators, such as morphodynamic equilibrium, adaptive landform persistence, and biogeomorphic succession.

To directly address the pedagogical objective of fostering resilience-oriented thinking, the Lusi Eco-Quest geo-game meticulously integrates these complex aspects of coastal resilience into its core gameplay mechanics, narrative, and environmental simulations. The development team translated the theoretical framework of coastal resilience into interactive scenarios, ensuring that players actively engage with the principles of adaptation and recovery.

3.4 Mangrove Conservation Education in Lusi Eco-Quest

Mangrove ecosystems constitute a vital biophysical component of coastal environments owing to their multifaceted roles in stabilizing coastlines, regulating sediment dynamics, and supporting diverse estuarine biodiversity. In the Lusi Eco-Quest, mangrove conservation is presented not merely as an ecological theme but as a spatially grounded imperative rooted in the geomorphological and ecological context of Pulau Lusi. Formed through sedimentation following the Lapindo mudflow, the island has experienced rapid ecological succession, with mangroves serving as pioneer vegetation that contributes to shoreline stabilization, organic matter accumulation and habitat structuring.

From a physical geography perspective, the presence and expansion of mangrove zones are key indicators of coastal-system resilience. Their root systems bind loose sediments, attenuate wave energy, and foster conditions for vertical land accretion, thereby acting as natural buffers against coastal erosion and the rise in sea level. In the game, these functions are integrated into the mission design, where players are tasked with identifying critical mangrove zones, evaluating ecosystem degradation, and executing restoration strategies. The game situates these challenges within a spatially explicit framework, simulating the biogeographic distribution of species, zonation patterns, and anthropogenic pressures, such as land reclamation and pollution.

By synthesizing these ecological and spatial dimensions, Lusi Eco-Quest transforms mangrove conservation from passive instructional content into an immersive and place-based environmental experience. This enables learners to analyze the spatial consequences of environmental degradation, assess restoration trade-offs, and engage in ethically grounded decision-making. By linking geospatial thinking with ecological awareness, the game fosters an integrated understanding of mangrove ecosystems as living infrastructure crucial to the sustainability of dynamic coastal landscapes such as Pulau Lusi.

3.5 UEQ-S and SUS Assessments

The UEQ-S was administered to 36 students and assessed two core dimensions of user experience: pragmatic quality, encompassing clarity, efficiency, and dependability, and hedonic quality, which includes emotional appeal, stimulation, and novelty. The results indicated an average score of 2.000 for pragmatic quality and 1.965 for hedonic quality, with a composite average of 1.983, all of which are classified as “excellent” according to the international UEQ benchmarks. These findings suggest that the Lusi Eco-Quest not only operates efficiently from a usability perspective but also provides an engaging and emotionally resonant experience for learners. A high hedonic score indicates that the game design effectively incorporates motivation-enhancing features, such as narrative engagement, interactivity, and thematic relevance, which are crucial for sustaining learner attention in educational gaming environments.

Conversely, the SUS assessment yielded an overall usability score of 69, placing it within the “marginal-high” acceptability range, which is acceptable. According to standard interpretation frameworks, this score corresponds to grade D and is categorized as "Good" in adjective rating. While this result reflects a generally positive reception, it also reveals certain usability limitations that require further refinement. Specifically, feedback obtained through student interviews highlighted challenges in navigating the game interface and understanding the gameplay mechanics, underscoring the need for enhanced scaffolding tools, such as guidebooks or video tutorials. These usability issues emphasize the importance of aligning interface design with the cognitive and technological readiness of the target demographic in the development of educational games.

4 Conclusion

This research successfully addressed key objectives in advancing geography education through digital innovation, focusing on ecological literacy and coastal resilience. The development of the Lusi Eco-Quest geo-game effectively created a contextual, place-based role-playing game for the Pulau Lusi mangrove environment. Utilizing a user-centered design methodology with RPG Maker MV for game development and Geospatial tool for precise spatial data integration, the game seamlessly incorporates local ecological content including Pulau Lusi's unique formation, coastal resilience principles, and mangrove conservation into an authentic and immersive learning environment. The evaluation revealed an excellent user experience through the UEQ-S across both pragmatic and hedonic dimensions, confirming the game's engaging and motivating design. Concurrently, SUS scores indicated good usability, validating the game's intuitive operation. These findings affirm Lusi Eco-Quest's significant educational value and strong user acceptance, although results also highlighted a need for enhanced in-game guidance to optimize user interaction. The game effectively facilitates ecological understanding and spatial awareness by immersing players in inquiry-driven, problem-based scenarios that translate complex environmental dynamics into actionable learning experiences. Ultimately, this study offers a replicable model for integrating real-world environmental scenarios into curriculum-aligned digital learning tools, advancing geography education as a transformative praxis. While Lusi Eco-Quest demonstrates strong educational potential, future research should focus on longitudinal impact assessments on learning outcomes and pro-environmental behaviors, comparative efficacy studies against traditional methods, refinement of in-game guidance mechanisms, scalability and adaptability of the framework to other ecological contexts, and integration with multi-modal learning approaches.

References

1. V. Hagger *et al.*, "Drivers of global mangrove loss and gain in social-ecological systems," *Nat. Commun.*, vol. 13, no. 1, 2022, doi: 10.1038/s41467-022-33962-x.
2. A. K. Putra, Sumarmi, I. Deffinika, and M. N. Islam, "The effect of blended project-based learning with stem approach to spatial thinking ability and geographic skill," *Int. J. Instr.*, vol. 14, no. 3, 2021, doi: 10.29333/iji.2021.14340a.
3. A. Amri, J. A. Lassa, Y. Tebe, N. R. Hanifa, J. Kumar, and S. Sagala, "Pathways to Disaster Risk Reduction Education integration in schools: Insights from SPAB evaluation in Indonesia," *Int. J. Disaster Risk Reduct.*, vol. 73, 2022, doi: 10.1016/j.ijdrr.2022.102860.
4. Y. Li, S. Krause, A. McLendon, and I. Jo, "Teaching a geography field methods course amid the COVID-19 pandemic: reflections and lessons learned," *J. Geogr. High. Educ.*, vol. 47, no. 2, 2023, doi: 10.1080/03098265.2022.2041571.
5. A. K. Putra, A. S. Sumarmi, A. Fajrilia, M. N. Islam, and B. Yembuu, "Effect of

Mobile-Augmented Reality (MAR) in Digital Encyclopedia on The Complex Problem Solving and Attitudes of Undergraduate Student," *Int. J. Emerg. Technol. Learn.*, vol. 16, no. 7, 2021, doi: 10.3991/ijet.v16i07.21223.

6. E. J. Slattery *et al.*, "Effectiveness of a minecraft education intervention for improving spatial thinking in primary school children: A mixed methods two-level cluster randomised trial," *Learn. Instr.*, vol. 94, no. July, p. 102003, 2024, doi: 10.1016/j.learninstruc.2024.102003.
7. D. Fernández Galeote, L. Gabrielaitis, G. Guillén, and J. Hamari, "Play, games, and gamification to support sustainability transitions: a scoping review and research agenda," *Environ. Innov. Soc. Transitions*, vol. 57, 2025, doi: 10.1016/j.eist.2025.101025.
8. W. . Borg and M. . Gall, "Educational Research an Introduction fourth edition," *Longman Inc*, vol. 1, no. 1, 1983.
9. ISO 9241-210, "Ergonomics of human system interaction Part 210 : Human-centred design for interactive systems," *Int. Organ. Stand.*, vol. 10406–1:20, 2019.
10. A. R. Soucy, P. Rahimzadeh-Bajgiran, S. De Urioste-Stone, K. Eitel, J. Jansujwicz, and M. Brownlee, "Impact of a place-based educational approach on student and community members' experiences and learning within a post-secondary GIS course," *J. Geosci. Educ.*, vol. 0, no. 0, pp. 1–16, 2024, doi: 10.1080/10899995.2024.2379194.
11. A. K. Putra, Sumarmi, S. Bachri, D. Al Khalidy, and I. A. S. Huda, "Spatial Acumen in the Digital Age: The Transformative Power of STEM-Enriched Mobile Textbooks in Geography," *Int. J. Interact. Mob. Technol.*, vol. 17, no. 23, 2023, doi: 10.3991/IJIM.V17I23.45949.
12. L. Rölfer, L. Celliers, and D. J. Abson, "Resilience and coastal governance: knowledge and navigation between stability and transformation," *Ecol. Soc.*, vol. 27, no. 2, 2022, doi: 10.5751/ES-13244-270240.
13. A. Kazemi, L. Castillo, and O. M. Curet, "Mangrove roots model suggest an optimal porosity to prevent erosion," *Sci. Rep.*, vol. 11, no. 1, pp. 1–14, 2021, doi: 10.1038/s41598-021-88119-5.

Traces of Submarine Volcanism and Tectonic Uplift : Evidence of Geological Interaction at Watudodol Beach

Muhammad Zaki Gymnastiar Putra[1], Melati Julia Rahma[2*], *and* Firman Sauqi Nur Sabila[3]

[1,2]Department of Geography, Faculty of Social Sciences, State University of Malang, 65145 East Java, Indonesia
[3]Ijen Geopark Geological Information Centre, 68465 East Java, Indonesia

Abstract. Watudodol Beach in Banyuwangi is an important location for understanding the interaction between undersea volcanism and tectonic uplift at the eastern tip of Java. The presence of pillow-structured basalt lava, fossilised limestone, and volcanic breccia records complex geological dynamics. This research was conducted through field observations, outcrop descriptions, sampling, petrographic analysis, and integration with regional geological literature studies. The results of the study show three main stratigraphic units: pillow-structured basalt lava and autobreccia with a thickness of >3 m, volcanic breccia ±1 m thick with basalt fragments and calcite veins, and clastic limestone associated with marine fossils (molluscs, corals, foraminifera) with a thickness of 20–50 cm. These outcrops are exposed at elevations ranging from 0 to 544 metres above sea level, indicating significant post-depositional uplift. Mineralogical analysis shows a dominance of plagioclase (36%), K-feldspar (11%), pyroxene (8%), and volcanic glass (24%), reflecting basaltic magma with a hypocrystalline texture. These findings confirm that the Watudodol area was formed through a sequence of submarine volcanism followed by tectonic uplift since the Early Pliocene. Regionally, these results reinforce our understanding of the geodynamic evolution of Banyuwangi and support the potential for developing geotourism based on geological heritage.

*Corresponding author: melatijuliarahma.fis@um.ac.id

1. Introduction

Watudodol Beach is one of the coastal areas at the eastern end of Java Island that has complex and interesting geological processes. The existence of diverse rock formations, such as old volcanic rocks in the form of basalt lava and breccia as well as limestone and rocks containing marine animal fossils, indicates intense geological activity in the past. One of the interesting geological activities is the trace of interaction between submarine volcanism and tectonic uplift that resulted in a distinctive stratigraphic condition.

Watudodol Beach on the south-eastern slope of the Ancient Ijen Volcano reveals old volcanic rocks in the form of basalt lava and limestone. The exposed lava was produced by the eruption of Mount Merapi east of the Ijen Crater which was thrown to the east to form pillow-structured basalt lava outcrops, indicating that the lava flow once entered the water or sea area [13].

Lava is an incandescent magma liquid that comes out of the earth through craters or fractures or faults on the surface which then undergoes a cooling process and freezes into various rocks. The cooling process of magma that comes out of the earth can occur in land and water or sea areas. Magma that comes out then undergoes a cooling process and freezes in the sea area generally forms a pillow lava structure.

Pillow lava is formed as a result of submarine volcanism activity which is formed when basaltic lava comes out and flows, then undergoes rapid cooling due to direct contact with water and creates a typical rounded layered structure resembling a pillow. This contact causes the crystallisation process of The contact causes the crystallisation process of the minerals in it to be incompletely formed, thus forming a geometry resembling a pillow structure. Pillow lava structures are usually formed when lava exits through submarine volcanic fissures or subglacial volcanoes or lava flows into the sea[4]. The pillow structure indicates that the formation process occurs when there is direct interaction between lava and water and indicates that the formation of the lava rock occurred in an underwater environment[5]. Recent studies in the archipelagic arc of Southeast Asia, such as in the Philippines and Sulawesi, further confirm the role of extensional tectonics in facilitating the rise of basaltic magma to form submarine volcanism, which is then uplifted and preserved in ophiolite complexes[6].

The limestone found exposed with the lava overlies unconformably old volcanic rocks dominated by lava. The limestone also contains fossils of marine animals, including molluscs, corals and foraminifera. The interaction between the basalt lava, limestone and marine animal fossils found in the rocks suggests that the area was once in a submarine environment. Then tectonic uplift occurred which caused the rocks to be lifted to the surface as they are today.

Limestone is a type of sedimentary rock whose constituent material is dominated by calcium carbonate (CaCO3) formed from the remains of marine animal organisms, such as shells, snails, and corals[3]. Limestone is formed from chemical and biological processes that occur in shallow marine environments. The chemical process of limestone formation involves dissolution, precipitation, and deposition of calcium carbonate (CaCO3) minerals. While the biological limestone formation process, usually known as biogenic limestone, involves the accumulation and solidification of the remains of marine organisms that are rich in calcium carbonate (CaCO3).

This research aims to identify and describe geological evidence that shows submarine volcanic activity in the form of pillowstructured basalt lava and tectonic uplift processes that cause the formation of geological structures in the area, such as the presence of limestone and fossils of ancient marine animals to the land surface. This research also explains the relationship between volcanism and tectonic activity with the current geological conditions in the Watudodol Beach area. It is expected that the results of this study can be used to update

 https://doi.org/10.1051/e3sconf/202565208002

data and information from previous studies as well as supporting data and information to conduct further exploration activities. This research is expected to update previous research data and information with a more integrative approach, as well as provide robust supporting data for further geological exploration activities in tectonically active coastal areas.

2. Method

2.1. Location Determination Method

Watudodol Beach is located in Bangsri Village, Kalipuro Subdistrict, Banyuwangi Regency. Geographically, this area is included in the north coastal area of Banyuwangi Regency or the east coast of Java Island and is directly adjacent to the Bali Strait. This location has coordinates 114°24'47.11 "E / 8°5'39.29 "S. Watudodol beach is located on the south-eastern slope of the caldera complex of Ancient Ijen Volcano, which now includes Pendil Volcano, Suket Volcano, Rante Volcano, Raung Volcano, Young Ijen Volcano, and Merapi Volcano. Geologically, this area is part of an active volcanic and tectonic belt that causes interaction between submarine volcanism and tectonic uplift.

Fig. 1. Research Location Map

2.2. Research Approach and Method

The method used in this research is a type of systematic approach in the form of descriptive analysis through field observation and field data collection to describe and analyse geological phenomena in the form of submarine volcanism activity and tectonic uplift in the Watudodol Beach area. Field observations and field data collection were carried out systematically, including the description of rock outcrops in the area. The study site and rock samples

https://doi.org/10.1051/e3sconf/202565208002

representing the lithological types found in the area were taken based on the regional geological map. The main focus of these field observations was to identify the geological structure, stratigraphic arrangement, inter-rock relationships of pillow lava, volcanic breccia and limestone associated with marine animal fossils. Field observations were conducted over five days on 23 May 2025 to ensure comprehensive coverage of the existing outcrops. A total of 10 representative fresh rock samples, consisting of 4 basalt lava samples, 2 limestone samples, and 4 volcanic breccia samples, were successfully collected for further analysis.

Data obtained from field observations were then analysed descriptively and integrated with literature studies that included previous research results and relevant geological theories. This literature study is in the form of the results of geochemical analysis, petrogenesis, and paleontology related to the Watudodol Beach area. Samples of basaltic lava and volcanic breccia were made into thin sections for petrographic and palaeontological analysis. This analysis aimed to identify the mineral composition and rock texture of the basaltic lava, which was key to confirming the underwater volcanic formation environment.

The integration of field observation data with literature studies is intended to strengthen the validation of geodynamic interpretations in the area. This approach supports the interpretation of regional geodynamic processes that represent the relationship between submarine volcanism activity and tectonic uplift processes in the Watudodol Beach area based on the integration of field observation results with literature studies and relevant and previous regional geological data.

3. Results and Discussion

3.1. Physiography and Morphology

Physiographically, Banyuwangi Regency is included in the South Mountain Range of Java. In this path stretches a coastline of approximately 175.8 km, where the eastern and southern regions are directly adjacent to the Bali Strait and the Indian Ocean and the north is directly adjacent to the Java Sea[12]. The southern part of the region has a coastal morphology that stretches from Sukamade and Rajegwesi Bay in the west to the Blambangan Peninsula in the east. This strip is composed of rocks of early Neogene age, dominated by clastic volcanic deposits, volcanic rocks, and intrusive rocks of the Batuampar Formation and Merubetiri Formation[11]. More fully, Banyuwangi Regency is classified into main physiographic units, including the Anticlinorium of Rembang, Randublatung Depression, Java Depression Centre Zone, Quaternary Volcanics, and part of the Southern Mountain Range of Java [15].

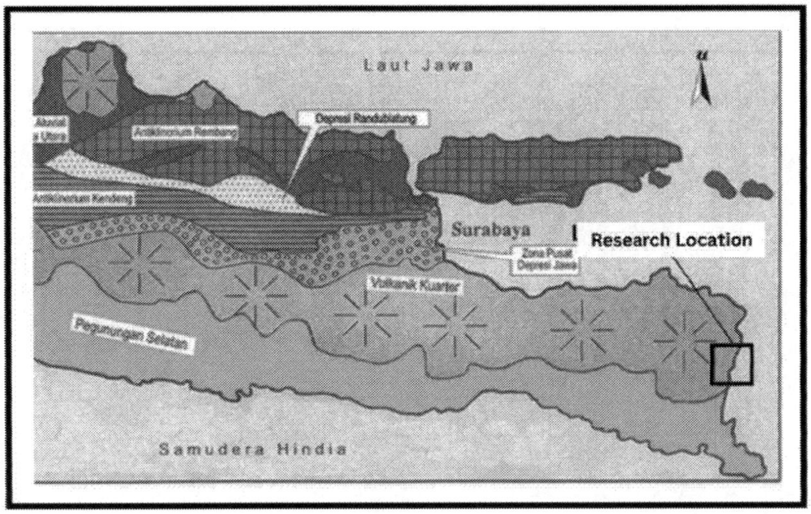

Fig. 2. Physiography of Watudodol Beach

In terms of physiography, Watudodol Beach exhibits the main physiographic features of volcanic or quaternary mountains. Generally, these quaternary mountains are located in the central part of Java Island. The quaternary volcanic belt stretches from the west to the east of Java Island.

3.2. Geological and Stratigraphic Conditions

The geological conditions of Banyuwangi Regency and its surroundings have been mapped by the Geological Survey Centre during the period 1990 to 1993. Regional geological maps covering the Banyuwangi area include the Banyuwangi Sheet Geological Map [14], Situbondo Sheet Geological Map, Jember Sheet Geological Map, and Blambangan Sheet Geological Map[1].

The stratigraphy of the Banyuwangi area is divided based on each physiographic zone, namely the Southern Mountain Zone and the northern part, namely the Kendeng Zone. The southern part is dominated by older volcanic rocks, while in the northern part younger rocks are distributed[11]. The stratigraphy of the Kendeng Zone is composed of deep-sea deposits at the bottom, which gradually transforms into deep-sea deposits. Gradually changed to shallow marine deposits and then to non-marine deposits at the top. Sediment types found in the Kendeng Zone include clastic turbidites, carbonates, and clastic volcanics.

The geological characteristics of the Banyuwangi area include formations that are Oligo-Miocene, Middle Miocene, Quaternary, and Alluvium in age. The oldest formations in the region consist of the Merubetiri Formation, Batuampar Formation, and Sukamade Formation. The OligoMiocene age is dominated by volcanic rocks consisting of volcanic breccia, lava, and tuff. Then there is the Middle Miocene formation generally dominated by limestone with clastic deposits and reefs that have undergone a karstification process. Quaternary rock formations are generally found to consist of volcanic rocks that began to form since the Pleistocene era. The youngest formation is alluvium which is the result of deposition from fluvial processes and coastal environmental activities.

 https://doi.org/10.1051/e3sconf/202565208002

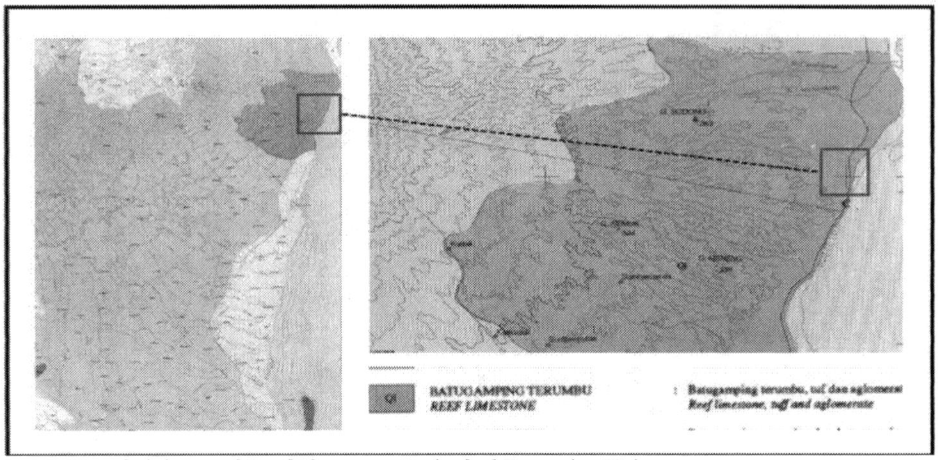

Fig. 3. Watudodol Coastal Reef Limestone Geological Formation Unit

The Watudodol Coastal Area shows a unit of Quaternary rock formations, composed of reef limestone, tuff, and agglomerate. This limestone unit is only exposed in certain locations and consists of reefs, layered clastics associated with conglomerates, agglomerates, lava, and a small amount of tuff. This unit is exposed in the north-eastern part of Banyuwangi, namely Watudodol Beach, and is estimated to be Holocene in age.

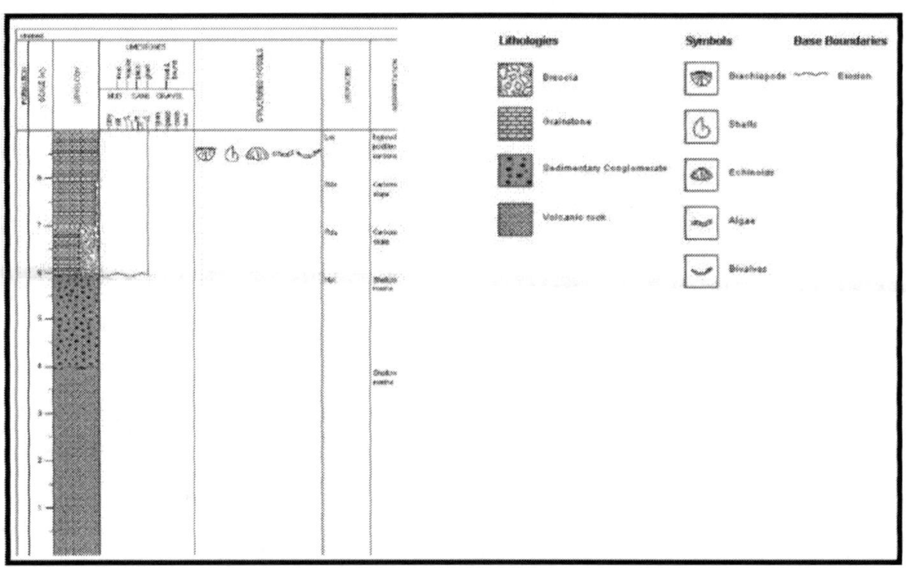

Fig. 4. Stratigraphic Column of Watudodol Beach

The stratigraphy of Watudodol Beach is divided into 3 outcrops or layers of rock formations. The lowest outcrop is composed of basalt lava which shows a pillow structure and autobrectification with fractures, afanitic texture and above it there are lava deposits with autobrectification texture. Above it are outcrops of brownish ash coloured breccia with black coloured basalt lava fragments. The uppermost layer is dominated by whitecoloured breccia

 https://doi.org/10.1051/e3sconf/202565208002

limestone with basalt fragments in which there are fossils of marine animals, such as gastropods and corals.

The uppermost layer is dominated by creamy white breccia limestone with basalt fragments containing marine animal fossils, namely gastropods and corals. This limestone layer is approximately 3 metres thick. Beneath it is a conglomerate deposit with a layer thickness of approximately 2 metres. The bottom layer consists of basaltic volcanic rock with a thickness of approximately 4 metres. This sequence and thickness indicate a geological cycle that began with undersea volcanic activity, followed by volcaniclast sedimentation, and ended with carbonate sedimentation in a shallow marine environment, before finally rising to the surface.

3.3. Pillow Lava and Volcanic Breccia

In the Watudodol Beach area, volcanic rocks in the form of lava produced by Mount Merapi were thrown eastward, forming lava outcrops with a pillow structure. Some of the pillow lava appears to have broken apart, forming lumps. The pillow structure indicates that the lava flow reached the coastal area and continued flowing into the sea. The lava formed by this volcanic activity occurred in shallow waters, allowing the solidified pillow lava to reach the sea surface[2]. This lava has a black basalt composition with a fine phaneritic to aphantic texture and shows a clear flow structure.

Fig. 5. Pillow lava on Watudodol Beach

Basaltic lava flowing out as a result of volcanic activity produces three types of lava flows, namely pahoehoe lava, aa lava, and blocky lava. The basaltic lava flow at Watudodol Beach has a lumpy structure resembling aa lava. This type of lava has high viscosity, so it flows slowly, while the inside remains in a liquid state. These lava fragments are formed as a result of rapid cooling, causing shrinkage and fragmentation. In the middle of the lava flow, because the cooling process is slower, a massive mass of igneous rock with a blocky structure is formed, which tends to develop into columnar or sheet joints [8]. The resulting lava layer is more than 3 metres thick. The presence of pillow lava and these flow structures provides critical physical evidence for reconstructing the submarine volcanic environment and understanding the nature of magma flow in the land-sea transition zone of island arcs.

https://doi.org/10.1051/e3sconf/202565208002

Fig. 6. Columnar Structured Lava on Watudodol Beach

The pillow lava at Watudodol Beach has a hypocrystalline texture, which is rock composed of crystals ranging in size from 0.025 to 2.6 millimetres[9]. This lava is composed of plagioclase crystals (36%), K-feldspar (11%), pyroxene (8%), palagonite (6%), and additional minerals in the form of black minerals (2%). These minerals range in size from 0.2 to 1.6 millimetres and are classified as fine fanerics with crystal shapes varying from subhedral to anhedral. The base mass consists of plagioclase microlites (13%) and volcanic glass material (24%). The relationship formed between rock fragments and minerals is inequigranular vitroveric[13]. The inequigranular vitroveric hypocrystalline texture and mineral composition indicate that the rock originated from magma that was basic or had low to medium acidity. The mineralogical composition and hypocrystalline texture indicate rapid magma solidification in an aquatic environment, while also providing clues to understanding the nature of the magma and the melting conditions of the mantle beneath the East Java volcanic arc.

Fig. 7. Thin section of pillow lava at Watudodol Beach (Nikol Sejajar). Ps = pyroxene, Kf = K-feldspar, Plag = plagioclase

Volcanic breccia rocks in the Watudodol Beach area were found exposed on the surrounding cliffs. These volcanic breccia rock outcrops were formed as a result of the eruption of Mount Merapi. These rocks are brownish-grey in colour, ranging in size from 10 cm to more than 256 cm, angular to rounded, composed of fragments of basaltic lava and have calcite veins that split the rock diagonally, composed of calcite (CaCO3) minerals. These calcite veins are characterised by their white absorption colour, interference colours varying from pink to green, and anhedral crystal shapes[7]. The hyaloclastite volcanic breccia is dark grey in colour with a jigsaw or puzzle texture. The sorting or uniformity of this outcrop is poorly separated and has an average thickness of 1 metre. The presence of hyaloclastite breccia and calcite veins is of significant geodynamic importance, as it not only records energetic volcanic fragmentation events beneath the sea, but also records phases of tectonic deformation and post-rock formation hydrothermal fluid circulation, indicating a complex tectono-magmatic history following the uplift of this region.

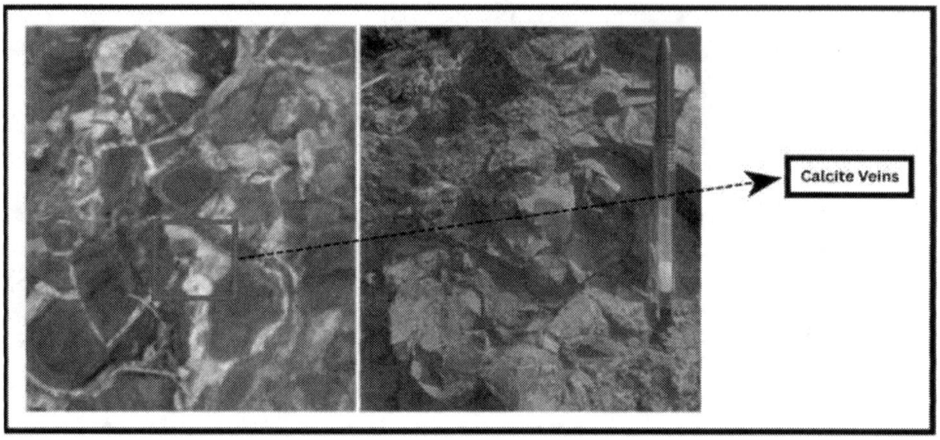

Fig. 8. Volcanic Breccia at Watudodol Beach

3.4. Association of Limestone and Marine Animal Fossils

Thick limestone formations exposed in the Watudodol Beach area, ranging in height from 0 to 544 metres above sea level, form hills on the eastern slopes of the Ijen Muda volcano, which has a maximum height of 504 metres above sea level, such as Mount Remuk, including other peaks such as Mount Sodong, and Mount Meneng, specifically expose limestone associated with coral reefs. This light brown limestone is partly brecciated or conglomerate due to its content of andesite, basalt, and a small amount of sedimentary rock material. This limestone is a clastic *rudstone* facies with the coarsest grain size, resulting from the breakdown of skeletal limestone that has undergone a process of transport and then been deposited and accumulated in a specific location, with a thickness of between 20 and 50 centimetres and generally well-layered. This facies is not included in the reef limestone facies but is associated with reefs[10]. This rudstone facies records high-energy events in a shallow marine environment, possibly due to storms or tectonic activity that uplifted ancient reefs, disrupted them, and redeposited their fragments. This indicates that this phase of carbonate sedimentation did not occur under calm conditions, but was influenced by the energy dynamics associated with surrounding tectonic activity.

Fig. 9. Clastic limestone containing marine animal fossils

Marine animal fossils found in the Watudodol Beach area and associated with limestone include molluscs (both shells and moulds), corals, and foraminifera. This light brown

https://doi.org/10.1051/e3sconf/202565208002

limestone is composed of a collection of marine biota, such as molluscs (mainly micro-sized gastropods commonly found in a fragmented state, ostracods), a small amount of planktonic foraminifera, and small benthic foraminifera, which are quite abundant.

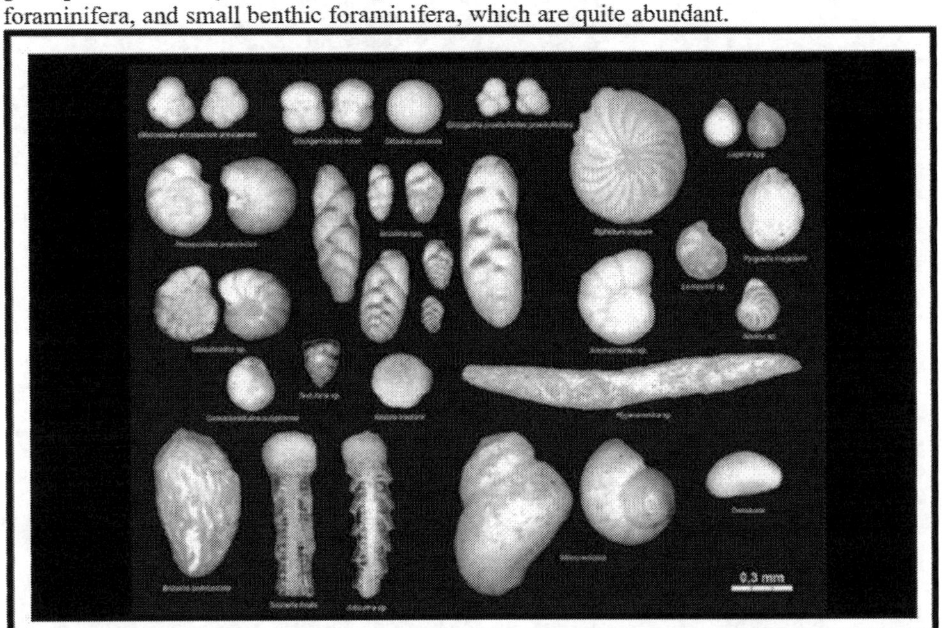

Fig. 10. Types of Fossils Associated with Limestone on Watudodol Beach

Planktonic foraminifera that can be identified using a 115-mesh sieve, or 125 microns, include Glorotalia acostaensis acostaensis Blow, Globigerinoides sp., Gs. rubber, Gs. praebulloides Blow, and Orbulina universa (d'Orbigny). Meanwhile, small benthic foraminifera species that were successfully identified include *Elphidium crispum* (Linnaeus), *Brizalina* sp., *Nodosaria* spp., *Neoeponides praecinctus* (Karrer), *Nonion* sp., *Cibicidoides cicatricosus*, *Lagena* spp., *Pyrgoella irregularis* (d'Orbigny), *Globocassidulina subglobosa* (Brady), *Textularia* sp., *Brizalina semicostata* (Cushman), *Rotalia beccarii* (Linne), *Elphidium* sp., *Anomalinoides* sp., *Marsipella* spp., *Tubinella finalis* (Brady), and *Articulina* sp [13]. The assemblage of benthic and planktonic foraminifera, particularly the presence of *Globorotalia acostaensis*, biostratigraphically confirms the Early Pliocene age and indicates a shallow to mid-neritic marine depositional environment. The geodynamic significance of these fossil findings is as strong evidence of palaeobathymetry; the elevation of these limestone outcrops, which today reach 544 metres above sea level, clearly demonstrates the magnitude of Quaternary tectonic uplift that has raised the ancient seabed hundreds of metres above the present sea level, recording the history of active compressional tectonics along the southern margin of Java.

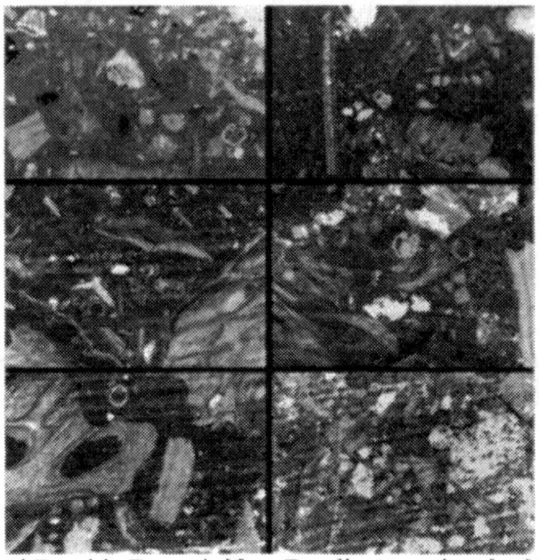

Fig. 11. Small Planktonic and Benthic Foraminifera Fossils Associated with Limestone on Watudodol Beach

3.5. Evidence of Geological Interaction

At Watudodol Beach, volcanic rocks in the form of basaltic lava produced by Mount Merapi form pillow lava outcrops, which are black in colour and have a rope-like structure on their surface due to solidification. These volcanic rocks are interbedded with basaltic lava. The presence of basaltic lava fragments in limestone layers similar to the exposed basaltic lava indicates that the limestone is relatively younger than the basaltic lava. The exposed limestone layers with lava have an elevation of 0 to 544 metres above sea level.

Fig. 12. Limestone Outcrop Containing Basalt Lava Components on Watudodol Beach

Based on the Banyuwangi Geological Map [14], the limestone unit in the Watudodol area extends northwest to the southeastern slopes of Mount Merapi. This Late Neogene (Early Pliocene) limestone is part of the Punung Formation limestone unit, which is widely exposed in the Blambangan Peninsula[1]. This young limestone unit also contains agglomerate, lava and tuff. The limestone unit in the Watudodol area unconformably overlies older volcanic rocks dominated by basalt lava and is regionally exposed as an open part of the Oligocene-Miocene or Early Miocene volcanic rock unit. This older volcanic rock unit is more dominant along the southern coast of the Banyuwangi region. The lower section of the Early Pliocene limestone, which contains a significant amount of lava components, can be interpreted as a base conglomerate.

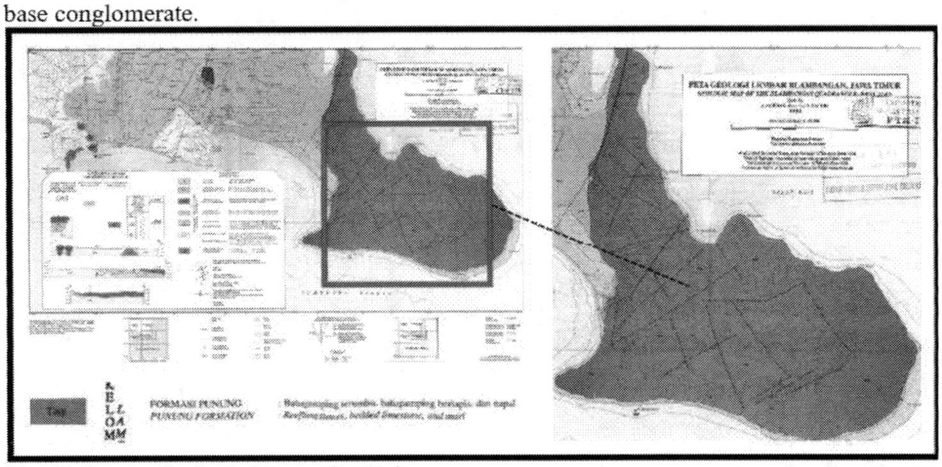

Fig. 13. Punung Geological Formation Unit

The discordant relationship between old volcanic rocks and young limestone that has been uplifted is not a unique phenomenon, but is also found in several other Indonesian island arcs, albeit with slightly different tectonic contexts. In Lombok, the Batu Putih Formation (Middle Miocene), which consists of limestone, also unconformably overlies Oligo-Miocene volcanic rocks. The main difference lies in the source of the volcanism; in Lombok, the ancient volcanism is associated with the Sunda magmatic arc, while in Watudodol, the source is the ancient Ijen volcano. Meanwhile, in the Yogyakarta area, limestone outcrops of the Wonosari Formation (Middle-Late Miocene) unconformably overlie the Nglanggran Formation, which is dominated by breccia and andesitic lava. This similar pattern indicates a geological cycle that is common in Java: a period of intense volcanic activity followed by a quiet period that allows for carbonate sedimentation in shallow seas. What makes Watudodol unique compared to these two locations is the presence of pillow lava, which clearly indicates an underwater volcanic environment, direct evidence that reinforces the paleo-environmental interpretation.

4. Conclusion

The geological phenomena at Watudodol Beach are the result of a unique interaction between undersea volcanic activity and intense tectonic uplift since the late Neogene or Early Pliocene. The formation of pillow lava structures, which are rarely found in Java, together with volcanic breccia, confirms that basaltic lava from Mount Merapi once flowed into the sea, underwent rapid cooling, and formed distinctive underwater structures. The presence of overlying limestone that is not in harmony with older volcanic rocks, as well as its association with shallow marine fossils such as foraminifera, molluscs, and corals, indicates a change in the environment from marine to terrestrial due to rapid and significant tectonic uplift. Watudodol's uniqueness compared to other locations in Java lies in the direct evidence of pillow lava as a record of the paleo-marine environment that was later uplifted hundreds of metres above the current sea level. To deepen our understanding, further research is needed on geochemistry, chronostratigraphy, and regional tectonic connections so that the geodynamic evolution of this area can be reconstructed more comprehensively. Additionally, the geological uniqueness of Watudodol has great potential to be developed as a geotourism site and geological heritage within the framework of the Ijen Geopark, which can provide educational value as well as sustainable economic benefits for the surrounding community.

Reference

1. Achdan, A. and Bachri, S., 1993. *Peta Geologi Lembar Blambangan, Jawa, skala 1:100.000*. Bandung: Pusat Penelitian dan Pengembangan Geologi.
2. Astjario, P. and Astawa, N., 2016. Proses Terbentuknya Pulau-pulau Wisata, Gili Trawangan, Meno dan Air, Akibat Aktifitas Gunungapi Bawah Laut di Pamenang, Kabupaten Lombok Barat. *Jurnal Geologi Kelautan*, *3*(1).
3. Beko, B.C.M., Angkie, J., Mahmud, D., Rawul, J.A., Guna, P.A. and Cahyono, Y.D.G., 2021. Analisis Kualitas Massa Batuan dengan Metode Rock Mass Rating (RMR) pada Batugamping. In *Prosiding Seminar Nasional Sains dan Teknologi Terapan* (Vol. 9, No. 1, pp. 153-160).
4. Dzakiya, N., Aditya, M.A.B., Jayadi, H. and Setiawan, D.G.E., 2020. *Analisis sebaran lava bantal Watuadeg yang didukung data Vertical Electrical Sounding (VES) di Lapangan 'AGSTY' Yogyakarta*. Jambura Physics Journal, 2(1), pp.44–53.
5. Eliezer, I., Winarno, T. and Ali, R.K., 2019. *Petrogenesis lava bantal Nampurejo di Dusun Kalinampu dan sekitarnya, Desa Jarum, Kecamatan Bayat, Kabupaten Klaten, Provinsi Jawa Tengah*. Jurnal Geosains dan Teknologi, 2(1), pp.33–41.
6. Guotana, J. M. R., Payot, B. D., Dimalanta, C. B., Ramos, N. T., Faustino-Eslava, D. V., Queaño, K. L., & Yumul Jr, G. P. (2018). Petrological and geochemical characteristics of the Samar Ophiolite ultramafic section: implications on the origins of the ophiolites in Samar and Leyte islands, Philippines. *International Geology Review*, *60*(4), 401-417.
7. Hasria, H., Dwiyanti, P., Azzaman, M.A., Okto, A. and Septiana, S., 2024. Pemanfaatan Batugamping pada Formasi Matano di Kecamatan Wiwirano Kabupaten Konawe Utara, Provinsi Sulawesi Tenggara. *Jurnal Geosains dan Teknologi*, *7*(2), pp.83-93.
8. Herho, S.H.S., 2018. *Volkanologi Selayang Pandang. Jilid 1.* Jakarta: Matahari.
9. Kristanto, A.T. and Sugarbo, O., 2020. Analisis petrogenesa batuan beku sebagai tinjauan keterdapatan mineral ekonomis di Daerah Wukirharjo. *Geoda*, *1*(1), pp.95-104.
10. Moreira, J.B.F., Yuwanto, S.H. and Mahardjo, E., 2019, September. Pemetaan Geologi dan Penentuan Lingkungan Pengendapan Batugamping Berdasarkan Analisis Petrografis diI Kecamatan Semanding dan Sekitarnya Kabupaten Tuban Provinsi Jawa Timr. In *Prosiding Seminar Teknologi Kebumian dan Kelautan (SEMITAN)* (Vol. 1, No. 1, pp. 154-163).
11. Prabowo, A., Permanadewi, S. and Samodra, H., 2021. Analisis Batuan Gunungapi di Daerah Banyuwangi dan Sekitarnya, Provinsi Jawa Timur Berdasarkan Aspek Geokimia. *Jurnal Geologi dan Sumberdaya Mineral*, *22*(4), pp.197-207.
12. Rohaendi, N., Setiawan, I.F., Budiyono, A., Harmoko, A.A. and Wahyudi, W., 2023. Geologi dan model kolaborasi pengembangan Geopark Ijen, di Kabupaten Banyuwangi [Geology and collaboration model for developing Ijen Geopark, in Banyuwangi Regency]. *Jurnal Geominerba*, *8*(2), pp.166-179.
13. Samodra, H., Prabowo, A., Permanadewi, S., Razak, J. and Yalin, S., 2019. *Inventarisasi keragaman geologi dan identifikasi warisan geologi daerah Kabupaten Banyuwangi, Jawa Timur*. Bandung: Pusat Survei Geologi, Badan Geologi, Kementerian Energi dan Sumber Daya Mineral, bekerja sama dengan BAPPEDA Kabupaten Banyuwangi.

https://doi.org/10.1051/e3sconf/202565208002

14. Sidarto, S.T. and Sudana, D., 1993. Peta Geologi Lembar Banyuwangi, Jawa, Scale 1: 100.000. *Bandung: Pusat Penelitian dan Pengembangan Geologi.*
15. Van Bemmelen, R.W., 1949. *The geology of Indonesia.* The Hague: Martinus Nijhoff.

Fluvial Marine Dynamics and Volcanic Sediment Supply in Barrier Beach and Mangrove Formation, Santen Island Banyuwangi

Putri Salma Shafira[1], *Melati Julia* Rahma[2*], and *Firman Sauqi Nur* Sabila[3]

[1,2]Department of Geography, Faculty of Social Sciences, State University of Malang, 65145 East Java, Indonesia
[3]Ijen Geopark Geological Information Centre, 68465 East Java, Indonesia

Abstract. The Banyuwangi Regency is distinguished by substantial volcanic activity, which has contributed to the formation of the region's coastal landforms. On Santen Island, the barrier island is a morphological feature formed primarily by fluvial–marine dynamics, with volcanic deposits serving as a significant sediment source. These processes also support the protection and growth of mangrove ecosystems. The present study analyses the interaction between volcanic sediment supply and fluvial–marine processes in barrier formation, and their relation to mangrove sustainability. An explanatory–descriptive approach was applied, integrating geological analysis with Sentinel-2 imagery from 2018 to recent monitoring. The results of the study indicate that the formation of barriers is influenced by volcanic materials derived from Mount Ijen, Mount Ranti, and Mount Merapi. These materials are transported through the Bendo watershed and deposited in the intertidal zone. Sentinel-2 imagery reveals a barrier extending longitudinally from north to south, with mangroves expanding both landward and laterally. Between 2018 and the most recent observations, mangrove cover has generally increased, although recent data suggest minor fluctuations that require further investigation. These findings underscore the combined role of volcanic sediment supply and fluvial–marine dynamics in shaping and sustaining vulnerable coastal ecosystems.

1 Introduction

Indonesia has the highest number of active volcanoes in the world, with 127 recorded [1]. The high volcanic activity in the Indonesian region is attributed to its location in the collision zone of three major tectonic plates: Indo-Australia, Eurasia, and the Pacific [2]. Based on data from the Centre for Volcanology and Geological Hazard Mitigation (PVMBG), out of a total of 127 active volcanoes in Indonesia, 19 are located in Java, with the majority of them situated in East Java. Active volcanoes in East Java include Mount Semeru, Mount Raung, Mount Bromo, Mount Kelud, Mount Ijen, Mount Arjuno-Welirang, and Mount Lamongan. The presence of volcanoes contributes to a variety of potential natural

* Corresponding author: melatijuliarahma.fis@um.ac.id

resources, such as geothermal energy, excavated materials from volcanic activity, and mineral deposits [3-4].

Various natural resource potentials and mineral deposits also occur in the eastern end of Java. Banyuwangi Regency was formed in the context of an active volcanic arc resulting from the subduction of the lower Indo-Australian Plate and Eurasian Plate. Where Neogene and Quaternary mammatism formed the constituent lithologies of the Batuampar and Merubetiri Formations since the Oligo-Mesozoic era [5]. Volcanic activity in the Banyuwangi Regency area occurred approximately 50,000 years ago, during the late Pliocene period, when Mount Ijen Purba experienced a large Plinian eruption that formed the Ijen Caldera [3]. The eruption produced by Ancient Mount Ijen was the first and oldest eruption in geological terms [5]. In the same period, namely the Pliocene, there was also an eruption originating from Mount Merapi Banyuwangi, where indications of ancient volcanic activity and old lava domes were found, indicating that it was considered an old eruption based on geological evidence [5].

However, the resulting eruption was not explosive. Then, the documented volcanic activity since the initial eruption includes the eruptive activity of Mount Raung in 1586, which was of an explosive type and claimed lives [1]. Until now, based on reports from the Centre for Volcanology and Geological Hazard Mitigation (PVMBG), volcanic activity in Banyuwangi is still ongoing, with the last eruption occurring on June 12, 2025, from Mount Raung, which formed an ash pond reaching 1,000 metres [1].

Volcanic activity can also produce a variety of landforms, including those found along coastal areas. In the context of Banyuwangi Regency, the influence of volcanic processes on coastal morphology is substantial[6]. The volcanic activities of the Ancient Ijen, Merapi, and Raung volcanoes have been observed to release materials with the potential to form new land and reshape coastal landscapes [3]. A notable geomorphological feature in such contexts is the barrier beach, which develops through the accumulation of sediment parallel to the shoreline. The presence of volcanic-derived sediments has been demonstrated to facilitate the formation of barrier islands, whilst concurrently supporting mangrove ecosystems through the provision of nutrient-rich substrates and the creation of conducive habitats for growth [7-10]. Santen Island serves as a prime example of a coastal environment where these interactions are clearly observable [6].

The formation of diverse landforms, including those observed in coastal areas, is often attributed to volcanic activity. In the context of Banyuwangi Regency, the influence of volcanic processes on coastal morphology is substantial. The volcanic activities of the Ancient Ijen, Merapi, and Raung volcanoes have been observed to release materials with the potential to form new land and reshape coastal landscapes [8]. A notable geomorphological feature in such contexts is the barrier beach, which develops through the accumulation of sediment parallel to the shoreline. The presence of volcanic-derived sediments has been demonstrated to facilitate the formation of barrier islands, whilst concurrently supporting mangrove ecosystems through the provision of nutrient-rich substrates and the creation of conducive habitats for growth [6]. Santen Island serves as a prime example of a coastal environment where these interactions are clearly observable.

Despite the significance of this finding, research linking volcanic sediment supply from upland sources to the dynamics and sustainability of mangrove ecosystems in Banyuwangi remains limited [9,11]. A plethora of studies have hitherto been conducted on either volcanic activity or coastal geomorphology in isolation. This has engendered a paucity of understanding regarding the integrated role of these phenomena in shaping the morphology of barriers and the associated ecosystems. The present study investigates how volcanic sediment contributions and fluvial–marine dynamics interact in the formation of barrier landforms on Santen Island in order to address this gap. The study further evaluates the role of barrier morphology in protecting mangrove ecosystems through the analysis of geological data and Sentinel-2 imagery, focusing on barrier orientation and mangrove distribution in the area[4-6].

2 Method

2.1 Location Determination Method

The present study was conducted in the coastal area of Banyuwangi, specifically on Santen Island in the eastern part, due to the paucity of geomorphological research on the island. Santen Island is located at the coordinates 8°13′5.11″ S and 114°22′41.53″ E, within Karangrejo Village, Banyuwangi District, Banyuwangi Regency (see Figure 1).

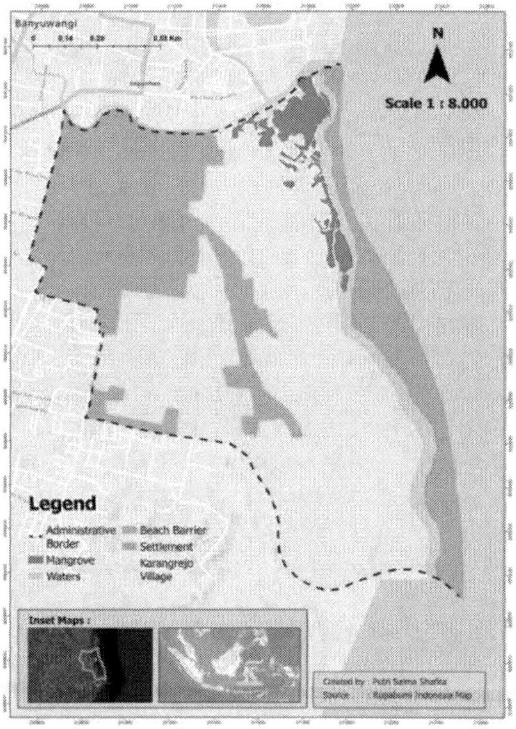

Fig. 1. Research Location Map

The total area of Karangrejo village is 172 ha. Santen Island is located in close proximity to the Bali Strait, which is situated at the easternmost extremity of Java Island. The Wuyu River is located to the north of Santen Island. The research was conducted between March and June 2025, encompassing both field surveys and spatial data processing.

2.2 Data Collection Method

Secondary data were obtained from literature studies related to geology and geomorphology, as well as from agency or institutional reports and previous research[3,5]. These included geological maps and previous geomorphological surveys, which supported the analysis of barrier shoal formation on Santen Island. Spatial data were derived from multispectral imagery provided by Sentinel-2A and Sentinel-2B satellites operated by the European Space Agency (ESA) as part of the Copernicus programme. The dataset spanned the years 2018–2025, with at least one representative image selected per year during the dry season (June–September) to minimise cloud cover. The specific acquisition dates of each image were documented to ensure temporal consistency.

Before analysis, the Sentinel-2 images underwent standard atmospheric correction using the Sen2Cor algorithm (Level-2A surface reflectance product). To reduce atmospheric and radiometric noise, cloud masking was applied using the QA60 band (cloud probability more than 60%), in combination with visual screening to ensure that only cloud-free pixels were retained [4]. Primary data consisted of direct field observations to validate the classification results of the satellite images. Field verification was conducted at selected points representing barrier shoals and mangrove stands. Additional primary data included semi-structured interviews with local stakeholders regarding environmental change, as well as sediment samples (pumice, sand, and soil) collected during surveys [5,11].

2.3 Data Analysis and Processing Methods

In analysing secondary and primary data, researchers used descriptive-explanatory methods. Descriptive analysis aims to describe the data as it is, while explanatory analysis aims to explain the cause-and-effect relationship between variables. In the descriptive stage, the researcher provides a factual and systematic description of the process of forming barrier scales on Santen Island, and then explains the causal relationship between this process and the distribution of mangroves on Santen Island[5].

For the spatial-temporal analysis, Sentinel-2 imagery spanning the period from 2018 to 2025 was processed using Google Earth Engine (GEE). The processing workflow included pre-processing, band combination, and temporal filtering to ensure accurate and consistent results across the dataset[4]. Two types of band composites were applied in this study to support the interpretation of coastal dynamics. The first was the True Colour Composite (Bands 4-3-2), which allowed the visualization of coastal sediment distribution in natural color, providing a realistic depiction of surface conditions. The second was the False Colour Composite (Bands 8-4-3), which utilized near-infrared reflectance to highlight chlorophyll content and assess the density of mangrove canopies, thereby offering deeper insights into vegetation health and spatial changes in coastal ecosystems[4].

To determine annual changes in barrier shoal and mangrove extent, a manual on-screen digitization technique was performed using ArcGIS 10.8. Digitization was conducted at a working scale of 1:3,000 – 1:5,000, which allowed precise delineation of geomorphic features. The criteria for delineating barrier shoals included tonal differences in sediment reflectance (light brown to gray) and linear accumulations parallel to the shoreline. For mangroves, digitization criteria included dense, dark red patches in the False Colour Composite and verified canopy cover during field surveys[4,9].

The digitized features were converted into vector polygons, and their areas were calculated using the Geometry Area function in ArcGIS[4,5,9]. This enabled quantification of annual changes in barrier shoal coverage and mangrove extent. To ensure replicability, each digitization process was performed twice by different operators, and discrepancies greater than 5% were re-examined through joint interpretation.

3 Results and Discussion

3.1 Geological Processes in the Formation of Barrier Beach

In order to comprehend the genesis of Berrier Beach on Santen Island, it is necessary to analyse the historical and morphological development of the coastline through geological data[14]. The stratigraphic arrangement of the Banyuwangi area indicates the presence of various rock formations, ranging from those dated to the Late Oligocene to the Early

Miocene, Middle Miocene, and Quaternary periods. The results of observations conducted through the Banyuwangi Sheet Geological map (Figure 2) demonstrate that Santen Island is situated within the Alluvium rock unit[3,5]. The Alluvium rock unit in Banyuwangi is the youngest of the rock units. Evidently, the rock formation is an accumulation of material from Mount Ijen, which is considered to be of ancient origin, as evidenced by the results presented on the geological map.

Evidence suggests that, as observed from the Ellene year, this mountain is recognised as the most ancient and progenitor of Mount Merapi, Mount Raung, and the surrounding mountains in Banyuwangi Regency. The formation of the Alluvium unit commenced with Quaternary rocks originating from the ancient Ijen volcano, dating back to the Pliocene, which then transformed clastic sedimentary rocks from the Merapi volcano of early Quaternary age [7]. Santen Island is located within a region that contains volcanic rocks from the Batuampar and Merubetiri Formations, which are of Oligo-Miocene age[5,15]. The Batuampar and Merubetiri formations in Banyuwangi are composed of basalt-type volcanic rocks, which are derived from the toleite magma series. These rocks are metalluminous and originate from depths ranging from 52.26 to 169.41 km[7]. Basalt rock is a type of extrusive igneous rock formed from the solidification of magma at the Earth's surface [15].

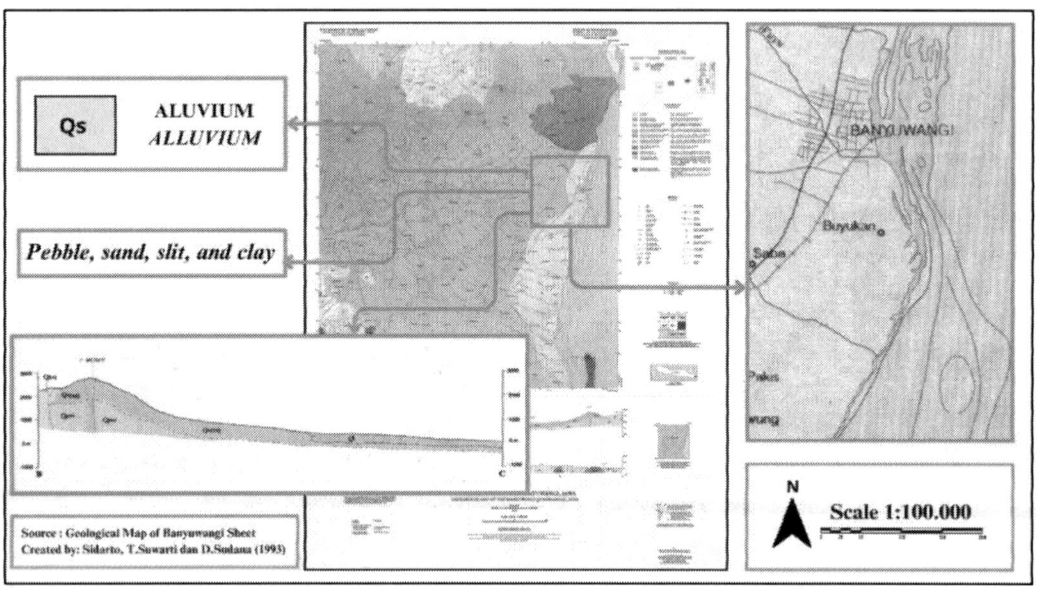

Fig. 2. Geological Map of Banyuwangi Sheet

The formation of alluvium sedimentary units, particularly barrier beach, is inextricably linked to a sequence of intricate geomorphic processes. These processes commence with the weathering of rock material on land and progress to the erosion and transportation of these materials by exogenous agents, such as river water, ocean currents, and waves. The resultant sediment is then deposited in the coastal zone through the process of sedimentation, influenced by tidal dynamics and wave energy. This process leads to the formation of deposits that extend parallel to the coastline[6,7]. The formation of basalt rocks, precipitated by volcanic activity at Mount Ijen Purba, Mount Merapi, and Mount Raung, also underwent the process of weathering through fault activity. Prior to the initiation of the process of erosion, a volcanic material must undergo a process of weathering. Banyuwangi is located along regional faults that extend from the Sukamade Plateau in the west to the Banyuwangi Low in the east[7]. With regard to the uplift fault movement process that has occurred in the western Banyuwangi area, it has had an effect on the sedimentation process in the coastal area. The presence of the fault has been shown to cause basalt rocks that were previously submerged to emerge in the coastal area through the weathering process[6,7].

3.2 Sedimentation Processes in the Formation of Barrier Beach

Following the weathering process, a series of geomorphic processes in the alluvium sediment unit occur, resulting in the formation of a barrier beach, which subsequently undergoes the erosion process. The role of water media is identified as the primary factor in this process. Sedimentary material is transported by the action of river flow from upstream to downstream, where sediments such as sand, silt, and clay are carried and gradually deposited on floodplains, deltas, or river mouths when the current speed decreases. This process forms a layer of alluvial deposits rich in loose material from weathering of rocks in the upstream area[11]. It is evident that the sedimentation activities that occur on Santen Island can reach the coast, and it is therefore not possible to separate these activities from the influence of the river flow that flows around it. The geographical location of the barrier beach on Santen Island is proximate to a number of streams. The northern part of the shoal area is characterised by the presence of the Wuyu River, while the southern part is marked by the Kalibendo River. When viewed in overview, the two streams are connected to the Bendo watershed (Figure 3).

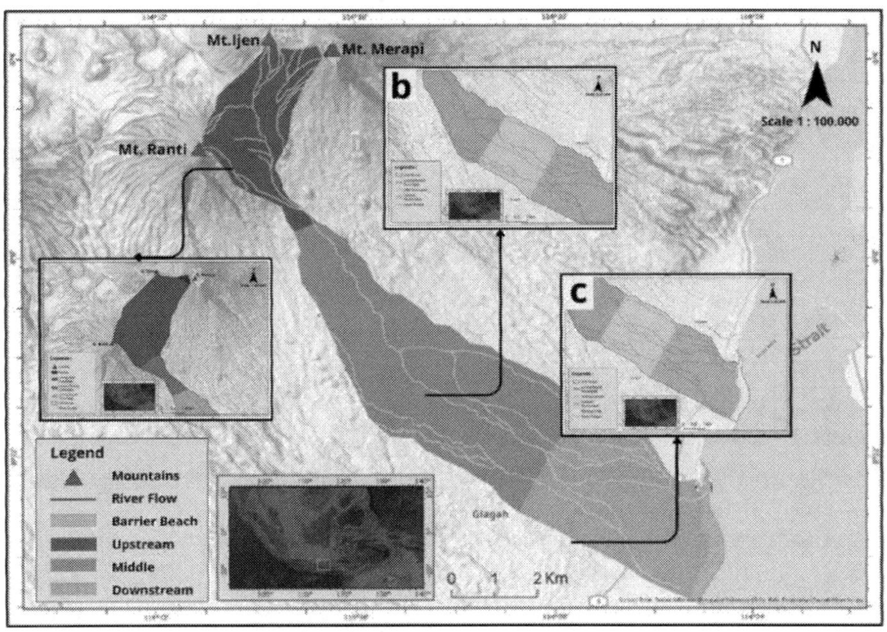

Fig. 3. The Bendo watershed becomes the transport medium for volcanic materials from Mt. Merapi, Mt. Ijen, and Mt. Ranti, Bendo watershed is divided into three parts, namely (a) Upstream, (b) Middle, (c) Downstream

The Bendo watershed is divided into three sections: upstream, middle, and downstream. The upstream part of the Bendo watershed originates from the Ijen Volcano complex, specifically Mount Ranti, Merapi, and Ijen [11]. The Bendo Watershed is characterised by a diverse array of landforms, which can be broadly categorised into three distinct types based on their formation processes: volcanic, fluvial, and marine landforms. In the upstream to middle zone of the Bendo watershed, landforms resulting from past volcanic activity predominate, while in the downstream area, land morphology is predominantly shaped by fluvial processes and the influence of marine processes [11]. This division is indicative of the variation in geomorphic processes that shape the overall landscape of the Bendo watershed.

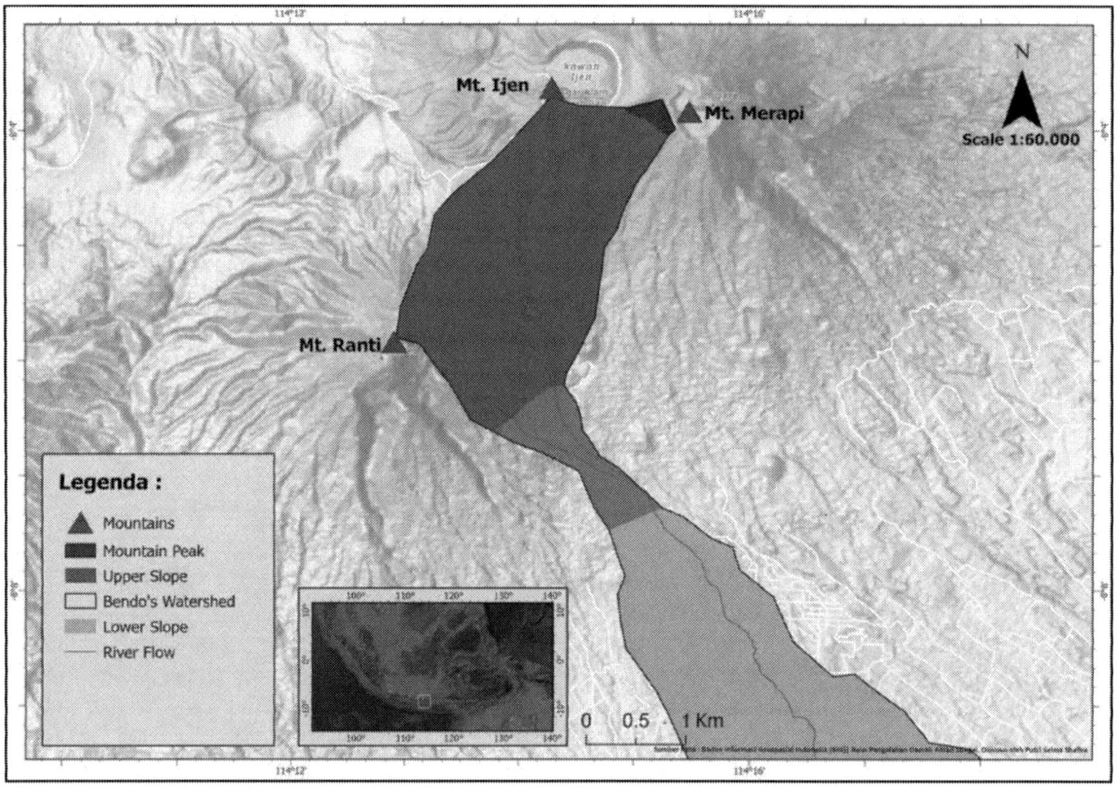

Fig. 4. Geomorphological Map of the Upper Bendo Watershed

The upper part of the Bendo watershed (see Figure 4) is characterised by landform units, including peaks, upper slopes, and middle slopes. These are composed of volcanic materials such as volcanic breccia, lava, and tuff, which formed from the accumulation of volcanic ash. From a morphological perspective, the area is characterised by undulating topography, with elevations ranging from mountains to hills and pronounced slopes. The predominant geomorphic processes in this area are erosion and mass movements, including landslides, which are of high intensity due to a combination of heavy rainfall and slope cutting activities, particularly during road construction.

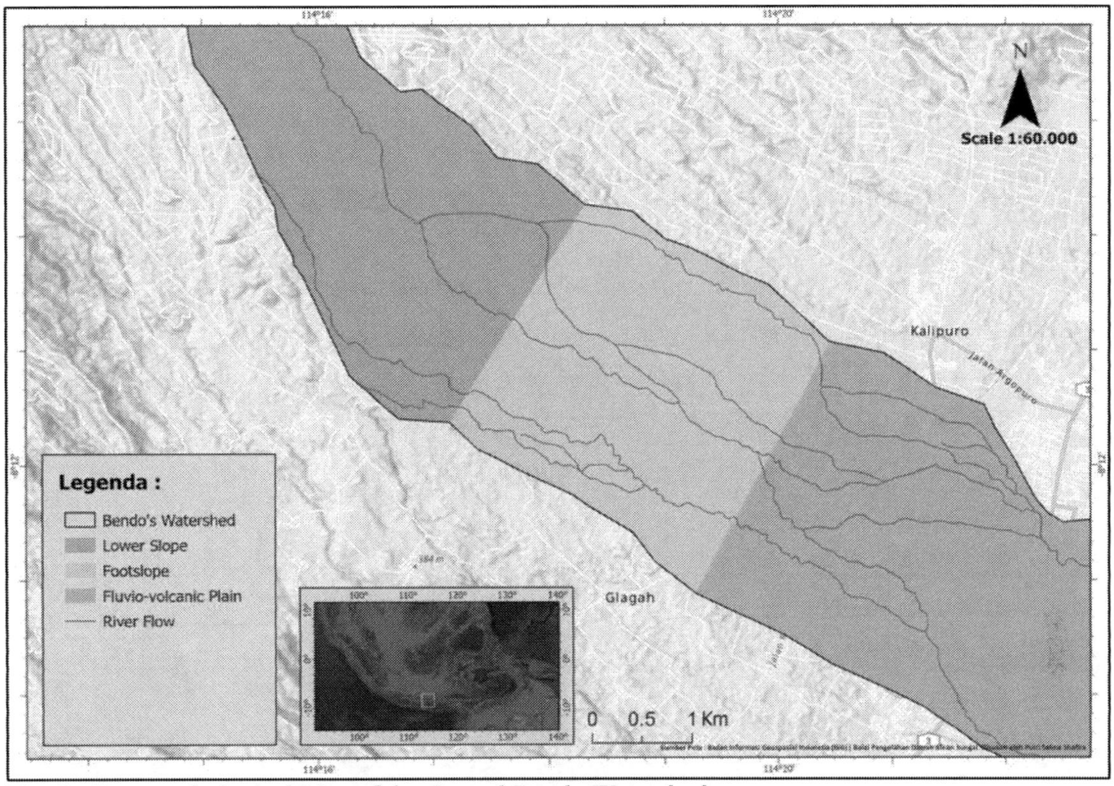

Fig. 5. Geomorphological Map of the Central Bendo Watershed

The middle zone of the Bendo watershed (see Figure 5) consists of landform units in the form of lower slopes and volcanic foothills. In this area, andesite lava material is present, which is the result of past volcanic eruption activities and has undergone natural deposition in the area (Figure 6a). The erosion process in the middle slope zone of the Bendo watershed is still ongoing, exhibiting relatively high intensity. The erosion rate is influenced by the type of land cover, especially vegetation in plantation areas located on the lower slopes. The plantation activities have been identified as a contributing factor to the observed increase in the erosion rate. The presence of waterfalls in this area suggests that changes in the rock layer structure have influenced the dynamics of water flow (Figure 6b). Concurrently, within the volcano's foot zone, the predominant geomorphic process is sediment transport, marking the transition phase from erosion to sedimentation. This is characterised by the presence of lava and pyroclastic material resulting from prior volcanic activity.

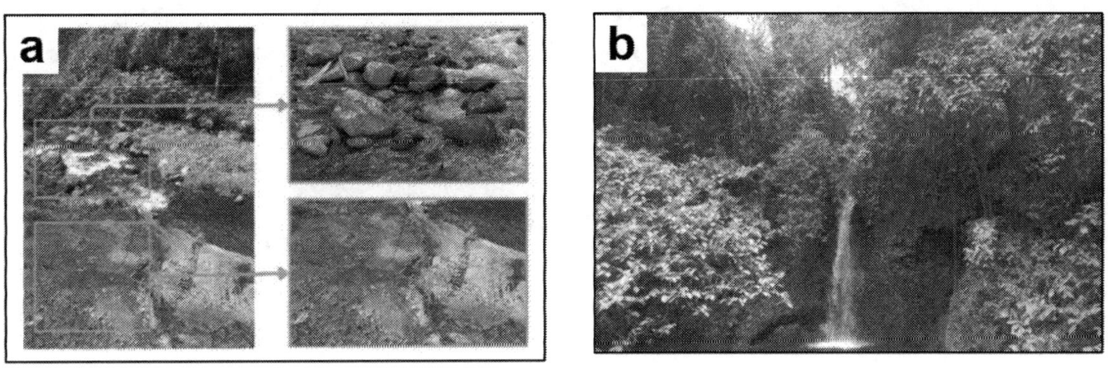

Fig. 6. Landfrom Bendo watershed centre (a) Andesite lava material and (b) waterfall as a volcanic material transfer route

https://doi.org/10.1051/e3sconf/202565208003

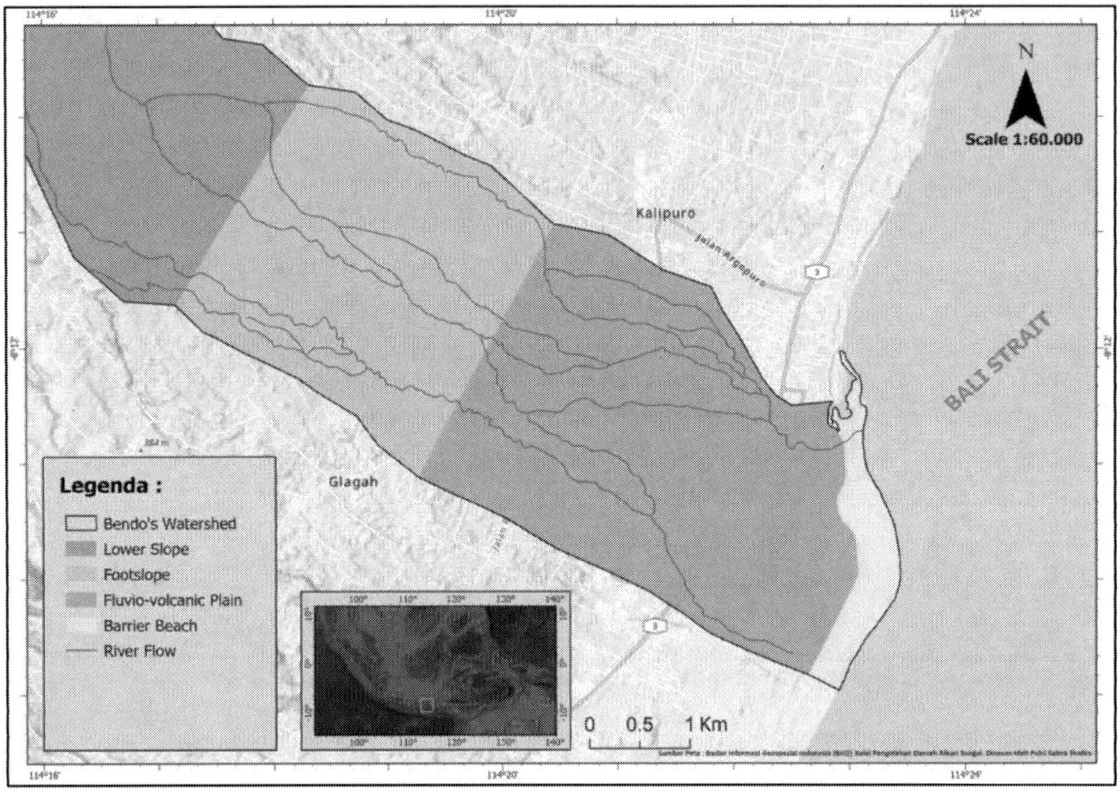

Fig. 7. Geomorphological Map of the Lower Bendo Watershed

The downstream section of the Bendo watershed (see Figure 7) comprises a variety of landform units, including fluviovolcanic plains, shoals, and lagoons. These features are the result of the interaction between fluvial, volcanic, and marine processes. The fluviovolcanic plains contain sedimentary deposits resulting from the accumulation of eroded material carried by river flow and volcanic material. From a genetic perspective, this plain thus constitutes a transition zone between landforms of volcanic and fluvial origin. Conversely, the beach is characterised by a preponderance of marine or maritime processes, as evidenced by the Geomorphology map of the Lower Bendo Watershed (Figure 7) and satellite imagery of the barrier beach on Santen Island (Figure 10).

A field analysis was conducted at Santen Island Beach, which revealed the presence of various basalt rock types exhibiting distinct colours and characteristics. Figure 8(a) is characterised by a darker grey colour with a porous texture, and it is sufficiently robust to exhibit a greater mass. Stones exhibiting such characteristics are commonly designated as lava stones[15]. While basalt stone (Figure 8(b)) possesses a lighter colour and weight, it is categorised as a floating stone based on its characteristics. The presence of basalt stones is indicative of a substantial volcanic process, encompassing the stages of weathering, erosion, transportation, and deposition[6].

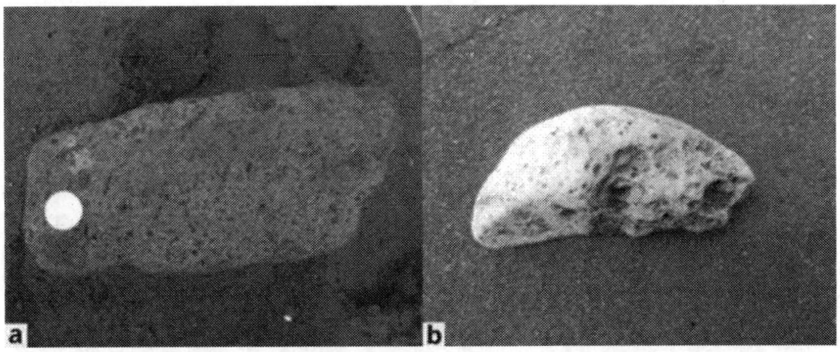

Fig. 8. Variations of Basalt Stone Types found on the coast of Santen Island (a) Lava Stone and (b) Pumice Stone

Fig. 9. There is a difference in texture between the coastal area (a) which tends to be finer while the closer to the settlement (b) the coarser the texture

The sand that accumulates to form a shoal becomes the primary material for the formation of the barrier beach. The barrier beach that has been formed on Santen Island is the result of a series of volcanic processes, through which a succession of weathering, erosion, transportation and sedimentation events have occurred. The material resulting from volcanic eruptions comprises volcanic ash, lava, and pyroclastics sourced from Mount Ijen Purba, Mount Merapi, and Mount Ranti. These materials are subject to erosion and are subsequently transported from the Bendo watershed through the Wuyu river to the coastal area. Thereafter, they undergo a deposition process influenced by river currents, coastal flows, and waves. It is a well-documented fact that, over time, the gradual accumulation of sediment results in the formation of a barrier structure that functions as a separator between sea and land[9]. Furthermore, the processes of land uplift and subsidence, consequent to volcanic activity, have the capacity to exert influence upon the patterns of ocean currents and sedimentation. In the event of an increase in land resulting from magma intrusion, the coastal zone may undergo changes that facilitate the formation of barrier beaches[6,9]. In addition to the impact of volcanic activity, the cycle of rising and falling sea levels, in conjunction with tectonic activity, has the capacity to exert influence on sedimentation patterns and the stability of barrier beaches. Should substantial changes occur, the barrier beach may experience erosion or even reform with different characteristics[6,9].

The formation of barrier beach as one of the coastal landforms plays a crucial role in supporting the existence of coastal ecosystems (Figure 10), including typical vegetation such as pioneer plants and other coastal species. Geomorphic processes, including weathering, erosion, and the transport of sand by river flows and ocean currents, contribute to the supply and distribution of sand sediments that comprise the dune body. The physical and chemical properties of these sand sediments, including grain size, porosity, and nutrient content,

https://doi.org/10.1051/e3sconf/202565208003

significantly influence the ability of this medium to support the growth of coastal vegetation[6,9].

Fig. 10. Satellite imagery of Gisik Pengalang on Santen Island

3.3 Analysis of Mangrove Distribution in the Formation of Barrier Beach

Santen Island is notable for its native ecosystem, which has existed since the island's formation. The mangrove forest on Santen Island is composed of the following species: Rhizophora mucronata, Avicennia marina, and Sonneratia alba [10] (see Figure 11). This phenomenon can be attributed to the geographical context of Santen Island, which is located at the mouth of a tranquil river. This unique geographical feature enables the natural growth of mangroves along the riverbanks and in the coastal waters. However, from the 1950s to the 2000s, several environmental and land-use changes, including the opening of ponds, the establishment of settlements and mangrove timber harvesting, began to be carried out, leading to a decline in the mangrove population. Since 2014, the local community and local government have engaged in systematic mangrove restoration activities, which have been carried out until 2022. In addition to the objective of restoring the mangrove ecosystem, the restoration activity also seeks to mitigate the effects of abrasion that occur on the coast.

Fig. 11. . Mangrove species, *Rhizophora mucronata* (a), *Avicennia marina* (b) and *Sonneratia alba* (c) comprise the mangrove ecosystem on Santen Island

The presence of a beach, in addition to its role as a natural barrier to wave energy and ocean currents, results in the creation of a more tranquil lagoon environment on the landward side. This renders the area an optimal transition zone for the growth of mangrove ecosystems, as it provides a stable substrate, a nutrient supply from sediments, and protection from abrasion and other hydrodynamic phenomena[9]. In order to ascertain the extent to which the barrier dune provides protection to the growth of the mangrove ecosystem, a spatial-temporal analysis must be conducted using Sentinel-2 satellite imagery. This paper utilised Google Earth Engine (GEE) and ArcGIS to analyse spatial data and undertake quantitative analysis of geometry features in order to generate a land cover change map for the period 2018 to 2025. The results of this analysis indicated significant area dynamics in the morphological unit of barrier dune and mangrove ecosystems in the north coastal area of Santen Island, Banyuwangi Regency.

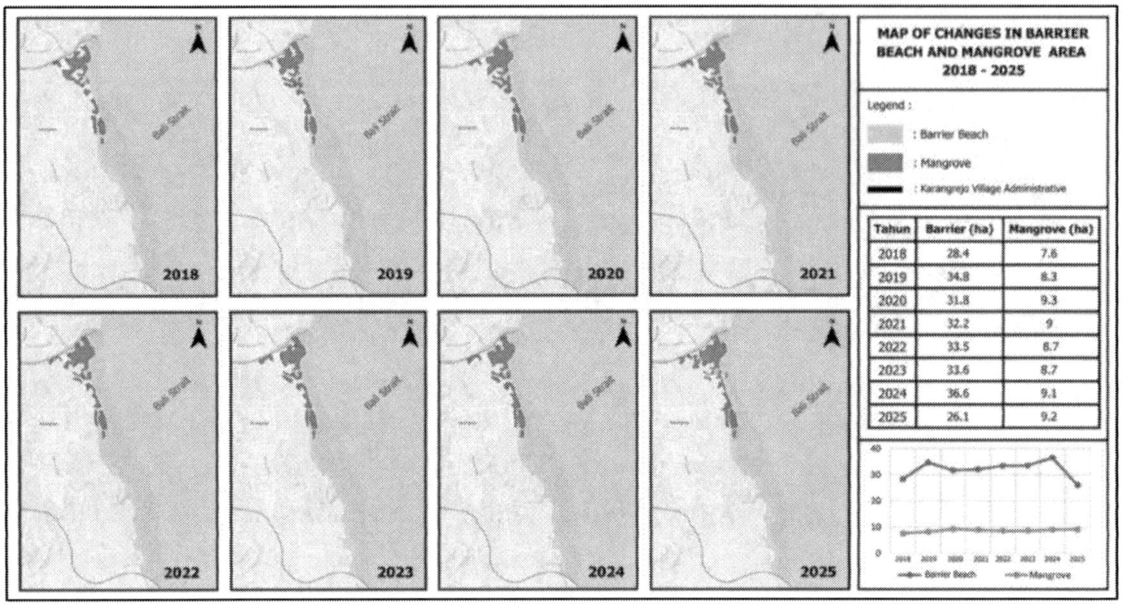

Fig. 12. Map of Changes in the Area of Gisik Barrier and Mangrove in 2018-2025

Spatial analysis of the barrier beach, as depicted on the map and symbolised by an orange hue, was conducted utilising Sentinel-2 satellite imagery (Figure 12). The analysis revealed a distinct pattern of shoal distribution, extending from north to south along the coastline directly adjacent to the Bali Strait. This distribution pattern is the result of sediment transport by the dominant longshore current flowing from the northeast towards the southwest, driven by ocean waves and tidal currents. Sediments derived from estuaries and local rivers are conveyed into the coastal zone, thereby forming longitudinally developed shoal bodies. A close examination of map images reveals that from 2018 to 2024, the dune area underwent a substantial expansion, increasing from 28.4 ha to 36.6 ha. This indicates a significant accumulation of sediment. However, in 2025, a significant decrease in area to 26.1 ha was observed, which may be attributable to hydrodynamic disturbances or coastal erosion processes that compromise sediment stability[4,9].

Conversely, the green hue of the mangrove ecosystem indicates a propensity for the growth direction of mangrove vegetation to deviate from the typical coastal trajectory, demonstrating an inclination to extend laterally in a southward direction along the barrier beach boundary. The distribution of these organisms is influenced by the presence of favourable substrate conditions, namely fine sediments such as mud and clay, which are formed in the intertidal zone behind the beaches. The mangrove root system has been demonstrated to play a role in stabilising sediments, capturing silt, and providing microhabitat conditions that support vegetative expansion. The area of mangrove growth increased from 7.6 hectares to 9.2 hectares between 2018 and 2025. This increase in area is relatively stable and sustained, suggesting that the environmental conditions behind the beach are relatively protected and supportive of ecological succession[9].

From a quantitative perspective, changes in barrier beach and mangrove areas demonstrate different dynamics. The Barrier Beach area demonstrated significant fluctuations in area, ranging from 28.4 hectares in 2018, peaking at 36.6 hectares in 2024, then dropping sharply to 26.1 hectares in 2025. Concurrently, the area covered by mangroves exhibited a gradual and consistent increase, from 7.6 hectares in 2018 to 9.2 hectares in 2025. This numerical contrast indicates that while barrier beach is more dynamic and vulnerable to sudden coastal processes, mangroves tend to respond more gradually and adaptively.

The concurrent proliferation of barrier beach and mangrove vegetation is of paramount importance in determining the natural protection system of the coastal area. Barrier beach is the primary geomorphic buffer, the function of which is to protect the backshore from abrasion and wave action. It is evident that the area of beach has undergone expansion between 2018 and 2024, thereby enhancing the coastal area's capacity to absorb wave energy and resist saltwater intrusion. This provides a more tranquil environment for mangroves to thrive, unencumbered by the pressures of extreme hydrodynamic factors. In this context, the beaches function in a dual capacity: as passive energy barriers and as active agents in the creation of ecohydrological conditions conducive to the growth of mangrove vegetation.

Concurrently, the presence of mangrove vegetation functions as an ancillary ecological safeguard, thereby augmenting the stability of the barrier beach. The structural characteristics of mangrove roots, including their ability to sluggish water currents and entrap sediments, serve to mitigate the erosion risk from the landward side, concurrently enhancing the structural resilience of the shoreline. Furthermore, the mangrove vegetative canopy exerts an additional function in reducing the strength of incoming waves, thereby acting as a two-way energy barrier between the sea and the land, and vice versa. The consistent growth of mangroves over the 2018–2025 period, despite a decrease in dune area in the previous year, suggests that mangroves are highly adaptable to changes in coastal morphology and can maintain their protective functions under conditions of environmental stress.

However, a substantial decrease in the area of beach was observed, from 36.6 ha in 2024 to 26.1 ha by 2025. This phenomenon has been linked to the impact of significant hydrodynamic events, such as intensified currents or erosion processes. However, without additional detailed supporting data, such as storm or wave records, the underlying causes remain uncertain. Concurrently, the mangrove area underwent continuous and consistent expansion. This finding suggests that mangroves possess a superior adaptive and mitigative capacity to environmental disturbances in comparison to barrier beaches, which are characterised by greater dynamism and vulnerability to rapid changes in coastal morphology.

4 Conclusion

It is evident from geological and geomorphological studies that the formation of barrier beaches on Santen Island is the result of complex interactions between volcanic, tectonic, fluvial, and marine processes. These processes have taken place gradually and continuously since the Quaternary period. The volcanic materials emanating from Mount Ijen Purba, Mount Merapi, and Mount Raung have undergone a process of weathering and erosion, subsequently being conveyed through the Bendo watershed system to the coastal region. The sedimentation process occurring in the intertidal zone, influenced by longshore currents and tidal dynamics, produced volcanic sand deposits that were naturally sorted and formed a barrier beach structure parallel to the coastline. This hypothesis is further substantiated by the presence of basaltic rocks and substantial black sand in the coastal region of Santen Island, suggesting a substantial contribution of past volcanic activity to the contemporary coastal morphology.

The presence of barrier beaches is of significance not only as a geomorphic entity, but also as an ecological function, acting as a natural barrier against marine dynamics. A spatial-temporal analysis reveals that the augmentation of the shoreline from 2018 to 2024 has engendered a more stable lagoon transition zone, which in turn has facilitated the proliferation of mangrove vegetation behind them. Mangrove species such as Rhizophora mucronata, Avicennia marina, and Sonneratia alba experienced consistent expansion, as the intertidal area was increasingly protected from wave action and abrasion. This synergistic relationship reflects the ecological interdependence between barrier beaches, which act as physical buffers, and mangroves, which act as biological buffers, thereby strengthening the coastal system as a whole.

However, the high environmental dynamics in the coastal area of Santen Island, particularly due to strong surface currents in the Bali Strait, have caused significant fluctuations in the extent of barrier beaches, including a decrease observed in 2025. The underlying drivers of this erosion event remain uncertain and cannot yet be conclusively linked to a single factor, representing one of the study's limitations. Despite the ongoing expansion of mangrove vegetation and its demonstrated capacity for adaptation to environmental fluctuations, it remains incapable of entirely supplanting the physical protective function of the barrier against large-scale wave energy.

It is recommended that future research address the limitations identified by incorporating more robust approaches. Such approaches may include quantitative modelling and long-term monitoring. It is recommended that future studies concentrate on simulating the potential impacts of sea-level rise and intensified hydrodynamic processes on the stability of barrier beaches and the resilience of associated mangrove ecosystems. The implementation of such endeavours would serve to provide a more robust scientific foundation for the prediction of coastal change and the support of integrated coastal management in Banyuwangi.

References

1. S. D. Andreastuti, E. T. Paripurno, Subandriyo, D. K. Syahbana, A. S. Prayoga, Volcanic Risk and Disaster Communication in Indonesia: Lessons from Recent Eruptions. J. Appl. Volcanol. 12, 1–15 (2023). https://doi.org/10.1186/s13617-023-00129-2
2. F. S. N. Sabila, M. Abdurrachman, A. Saepuloh, I. A. Kurniawan, A. Baraas, D. F. Yudiantoro, H. Kusdaryanto, Geology, Archaeology, and Historical Studies of the Late 16th Century Plinian Eruption of Raung Volcano: A Potential Case for Disaster Geotourism in Ijen UNESCO Global Geopark, East Java, Indonesia. Geosciences 14, 284 (2024). https://doi.org/10.3390/geosciences14110284
3. A. Prabowo, S. Permanadewi, H. Samodra, Volcanic Lithology and Geomorphology of Banyuwangi Regency. J. Geol. Sumberdaya Min. 22, 197–207 (2021). https://doi.org/10.33332/jgsm.geologi.v22i4.658
4. M. Irham, I. Rusydi, H. A. Haridhi, I. Setiawan, Y. Ilhamsyah, A. Deli, M. Rusdi, A. M. Siregar, Mangrove Ecosystem Dynamics from Sentinel-2 Data. J. Mar. Sci. Eng. 9, 815 (2021). https://doi.org/10.3390/jmse9080815
5. L. H. Trinh, T. G. Le, X. B. Tran, Q. V. Tran, V. P. Le, T. P. To, Barrier Island Morphodynamics under Coastal Forcing. Bull. Geogr. Phys. Geogr. Ser. 26, 77–87 (2024). https://doi.org/10.12775/bgeo-2024-0006
6. Z. Xue, S. Qian, Monitoring Barrier Dynamics with Sentinel-2 and SAR Data. IEEE J. Sel. Top. Appl. Earth Obs. Remote Sens. 15, 5131–5146 (2022). https://doi.org/10.1109/jstars.2022.3185078
7. N. R. Permana, D. F. Fadrian, B. Gunawan, A. Wulandari, Distribution of Geological Structures on Java Island Based on Derivative Analysis of Satellite Gravity Data. J. Phys. Fundam. Appl. 14, 154–168 (2024). https://doi.org/10.26740/jpfa.v14n2.p154-168
8. K. Salgado, M. L. Martínez, L. L. Álvarez-Molina, P. Hesp, M. Equihua, I. Mariño-Tapia, Coastal Dune and Vegetation Interactions. Ecoscience 29, 149–161 (2022). https://doi.org/10.1080/11956860.2021.1934299

9. A. Di Capua, R. De Rosa, G. Kereszturi, E. Le Pera, M. Rosi, S. F., From Volcanoes to Sedimentary Systems. Geol. Soc. Lond. Spec. Publ. 520, (2022). https://doi.org/10.1144/SP520-2022-303
10. Y. I. Rahmila, L. B. Prasetyo, C. Kusmana, Suyadi, M. Basyuni, B. Pranoto, Spatial Analysis of Mangrove Ecosystem Dynamics in Banyuwangi: A Geographically Weighted Regression Approach. Trop. Ecol. 64, (2024). https://doi.org/10.1080/21580103.2024.2438602
11. J. J. Major, Subaerial Volcaniclastic Deposits – Influences of Initiation Mechanisms and Transport Behaviour on Characteristics and Distributions. Geol. Soc. Lond. Spec. Publ. 520, 29–100 (2021). https://doi.org/10.1144/SP520-2021-142
12. R. A. Davis, D. FitzGerald, *Beaches and Coasts*, 2nd ed.; Wiley-Blackwell: Oxford, 2020. https://doi.org/10.1002/9781119334566
13. S. J. Dadson, Geomorphology and Earth System Science. Geol. Soc. Lond. Spec. Publ. 520, (2021). https://doi.org/10.1144/SP520-2021-142.
14. F. Yuliansyah, C. Sujatmiko, Petrological Characteristics of Basalt in Java. Teknika Sains 4, 1–10 (2019). https://doi.org/10.24967/teksis.v4i2.639
15. 14. A. Ashari, S. Purwantara, *Bentanglahan Vulkanik Indonesia: Aspek Fisik dan Kultural*; UNY Press: Yogyakarta, 2022. ISBN: 978-602-498-423-2

Stratigraphic Variations And Structural of Autobreccia: Implications Of Edaphic Potential in The Merapi-Ranti Volcanic Transition Zone, Geopark Ijen

Avanza Iqbal Ar Rafi[1], *Melati* Julia Rahma[2*], and *Firman* Sauqi Nur Sabila[3]

[1,2]Departemen Geografi, Fakultas Ilmu Sosial, Universitas Negeri Malang, Jalan Semarang No. 5, Malang, East Java
[3]Geology Geopark Ijen, Jalan Gajah Mada, Banyuwangi, East Java

Abstract. The transition zone between the Merapi Formation and the Ranti Formation in Kalibendo, Ijen Geopark, provides information in the form of rock outcrops that are used as parameters to understand and reconstruct geological phenomena in the past. The main problem in this study is the lack of clear stratigraphic information regarding the boundary between the two formations on geological maps and in previous studies. This study aims to reveal the stratigraphic sequence, identify autobreccia structures, and assess the edaphic implications in the Merapi–Ranti transition zone. This study uses a systematic method in the form of field observations and edaphic potential analysis as well as reconstructing geological phenomena and their potential for the surrounding environment. Based on rock outcrops at the research site, the stratigraphic sequence shows lava rocks from the Merapi Formation stacked on lava rocks from the Ranti Formation, indicating sequential formation. The discovery of autobreccia structures between the two formations is evidence of the interaction between molten lava from the Ranti Volcano and lava that had solidified from the Merapi Volcano. In addition, the diversity of rock and mineral structures is an indicator in reconstructing geological phenomena. Mineral diversity also correlates with edaphic potential, where soil resulting from the weathering of volcanic rocks is rich in minerals that can release nutrients, contributing to the formation of soil with high cation exchange capacity (CEC) that has long been utilized by the Osing ethnic for agriculture.

1 Introduction

Ijen Geopark is an area in Banyuwangi and Bondowoso that has geological, biological, and cultural diversity. This geopark features various natural landscapes such as volcanoes, rivers, lakes, karsts, and beaches, which were formed through long geological processes. Ancient volcanic activity left traces in the form of volcanic breccia, lava, and Pleistocene tuff [1]. The existence of volcanoes in this region is influenced by Indonesia's position at the meeting point of the Pacific, Eurasian, and Indo-Australian plates [2]. The collapse of the ancient Ijen Volcano around 50,000 years ago triggered the formation of new volcanoes, including

* Corresponding author: melatijuliarahma.fis@um.ac.id

Raung, Ijen, Merapi, Ranti, Pendil, and Suket [3]. This activity resulted in the formation of adjacent Quaternary volcanic rocks, creating a complex stratigraphy.

Although Ijen Geopark has been extensively studied from a general geological perspective, no specific studies have been conducted on the relationship between stratigraphy and autobreccia structures in the Merapi–Ranti transition zone. This research gap is an important basis for our study. The relevance of this research is further strengthened by the fact that Ijen Geopark is an area with geological diversity recognized by UNESCO. Detailed studies of stratigraphy and autobreccia not only enrich scientific information and support geological conservation, but also provide added value to edaphic utilization for local communities.

This diversity is inseparable from a long geological process, in which there were once major eruptions of ancient volcanoes that are evidenced by the presence of volcanic rocks such as volcanic breccia, lava, and tuff dating back to the Pleistocene era [1]. The presence of volcanoes in the Ijen Geopark area is the result of a geological process as Indonesia lies at the convergence of the the Pacific Plate, the Eurasian Plate, and the Indo-Australian Plate [2]. The geological phenomena that have been occurring since 50,000 years ago began when the ancient Ijen volcano collapsed, leading to a series of volcanic activities that formed new volcanoes, including Raung, Ijen, Merapi, Ranti, Pendil, and Suket [3]. The emergence of younger volcanoes led to the formation of various Quaternary volcanic rock formations were formed, which then formed the stratigraphy.

Stratigraphy refers to the study of the composition, relative age, distribution, and historical interpretation of rock layers. Stratigraphically, the Banyuwangi region and its surroundings are composed of various types of rocks dating from the Late Oligocene (28–23 million years ago) to the Early Miocene (23–16 million years ago), Middle Miocene rocks (11.6–16 million years ago), and Quaternary rocks (2.58 million years ago to present). The Oligo-Miocene rocks, dominated by volcanic rocks and lava, are the oldest exposed rocks. Additionally, shallow marine sedimentary rocks and volcanic clastic rocks were formed, including intrusive rocks such as granodiorite and andesite porphyry. These data are based on geological mapping by the Geological Research and Development Center (now the Geological Survey Center) from 1990 to 1993 [3].

The description of rocks based on their physical characteristics and composition is called lithology. Lithological studies are conducted to reconstruct past geological processes and their relationship with the surrounding environment. One of the lithological structures that can be found in the Ijen Geopark area is the autobreccia structure. Autobreccia is a type of rock that forms fragments on its own [4]. This structure is formed as a result of lava cooling, which causes the formation of angular rock fragments [5]. Its formation process is influenced by varying cooling times, resulting in the appearance of a arrangement of fragment components and matrix with the same composition [5]. The lithology and structure of autobreccia can serve as indicators for understanding lava eruption dynamics, volcanic deposition environments, and the volcanic activity history of a region.

Understanding magmatism and volcanoes requires exploration of the mineral diversity of a region. Minerals commonly found in igneous rocks, which are the product of volcanic activity, are pyroxene and plagioclase, formed in magma chambers under high temperatures and pressures. Pyroxene and plagioclase are generally found in oceanic crust undergoing subduction [6]. In addition to mineral diversity, the rock structure found at a site is necessary to understand the formation scheme of a region. The mineral diversity present in a region can contribute to edaphic potential, which is the capacity of soil to support vegetation growth through its physical, chemical, and biological characteristics.

Volcanic rocks contain minerals that can weather, releasing nutrients that are important for plants. Minerals such as feldspar, pyroxene, and olivine found in volcanic rocks will begin to weather over time, producing weathered minerals such as kaolin and clay that increase the soil's capacity to retain water and nutrients, thereby enhancing soil fertility [7]. This mineral weathering process naturally increases soil fertility, but its speed and efficiency depend on the type of mineral and environmental conditions. Therefore, mineral-rich volcanic rocks that are easily weathered have great potential as a source of natural nutrients for the soil, supporting plant growth and soil ecosystem sustainability.

The 1993 Banyuwangi geological map, published by the Center for Geological Research and Development (now the Geological Survey Center), does not provide definitive information about the stratigraphy of the Merapi and Ranti formations. Both formations have the same code, Qv, which indicates volcanic rock. However, no studies have been conducted yet to explain the stratigraphic sequence or cross-section of these formations. The primary objective of this research is to reveal the stratigraphic sequence between the two formations to determine which formed first. Understanding the interaction between rock formations at the research site can indicate the diversity of rocks and minerals and their contribution to soil formation in a particular area. The geological features found at the research site can be systematically organized to reconstruct past geological phenomena.

2 Location and Methodology

2.1. Location

The research was conducted in the Kalibendo Waterfall area (8°09′37.854″ S; 114°16′18.196″ E) located within the Ijen Geopark, Banyuwangi Regency, East Java. This location is a relatively unspoiled natural tourist area, located at the foot of Mount Merapi and Mount Ranti, making it strategic for volcanic geological observation. The surrounding area consists of coffee plantations and agricultural land owned by the Osing tribe, the indigenous ethnic group of Banyuwangi. This area was chosen because it is located precisely at the transition zone between the Merapi Formation and the Ranti Formation, as shown on the Banyuwangi Geological Map **Fig.1**. The location's position on the contact line between the two volcanic rock units allows for direct observation of the differences in lithology, geological structure, and stratigraphic relationships between the two in the field.

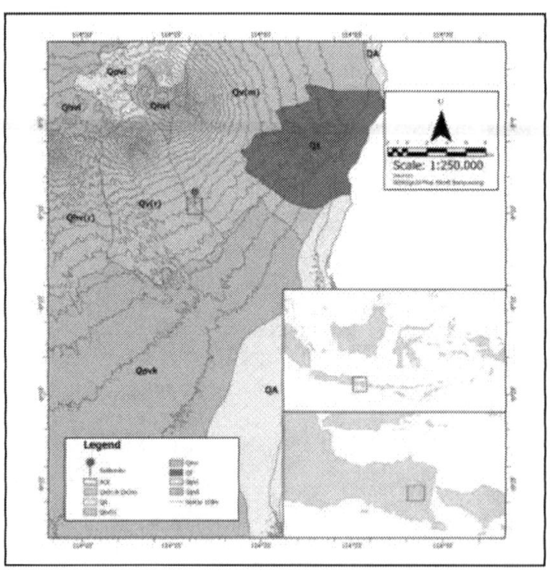

Fig. 1. Research Location Map

2.2. Data Analysis Techniques

This study used a systematic method [6], with the initial stage being the determination of the research location based on the Banyuwangi Geological Map. The location was then validated through field observations to identify the stratigraphic arrangement between the Merapi Formation and the Ranti Formation and determine the order of their formation [6]. Additionally, an analysis of rock characteristics was conducted based on igneous rock classification, by matching the physical and mineralogical characteristics of the rocks [5] .

Field data collection was carried out using various geological equipment. Geological hammers were used to collect rock samples, geological compasses to measure the orientation of structures, GPS to record location coordinates, and magnifying glasses to observe mineral details macroscopically. In addition, HCl solution was used to test for the presence of carbonate minerals, while cameras were used to document rock outcrops and geological structures. All of these activities were based on geological maps as the main reference for field mapping.

The field data obtained was then analyzed descriptively and comparatively, focusing on the differences in lithology, geological structure, and stratigraphic relationships between the two formations. Observations of rock outcrops around Kalibendo Waterfall were used as the main data to determine the stratigraphic sequence, rock structure, and mineral identification. All collected data were then compiled and analyzed to reconstruct the sequence of geological events in the transition zone between the Merapi Formation and the Ranti Formation. This interpretation was based on observations of characteristic geological structures such as autobreccia, xenoliths, sheeting joints, and vesicular structures found in outcrops at the research site.

3 Result and Discussion

3.1. Rock Outcrops and Stratigraphy

The rock outcrop around Kalibendo Waterfall, with a height of about 8 meters, reveals two rock units exposed vertically. The lower part of the outcrop is composed of rocks from the Merapi Formation, while the upper part is part of the Ranti Formation. The Merapi Formation is dominated by dark-colored andesitic-basaltic lava with a massive texture, while the Ranti Formation is composed of younger lava with volcanic breccia fragments and vesicular structures. These two formations are separated by an autobreccia structure formed by the interaction of young lava from the Ranti Volcano with lava that had solidified from the Merapi Volcano, so that the boundary between the two can be clearly identified.

Identification of the boundary between the two formations in **Fig.2** is based on the principle of superposition, whereby the lower layer tends to be older [8], and is reinforced by the presence of autobreccia formed by the interaction of young lava from Mount Ranti with lava that had already solidified from Mount Merapi. In addition to clear lithological differences, this outcrop also reveals other geological structures in the form of autobreccia and sheeting joints, which are important indicators in reconstructing the geological processes in the Merapi–Ranti transition zone.

Fig. 2. Rock Formation Exposure

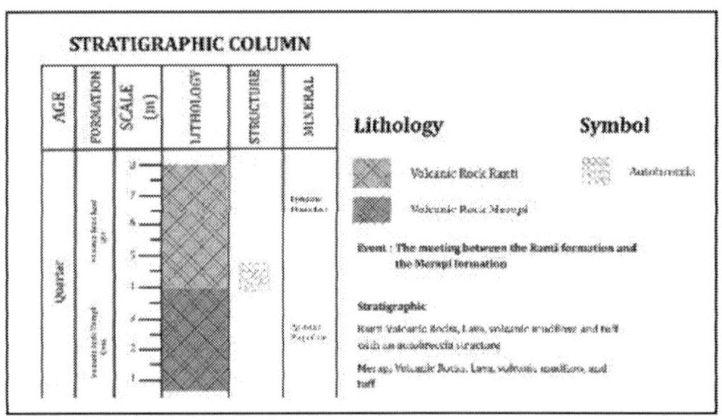

Fig. 3. Stratigraphic Column, Form Field observation

Stratigraphy in a region provides important clues about the geological phenomena that shaped the current landscape. In the Banyuwangi geological map sheet, cross sections show variations in rock composition around Banyuwangi, with each cross section cutting through a different formation. However, there are no sections that directly connect the Merapi Qv(m) Formation and the Ranti Qv(r) Formation, so the stratigraphic relationship between the two cannot be confirmed. Kalibendo, located between the two formations, shows rock outcrops that reveal the vertical stratigraphic sequence **Fig.2**. Based on the law of superposition, rock layers at the bottom in a normal position are generally older than the layers above them [8]. Thus, the stratigraphic sequence in Kalibendo consists of the Merapi Qv(m) Formation at the bottom and the Ranti Qv(r) Formation at the top **Fig.3**.

This formation structure forms the basis for assumptions regarding geological processes in the past, where the Merapi formation at the bottom, or older rock, was formed first, followed by the formation of the Ranti formation at the top, which is younger rock. The determination of these two formations is based on the surrounding geomorphology and the results of geological mapping conducted by the Geological Research and Development Center (now the Geological Survey Center). The exposed formations are of the same type, namely the Quaternary Volcanic Formation (Qv), but originate from two different volcanoes.

3.2. Rock Structure

The rock outcrops at Kalibendo not only reveal two rock formations but also show a rock structure in the form of an autobreccia **Fig.4a**. This structure was formed as a result of the interaction between hot lava and previously cooled rock. An autobreccia is a rock structure that is the product of volcanic activity [9]. The presence of this autobrexial structure indicates the formation process of these two formations, where the Merapi Formation formed first, followed by hot lava, which is a product of the Ranti volcano. The hot lava then interacted with the Merapi volcanic rock, resulting in the formation of an autobreccia structure consisting of rock fragments of similar size.

This finding reinforces the assumption of a sequence of geological events that indicate a gradual process between the two volcanic systems. Another structure found in the Kalibendo area is a xenolith structure, which is a fragment of older rock originating from an intrusion that was later trapped in a younger rock body [1]. These structures are commonly found in andesite rocks formed within the magma chamber **[8]**. In addition to autobreccia, other rock structures were also found and were quite diverse **Table 1**.

Fig. 4. Rock structure diversity a.) Autobrexial structure. b.) Xenolith structure. c.) Sheeting joint structure. d.) Vesicular structure. e.) Rock oxidation

Table 1. Rock structure diversity

No	Structure	Description
1	Autobreccia	This structure is not only found at one outcrop location, but also at several other outcrops in the study area. This indicates that the interaction process between hot lava and rocks that had already cooled down occurred repeatedly [9] and was evenly distributed across various points. This finding reinforces the assumption that the phenomenon of autobrecciation is an important part of the volcanic dynamics in this region.
2	Xenolith	Xenolith structures were found among rocks along the riverbed that had undergone erosion. The presence of older rock fragments trapped within younger rock masses indicates a process of young

https://doi.org/10.1051/e3sconf/202565209001

No	Structure	Description
		magma intrusion into previously formed rocks. This finding indicates that the geological processes at the site involved the intrusion of magma carrying older rock fragments, which were then trapped and solidified within the younger rock body [3].
3	Sheeting joint	This structure was found in the same outcrop as the autobreccia structure, indicating that the rock formation process occurred gradually through repeated lava flows. This structure was formed by the gradual cooling of flowing lava, which formed layers parallel to the surface. Over time, these layers were eroded by surface water flow, exposing the outcrop and revealing these structures in the field. This finding supports the interpretation that the Ranti Formation was formed through repeated volcanic processes with the contribution of lava deposition and gradual mechanical weathering [5].
4	Vesicular	Vesicular structures are found in many locations along the Kalibendo stream, characterized by the presence of holes or cavities on the surface of volcanic rocks formed by the release of volatile gases during the cooling and solidification of lava [10].
5	Rock oxidation	Oxidation in rock samples shows contrasting black and red colors, with the red color indicating zones that have cooled more quickly and undergone intense oxidation. These conditions are commonly found on the surface of autobreccia fragments or lava exposed to air, and are important indicators in the interpretation of the thermal and environmental conditions of lava solidification [13].

Data : Field observation

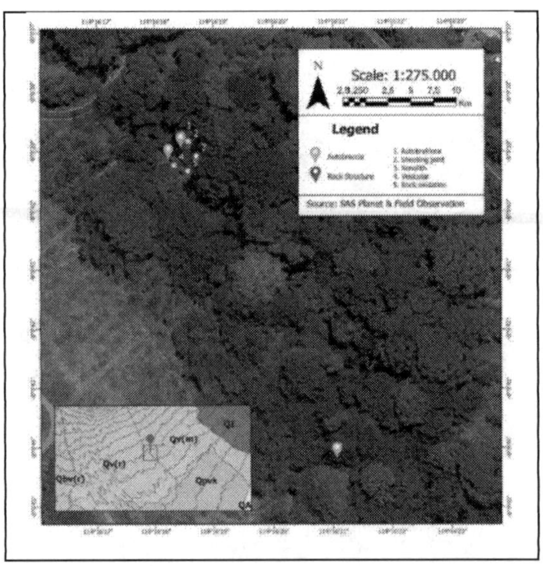

Fig. 5. Map of Rock Structure Sample Distribution

In rock findings in the area around the Kalibendo waterfall, there are xenolith structures in the form of old rock fragments that are thought to be volcanic products of Mount Merapi found trapped in young rocks from Mount Ranti **Fig 4b**. This finding indicates that during the formation of Mount Ranti, magma penetrated older rocks from Mount Merapi. This supports the interpretation that volcanic activity at Merapi occurred earlier than the development of the Ranti volcanic system.

The presence of these xenoliths does not stand alone, but is consistent with the findings of other rock structures located nearby at coordinates 8°09′37.854″ S and 114°16′18.196″ E. Based on the rock distribution map, this point is an important location for sample collection and also shows the distribution of autobreccia structures **Fig 5**. This distribution indicates that the interaction between young lava and old rocks occurred repeatedly in the Merapi–Ranti transition zone. Thus, both xenoliths and autobreccia provide strong evidence for reconstructing the complex volcanic dynamics in the Ijen Geopark area.

In addition to these two structures, field observations also revealed the presence of another rock structure in the form of sheeting joint **Fig 4c**. This structure develops parallel to the natural surface of the rock and is commonly found in massive or igneous rocks [5]. The presence of sheeting joints in Kalibendo usually forms on steep slopes and plays an important role as the main pathway for water movement, as seen in the Kalibendo Waterfall stream. Sheeting joints formed gradually following topographic stress conditions, creating layers of rock [5]. These layers are traversed by water flow, causing the rocks to undergo weathering and deposition processes that expose these layers. Another structure found is a vesicular structure **Fig 4d**, which consists of porous rocks resulting from the release of trapped gas during cooling. The released gas is volatile gas contained in the magma and can be released during the magma's ascent to the surface [10]. This gas is more soluble in magma at high pressure and will undergo exsolution to form bubbles (vesiculation) when pressure decreases, affecting the behavior and properties of the magma during eruption [10].

There are rocks with different colors **Fig 4e**, the presence of these rocks indicates that there were different oxidation and cooling processes between parts of the rock. As explained by Morrison-Evans (2024), color variations such as dark gray to reddish or brownish represent different levels of oxidation in andesitic basaltic lava. These color changes reflect differences in temperature, exposure time to air, and oxygen diffusion during the cooling and weathering of the lava. The presence of these rocks allows us to make assumptions about the rock formation process. When lava flows and begins to solidify, the parts closer to the air or surface are more easily exposed to oxygen, resulting in more intense oxidation and a reddish-brown color, indicating a high content of oxidized iron. Meanwhile, the inner parts, which are less exposed to oxygen, remain dark gray, indicating minimal oxidation [13].

3.3. Geological Reconstruction

Based on the rock structures found at the research site, past geological reconstructions can be made by linking the relationships between structures and stratigraphic positions. After the formation of Mount Merapi, volcanic activity at Mount Ranti produced magma that broke through the Merapi rock layer, forming xenolith structures. Subsequently, lava flows from Ranti covered the cooled Merapi rocks, and the interaction between the hot lava and the cold rocks formed autobreccia structures.

In addition, lava flows also form layers that follow the contours of the local topography, which over time undergo weathering due to water flow. This process causes material deposition and produces sheeting joint structures in the form of fractures parallel to the rock surface. The discovery of vesicular structures in the outcrop also indicates an explosive eruption phase, in which high vesicularity affects the behavior of magma as it rises to the surface [10].

Although there are no written records of historical eruptions of Mount Merapi and Mount Ranti, stratigraphic observations and rock structures in Kalibendo show evidence of repeated volcanic activity during the Quaternary period. This is in line with the geological reconstruction of the Ijen region, which explains that the collapse of the ancient Ijen volcano triggered the formation of younger volcanoes, including Merapi and Ranti, which produced lava flows, breccias, and volcanic structures that can still be observed in the current outcrops [3,2]. Thus, field observations can be considered a historical representation of the volcanic activity of these two volcanoes.

3.4. Mineral Diversity

The long history of volcanic activity in the Merapi and Ranti volcanoes is rich in minerals resulting from magma crystallization. At the research site, particularly around Kalibendo Waterfall, macroscopic minerals such as plagioclase and pyroxene were found **Fig 6**. The presence of plagioclase, commonly found in continental rocks, as explained by [12], indicates that plagioclase is one of the main rock-forming minerals in the continental crust.

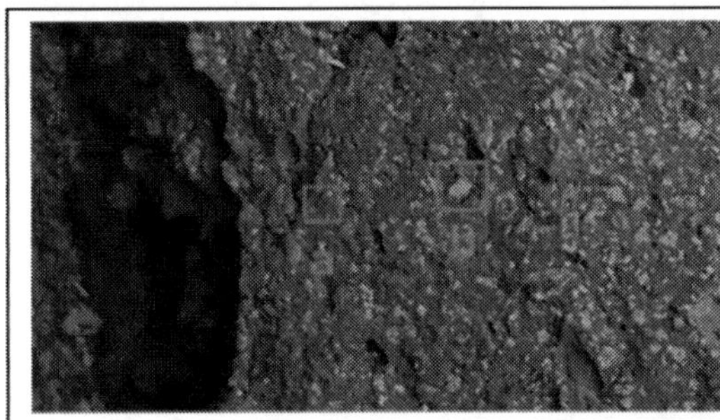

Figure 6. Macroscopic Minerals (a) Pyroxene, (b) Plagioclase

Pyroxene, which is commonly found in oceanic rocks is a common component found in basaltic rocks in marine environments [6]. These two minerals support the research assumption that magmatic activity in this area is related to the subduction zone, which is the meeting point between oceanic and continental plates. Although volcanic processes from volcanoes can produce various types of minerals, macroscopic observations in the field show the dominance of plagioclase and pyroxene as the main minerals present.

Tabel 2. Mineral Diversity

No.	Rock	Mineral
1	Andesite	Pyroxene Plagioclase Magnetite Amphibole Biotite Feldspar
2	Basalt	Pyroxene Plagioclase Olivine Magnetite
3	Scoria	Pyroxene Plagioclase Olivine Magnetite

Data : field observations & Toderaş et al. (2017) [13]

The abundance of minerals can also be seen from the rocks found at the research site. The results of the analysis and matching of rock and mineral characteristics show the diversity of minerals in each rock **Table 2**. Volcanic rocks such as andesite, basalt, and scoria match the results of previous research by Toderaş et al. (2017) [13], stating that the igneous rocks found at the research site are dominated by plagioclase, pyroxene, olivine, feldspar, and magnetite minerals. The minerals from these rocks will eventually undergo weathering and contribute to soil formation, becoming an important factor in soil fertility.

3.5. Edaphic Potential

Field studies show that the area around Kalibendo Waterfall, located in the volcanic transition zone between Mount Merapi and Mount Ranti, is dominated by land use in the form of coffee plantations and agricultural land. This land use pattern reflects high edaphic and environmental potential, influenced by local geological conditions, particularly the presence of mineral-rich volcanic rocks. These findings are consistent with another research, which states that volcanic lava rocks, such as those found in this area, generally contain primary minerals such as plagioclase and pyroxene [9]. When weathered, these minerals can release important nutrients such as K^+, Ca^{2+}, and Mg^{2+} into the soil.

The diversity of minerals in volcanic rocks in the Mount Merapi and Ranti volcanic areas is the result of active geological processes that produce primary minerals rich in nutrients. These minerals, such as plagioclase, pyroxene, feldspar, and olivine, play an important role in supporting soil fertility, but must first undergo weathering. Weathering is a series of chemical and physical processes that transform primary minerals into more stable forms that are easily absorbed by plants. These minerals fall into the category of primary minerals that are essential for plants when they undergo weathering and decomposition [17]. Once primary minerals have undergone weathering, they form simple charged compounds.

Environmental factors have a major influence on the intensity and direction of weathering. Parameters such as soil pH, temperature, humidity, redox conditions, and the presence of microorganisms will determine the rate of mineral decomposition and nutrient release. Humid and warm tropical conditions such as those found in the Kalibendo area of Ijen Geopark, tend to accelerate the weathering process and form soils that are richer in nutrients [17].

Previous studies have also shown that soil modification through the addition of volcanic materials such as basalt powder can increase macro and micro nutrient content. However, this effect can vary depending on the physicochemical properties of the rock, soil dynamics, and the complexity of the nutrient cycle [9]. Primary mineral weathering not only provides important nutrients such as K^+, Ca^{2+}, and Mg^{2+}, but also forms soil structures with high cation exchange capacity (CEC), which is very beneficial for plant growth. Soils developed from volcanic rocks tend to have high natural fertility due to the direct contribution of mineral weathering and intense biological activity [9].

3.6. Implications of Minerals on Edaphic Potential

The rocks found in the Kalibendo area are dominated by igneous rocks in the form of basalt and andesite, which are classified as basic rocks. These rocks are generally weathered, fine-grained, and produce clay with a predominance of 2:1 montmorillonite and high base saturation, making them relatively fertile [14]. Mineralogically, the rock contains primary minerals such as plagioclase, pyroxene, feldspar, and olivine, which play an important role as providers of nutrients such as K^+, Ca^{2+}, and Mg^{2+}. In addition, secondary minerals such as calcite, apatite, chlorite, and quartz were also found. According to Hanafiah (2014) in his book he mentions, secondary minerals play a role in determining the chemical and physical properties of soil and serve as indicators of the intensity of primary mineral weathering. The greater the proportion of secondary minerals, the more intense the weathering process and the more mature the soil genesis process. Hanafiah (2014) in his book he mentions, some non-silicate minerals, such as calcite, have relatively high nutrient content and function as sources of Ca and OH^-, so they are often used as liming agents for acidic soils. Basically, plants need macro nutrients such as N, P, K, Ca, and Mg, as well as other micro nutrients that generally come from parent rocks and their constituent minerals [15].

4 Conclusions

This study reveals the stratigraphic sequence between the Merapi Formation and the Ranti Formation in the Kalibendo transition zone, Ijen Geopark, Banyuwangi. This research was motivated by the lack of clear information regarding the stratigraphic cross-section of the two formations, both in regional geological maps and previous studies. Field observations show that the stratigraphic sequence at the research site consists of Merapi Formation lava at the bottom and Ranti Formation lava at the top. The presence of autobreccia structures in the contact zone is evidence of interaction between hot lava from the Ranti Volcano and solidified lava from the Merapi Volcano, reinforced by the discovery of other geological structures such as xenoliths, vesicular textures, and evidence of differential oxidation, which indicate a gradual and complex volcanic process. In addition, the diversity of primary minerals (plagioclase, pyroxene, olivine, feldspar, magnetite) and secondary minerals (calcite, chlorite, quartz) plays an important role in releasing nutrients such as K^+, Ca^{2+}, and Mg^{2+}, thereby supporting the formation of soil with high cation exchange capacity (CEC). These conditions have implications for the edaphic potential that has long been exploited by the Osing community, for example for coffee plantations and agriculture around Kalibendo. Thus, this study confirms that stratigraphy, autobreccia structure, and mineral diversity are not only important for geological reconstruction, but also have real implications for edaphic potential and land use. Further studies using petrographic and geochemical approaches can deepen our understanding of the relationship between rock mineralogy and edaphic potential in volcanic areas.

References

1. A. Prabowo, H. Samodra, Analisis batuan gunungapi di daerah Banyuwangi dan sekitarnya berdasarkan aspek geokimia. *Jurnal Geologi dan Sumberdaya Mineral* 21(4), 197–207 (2024). https://doi.org/10.33332/jgsm.geologi.v22.4.197-207
2. D. Bagus, V. Dosen, J. Tadris, I. P. Sosial, F. Tarbiyah, I. Keguruan, Ancaman erupsi Kawah Ijen terhadap masyarakat lereng Ijen (Kabupaten Bondowoso). (2020). http://www.vsi.esdm.go.id
3. H. Samodra, A. Prabowo, S. Permanadewi, J. Razak, S. Yalin, Bappeda Banyuwangi, Inventarisasi keragaman geologi dan identifikasi warisan geologi daerah Kabupaten Banyuwangi (Jawa Timur) (2019). https://doi.org/10.33332/jgsm.geologi.v22i4.658
4. J. P. Wiersma, E. M. Roberts, P. H. Dirks, Formation of mud clast breccias and the process of sedimentary autobrecciation in the hominin-bearing (Homo naledi) Rising Star Cave system, South Africa. *Sedimentology* 67(2), 897–919 (2020). https://doi.org/10.1111/sed.12666
5. J. Martí, G. Groppelli, A. Brum da Silveira, Volcanic stratigraphy: A review. *Journal of Volcanology and Geothermal Research* 357, 68–91 (2018). https://doi.org/10.1016/j.jvolgeores.2018.04.006
6. W. Ning, T. Kusky, L. Wang, B. Huang, Archean eclogite-facies oceanic crust indicates modern-style plate tectonics. *Proceedings of the National Academy of Sciences of the United States of America* 119, e2117529119 (2022). https://doi.org/10.1073/pnas.2117529119
7. A. L. Lewis, B. Sarkar, P. Wade, S. J. Kemp, M. E. Hodson, L. L. Taylor, K. L. Yeong, K. L. Davies, K. Nelson, M. I. Bird, I. B. Kantola, M. D. Masters, E. DeLucia, J. R. Leake, S. A. Banwart, D. J. Beerling, Effects of mineralogy, chemistry and physical properties

of basalts on carbon capture potential and plant-nutrient element release via enhanced weathering. *Applied Geochemistry* 132, 105023 (2021). https://doi.org/10.1016/j.apgeochem.2021.105023
8. R. Daniswara, Y. Firmansyah, R. M. Ganjar Gani, I. Syafri, Geologi daerah Desa Karangtengah dan sekitarnya, Babakan Madang, Bogor. (2020). https://doi.org/10.24198/pgj.v4i6.32199
9. X. Dupla, R. Claustre, E. Bonvin, I. Graf, R. C. Le Bayon, S. Grand, Let the dust settle: Impact of enhanced rock weathering on soil biological, physical, and geochemical fertility. *Science of the Total Environment* 954, 176297 (2024). https://doi.org/10.1016/j.scitotenv.2024.176297
10. K. V. Cashman, R. S. J. Sparks, How volcanoes work: A 25 year perspective. *Geological Society of America Bulletin* 125, 664–690 (2013). https://doi.org/10.1130/B30720.1
11. B. Morrison-Evans, E. Melekhova, J. Blundy, Experimental insight into basaltic andesite lava dome oxidation textures at La Soufrière, St Vincent. (2024). https://doi.org/10.1144/SP539-2022-298
12. M. Baïsset, L. Labrousse, A. Schubnel, J. Gasc, K. Béneut, M. Guillaumet, Rheology of hydrated plagioclase at lower crustal conditions: Cataclasis, creep and transformational plasticity. *Journal of Structural Geology* 178, 105010 (2024). https://doi.org/10.1016/j.jsg.2023.105010
13. M. Toderaş, R. I. Moraru, C. Danciu, G. Buia, L. I. Cioca, Model to assess the quality of magmatic rocks for reliable and sustainable constructions. *Sustainability* 9(10), 1925 (2017). https://doi.org/10.3390/su9101925
14. A. Andreas, A. Putra, Perbandingan karakteristik batuan beku erupsi Gunung Gamalama dan Gunung Talang. *Jurnal Fisika Unand* 7, 293–298 (2018). https://doi.org/10.25077/jfu.7.4.293-298.2018
15. I. Bali, A. Ahmad, C. Lopulisa, Identifikasi mineral pembawa hara untuk menilai potensi kesuburan tanah. *Jurnal Ecosolum* 7(2), 81–100 (2018). https://doi.org/10.20956/ecosolum.v7i2.5126

 https://doi.org/10.1051/e3sconf/202565209002

Quantitative Morphometric Characteristics in the Geomorphic Setting of Southern Malang, Indonesia

Heni Masruroh[1,2*], *Alfi* Sahrina[1,2], *Choirul* Anam[2,3], *Zukhrufia* Rahmi[4], *Zafira* Fatimah Azzahra[4], *K*resno Sastro Bangun Utomo[1], *B*intang Harfian[4]

[1]Geography Departmen, Faculty of Social Science, Universitas Negeri Malang
[2]Center for Environment, Mitigation and Disaster, Research Institute and Community Engagement, Universitas Negeri Malang
[3]Management Department, Faculty of Economics and Business, Universitas Negeri Malang
[4]Geography Student, Faculty of Social Science, Universitas Negeri Malang

Abstract. Geomorphometry serves as a vital tool for providing quantitative data on watershed characteristics. Morphometric analysis is essential for understanding the relationship between landform features and hydrological responses. This research aimed to examine geomorphic characteristicnahs in the South Malang area. The methodology involved utilizing remote sensing data through a morphometric approach. Data collection utilized DEMNAS with 8-meter spatial resolution and field observations. Processing was conducted using ArcGIS 10.8, incorporating geomorphometric parameters including hydrological morphometric categories with 12 derived parameters and morphological morphometric categories with 16 derived parameters. The Wang and Liu fill algorithm was implemented for its benefits in filling depressions from edges to center while preserving original topographic integrity. Hydrological processing followed Strahler's formula for river order determination. Results indicate that the Gurah Watershed exhibits elongated basin morphometric characteristics. This configuration affects complex geomorphological process dynamics. The watershed's elongated morphometry combined with steep gradients and intermediate geomorphic development suggests potential landslide vulnerability from stream bank erosion. The elongated shape influences hydrological response patterns, causing slower downstream flow arrival and concentrating water flow in the main channel. These conditions are demonstrated through quantitative analysis of hydrological and morphological morphometric categories, providing insights into watershed behavior and potential hazards.

* Corresponding Author: heni.masruroh.fis@um.ac.id

1 Introduction

Geomorphology has experienced rapid development in the past five decades, both in fundamental aspects such as methods and approaches, and shifts in research interests [1]. This development includes the emergence of geomorphometric approaches in geomorphological analysis. Geomorphometrics serves as a quantitative approach for the numerical analysis of landforms. Furthermore, geomorphometry plays a crucial role in interpreting land surface parameters within the scope of geomorphological studies and enables the integration of spatial data in modeling geomorphic processes. This represents progress at a time when geomorphologists need a mathematical theoretical framework to answer systematic, structured questions about the geomorphic processes occurring on Earth [2], [3]. Geomorphometry can be applied to identify active erosion zones, drainage patterns, or the distribution of depressions in mountainous landscapes [2]. Advances in remote sensing technology and high-resolution digital elevation models (DEMs) strengthen this capacity, enabling quantitative analysis that links form and process within a single, interconnected framework [3]. This provides insight into how digital topographic data can reveal the dynamics of landform formation and evolution. In more complex conditions, the geomorphometric approach helps identify morphological patterns and unravels the spatial relationships between landform elements and active processes on the Earth's surface [4]. This capability becomes very relevant when applied to areas that are ecologically and geomorphologically vulnerable to change.

Watershed morphometry plays a crucial role in environmental sustainability management as it is a key instrument for understanding the interaction between landform characteristics and hydrological responses [5]. Most studies on morphometric analysis are applied to homogeneous landscapes, such as volcanic landscapes or karst landforms. Most previous studies have focused on interpreting morphometric data in disaster-prone areas, particularly in hydrometeorological disasters. Studies on detailed analyses integrating morphometric characteristics based on morphological and hydrological perspectives into a unified whole to predict potential disaster threats are still rare [6]. In Indonesia, particularly in the South Malang region, most studies still focus on disaster modelling, land use changes, coastal sedimentation, and disaster vulnerability mapping [7]. Studies on applying morphometry by linking landforms and hydrological responses are still rare.

The Gurah Watershed is an area with complex landscape conditions indicated by the diversity of lithology. This area is included in the Wonosari Formation (Tmwl) with lithology in the form of coralline limestone, tuffaceous, sandy limestone, marl, black claystone and is included in the Mandali Formation (Tamm) with lithology in the form of andesite-basaltic, latite phorphyry, and dacite. Based on geological formation, the Gurah Watershed is included in the transition zone, which meets the karst and volcanic landscapes [8]. The existence of these two main geomorphological systems makes the Gurah Watershed, despite its relatively small size, representative of the geomorphic complexity typical of South Malang. Karst and volcanoes are also known as areas with high levels of geomorphological and ecological sensitivity [8]. This high sensitivity is the impact of desertification caused by the structural, ecological, and geomorphological characteristics of the Karst region [8].

Over the last two centuries, impacts and hazards have increased drastically in karst and volcanic areas [8]. This increase is triggered by exponential population growth that impacts anthropogenic activities that are considered the main geomorphic agents in landscape evolution. The complexity of causal factors originating from the interaction between natural processes and anthropogenic activities creates dynamics that ultimately determine the

https://doi.org/10.1051/e3sconf/202565209002

character of karst and volcanic landscapes [8]. Geomorphometrically, karst and volcanic landforms are highly dynamic and responsive to environmental disturbances, especially in forested areas that are vulnerable to vegetation degradation, land cover changes, and accelerated erosion. The complexity of processes and physical characteristics of karst and volcanic areas not only shape distinctive landscapes but also determine their level of vulnerability to environmental change [8]. Therefore, quantitative analysis is needed to understand these dynamics and support mitigation efforts against rapid landscape change. This study conducted a quantitative morphometric analysis to interpret the morphological and hydrological characteristics of the Gurah Watershed. The results provide essential baseline information for sustainable Gurah Watershed management.

2 Method

2.1 Study Area

The Gurah Watershed in this study is located on the Southern Cross Route of Java Island and is located in the downstream region south of Mount Kawi. Based on its latitude and longitude, this watershed is located between 8°18'S and 08°21'S and 112°20'E and 112°22'E. Stretching 6 km in length and covering an area of approximately 11 km2, the Gurah Watershed has an elevation ranging from 0 to 378 m. Administratively, it is located within Tugurejo Village, Wates District, Blitar Regency, East Java Province (Fig. 1). Most of the area has a linear settlement pattern along the road due to the influence of the hilly area. Geologically, the layers formed are Mandalika formations, with volcanic rock types such as andesite-basalt, latite porphyrite, rhyolite, and dacite. The Gurah Watershed area has a tropical climate with an average temperature of 24°C and an average annual rainfall of 1478.8 mm . This tropical climate is divided into two seasons: the rainy season in December-June and the dry season in July-November. Climatologically, it is actively influenced by the movement of the west monsoon and the east monsoon with more humid conditions occurring during the west monsoon.

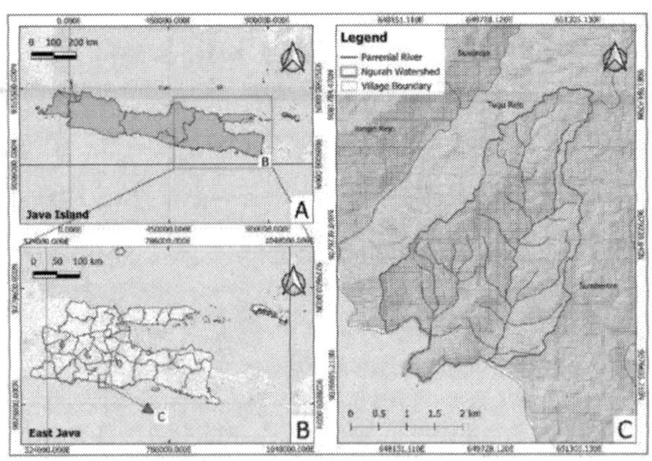

Fig. 1. Research Area

2.2 Geospatial Data Acquisition

https://doi.org/10.1051/e3sconf/202565209002

In this study, the primary data was obtained from DEMNAS, namely the Digital Elevation Model data created by the Geospatial Information Agency (BIG) with a spatial resolution of 8 m. In water resource studies, information such as the stream network and its impact on the surrounding environment is crucial, one of which is identifying drainage system characteristics. The study used geomorphometric methods for this analysis, divided into several factors. First, hydrological factors include stream order, stream number, stream length, mean stream length, stream length ratio, drainage density, stream frequency, overland flow length, and main channel length. Second, morphological factors include basin area, perimeter, relief ratio, hypsometric curve, hypsometric integral, slope, aspect, TWI, TRI, TPI, form factor, elongation ratio, circularity ratio, compactness coefficient, and basin length. Table 1 shows the parameters used in this study.

Table 1. Geomorphometric Factor

Category	Parameter	Equation	References
Hidrologi	Stream Order	Hierarchical rank	Strahler, 1964
	Stream Number (Nu_t)	Nu_t= Nu1+Nu2…Nun Where Nu_t is total stream number of order	Horton, 1945
	Stream Length (Lu_t)	Lu_t = Lu1+Lu2+Lu3…Lun	Strahler, 1964
	Mean Stream Length (Lurm)	Lu / N	Horton, 1945
	Stream Length Ratio (Lur)	Lur = Lu(x+1)/Lu(x)	Strahler, 1964
	Bifurcation Ratio (Rb)	Rb = Nu(x) / Nu(x+1)	Horton, 1945
	Drainage Density (Dd)	Dd = Lu_t/A where Lu_t is total stream length of order	Horton, 1932
	Stream Freaquency (Fs)	Fs = Nu_t/A	Horton, 1931
	Overland Flow Length (Lg)	Lg = 1/2Dd	Horton, 1945
	Constant of Channel Maintenance (C)	C = 1/Dd	Horton, 1932
	Stream Power Index (SPI)	SPI = $A_s * \tan\beta$ Where As indicates the specific catchment area, and β is the slope expressed in radians	Moore et al, 1993
	Main Chanel Length (Cl) km	Arcgis Analysis	
Morphology	Form Factor (Ff)	Ff= A/Lb^2	Horton, 1932
	Elongation Ratio (Re)	Re = $2 * (A/3.14)^{1/2}$ /Lb	Schumm, 1956
	Circularity Ratio (Rc)	Rc = $12.57 * (A/P^2)$	Miller, 1953
	Compactness Coefficient (Cc)	Cc = $0.2841 * P/A^{0.5}$	Gravelius, 1914
	Length of Basin (Lb) km	Arcgis Analysis	Schumm, 1956

Category	Parameter	Equation	References
	Basin Area (A) km^2	Arcgis Analysis	Schumm, 1956
	Perimeter (P) km	Arcgis Analysis	Schumm, 1956
	Relief Ratio (Rr) m/km	Rr = H/Lb Where H is the difference between the maximum and minimum elevations in the watershed	Schumm, 1956
	Hypsometric Curve	X= elevasi Realtif Y= area relative Where x and y calculate from: $X = (\dot{h} - h_{min}) / (h_{max} - h_{min})$ $Y = A_h / A_t$	Strahler, 1952
	Hypsometric Integral	$HI = (\dot{h} - h_{min}) / (h_{max} - h_{min})$ $\dot{h}$ = average watershed elevation h_{max} = maximum elevation h_{min} = minimum elevation	Strahler, 1952
	Slope	f(x)=a_0+(√(σz/σx))+(√(σz/σy))	Zevenberg & Thorne, 198
	Aspect	arctangent [(∂z/∂y)/(∂z/∂x)]	Zevenberg & Thorne, 1987
	Topography Wetness Index (TWI)	$TWI = \ln(A_s/\tan\beta)$ Where As represents the specific catchment area, and β denotes the slope gradient expressed in radians	Moore et al.,1993
	Topography Roughness Index (TRI)	$TRI = \sqrt{[\Sigma (z_i - z_a)^2]}$	Riley et al., 1999
	Topography Position Index (TPI)	$TPI = Z_0 - (\sum Z_n / n)$ where Z_0 is the elevation of the point under evaluation, Z_n are the elevations of the surrounding points, and n is the total number of surrounding points considered	Weiss, 2001; Gallant & Wilson, 2000

2.3 Data Pre-Processing

All parameters were processed using Geographic Information System (GIS) software, such as ArcGIS 10.8. The study area is in the UTM 49S zone, with a watershed obtained through several processing stages. The DEM input was first filled, which is a crucial procedure, particularly in hydrological analysis [2]. The Wang and Liu fill algorithm was used because it offers advantages such as filling sinks from the edge toward the center without disrupting the original topographic structure [2]. This results in the final result that water flows from the

https://doi.org/10.1051/e3sconf/202565209002

highest to the lowest elevation cells. The results of this process are then followed by flow direction and flow accumulation. The raster provides values to indicate the direction of water flow and accumulation, which can then be used to determine stream network extraction. The formula proposed by Strahler is used as a guideline in determining river order. Firstorder is identified by the criterion of a network that has no tributaries. If two first-order networks meet and merge, it is considered second-order, and the rules for subsequent orders apply [9]. Further analysis of other parameters was conducted using various tools available in ArcGIS 10.8 software. Fig. 2 shows the research flowchart.

2.4 Data Processing Techniques

Data was processed by obtaining morphometric and geomorphometric parameters for the Gurah Watershed (DAS). The analysis used DEMNAS data as the primary input data with a resolution of 8 meters, which was then resized to 10 meters. This DEMNAS data was also supported by Gurah Watershed boundary data obtained through delineation using a hydrological approach. The entire processing of both data sets was performed using QGIS 3.40 software with the support of the SAGA GIS toolbox. The first stage focused on correcting the DEM data using the fill sink method. This process aimed to remove artificial depressions in the DEM, ensuring the validity of the resulting hydrological parameters. This process was followed by river network extraction through flow direction, flow accumulation, and channel extraction. The main rivers in this study were determined based on the Strahler order classification results. The results of this classification will later be used in calculating other linear parameters such as total river length (Lut), number of rivers per order (Nu), bifurcation ratio (Rb), drainage density (Dd), stream frequency (Fs), and overland flow length (Lg).

Meanwhile, parameters such as perimeter (P), watershed area (A), and main watershed length (Lb) were obtained through calculations on shapefile data. These three parameters will be used in further processing to obtain several parameters such as the form factor (RF), constant of channel maintenance (C), and relief ratio (Rh). All of these parameters were calculated using standard formulas from Gravelius (1914), Horton (1932, 1945), and

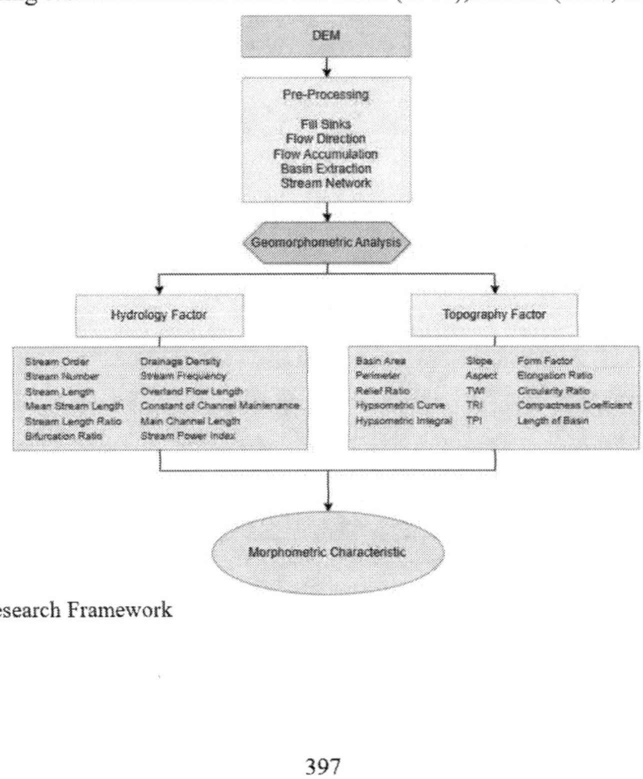

Fig. 2. Research Framework

Schumm (1956). Several parameters were obtained through direct processing of DEM data. These parameters include slope, aspect, and curvature, which were processed using QGIS tools as an application of the second-order polynomial method developed by Zevenbergen & Thorne (1987). The resulting indices such as the Topographic Wetness Index (TWI), Terrain Ruggedness Index (TRI), and Topographic Position Index (TPI) were analyzed to obtain results on surface morphological diversity. The TWI index was calculated through slope and flow accumulation. Meanwhile, the raster algorithm calculated the TRI and TPI based on the pixel radius. SAGA. Parameter Hypsometric Curve is obtained through the SAGA Morphometry – Hypsometry tool with an elevation interval class of 100. Furthermore, the Hypsometric Integral value is obtained based on the cumulative area of relative elevation and the relative area mapped in the Hypsometric Curve.

3 Result and Discussion

3.1 Morphology Characteristics

Morphometry is an approach in geomorphological studies that provides a quantitative understanding of landform structure, utilizing numerical data based on a Data Elevation Model (DEM). The application of morphometry to morphological characteristics is crucial for quantifying the morphological aspects of a region, particularly in areas with heterogeneous surface features. [5], [9] indicate that the application of morphometry to morphology is categorized into basic morphometry parameters, such as elevation, slope, and aspect; derived parameters, such as TWI, SPI, and TRI; and shape and form descriptors, such as curvature and topographic position index, which describe the configuration of the Earth's surface in greater detail.

The area in this study is a transitional region with karst and volcanic systems, with a highly complex and dynamic landscape zone geomorphologically. The complexity of this zone is due to the region being formed by two morphometric systems: the karstic system and the volcanic system. Studying morphological characteristics using a morphometric approach will provide important information for systematic identification and classification of land units and as a basis for determining zoning for ecological vulnerability and disaster threats [9].

This study used a morphometric approach to morphological characterisation based on basic morphometric parameters, derived parameters, and form descriptions (Fig. 3). The results indicate that the study area is dominated by strong-very strong slope classes, covering 65,610 pixels or 60.67% of the total area. This slope formation in the transition zone is caused by past tectonic and volcanic activity. The morphometric type for the SPI provides information on the potential strength of surface flow in erosion. The steep and fragmented slope conditions cause this area to be categorized as very high to high erosion. This high erosion is strongly influenced by surface material in the form of volcanic weathering [10]. Aspect is a DEM derivative that provides information on terrain orientation to the cardinal points, so aspect significantly influences soil moisture conditions. Characterization of aspects in the research area shows that most of the slopes face southwest (19.35%), west (18.63%), and northwest (17.48%).

Curvature provides information about the shape of the earth's surface, namely concave, flat, and convex. The dominance of flat curvature values indicates that most areas do not experience extreme flow convergence or divergence changes. Curvature characterization is very important because it influences surface flow that can increase the risk of local inundation and soil saturation. Convex and concave curvature show a small amount, but this can be used as a formation of weathering processes or rock structural deformation [10]. To determine the

 https://doi.org/10.1051/e3sconf/202565209002

vertical landscape characterization, this study conducted a TPI analysis. The land unit that dominates this area is the upper slope which has an important role in the hydrological system because it is the initial process of erosion, cracking, and the beginning of the start of land mass movement. On the other hand, the lower slope is a zone of accumulation of material resulting from erosion and land mass movement (landslides). TRI provides information regarding the characterization of the study area's complexity or diversity of elevation. Geomorphological processes greatly influence elevation diversity. A total of 86.41% of the study area exhibited smooth geological surfaces, indicating advanced weathering and landform homogenization. Meanwhile, 5.28% and 0.67% exhibited rough and very rough surfaces, with surface conditions characterized by geomorphological processes such as gully erosion and hills with hard rocks.

The TWI provides information characterizing the study area's soil fertility. Data processing results indicate that the Ngorah Subwatershed is predominantly classified as very dry, with 52,106 hectares (49.22%) and dry with 50,375 hectares (47.59%). The dominance of these two classes indicates that the transitional karst-volcanic physiographic characteristics influence water content. Based on the morphometric approach, the distribution of TWI is strongly influenced by slope gradient and topographic position. Steep slopes and topographic positions, such as upper to middle slope units, have a low TWI index, while lower areas with foot slope units, which are deposition areas, have a high TWI index. Table 2 shows the morphometric categories in the study area.

Tabel 2. Characteristics of the Research Area Based on Morphometric Type

Morphometry Type	Category	Number of Pixels	Percentage
Slope	Near level – Very gentle	5296	4.90
	Gentle slope	8294	7.67
	Moderate slope	18042	16.68
	Strong – Very strong slope	65610	60.67
	Extreme – Steep slope	10436	9.65
	Very steep slope	462	0.43
Stream Power Index (SPI)	Very low erosion potential	1	0.0009
	Low erosion potential	797	0.75
	Moderate erosion potential	42428	40.08
	High erosion potential	60434	57.09
	Very high erosion potential	2192	2.07
Aspect	Southeast	2584	2.38
	South	7447	6.88
	Southwest	20925	19.35
	West	20153	18.63
	Northwest	18906	17.48
	North	17467	16.15
	Northeast	15241	14.09
	Southeast	5395	4.98
Curvature	Strong concave	18	0.02
	Medium concave	60	0.06
	Flat/neutral	107865	99.74
	Medium convex	182	0.16
	Strong convex	15	0.01
Topographic Position Index (TPI)	valley	2362	2.23
	lower slope	24910	23.53

	middle slope	7438	7.02
	upper slope	71142	67.20
Topographic Roughness Index (TRI)	very smooth	8073	7.62
	smooth	91477	86.41
	rough	5589	5.28
	very rough	713	0.67
Topographic Wetness Index (TWI)	very dry	52106	49.22
	dry	50375	47.59
	wet	2974	2.80
	very wet	397	0.37

Based on Table 3 the analysis of morphological characteristics with a form factor (Rf) value of 0.297 and an elongation ratio (Re) of 0.616, the Ngorah Watershed has an elongated basin shape. The elongated morphometric shape of the Ngorah Watershed impacts geomorphological processes because in an elongated watershed, the hydrological response is slower to reach downstream. Typically, in watersheds with an elongated morphometric shape, the river network pattern is subdendritic or trellis whose flow pattern follows the geological structure. This elongated basin also has implications for the potential for high landslide vulnerability due to its steep slope gradient [11]. Other morphometric parameters such as The main river length (Lb) is recorded at 6,034 km, with a watershed area (A) of 10,845 km^2 and a watershed circumference (P) of 21,951 km. The calculated relief ratio (Rh) reached 55.328. The analysis of the relief ratio (Rh) of 55.328. These morphometric parameters were obtained from DEM analysis using QGIS hydrology tools. The watershed boundary was delineated, the river network was extracted, and subsequently the main river length (Lb), watershed area (A), and perimeter (P) were calculated using the geometry functions. The relief ratio (Rh) was then derived from these values following the standard morphometric formula.

The results of these four parameters indicate that the Ngorah Watershed is characterized by steep slopes, which impact geomorphological processes such as high erosion potential, sediment transport, and landslide potential due to stream bank erosion and gully erosion. Furthermore, the high Rh value also impacts the morphometry of the V shaped valley with narrow alluvial terraces. This condition is further strengthened by the hypsometric integral value of 0.48, obtained from DEM analysis using the area–altitude method in QGIS. The hypsometric integral (HI) value is calculated by normalizing the elevation and area of the watershed (DAS), then determining the proportion of the area under the hypsometric curve. Based on Strahler's (1952) classification as cited in Singh et al. (2008) [12], an HI value between 0.35–0.60 indicates that the watershed is in the intermediate geomorphic stage (mature stage), which is in accordance with the development conditions of the Ngorah Watershed.

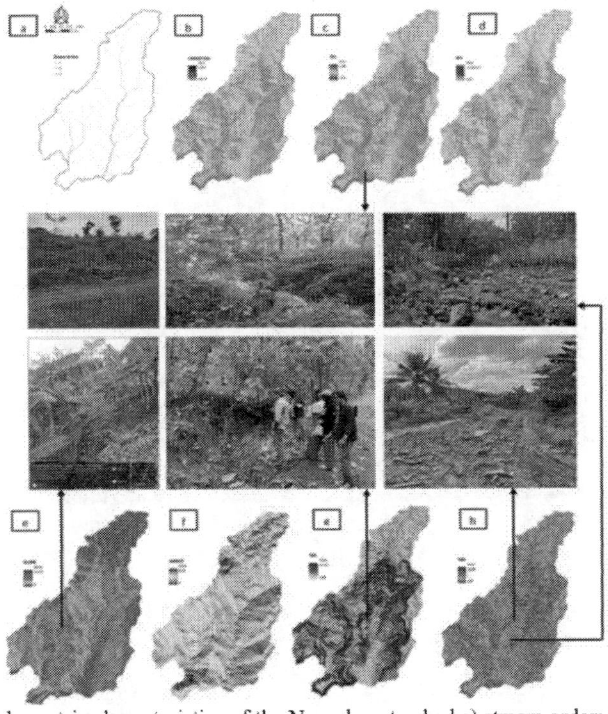

Fig. 3. Morphometric characteristics of the Ngorah watershed a) stream order; b) curvature; c) TPI; d) SPI; e) Slopes; f) Aspect; g) TRI; h) TWI

Tablel 3. Morphology Characteristic

Rf	Re	C	Cc	Lb	A	P	Rh	Hypsometric Integral
0.297	0.616	0.28	1.894	6.034	10.845	21.951	55.328	0.48

Overall, the morphometric characteristics of the Ngorah Watershed are shaped, indicating the dynamics of complex geomorphological processes. The elongated morphometric shape of the watershed with steep slopes and its intermediate geomorphic development indicates that the Ngorah watershed is potentially vulnerable to landslides due to stream bank erosion. The elongated morphometric shape of the Ngorah watershed results in a slower hydrological response reaching downstream and water flow being focused on the main river. Fig. 3 shows the characteristics of the Ngorah watershed's appearance.

https://doi.org/10.1051/e3sconf/202565209002

3.2 Hydrology Characteristics

River network order analysis is a key component in conducting morphometric analysis using a hydrological approach [11]. The Strahler method was chosen as the reference, where the first-order network is sorted based on its position without tributaries. Furthermore, when these orders meet, they form new branches, which are then referred to as second-order. This rule applies to the following orders. Based on the analysis, it was found that the Ngorah Watershed has three orders, with complete information shown in Table 4.

Table 4. Hydrology Characteristics

<table>
<tr><th>Stream Order</th><th>Nu</th><th>Lu</th><th>Lurm</th><th>Lur</th><th>Rb</th><th>Cl</th><th>Dd</th><th>Fs</th><th>Lg</th><th>C</th><th>Nu_t</th></tr>
<tr><td>1</td><td>25</td><td>18.493</td><td>0.740</td><td>0.230</td><td>30</td><td rowspan="3">9.25</td><td rowspan="3">2.3</td><td rowspan="3">2.95</td><td rowspan="3">0.21</td><td rowspan="3">0.42</td><td rowspan="3">32</td></tr>
<tr><td>2</td><td>5</td><td>4.256</td><td>0.851</td><td>0.745</td><td>7</td></tr>
<tr><td>3</td><td>2</td><td>3.170</td><td>1.585</td><td></td><td></td></tr>
</table>

The highest order is order 3 with an accumulated length of approximately 3.1 kilometers. This figure indicates that the complexity of the river network in the Ngorah Watershed is relatively low. The relatively simple topography also influences this condition. Furthermore, by identifying the largest order, it is also known that this order represents the main river flow [13] of the Ngorah Watershed. Morphometric analysis of the Ngorah Watershed (DAS) shows river network characteristics that reflect local hydrological and physiographic conditions. The total number of identified rivers reached 32 (Table 4), with a decreasing pattern with increasing order as described by Horton [13].

The high number of rivers in the low order indicates relatively low surface runoff during the rainy season [13], while the dendritic pattern formed represents topography with relatively low slope variations and the natural development of the river network following the slope. These results indicate a regular flow system that supports a uniform distribution of runoff. Table 4 shows that the total length of the river network reached 25.9 km with an average of 3.1 km, with the first order having the largest cumulative length of approximately 18.4 km. In line with the number of rivers, river length shows an inverse relationship with order, indicating a positive relationship between length (Lu) and the number of channels (Nu) [14]. Low order rivers are generally located in areas with steep slopes, contributing to greater accumulation of length, while downstream sections tend to be gentler. The river length ratio (<1 across all orders) indicates steeper headwater conditions, accelerating flow concentration downstream and resulting in short concentration times [15]. Bifurcation (Rb) values of 5 for order 1 and 2.5 for order 2 indicate a moderate degree of branching upstream, with a simpler network downstream. The relatively low Rb values also indicate minimal structural disturbance and geological stability within the watershed [15].

The characteristics of flow density and frequency reinforce the hydrological picture of the Ngorah Watershed. The drainage density (Dd) of 2.3 indicates moderate conditions with relatively low erosion, in line with vegetation cover and permeable soil material [15]. The Dd value per order ranges from 0.2 to 1.7, indicating the influence of river network length, lithology, and vegetation conditions. The river frequency (Fs) of 2.9 is also in the moderate category, with a positive relationship to drainage density indicating a drainage texture that is not too coarse [15]. The length of the surface flow (Lg) is only 0.21, indicating a short flow

https://doi.org/10.1051/e3sconf/202565209002

distance so that rainwater is quickly concentrated in the main channel. The negative relationship between Lg and Fs confirms that the denser the river network, the shorter the distance traveled by the surface flow. Other supporting parameters strengthen the conclusions of the watershed hydrology. The constant channel maintenance (C) value of 0.42 indicates compliance with Schumm's theory of a negative correlation with drainage density, indicating a relatively low maintenance requirement to maintain the stability of the flow network. The length of the main channel (Cl) is recorded at 9.25 km, representing the longest path from the upstream to the watershed outlet. Meanwhile, the Stream Power Index (SPI) generated through SAGA analysis indicates high flow strength within the main river network, which impacts sediment transport capacity. Variations in SPI values emphasize the role of river cross-sectional shape in determining flow power and sediment transport stability.

4 Conclussion

The morphometry of the Ngorah watershed shows a complex geomorphological karst-volcanic transition basin with steep and heterogeneous terrain. Basic and derivative parameters indicate that strong to very strong slopes cover more than 60% of the area, the relief ratio reaches 55.3, and the hypsometric integral of 0.48 places the watershed in a mature geomorphic stage. The values of the curvature and TRI parameters indicate that most of the surface is smooth but with localized rough zones formed by gully erosion and structural deformation. The TWI parameter shows a predominance of very dry to dry classes, reflecting the influence of steep slopes and transitional lithology on soil moisture. The elongated basin shape (Rf = 0.297; Re = 0.616) will slow down the hydrological process in the downstream section while concentrating the flow in the main channel, which is 9.25 km long and capable of producing strong flow forces and localized bank erosion. The drainage density (2.3 km^2) and flow frequency (2.9 km^2) identify the Ngorah watershed network as somewhat coarse, supporting controlled surface runoff, but high erosion potential still exists in areas on steep slopes. Overall, these morphological and hydrological attributes indicate that the Ngorah watershed is in a mature stage of landscape evolution with moderate to high vulnerability to erosion and landslides, especially along steep upper slopes and cliff zones. Although administratively part of Blitar Regency, its morphometric characteristics represent a small basin in the southern Malang karst-volcanic transition zone, which provides important basic information for ecological zoning, erosion management, and disaster mitigation planning.

Acknowledgments

The researcher would like to thank the LPPM State University of Malang for funding this research through the PNBP Grant 2025 and Geoenviroster Research Group

Referrences

1. S. Mukherjee, "A Review of the Contemporary Approaches to Geomorphology," Indian Journal of Spatial Science, vol. 13.0, no. 4, pp. 78–83, Winter 2022.
2. P. Costabile, C. Costanzo, C. Gandolfi, F. Gangi, and D. Masseroni, "Effects of DEM Depression Filling on River Drainage Patterns and Surface Runoff Generated by 2D Rain-on-Grid Scenarios," Water, vol. 14, no. 7, pp. 997–997, Mar. 2022.
3. L. Xiong, S. Li, G. Tang, and J. Strobl, "Geomorphometry and terrain analysis: data, methods, platforms and applications," Earth-Science Reviews, vol. 233, p. 104191, Oct. 2022.

4. L. F. L. Garzon, M. F. Johnson, N. Mount, and H. Gomez, "Exploring the effects of catchment morphometry on overland flow response to extreme rainfall using a 2D hydraulic-hydrological model (IBER)," J. Hydrol., vol. 627, no. October, 2023.
5. N. Sharma et al., "Prioritization of erosion susceptible watersheds using morphometric analysis and PCA approach: A case study of lower Sutlej River basin of Indian Punjab," Watershed Ecol. Environ., vol. 5, no. March, pp. 209–224, 2023.
6. M. H. AlRifai, A. Al Kafy, and H. A. Altuwaijri, "Quantitative Assessment of Flood Risk Through Multi Parameter Morphometric Analysis and GeoAI: A GIS-Based Study of Wadi Ranuna Basin in Saudi Arabia," Water (Switzerland), vol. 17, no. 14, pp. 1–31, 2025.
7. N. Hamid, "Natural Disaster Risk Mapping in Realizing Sustainable Environmental Management in the East Rembang Region, Indonesia," Int. J. Environ. Impacts, vol. 7, no. 3, pp. 495–504, 2024.
8. L. Pisano, V. Zumpano, M. Pepe, I. S. Liso, and M. Parise, "Assessing Karst Landscape Degradation: A Case Study in Southern Italy," Land, vol. 11, no. 10, pp. 1–18, 2022.
9. A. Mahala, "The significance of morphometric analysis to understand the hydrological and morphological characteristics in two different morpho-climatic settings," Applied Water Science, vol. 10, no. 1, Dec. 2019.
10. H. Deng et al., "Slope-Unit Scale Landslide Susceptibility Mapping Based on the Random Forest Model in Deep Valley Areas," Remote Sensing, vol. 14, no. 17, pp. 4245–4245, Aug. 2022, doi: https://doi.org/10.3390/rs14174245.
11. C. Kant, G. Kumar, and R. S. Meena, "Modeling morphometric and geomorphological parameters of mountainous river basin for water resource management using remote sensing and GIS approach," Modeling Earth Systems and Environment, vol. 9, no. 2, pp. 2151–2163, Dec. 2022.
12. O. Singh, A. Sarangi, and M. C. Sharma, "Hypsometric Integral Estimation Methods and its Relevance on Erosion Status of North-Western Lesser Himalayan Watersheds," Water Resources Management, vol. 22, no. 11, pp. 1545–1560, Feb. 2008, doi: https://doi.org/10.1007/s11269-008-9242-z.
13. M. S. Bhat, A. Alam, S. Ahmad, H. Farooq, and B. Ahmad, "Flood hazard assessment of upper Jhelum basin using morphometric parameters," Environmental Earth Sciences, vol. 78, no. 2, Jan. 2019.
14. D. Asfaw and G. Workineh, "Quantitative analysis of morphometry on Ribb and Gumara watersheds: Implications for soil and water conservation," International Soil and Water Conservation Research, vol. 7, no. 2, pp. 150–157, Jun. 2019.
15. S. Gajbhiye, S. K. Mishra, and A. Pandey, "Prioritizing erosion-prone area through morphometric analysis: an RS and GIS perspective," Applied Water Science, vol. 4, no. 1,pp. 51–61, Sep. 2013.

Volcanic Ash Weathering and Pedogenesis in Mount Penanjakan: Soil Evolution Supporting Pioneer Vegetation

Melati Julia Rahma[1*], *Jati* Batoro[2], *Danniary Ismail* Faronny[3], *Rivandi Pranandita* Putra[4], *Avanza Iqbal Ar* Rafi[5], *Muhammad Zaky Gymnastiar* Putra[6], *Eka Prasetyo* Nugroho[7], *Roy Juan* Akbar[8], *Ahmad* Habibi[9], *Erisa* Rahmawati[10], and *Novi* Silvia[11]

[1,4,5,6,7,8,9,10,11]Geography Department, Faculty of Social Science, State University of Malang
[2]Biology Department, Faculty of Mathematic and Science, Universitas Brawijaya
[3]Doctoral Program of Environmental Studies, Graduate School, Universitas Brawijaya

Abstract. Volcanic ash deposits play a fundamental role in soil formation processes and the establishment of pioneer vegetation in highland ecosystems. This study investigates soil evolution and pedogenesis derived from volcanic ash in Mount Penanjakan, Bromo Tengger Semeru National Park, East Java, Indonesia, focusing on its role in supporting Edelweiss (Anaphalis spp.) as a pioneer species. Field surveys were conducted at three sites—Lemah Pasar, Bukit Prau, and Tikungan S using transect-based soil profile observations up to 2 m depth. Soil samples were analyzed for organic carbon (Walkley and Black), texture (pipette method), soil pH (pH meter), and porosity, bulk density, and aggregate stability. Results indicated that soils were predominantly silty clay loam, with moderately acidic to neutral pH (5.5–7.5) and organic carbon content ranging from 1.2% to 2.1%. Taxonomic classification using USDA Soil Taxonomy (2014) identified variations from Mollic Hapludands (Andisols) at Lemah Pasar, Typic Dystrudepts (Inceptisols) at Bukit Prau, to Typic Hapludands (Andisols) at Tikungan S. These findings reflect a pedogenic gradient from weakly developed Inceptisols to more advanced Andisols influenced by volcanic ash weathering. The edaphic conditions, particularly high porosity, moderate organic matter, and andic properties, provide favorable habitats for Edelweiss species, each occupying different soil development stages. This study highlights the importance of volcanic ash pedogenesis in shaping soil-vegetation interactions and provides a scientific basis for designing ex-situ conservation strategies that mimic natural soil conditions.

* Corresponding author: melatijuliarahma.fis@um.ac.id

1 Introduction

Volcanic ash weathering and pedogenesis represent fundamental processes that shape ecosystem development in tropical mountainous regions, particularly where volcanic activity creates unique soil formation pathways and vegetation establishment patterns. Indonesia, as part of the Pacific Ring of Fire, contains extensive volcanic landscapes where these processes are critically important for understanding ecosystem dynamics and supporting biodiversity conservation [1]. Volcanic ash soils retain the largest and most persistent soil carbon pools of any ecosystem, yet the mechanisms governing soil carbon accumulation and weathering during initial phases of ecosystem development remain poorly understood, especially in high-elevation tropical environments [2].

The pedogenesis of volcanic ash follows distinct developmental trajectories influenced by climate, topography, and biological factors. Recent research has demonstrated that weathered volcanic ash provides essential nutrients for plant establishment, creating improved nutrient and water balance that accelerates vegetation recovery in disturbed volcanic environments [3]. Pedogenesis serves as a key factor driving nutrient immobilization and mineralization, particularly relevant in fertility management of volcanic soils, affecting both plant production and soil microbial functionality. However, the specific environmental conditions at high elevations result in unique combinations of soil properties and development patterns that differ significantly from lower elevation volcanic soils [4].

Pioneer species play a crucial role in initiating ecological succession following volcanic disturbances, serving as the first colonizers of newly formed volcanic substrates. These species must possess exceptional physiological adaptations to survive in environments characterized by nutrient-poor volcanic ash deposits, extreme temperature fluctuations, and harsh chemical conditions. Pioneer species are defined as the first group to colonize new habitats created by disturbances, including volcanic activity that creates very fine soils, high heat conditions, or water scarcity [5]. Understanding their establishment patterns is essential for comprehending ecosystem recovery processes and developing effective conservation strategies.

Anaphalis spp, endemic to the mountainous regions of Java, represents a remarkable example of evolutionary adaptation to volcanic environments. This species functions as a pioneer in recent volcanic lands, demonstrating exceptional capability to colonize and establish in freshly deposited volcanic materials. Recent research has revealed that edelweiss species develop mono-specific forb layers on young volcanic soils rich in sulfur, indicating unique adaptation mechanisms to sulfur-rich volcanic environments and their critical role in establishing initial vegetation communities. Traditional ecological knowledge has been successfully applied to monitor Anaphalis javanica distribution, abundance, and population dynamics in Bromo Tengger Semeru National Park, providing valuable insights into the species' ecological requirements and habitat preferences [6].

Mount Penanjakan, located within Bromo Tengger Semeru National Park at elevations of 2,500-2,700 meters above sea level, presents a unique natural laboratory for studying volcanic ash weathering and pedogenesis processes. The site's volcanic ash and tuff parent materials, combined with high elevation climate conditions and the presence of Anaphalis javanica as a dominant pioneer species, create ideal conditions for investigating soil-vegetation interactions in volcanic systems. The elevation-dependent influence of climate on pedogenesis has been recognized as a critical factor in soil development, though specific relationships between weathering processes, soil evolution, and pioneer vegetation establishment remain inadequately understood in tropical high-mountain volcanic environments [1, 2, 7].

Despite the ecological and economic importance of volcanic soil systems, significant knowledge gaps exist regarding the specific pedological processes that support pioneer vegetation in high-elevation volcanic environments. Most previous research has focused on agricultural volcanic soils at lower elevations, leaving critical gaps in understanding soil development processes in conservation areas where natural succession occurs. The relationship between volcanic ash weathering rates, soil property development, and Anaphalis javanica establishment patterns has not been systematically investigated, limiting our ability to develop science-based conservation strategies for this endemic species [8].

This research is critically important for several interconnected reasons that span scientific understanding, conservation applications, and sustainable land management. From a scientific perspective, this study addresses fundamental knowledge gaps in volcanic soil science by investigating pedogenesis processes in high-elevation tropical environments where climate, parent material, and biological factors interact in unique ways. Understanding these processes contributes to global knowledge of volcanic soil development and provides insights applicable to similar volcanic systems worldwide [5, 8].

Finally, this research addresses the urgent need for science-based land management in protected volcanic ecosystems where traditional approaches may be insufficient for addressing contemporary conservation challenges. By providing detailed understanding of soil-vegetation relationships, the study supports evidence-based decision making for ecosystem management, species conservation, and sustainable land use planning in one of Indonesia's most iconic volcanic landscapes.

2 Methodology

2.1 Study Area

The research was conducted at Resort Gunung Penanjakan, TNBTS, Pasuruan Regency, East Java, Indonesia. The site is located near Wonokitri Village, which hosts the only ex-situ Edelweiss conservation site in East Java. The study period was from March to November 2025 (Fig. 1).

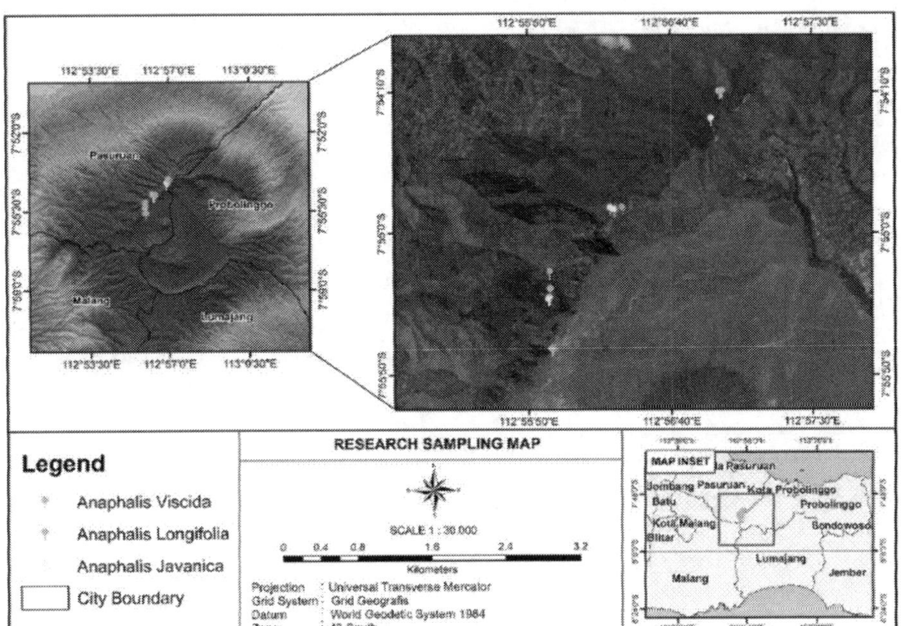

Fig. 1. Map of Site Location

2.2 Determination of Observation Points

Observation points were determined using the transect method along cliff edges where Edelweiss naturally grows. Four observation sites were selected. At each site, soil profiles were excavated to a depth of 2 meters, and morphological as well as physiographic observations were conducted [9].

2.3 Soil Sampling Method

Soil samples were collected from each horizon within the soil profile. Horizons were identified and measured using a soil profile tape and field meter. Samples were stored separately in plastic bags for laboratory analysis [3].

2.4 Laboratory Analysis

Laboratory analyses were carried out at the Soil Physics and Chemistry Laboratory to determine the physical and chemical properties of soil samples collected from each horizon. The parameters analyzed included organic carbon (C-organic), which was measured using the Walkley and Black method to estimate the organic matter content in the soil. Soil texture was determined by the pipette method, providing the proportion of sand, silt, and clay fractions. Soil reaction was assessed through pH measurement using a pH meter in H_2O solution, allowing evaluation of soil acidity. In addition, soil porosity, bulk density, and aggregate stability were analyzed to obtain information regarding soil structure, compaction, and aeration conditions [5].

2.5 Data Analysis

Soil characteristics were matched with the USDA Soil Taxonomy (2014). The classification process identified diagnostic horizons, epipedon, endopedon, and determined the soil order, suborder, great group, and subgroup [7].

3 Result and Discussion

The results of soil profile observation in three locations showed differences in morphology and characteristic horizons. It is listed in detail in Table 3.

Table 3. Soil Profile Identification Results

Soil Profile Photo and Physiographic Conditions	Morphological Characteristics of Each Horizon
Physiographs : **Survey and observation area in Lemah Pasar, Bukit Cinta** (southern part), Wonokitri Village, Tosari District, Pasuruan Regency; Observation date January 16, 2024; Coordinates 7.882821 LS and 11.91243 BT; Elevation 2,531 meters above sea level; Flat macro relief and termite hill micro-relief; slope slope 15%; The surface flow is moderate; Soil permeability is	**Layer 2.1:** A Depth 0-18/20cm; Humid soil color (7.5 Y 3/1); Texture of silty clay loam (Sicl, L); Granular structure (G); Moist consistency (G) and slightly adherent wet conditions (aL); Field pH 5.7; H2O soluble pH 6.72; C-Organic 2.12%; N-Total 0.103%; C/N ratio 20.58; P_2O_5 14.92 mg/100g; K-dd 0.38 cmol(+)/kg.

Soil Profile Photo and Physiographic Conditions	Morphological Characteristics of Each Horizon
relatively slow; The form of erosion is in the form of grooves with sufficient erosion hazard; Natural vegetation is dominated by Edelweiss specific *Anaphalis viscida*; Rainwater source. 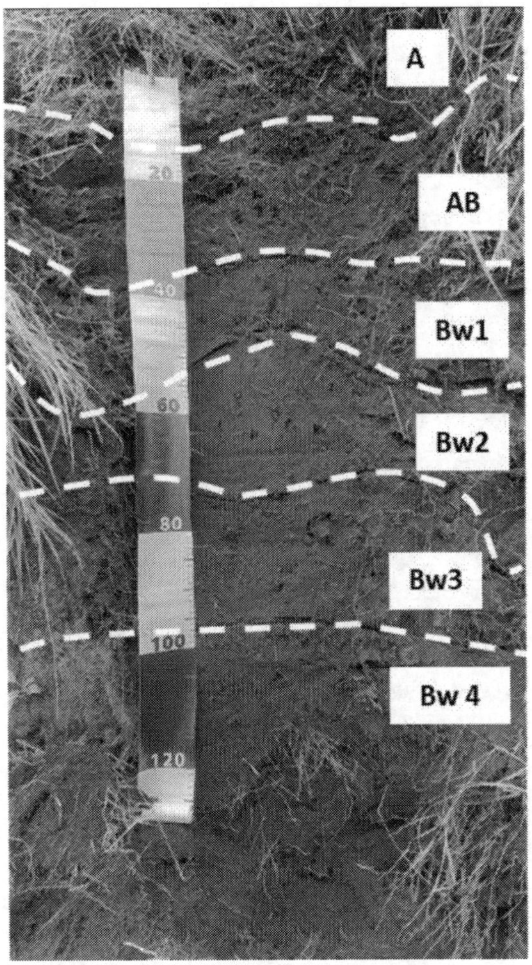 **Fig. 2.** Soil Horizon at Lemah Pasar	**Layer 2.2 :** AB Depth 18/20-38/40 cm; Moist soil color (10 YR 3/1); Texture of silty clay loam (Sicl, L); Granular structure (G); Humid consistency (G) and slightly adherent wet conditions (aL); Field pH 5.6; H2O soluble pH 6.48; C-Organic 1.41%; N-Total 0.102%; C/N ratio 13.82; P_2O_5 15.61 mg/100g; K-dd 0.36 cmol(+)/kg; Field records there are a lot of mosses. **Layer 2.3 :** Bw1 Depth 38/40-50/60 cm; Color of moist soil (2.5 Y 3/1); Texture of silty clay loam (Sicl, L); Granular structure (G); Moist consistency (G) and slightly adherent wet conditions (aL); No field pH test was performed; Soluble pH H2O 7.02; C-Organic 1.36%; N-Total 0.104%; C/N ratio 13.08; P_2O_5 14.34 mg/100g; K-dd 0.41 cmol(+)/kg. **Layer 2.4 :** Bw2 Depth 50/60-75/98 cm; Moist soil color (2.5 Y 2.5/1); Texture of silty clay loam (Sicl, L); Subangular blocky (SB) structure; Moist consistency (G) and slightly adherent wet conditions (aL); No field pH test was performed; H2O soluble pH 6.88; C-Organic 1.56%; N-Total 0.101%; C/N ratio 15.45; P_2O_5 15.63 mg/100g; K-dd 0.46 cmol(+)/kg. **Layer 2.5 :** Bw3 Depth 75/98-99/102 cm; Color of moist soil (5 Y 2.5/2); Texture of silty clay loam (Sicl, L); Subangular blocky (SB) structure; Gummy moist consistency (G) and slightly adherent wet conditions (aL); No field pH test was performed; H2O soluble pH 7.01; C-Organic 1.56%; N-Total 0.103%; C/N ratio 15.15; P_2O_5 15.77 mg/100g; K-dd 0.40 cmol(+)/kg. **Layer 2.6 :** Bw4 Depth 99/102-200 cm; Moist soil color (5 Y 2.5/1); Texture of silty clay loam (Sicl, L); Subangular blocky (SB) structure; Moist consistency (G) and slightly

Soil Profile Photo and Physiographic Conditions	Morphological Characteristics of Each Horizon
	adherent wet conditions (aL); No field pH test was performed; H2O soluble pH 5.97; C-Organic 1.38%; N-Total 0.201%; C/N ratio 15.15; P_2O_5 6.87 mg/100g; K-dd 0.40 cmol(+)/kg.
Physiographs : Survey and observation area near **Bukit Prau, Wonokitri Village**, Tosari District, Pasuruan Regency; Observation date January 16, 2024; Coordinates -7.882821 LS and 112.91245 E; Elevation 2,743 meters above sea level; Flat macro relief and termite hill micro-relief; slope slope 10%; The surface flow is moderate; Good drainage; Soil permeability is very slow; Surface erosion forms with sufficient erosion hazard; The natural vegetation is dominated by Edelweiss specific *Anaphalis longifolia*; Rainwater source. 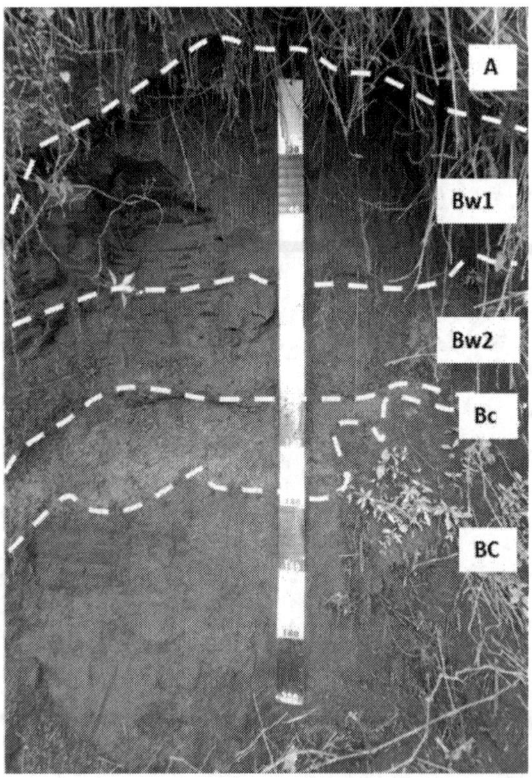 **Fig. 3.** Soil Horizon at Bukit Prau	**Layer 3.1 :** A Depth 0-15/40 cm; Humid soil color (2.5 Y) 2.5/1; Silty loam (SiL) texture; Granular structure (G); Very loose moist consistency (sG) and wet conditions are adherent (L) and slightly plastic (aP); Field pH test 5.5; H2O soluble pH 6.57; C-Organic 1.21%; N-Total 0.202%; C/N ratio 5.99; P_2O_5 17.81 mg/100g; K-dd 0.41 cmol(+)/kg. **Layer 3.2 :** Bw1 Depth 15/40-40/60 cm; Humid soil color (5 Y) 2.5/1; Texture silty clay loam (SiCL); Granular structure (G); Very loose moist consistency (sG) and wet conditions are adherent (L) and slightly plastic (aP); Field pH test 5.3; H2O soluble pH 5.91; C-Organic 1.34%; N-Total 0.105%; C/N ratio 12.76; P_2O_5 13.52 mg/100g; K-dd 0.37 cmol(+)/kg. **Layer 3.3 :** Bw2 Depth 40/60-80/120 cm; Humid soil color (10 Y) 2/1; Silty loam (SiL) texture; Angular blocky structure (AB); Very loose moist consistency (sG) and wet conditions are adherent (L) and slightly plastic (aP); No field pH test was performed; Soluble pH H2O 6.36; C-Organic 1.29%; N-Total 0.104%; C/N ratio 12.40; P_2O_5 14.18 mg/100g; K-dd 0.43 cmol(+)/kg. **Layer 3.4 :** Bc Depth 80/120-120/140 cm; Moist soil color (10 Y) 3/3; Texture of sandy loam (Sl); Angular blocky structure (AB); Very loose moist consistency (sG) and wet conditions are adherent (L) and slightly plastic (aP); No field pH test was performed; H2O soluble pH 6.39; C-

Soil Profile Photo and Physiographic Conditions	Morphological Characteristics of Each Horizon
	Organic 1.39%; N-Total 0.103%; C/N ratio 13.50; P_2O_5 15.01 mg/100g; K-dd 0.43 cmol(+)/kg. There is volcanic tuff concrete. **Layer 3.5 :** BC Depth 120/140-200 cm; Humid soil color (10 Y) 2/1; Texture silty clay loam (SiCL); Subangular blocky (SB) structure; Very loose moist consistency (sG) and wet conditions are adherent (L) and slightly plastic (aP); No field pH test was performed; H2O soluble pH 6.56; C-Organic 1.43%; N-Total 0.105%; C/N ratio 13.62; P_2O_5 14.72 mg/100g; K-dd 0.46 cmol(+)/kg.
Physiographs : The survey and observation area is at **the Tikungan S**, near the sea of sand, Bukit Cinta, Wonokitri Village, Tosari District, Pasuruan Regency; Coordinates -7.928255 LS and 112.932886 E; Elevation 2,173 meters above sea level; Observation date January 16, 2024; Macro physiography is in the form of choppy shape; Microphysiography in the form of Termite Hill; slope slope 35%; Fast surface flow; The form of erosion is in the form of surface; Natural vegetation is dominated by Edelweiss specific *Anaphalis javanica*; Land use in the form of shrubs; The water source comes from rain; Water management is in the form of rainfall	**Layer 4.1 :** A1 Depth 0-50/60 cm; Color of moist soil (10 YR) 4/4; Texture silty clay loam (SiCL); Subangular blocky (SB) structure; The moist consistency is very loose (sG) and wet conditions are slightly adherent (aL) and slightly plastic (aP); Field pH test 5.5; H2O soluble pH 5.72; C-Organic 1.37%; N-Total 0.106%; C/N ratio 12.92; P_2O_5 15.61 mg/100g; K-dd 0.31 cmol(+)/kg. **Layer 4.2 :** A2 Depth 50/60-80/100 cm; Color of moist soil (10 YR) 3/3; Texture silty clay loam (SiCL); Subangular blocky (SB) structure; The moist consistency is very loose (sG) and wet conditions are slightly adherent (aL) and slightly plastic (aP); Field pH test 5.6; H2O soluble pH 5.91; C-Organic 1.41%; N-Total 0.101%; C/N ratio 13.96; P_2O_5 18.82 mg/100g; K-dd 0.38 cmol(+)/kg. **Layer 4.3 :** Bw1 Depth 80/100-120/130 cm; Soil color moist (2.5 Y) 3/2; Texture silty clay loam (SiCL); Subangular blocky (SB) structure; The moist consistency is very loose (sG) and wet conditions are slightly adherent (aL) and slightly plastic (aP); No field pH test was performed; Soluble pH H2O 7.26;

https://doi.org/10.1051/e3sconf/202565209003

Soil Profile Photo and Physiographic Conditions	Morphological Characteristics of Each Horizon
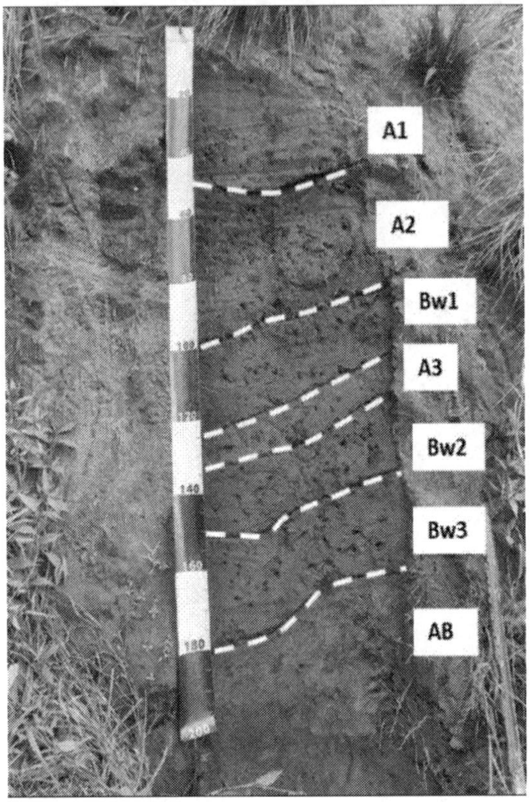 **Fig. 4.** Soil Horizon at Tikungan S	C-Organic 1.45%; N-Total 0.105%; C/N ratio 13.81; P_2O_5 13.27 mg/100g; K-dd 0.29 cmol(+)/kg. **Layer 4.4 :** A3 Depth 120/130-135/138 cm; Color of moist soil (10 YR) 3/4; Texture sandy clay loam (SCL); Angular blocky structure (AB); The moist consistency is very loose (sG) and wet conditions are slightly adherent (aL) and plastic (P); No field pH test was performed; H2O soluble pH 6.82; C-Organic 1.25%; N-Total 0.105%; C/N ratio 11.90; P_2O_5 13.01 mg/100g; K-dd 0.22 cmol(+)/kg. **Layer 4.5 :** Bw2 Depth 135/138-150/160 cm; Moist soil color (2.5Y) 3/3; Silty loam (SiL) texture; Subangular blocky (SB) structure; The moist consistency is very loose (sG) and wet conditions are slightly adherent (aL) and plastic (P); No field pH test was performed; H2O soluble pH 7.01; C-Organic 1.56%; N-Total 0.106%; C/N ratio 14.72; P_2O_5 12.31 mg/100g; K-dd 0.35 cmol(+)/kg. **Layer 4.6 :** Bw3 Depth 150/160-160/180 cm; Soil color moist (10 YR) 3/6; Texture silty clay loam (SiCL); Subangular blocky (SB) structure; Very loose (sG) moist consistency and adherent wet (L) and plastic (P) conditions; No field pH test was performed; H2O soluble pH 7.52; C-Organic 1.34%; N-Total 0.106%; C/N ratio 12.64; P_2O_5 11.92 mg/100g; K-dd 0.39 cmol(+)/kg. **Layer 4.7 :** AB Depth 160/180-200 cm; Moist soil color (2.5Y) 3/3; Texture sandy clay loam (SCL); Subangular blocky (SB) structure; The moist consistency is very loose (sG) and wet conditions are slightly adherent (aL) and plastic (P); No field pH test was

Soil Profile Photo and Physiographic Conditions	Morphological Characteristics of Each Horizon
	performed; H2O soluble pH 5.41; C-Organic 1.48%; N-Total 0.106%; C/N ratio 13.96; P_2O_5 15.71 mg/100g; K-dd 0.44 cmol(+)/kg

3.1 Soil Morphology and Horizon Characteristics

The results of soil profile observation in three locations showed differences in morphology and characteristic horizons. In Lemah Pasar Bukit Cinta, the soil horizon consists of A to Bw4 with a dominant texture of silty clay loam. The soil color ranges from 7.5Y 3/1 in the top layer to 5Y 2.5/1 in the bottom layer. The soil structure changes from granular on the surface horizon to subangular blocky on the lower horizon, with a depth of up to >200 cm. Natural vegetation is dominated by Anaphalis viscida, showing adaptation to moist volcanic soil conditions with moderate erosion hazards. The change in structure from granular (upper layer) to subangular blocky (lower layer) reflects the presence of a camic horizon, which is an early indication of the pedogenesis process. The depth of the profile of up to >200 cm shows the potential of the soil to store water and support long-term vegetation. The presence *of Anaphalis viscida* as the dominant vegetation shows adaptability to soils with high humidity but at moderate risk of erosion, in accordance with sloped physiographic conditions [10].

On Bukit Prau, the soil profile consists of an A–BC horizon with a silty loam to sandy loam texture. The soil color is lighter (2.5Y 2.5/1 to 10Y 3/3), with a very loose consistency. The very loose consistency of the soil allows for good aeration, but the slow permeability of the underlayers (Bc) is characterized by the presence of volcanic tuff concretions. The presence of this cremation is direct evidence of the influence of volcanic parent material that is still strong, so that the kambik horizon in this location has not developed optimally. The *dominant vegetation of Anaphalis longifolia* shows tolerance to soil conditions with good drainage but relatively low fertility. This is consistent with the character of pioneer vegetation that was able to adapt to nutrient-poor substrates [11].

At the Tikungan S, the soil profile consists of A1 to AB horizon, with a dominant texture of silty clay loam to sandy clay loam. The color of the soil varies from 10YR 4/4 in the upper horizon to 10YR 3/6 in the lower horizon. The thick surface horizon (0–60 cm) has a high volcanic ash content that shows the nature of the andik. The vegetation is dominated by Anaphalis javanica, which corresponds to the properties of high porosity soil and good humidity in the udic zone. The relationship between soil morphology and vegetation Edelweiss pioneers confirmed that the characteristics of young volcanic soils (andik, loose, high-porosity, medium organic) were key factors for the successful colonization of *Anaphalis spp.* in the Montana area TNBTS [12].

3.2 Soil Physicochemical Properties

In general, all three sites have a slightly acidic to neutral soil pH (5.5–7.5 H_2O). The C-organic content ranges from 1.2–2.1%, with the highest value on the Weak Market surface horizon (2.12%). This value decreases as depth increases, in line with the pattern of accumulation of organic matter in the upper layers. The C/N ratio was relatively variable, from low (5.99 in Bukit Prau) to high (20.58 in Lemah Pasar), indicating differences in the level of decomposition of organic matter between locations. These findings show that differences in C/N ratios between sites are closely related to the microclimate, drainage, and physical properties of the soil. A high C/N ratio value favors the accumulation of organic matter, while a low ratio indicates faster mineralization, which in the long run can affect nutrient availability [13].

3.3 Soil Taxonomic Classification

Based on taxonomic analysis using USDA 2014, soils in three locations are classified differently, detailed in Table 2.

Table 2. Soil Taxonomy

Soil Location and Taxonomy	Detailed description
Lemah Pasar Bukit Cinta: Mollic Hapludands	**Epipedon : Mollic** (Dark color, thick >25 cm (horizon 2.1–2.2); C-organic >1%; pH >6; granular structure.) **Endopedon : Cambic** (Smooth, subangular blocky texture from horizon 2.4 and down; however, there is no clear evidence of clay accumulation.) **Ordo : Andisol** (volcanic parent material (tephra/ash), high porosity & low BD on the surface, moderate C-organic) **Board : Udands** (Humid tropical Highland location (UDIC moisture regime is expected). **Great Group : Hapludands** (signifying no additional diagnostic horizon) **Subgroup : Mollic Hapludands** (because epipedons exhibit *a mollic* character, so subgroups can be designated *Mollic Hapludands)*
Bukit Prau: Typic Dystrudepts	**Epipedon : Ochric** (mineral A horizon, thin (0–15/40 cm) and low C-organic (~1.2%) **Endopedon : Cambic** (two cambic subhorizons (Bw1, Bw2) are visible; Bc shows concretions but there is no evidence of clay illusionation (argillic). **Ordo : Inceptisol** (The presence of a cambic horizon (Bw) indicates the beginning of the formation of a subsurface horizon but no diagnostic horizon has developed yet)

Soil Location and Taxonomy	**Detailed description**
	Subordo : Udepts (UDIC Land Regime) **Great Group : Dystrudept** (indicates that the soil is slightly acidic/low base saturation (pH H_2O 5.9–6.6, field pH 5.3–5.5) and has an endopedon cambic) **Subgroup : Typic Dystrudept**
Tikungan S: Typic Hapludands	**Epipedon : Andik** (Surface horizon (0–50/60 cm) is mineral soil with andic properties (dominant texture of volcanic dust/ash, C-organic ±1.3%) **Endopedon : Cambic** (The lower horizon (Bw1–Bw2) shows a change in structure/color, without obvious clay accumulation (not argillic), but the andic properties are still dominant) **Ordo : Andisol** (Dominated by volcanic ash/tuff parent material, andic properties >60% thickness of the first 60 cm) **Board : Udand** (Soil moisture regime *udic* (Bromo-Tengger location with high rainfall and good drainage) **Great Group : Hapludans** (There are no other specific traits (e.g. gelic, cryic, spodic) other than the basic traits of Hapludands) **Subgroup : Typic Hapludans** (The "typical" form of Hapludands)

At the location of "Lemah Pasar, Bukit Cinta" the soil taxonomy is Mollic Hapludands (order Andisol, suborder Udands, subgroup Mollic Hapludands). Epipedon mollic is identified from a thickness of >25 cm, dark color, and C-organic >1%. Cambic endopedon is characterized by a blocky subangular structure without obvious clay accumulation. While in Bukit Prau are Typic Dystrudepts (order of Inceptisol, suborder of Udepts). Epipedon ochric with low C-organic content (~1.2%) and cambric horizon at Bw1–Bw2 shows soil at an early stage of development. At the "Tikungan S" are found Typic Hapludands (order of Andisol, suborder of Udands). Thick surface horizons with pronounced andebris properties (high volcanic dust, moderate C-organic, high porosity), are classified as typical Hapludands.

The P_2O_5 content is relatively stable between 12–18 mg/100 g of soil, which is sufficient to support the growth of typical montana vegetation. K-dd values (0.29–0.46 cmol(+)/kg) indicate low to moderate cation exchange capacity, characteristic of young volcanic soils with limited horizon development [2, 3, 12]

3.4 Implications for Pioneer Vegetation (Anaphalis spp.) Habitat and Conservation

The results of the study showed that Edelweiss in TNBTS grows on young volcanic soils with prominent andean characteristics. The presence of Mollic Hapludands in Lemah Pasar provides an indication of fertile soil with a fairly high content of organic matter, supporting Anaphalis viscida. On the other hand, Typic Dystrudepts in Bukit Prau show relatively nutrient-poor soil conditions, but are still able to support the growth of Anaphalis longifolia which is more tolerant. Typic Hapludands in Tikungan S affirm the importance of andic properties (high porosity, good humidity) for Anaphalis javanica [14].

These findings confirm that Edelweiss ex-situ conservation strategy in Wonokitri must adapt the planting medium to the edapical conditions of the native habitat. The properties of sandy soils with a smooth texture, high porosity, and medium-high organic matter content need to be maintained to support the sustainability of the Edelweiss population [10, 15].

References

1. Fitriana KN, Ashari A, Olivia RA, Ardhana A-Z, Syarifudin B (2025) Progress research on pedological investigation of the volcanic landform in the western part of the Ring of Fire: A review
2. Beckstrom TB, Maaz TM, Deenik JL, Peter-Contesse H, Koch A, Tallamy Glazer C, Rivera-Zayas J, Crow SE (2025) From volcanic ash to abundant earth: understanding Andisol organic matter dynamics in relation to soil health on Hawaiʻi Island. Biogeochemistry 168:22. https://doi.org/10.1007/s10533-025-01216-9
3. Maghiar LM, Stoica IA, Tanentzap AJ (2021) Integrating demography and distribution modeling for the iconic *Leontopodium alpinum* Colm. in the Romanian Carpathians. Ecology and Evolution 11:12322–12334. https://doi.org/10.1002/ece3.7864
4. Staß S, Dorau K, Aguirre P, Schwiebert P, Mansfeldt T (2025) Land Use Effects on Organic Carbon in Andean Volcanic Ash Soils. J Plant Nutr Soil Sci 188:196–208. https://doi.org/10.1002/jpln.202400010
5. Yokobe T, Hyodo F, Tokuchi N (2020) Volcanic deposits affect soil nitrogen dynamics and fungal–bacterial dominance in temperate forests. Soil Biology and Biochemistry 150:108011. https://doi.org/10.1016/j.soilbio.2020.108011
6. Susanto D, Nugroho P, Numata S (2024) Traditional ecological knowledge for monitoring Anaphalis javanica (DC.) Sch.Bip. (Asteraceae) in Bromo Tengger Semeru National Park, Indonesia. Environ Monit Assess 196:717. https://doi.org/10.1007/s10661-024-12869-8
7. Clunes J, Hood-Nowotny R, Férnandez-Balado C, Valle S, Huber S, Hartmann F, Bruhn N, Rosinger C, Pinochet D, Lagos L, Keiblinger K (2025) Pedogenesis of volcanic ash soils determines the soil microbial pool and its contribution to nutrient availability. European Journal of Soil Biology 126:103756. https://doi.org/10.1016/j.ejsobi.2025.103756
8. Rolfes S, Longman J, Pahnke K, Engelen B (2025) Unique microbial communities in ancient volcanic ash layers within deep marine sediments are structured by the composition of iron phases. Front Microbiol 16:1526969. https://doi.org/10.3389/fmicb.2025.1526969
9. Gutiérrez-Patricio S, Gómez-Árias A, Nolasco-Jiménez P, Matáix-Solera J, Martínez-Martínez J, Martínez-Haya B, Vegas J, Jiménez-Morillo NT, Miller AZ (2025) Fresh tephra deposits from the Tajogaite Volcano boost thermophile proliferation and soil organic matter recovery

10. Pozumentshtikov A, Boycheva I, Bonchev G, Kitanova M, Zhiponova M (2024) The secrets of the edelweiss (Leontopodium R. Br. ex Cass.): a review on this symbol, scientific object, and natural treasure. nnual of Sofia University "St Kliment Ohridski", Faculty of Biology, Book 2 - Botanics 108:133–158. https://doi.org/10.60066/GSU.BIOFAC.Bot.108.133-158
11. Yang HI, Baek N, Kwak J-H, Lim S-S, Lee Y-H, Lee S-M, Choi W-J (2023) Microbial contribution to organic carbon accumulation in volcanic ash soils. J Soils Sediments 23:866–879. https://doi.org/10.1007/s11368-022-03349-y
12. Gerzabek MH, Poch RM, Álvarez D, Ploszczanski L, Zehetner F (2025) Strong Differences in Pedogenesis on Lava Vs. Scoria Along a Hydroclimatic Gradient on Santa Cruz Island/Galápagos Archipelago (Ecuador) – Insights From Micromorphology. Span J Soil Sci 15:14219. https://doi.org/10.3389/sjss.2025.14219
13. Huang Q, Yang F, Cao H, Cheng J, Jiang M, Li M, Ni H, Xie L (2024) Comparison of Microbial Diversity of Two Typical Volcanic Soils in Wudalianchi, China. Microorganisms 12:656. https://doi.org/10.3390/microorganisms12040656
14. Chen J, Cui Y, Xiao Q, Lin K, Wang B, Zhou J, Li X (2024) Difference in microbial community structure along a gradient of crater altitude: insights from the Nushan volcano. Appl Environ Microbiol 90:e00753-24. https://doi.org/10.1128/aem.00753-24
15. Rahma MJ, Soemarno S, Batoro J (2023) "Hulun Hyang" Farmer Group Asset Mapping Using Asset-Based Community Development Methode as an Ex-situ Edelweiss Flower Conservation in the Bromo Tengger Semeru National Park Area (Case Study: Edelweiss Park, Wonokitri Village). IOP Conf Ser: Earth Environ Sci 1131:012013. https://doi.org/10.1088/1755-1315/1131/1/012013

Analysis of The Natural Physical Environment And Artificial Physical Environment To Determine The Effective Defense Space In The Area Around Yogyakarta International Airport

Agoeng Koesoemo Arjanto Agustinus[1]*, *Sigit* Heru Murti.[2], and *Suratman*[3]

[1] Doctoral Program in Geography, Universitas Gadjah Mada, Yogyakarta, Indonesia
[2] Department of Geographic Information Science, Universitas Gadjah Mada, Yogyakarta, Indonesia
[3] Department of Environmental Geography, Universitas Gadjah Mada, Yogyakarta, Indonesia

Abstract. From a national development perspective for the advancement of the Indonesian nation, the construction of Yogyakarta International Airport on the southern coast aims to create a regional development center, transforming it into an airport city. From a national security and protection perspective, the airport's vital location in an open area facing other countries represents a center of gravity, containing both strengths and vulnerabilities. Therefore, managing and organizing space toward a safe and secure environment in the long term requires an effective defense zone to maintain the stability of defense and security protection in the nation's front yard. Based on spatial and environmental awareness, this study aims to determine the extent and distribution of effective protection zones in the area surrounding Yogyakarta International Airport. This study utilizes terrain analysis using the GIS-based Analytical Hierarchy Process (AHP) method, guided by the assessment of operational agility, namely the ability to exploit opportunities for movement and maneuver in military operations. Based on the combined physical aspects of the land, consisting of the natural physical environment of the earth's surface and the artificial physical environment of the earth's surface, using overlay union analysis, the area is 3,873.43 hectares, or 61.45% of the total area.

1 Introduction

National defense is inseparable from efforts to maintain the nation's integrity, especially in Indonesia [1]. Therefore, the control and utilization of territorial space in Indonesia aims to realize the integrity and sovereignty of the state against various potential threats. Indonesia is currently on the path to becoming a developed nation; thus, multiple uses of territorial space have been filled with strategic development projects in recent years. One such development project is the Yogyakarta International Airport. The presence of the Yogyakarta International Airport aims to accelerate the development of the surrounding area, leading to an airport city in Kulon Progo Regency.

Since the airport's operation in 2019, the surrounding environment has undergone a change in appearance, from rice fields and dry fields to residential areas and business and trade centers. Uncontrolled changes and developments in the environmental space that are not synergistic with the interests of the national defense sector will undoubtedly impact the preparation of the regional defense system to face the threat of enemy attacks/terrorism that

*Corresponding author : agustinus.sge@mail.ugm.ac.id

can occur at any time. Therefore, the state must be prepared to face threats that are more strategic, large-scale, and have long-term but invisible impacts [2].

Based on the objectives of national spatial planning as stated in the Republic of Indonesia Law on Spatial Planning, namely to create a safe, comfortable, productive, and sustainable space based on the Archipelago Insight, the Environment, and National Resilience [11]. Every regional spatial plan must not only be followed up with development directed at changing land use, but must also consider the appearance of the earth's surface as a space that can support the implementation of a regional defense system [5]. The area around Yogyakarta International Airport is a vital national strategic area, and its security and defense must be guaranteed. Therefore, environmental changes around the airport will change the potential of the space, one of which is as a defense area. Furthermore, changes in land use in an area are sometimes not in line with the established spatial plan [3]

Over time, the existence of a vital international airport has triggered changes in the use of space for various development activities in the surrounding area. Based on awareness of changing environmental conditions, perceptions need to be addressed to maximize defense preparedness. Therefore, to meet the need for defense space, a container (place) with broad dimensions and a form of defense space, the need for defense space is crucial. Therefore, a study is needed to determine effective defense spatial zones as operational defense fields. This is in line with the Decree of the Minister of Defense concerning Defense Sector Guidelines for the Preparation of Spatial Planning Regulations, which states that international airports are vital national strategic objects [12]. Just as the capital city acts as a CoG [6], so too the existence of airports in the southern coastal areas of Java Island functions as a center of balance that contains both strengths and vulnerabilities.

The selection of effective defense areas is analyzed based on the level of operational agility, which relates to the ability to seek opportunities, ease, and smoothness in movement and maneuvering in the face of obstacles posed by physical terrain, both natural and artificial. In other words, in military operations, the possibility of conducting certain military maneuvers is often limited by terrain limitations [10].

In line with the development and utilization of geospatial technology, the determination of effective defense areas uses the Analytical Hierarchy Process (AHP) method, which is based on Geographic Information Systems. AHP is a decision support model that breaks down complex multi-factor or multi-criteria problems into a hierarchy [4]. Therefore, the results of the analysis, in the form of information regarding the location, area, and distribution of effective defense zones/areas, can serve as a reference for policymakers in spatial management at Yogyakarta International Airport.

2 Materials and Methods

2.1 Research Location

This research was conducted in the Yogyakarta International Airport area, which has been designated by the Regent of Kulon Progo as the YIA International Airport Aerotropolis Strategic Area according to the Kulon Progo Regent's Decree in 2020 and the Kulon Progo Regent's Regulation in 2023[13,14]. Astronomically, the research area is located at coordinates 07° 51' 43.30"– 07° 55' 49.76" South Latitude and 110° 02' 28.60"-110° 06' 32.97" with an area of 5,548.62 ha, an average elevation of 12.2 m above sea level, and a very gentle average land slope of 3-7%.

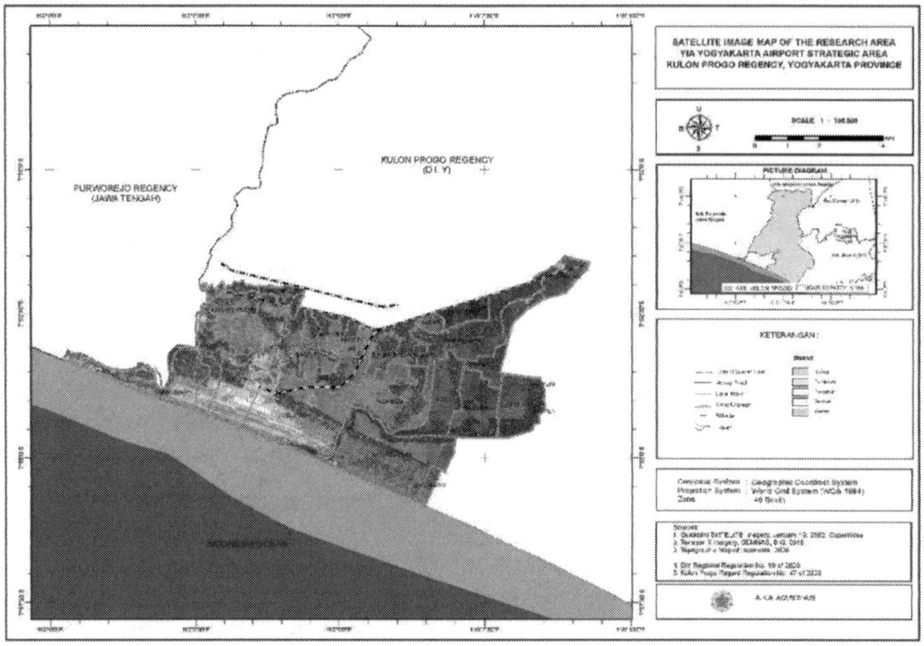

Fig. 1 Satellite Image Map of Yogyakarta International Airport Strategic Area

2.2 Research Variables and Materials

There are two variables in physical terrain analysis used in determining defense space: the natural physical environment appearance variable, namely the state of the natural profile with all natural objects, and the man-made physical environment appearance variable, namely the state of land cover/land use with all human cultural objects on the earth's surface related to the ability and opportunity to carry out movement and maneuver activities in military operations, as follows:

2.2.1 Physical Appearance of the Earth's Surface (Natural Physical Environment)

Table 1 Parameters and Research on Natural Physical Environment

Parameters	Source	Data Format
Land Slope	SRTM	Raster /Vektor
Soil Surface	Soil Type Map	Vektor
Earth Surface Shape	DEM Map	Raster/Vektor
Surface Water	Groundwater Basin Map	Vektor
Weather Conditions (Rainfall Intensity)	Rainfall Map from Rainfall Station	Raster/Vektor

2.2.2 Land Cover Condition (Artificial Physical Environmental)

Table 2.2 Parameters and Research on the Environmental Conditions of Medan as a Result of Human Culture

Parameters	Source	Data Format
Built-up Areas	Google Earth	Raster/Vektor
Disaster Areas	Google Earth	Raster/Vektor
Vegetation Areas	Google Earth	Raster/Vektor
River Networks/Irrigation and Drainage Channels	Google Earth	Raster/Vektor
Road-Railway Networks	Google Earth	Raster/Vektor
High-Voltage Power Networks	Google Earth	Raster/Vektor
Telekomunication Networks	Google Earth	Raster/Vektor

2.3 Methods

2.3.1 Land Defense Operations Field Assessment Techniques

For land defense operations, terrain information is one of the main focuses and is important to know [7]. So, terrain assessment will provide maneuver control advantages for the troops occupying it. Maneuvering activities must be accompanied by an accurate interpretation of the Earth's surface configuration and any threats [8]. The process of identifying effective defense space is one of the challenges in land defense planning in this Strategic Area. The following is an operational assessment technique based on the level of maneuverability of personnel, equipment, and material, namely the creation of criteria based on the results of satellite image interpretation and detailed and accurate field identification of the physical terrain conditions as a classification level for the form of obstacles.

Table 2 Criteria for assessing the military operation terrain based on maneuverability

Terrain Parameters	Assessment
Slope Gradient	The flatter the area, the easier it is to maneuver, and vice versa
Land	The looser the soil, the easier it is to maneuver, and vice versa
Earth Surface Shape	The more terrain profiles, the more difficult it is to maneuver, and vice versa.
Clean Water	The easier it is to get clean water, the easier it is to maneuver, and vice versa
Rainfall	The lower the rainfall, the easier it is to maneuver, and vice versa
Buildings	The closer the distance to the building, the easier it is to maneuver, and vice versa
Vegetation	The closer the distance to the vegetation, the more difficult it is to maneuver, and vice versa
Disaster Areas	The closer the distance to the disaster area, the more difficult it is to maneuver, and vice versa

Terrain Parameters	Assessment
Water Bodies	The closer the distance to the water body, the easier it is to maneuver, and vice versa
Road/Railway Networks	The closer the distance to the road/railway network, the easier it is to maneuver, and vice versa
High-Voltage Electricity Networks	The closer the distance to the high-voltage electricity network, the more difficult it is to maneuver, and vice versa
Telecommunication Networks	The closer the distance to the telecommunications network, the easier it is to maneuver, and vice versa

Source: Department of the Army Headquarters [9] (analyzed), 2024

2.3.2 Multicriteria Analysis with Analytical Hierarchy Process (AHP)

This study uses a quantitative approach. The method used in this study is spatial analysis using Geographic Information Systems with the application of the Analytical Hierarchy Process (AHP) to evaluate and make multi-criteria decisions from various configurations of the Earth's surface

Complex unstructured problems are broken down into their parts and then arranged in a hierarchical order [15]. Then, each terrain criterion will be given a relative score, which will then be added up to obtain a hierarchical level of interrelationship. The final result of the scoring system is to classify the level of spatial effectiveness needed in military operations as an output parameter. The classification is based on the total score value of each input parameter. The output parameter classification range is determined based on the range of the lowest (X_{min}) to the highest (X_{max}) values divided by the desired number of classes.

$$Score\ Range = \frac{(X_{max} - X_{min})}{n}$$

2.3.3 Spatial Analysis Using Geographic Information Systems (GIS)

The spatial model analysis uses the AHP technique through overlay-union analysis, namely the combination of scores from each field criterion according to the criteria in the physical field unit and the criteria in the land cover unit, which is continued by creating a score range that shows the scale of the effectiveness zone of the land defense operation space. The hierarchy of the highest scores will occupy the top position, and the lower or smaller scores will occupy the next position.

2.4 Determination of Effective Defense Space

The configuration of the Earth's surface influences military operations, so the natural physical environment and man-made land cover are determined by the parameters that influence them. The scoring process for each physical terrain and land cover parameter indicates the classification of the level of obstacle.

https://doi.org/10.1051/e3sconf/202565210001

2.4.1 Classification Of Maneuverability Of Obstacles Based On Natural Physical Environment

Table 2.4 Classification of Maneuver of Obstacle Based on Natural Physical Environment

Parameter	Criteria	Classification of Maneuver of Obstacle	Score	Weight
Slope (%)	0 -2	not an obstacle	3	30
	3 – 7	not an obstacle	3	
	8 - 13	not an obstacle	3	
	14 - 20	obstacle	2	
	21 - 55	very obstacle	1	
	>56	very obstacle	1	
Earth Surface Shape (Morphology)	Alluvial Plain	not an obstacle	3	20
	Natural Embankment	somewhat an obstacle	2	
	Foothill Slope	somewhat an obstacle	2	
	Coastal Shoal	obstacle	1	
	Sand Dune	very obstacle	1	
Soil consistency and plasticity level	Low	not an obstacle	3	30
	Medium	obstacle	2	
	High	very obstacle	1	
Buffer distance (m) of well/fresh water source to the soldier's position (Ground Water Basin)	1 - 50	not an obstacle	3	10
	51 - 100	obstacle	2	
	>100	very obstacle	1	
Rainfall intensity (mm/rainy day)	Up to 13.6	not an obstacle	3	10
	13.7 – 20.7	not an obstacle	3	
	20.8 – 27.7	obstacle	2	
	27.8 – 34.7	obstacle	2	
	>34.8	very obstacle	1	

Source: Analysis Data 2024

https://doi.org/10.1051/e3sconf/202565210001

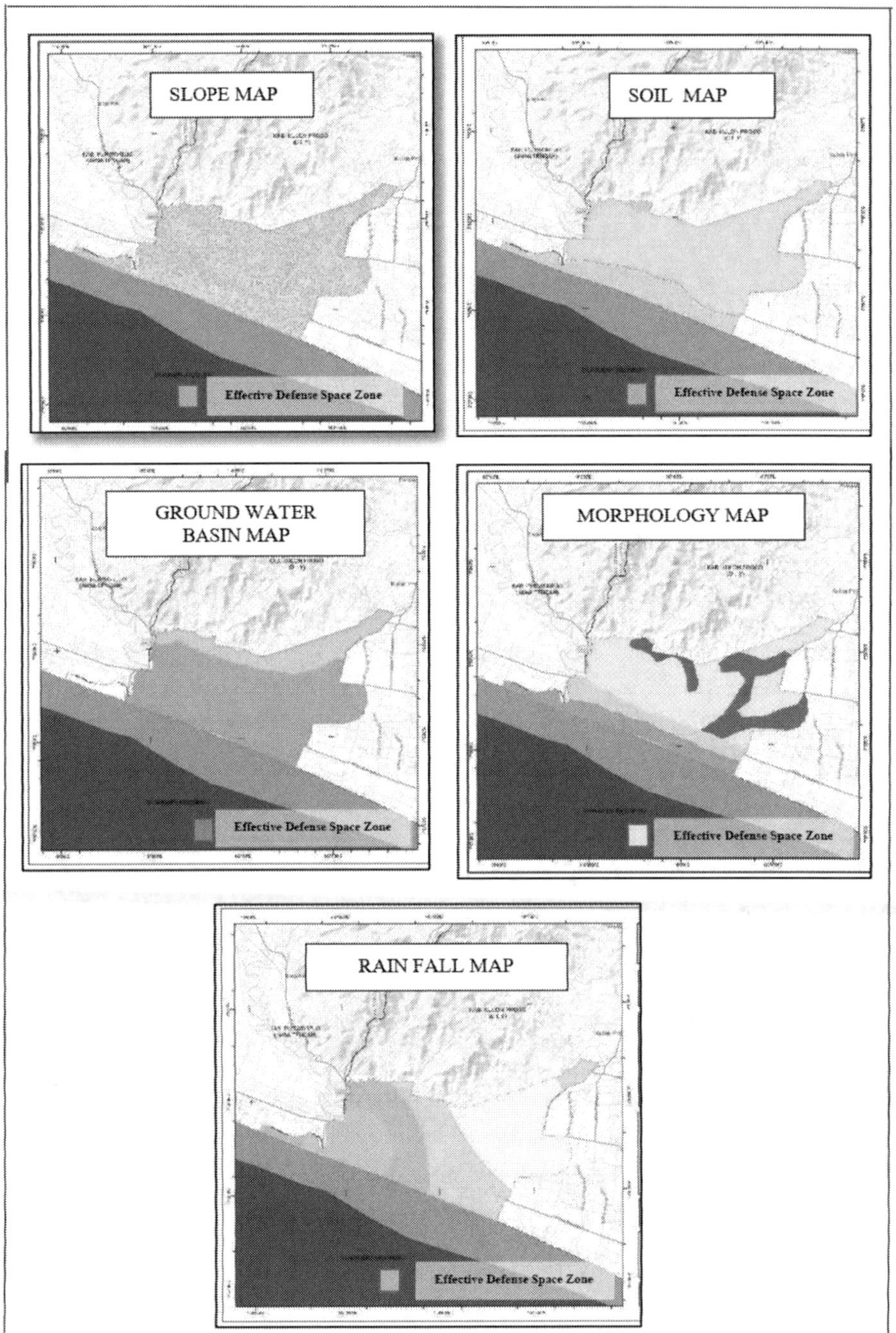

Fig. 2 Map of Effective Defense Zones according to Built Physical Environment Appearance (Land Cover Condition)

2.4.2 Classification Of Maneuverability Of Obstacles Based On Artificial Terrain Environment

Table 2.5 Classification of Maneuver of Obstacle Based on Artificial Physical Environment

Parameter	Criteria	Classification of Maneuver	Score	Weight
Effective distance of the settlement area to the soldier's position (m)	1 -50	not an obstacle	3	30
	51 - 100	obstacle	2	
	>101	very obstacle	1	
Effective distance of vegetation area to soldier's position (m)	1 - 100	not an obstacle	1	10
	101 - 250	obstacle	2	
	>251	very obstacle	3	
Effective distance from Disaster Prone Areas to soldiers' positions (m)	>251	not an obstacle	3	10
	101 - 250	obstacle	2	
	1 - 100	very obstacle	1	
Distance of river/drainage network to soldier's position (m)	1 - 25	not an obstacle	3	20
	26 -50	obstacle	2	
	>51	very obstacle	1	
Distance of road/railway network to soldier's position (m)	1 - 25	not an obstacle	3	10
	26 -50	obstacle	2	
	>51	very obstacle	1	
Distance of high voltage electric network to soldier's position (m)	1 - 20	not an obstacle	1	10
	20 - 40	obstacle	2	
	>41	very obstacle	3	
Distance of telecommunication network buffer to soldier's position (m)	1 - 20	not an obstacle	3	10
	20 - 40	obstacle	2	
	>41	very obstacle	1	

Source: Analysis Data 2024

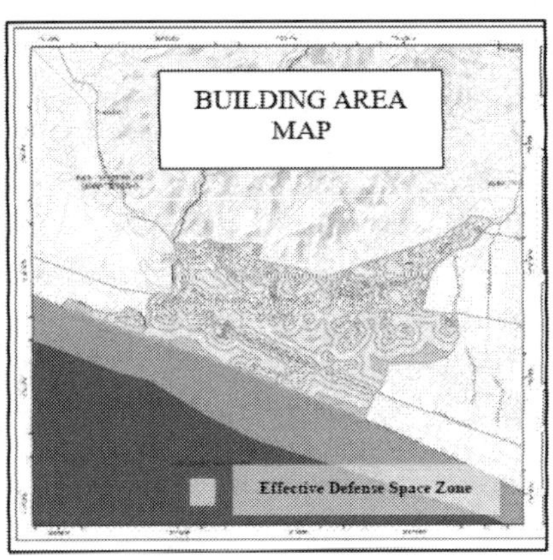

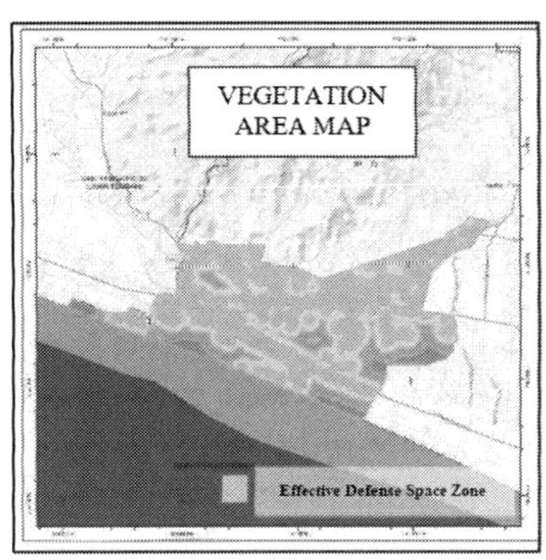

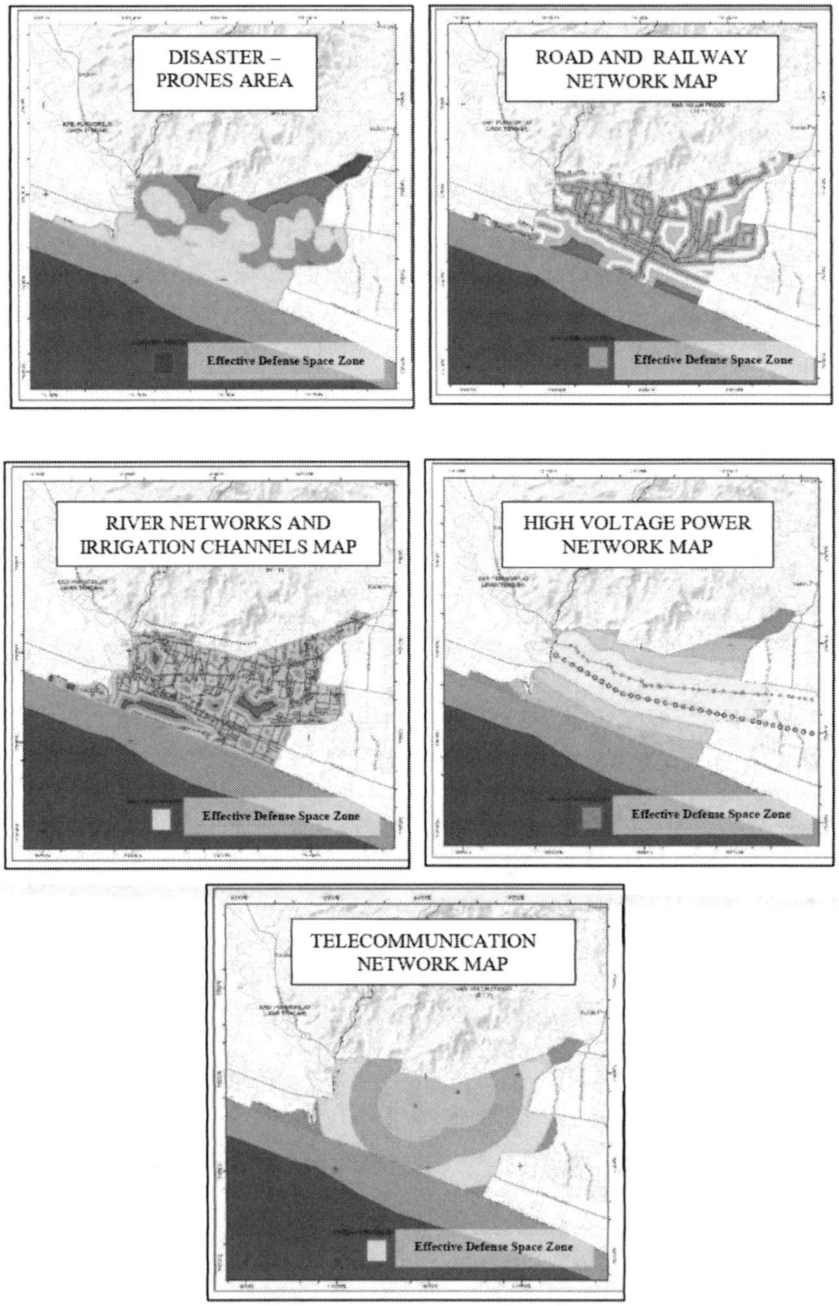

Fig. 3 Map of Effective Defense Zones according to Artificial Physical Environment Appearance (Land Cover Condition)

2.5 Determining the Effectiveness of the Land Defense Operations Space

The weighting step to determine the score range is to add up the total score from the results of the overlay union analysis process. The difference between the lowest and highest total scores will determine the score range of the land defense operations space effectiveness zone.

The score range will be classified into five scales (groups) according to the level of the most effective land defense operations space zone to the least effective land defense operations space zone, both in the physical field unit and the land cover unit. The scale of the effectiveness of the defense operations space in the physical field unit and the land cover unit can be seen in the following table:

Table 2.6 Scale of effectiveness of the land defense

Range Score	Land Defense Operations Space Effectiveness Zone
68 - 100	Effective
35 - 67	Quite Effective
1 - 34	Not Effective

3 Results and Discussion

3.1 Effective Defense Area Based on Natural Physical Environment

The map of the effective defense area based on physical conditions/earth surface profile is compiled based on the results of a spatial analysis using an overlay with the sum of land slope, surface soil type, landform, surface water conditions, and rainfall using the Analytical Hierarchy Process (AHP). The total score for each natural physical appearance/earth surface profile is Total Score = (land slope x 30) + (surface soil type x 20) + (landscape form x 30) + (surface water conditions x 10) + (rainfall intensity x 10).

The sum of each calculation will yield the highest and lowest scores, allowing for the range. The following table shows the implementation of the overlay and map of the effective defense area based on physical conditions/earth surface profile, as presented in Fig. 4

 https://doi.org/10.1051/e3sconf/202565210001

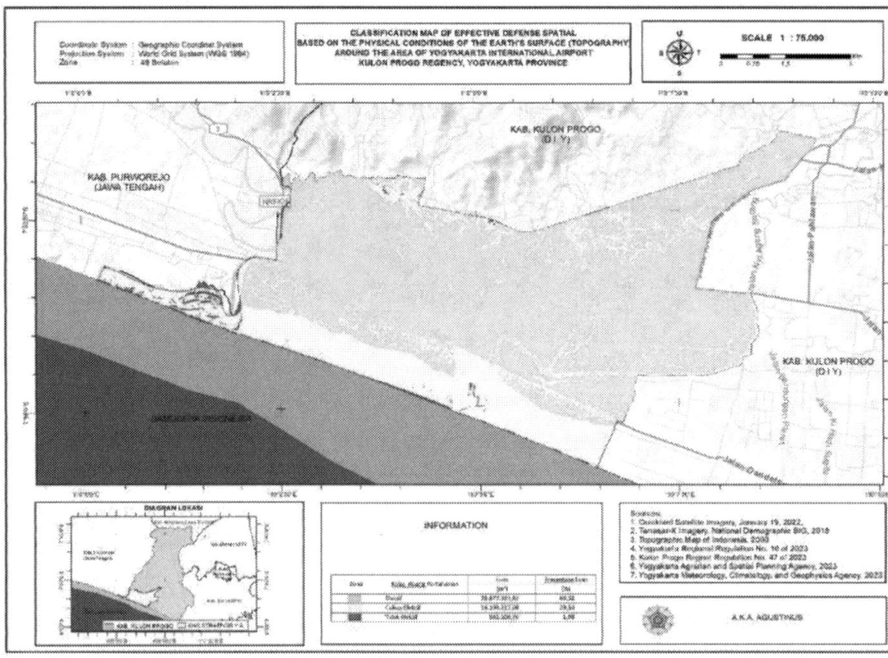

Fig. 4 Effective Defense Space Map by Natural Physical Environment Unit

The analysis of the physical conditions/surface profiles of the natural conditions of the earth shows that the majority of the strategic area of YIA Airport constitutes an effective defense area, at 69.31%. Therefore, the majority of this area is highly supportive for operational, tactical, and strategic purposes, particularly concerning the implementation of capabilities, maneuvering opportunities, and troop mobilization.

3.2 Effective Defense Area Based on Artificial Physical Environment

The map of effective defense areas according to land cover conditions is compiled based on the results of a spatial analysis using an overlay with a sum. Total Score = (built-up area x 30) + (vegetation area x 10) + (disaster area x 10) + (river network (drainage/irrigation) x 20) + (road and railway network x 10) + (electricity network (SUTET) x 10) + (telecommunication network x 10).

From each calculation, the highest and lowest scores are obtained, allowing for the range of differences. The sum of each calculation yields the highest and lowest scores, allowing for the range of differences. The following table shows the implementation of the overlay and the map of effective defense areas according to land cover conditions, as presented in Fig. 5

https://doi.org/10.1051/e3sconf/202565210001

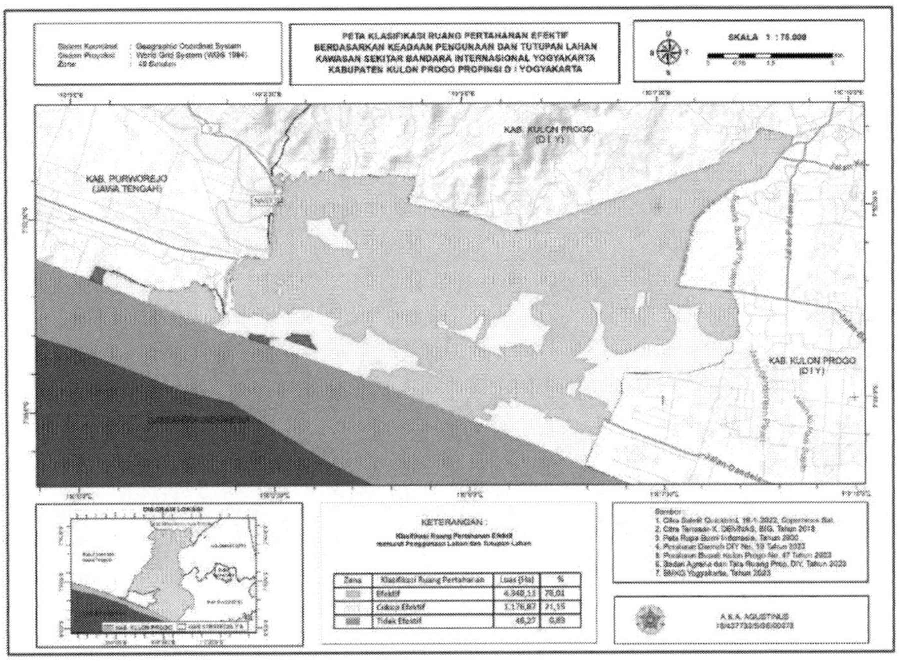

Fig. 5 Effective Defense Space Map by Artificial Physical Environment Unit

The analysis of the physical condition/profile of the Earth's surface, which includes land cover, shows that 78.01% of the area is effectively defensive. Therefore, the majority of this area is highly supportive of operational, tactical, and strategic operations, particularly in relation to the implementation of capabilities, maneuvering opportunities, and troop mobilization.

3.3 Effective Defense Area Based on Artificial Physical Environment

The determination of effective defense areas in the strategic area of Yogyakarta International Airport was carried out by overlaying the weighted values of the effective defense areas based on physical features/surface profiles with the effective defense areas based on land cover conditions. This resulted in a total score for each zone to determine the effective defense area category.

The overlay analysis yielded 12,019 alternative zones, of which 3,006 zones were categorized as having the maximum total score (100), and 9,013 zones with a total score less than 100. This means that 3,006 selected zones are considered effective defense areas, and the remaining 9,013 zones are regarded as ineffective defense areas based on the physical terrain. The total area of these 3,006 effective defense areas is 3,873.47 ha, or 69.45% of the total strategic area of YIA Airport. The description of the physical aspects of the terrain has a very gentle slope (1,167 zones), alluvial soil type (2,709 zones), has a landform in the form of an alluvial plain (1,899 zones), easily accessible (obtained) water sources because it is located in a discharge area (1,869 zones), rainfall < 150 (mm/month) (534 zones), is 0 - 165 m from the built-up area (1,901 zones), is 0 - 45 m from the vegetation area (1,901 zones), is 0 - 330 m from disaster-prone areas (1,220 zones), is 0 - 170 m from the river network (1,553

https://doi.org/10.1051/e3sconf/202565210001

zones), is 0 - 70 m from the road and railway network (1,468 zones), is 1 - 778 m from the high-voltage electricity network SUTT and SUTET (1,243 zones) and is 0 - 1850 m from disaster-prone areas (1,468 zones). m of telecommunication network (towers) BTS) (840 zones). Thus, of the 3,006 effective defense areas, the majority strongly support the implementation of defense operations.

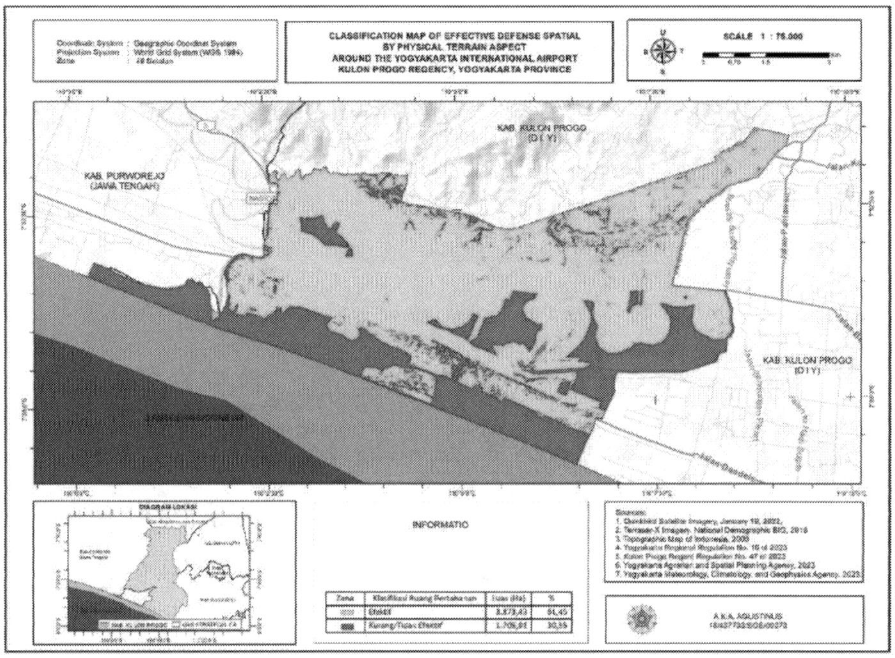

Fig. 6 Effective Defense Space Map by Physical Aspect

5 Conclusion

Based on the combination of physical aspects of the terrain, consisting of the natural physical environment of the earth's surface and the artificial physical environment of the land on the earth's surface, using overlay union analysis, the effective defense space in the strategic area of Yogyakarta International Airport is 3,873.43 ha or 61.45% of the total area. This means that based on the physical aspects of the terrain, most of this area is very supportive for planning and implementing capabilities, opportunities for maneuver and troop mobilization, especially in areas near the arterial road network and local roads and river and irrigation canal networks, especially in urban areas. Meanwhile, areas near the airport building and coastal areas are not included in the effective defense space.

References

1. Y. Sutrasna, Indonesia's defense strategy in facing military and non-military threats through an economic perspective. Indonesian Scientific Journal. **8**, 7 (2023)
2. Pratiwi, Agrenisa, Putri, A. Lutfiiah, Bahirah, H. Indah, Syahnanda, Restu, The Role of civil society in overcoming COVID-19 as a non-military threat in Indonesia. Publiciana. **12**, 2 (2022)
3. Alwan, Land use changes and spatial planning alignment in Kendari City. Urban and Regional Studies Journal. **3**, 1 (2021)
4. A. Afandi, D. Soedianto, E.I. Bangun, Utilization of the analytical hierarchy process to determine combat aircraft characteristics in support of the national defense industry. Rekayasa Journal of Science and Technology. **15**, 2 (2022)
5. A. Fahmi, A study of defense area planning in a legal anthropology review (The Readiness of Indonesian Army Legal Officers to Carry Out Legal Support Duties), (CV. Penerbit Qiara Media, JawaTimur, 2021)
6. Maharani C, Gandarsah I, Santiko U, and Lebang G, Defense of the National Capital Strategy and Military Implementation, (LAB 45 Monograf Jakarta: Laboratorium Indonesia, 2022)
7. Bateman M and Riley R., The Geography of Defence, (Routledge, London, 1987)
8. Clausewitz, C., edited and translated by Howard M and Paret, P.. On War, (New Jersey: Princeton University Press, 1989)
9. Department of the Army Headquarters, FM 5-33. Terrain Analysis (Washington, DC, 1990)
10. Erdeniz R., Military Operations Planning and Methodology: Thoughts on military problem-solving, (Swedish Defence University (SEDU), Stockholm, 2017)
11. State Secretary of the Republic of Indonesia, Law of the Republic of Indonesia Number 26 of 2007 Concerning National Spatial Planning, (Ministry of Law of the Republic of Indonesia, Jakarta, 2007)
12. Ministry of Defense of the Republic of Indonesia, Defense Sector Content Guidelines For The Formulation Of Spatial Planning Regulations, (Directorate of Defense Areas, Ministry of Defense of the Republic of Indonesia, Jakarta, 2023)
13. Regional Secretary of Kulon Progo Regency, Kulon Progo Regent Regulation Number 15 of 2013 concerning Detailed Spatial Planning for Urban Areas Around the Airport, (Legal Bureau of the Kulon Progo Regency Regional Government, 2013)
14. Regional Development Planning Agency of the Special Region of Yogyakarta Province 2023 Regional Spatial Planning Plan Special Region of Yogyakarta Province 2023-2043, (Legal Bureau of the Kulon Progo Regency Regional Government, 2023)
15. Marx J J.T.,. Military Integrated Environmental Management at The Donkergat Military Training Area, Ph.D. thesis, Stellenbosch University, South Africa (2014)

The Land Cover and Ecosystem Services Identification in the Special Purpose Forest Area (KHDTK) of Gurah, Blitar-Indonesia

Alfi Sahrina[1*], *Heni Masruroh*[1], *Fatchur Rohman*[2], *Choirul Anam*[3], *Mohammad Ainul Labib*[4], *Muhammad Rafif Fadihilah*[1], *Muhammad Rafli Tegar Hati*[1], *Muhammad Yusril Abidin*[1],

[1]Department of Geography, Faculty of Social Sciences, Universitas Negeri Malang, Malang, Indonesia
[2]Departement of Biology, Faculty of Mathemathics and Natural Sciences, Universitas Negeri Malang, Malang, Indonesia
[3]Departement of Management, Faculty of Economics and Business, Universitas Negeri Malang, Malang, Indonesia
[4]Social Sciences Education, Faculty of Social and Political Sciences, Universitas Negeri Surabaya, Surabaya, Indonesia

Abstract. The Special Purpose Forest Area (KHDTK) of Gurah is located in coastal and hilly area of Southern part of Blitar Regency. KHDTK of Gurah is located in protected and productive forests, which function as forestry education and training areas . The existence of these functional areas have a characteristic land cover that can be utilized, so zoning the area that had been managed, should be done. The aim of this search is to identify the land cover and ecosystem services in the Special Purpose Forest Area of Gurah. The method used in this research is by mapping the KHDTK of Gurah using drone. Mapping was conducted to aim the result in the form of aerial photograph that used in identifying the land cover in the area of Gurah KHDTK. Therefore, a ground check was conducted to see the suitability of land cover in Gurah KHDTK. Ground check location were determined randomly based on the aerial photograph obtained. The ecosystem services were analyzed descriptively based on the developing of land cover in Gurah KHDTK. The results of the ,research spatially show that the protected forest area in Gurah KHDTK has changed into dry fields or farmland. This has also been observed in several productive forest locations, where there have been changes in the area's designation. The land cover that has developed in Gurah KHDTK includes forest, fields, open land, rice fields and shrubland.

1 Introduction

The KHDTK of Gurah is located in Tugurejo Village, Wates District, Blitar Regency. This area will be used as training, research, and scientific development area [1]. Results of visual field identification, currently several areas of the land cover and land use in KHDTK Gurah

* Corresponding author: alfi.sahrina.fis@um.ac.id

have been utilized by he local community.This understandable given that the area is still relatively new and is in the process of being used according to its designation. A study on the forms of land use that have taken place is important. he results of this study are expected to provide an overview of the area's current condition, as well as become the base of formulating a sustainable utilization strategy.

To get an accurate description of the area's condition, it is necessary to identify the land cover exist in KHDTK Gurah. This can be done by applying geo-spatial technology, by using GIS (Geographic Information System) in particular. Studies on the use of this technology have been widely conducted, both with remote sensing data [2], [3], [4], the use of drones [5], [6], [7], and through other media that can suport the identification process. The application of these various technologies aims to increase the accuracy, efficiency, and ease the land cover identification.

This identification is getting more relevant because the KHDTK of Gurah functions as both a protected and a permanent production forests [1]. These two functions make the area vulnerable to changes in land use that are not in line with its designated purpose. Land use by the community is essentially a form of direct interaction between human and their environment. The community carries out this interaction by cultivating seasonal agricultural land such as corn, cassava, and banana plantations. However, if this use is not controlled, it has the potential to cause environmental damage. Therefore, further study is needed on land use patterns and their reciprocal relationship with environmental conditions, in order to obtain strong base for the sustainable management of the area.

In addition to the land use, ecosystem services also need to be considered in the management of KHDTK Gurah. KHDTK Gurah previously had a function as a protected forest and production forest, of course the ecosystem services that support this area can be in the form of habitat protection services, food services, production services. Ecosystem services play vital role in understanding the reciprocal relationship between humans and the environment. As a form of production, provision, and service, ecosystem services function to control and balance every change caused by human activities. Various studies show that changes in land cover can affect the value of ecosystem services studies on land changes that have resulted a change of ecosystem services value that had happened in several areas, such as in Sleman Regency, where the ecosystem services decrease [8]. Based on this, this research focuses on the identification of land cover and the analysis of ecosystem services found in the KHDTK of Gurah area.

2 Method

The method used was mapping the KHDTK of Gurah area by using a drone, similar to the previous studies on using GIS for land identification [5], [6], [7]. The result of aerial photographs were then processed and used as the base of land cover identification process. To improve the accuracy of the mapping results, a ground check was conducted in the field. The ground heck location were determined randomly based on the initial interpretation of the aerial photographs and the field conditions encountered, areas that have easy access can be ground checked, but areas that are difficult to reach can use aerial photography. The ground check data were then used to validate the interpretation results and served as reference for the digitization process of land se areas in KHDTK Gurah. Additionally, descriptive analysis related to ecosystem services in conducted by reviewing the existing forms of area management. This analysis was done descriptively to provide an overview of the land use forms by the community and the implications to the ecosystem services in KHDTK Gurah.

3 Results and Discussion

3.1 Condition and Aerial Photographs

The KHDTK of Gurah is located in the Southern part of Blitar Regency, precisely in the coastal area with hilly topography. Commonly, the land slope in this area varies, ranging from flat to gentle slopes around the coast, and moderately steep to extreme steep slopes in the hills and valleys. Figure 1 shows the slope conditions with a percentage of flat 6%, gentle 13%, slightly steep 20%, steep 26%, and very steep 35%. With in the valley area is the Gurah River, which flowstone Gurah Beach. This river is part of the Gurah Sub-watershed, which is seasonal and supported by several surrounding springs.

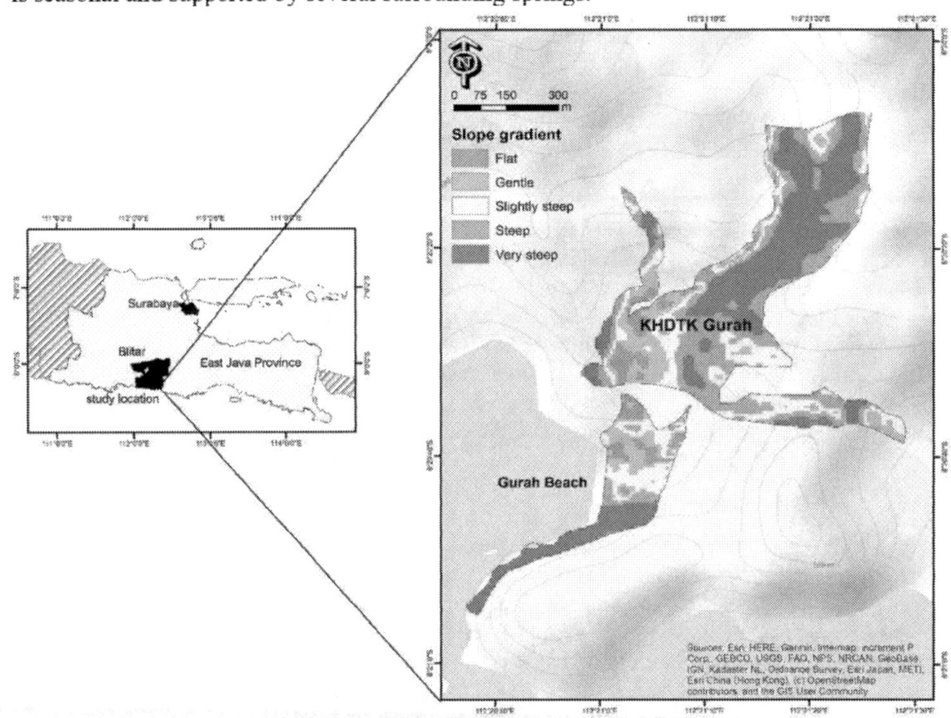

Fig. 1. Slope Gradient at KHDTK Gurah

Administratively the Gurah KHDTK is located within Tugurejo Village, Wates District, which is in the border of Malang Regency. Based on the visual observations in Figure 2a, most of the landscape is covered by vegetation. This aligns with its primary function as protected and productive forests. Figure 2b shows pristine protected forest area, while Figure 2c depicts a production forest area that has become land conversion. Furthermore, Figure 2d shows the land cover transition from protected forest to fields utilized by the community. This phenomenon demonstrates the dynamics and land use that have the potential to impact the ecological function of the area.

https://doi.org/10.1051/e3sconf/202565210002

Fig. 2. The Condition of Gurah KHDTK

To support further research, aerial photographs were taken in August 2024. These photographs were then processed and combined to produce a comprehensive representation of Gurah KHDTK area (Figure 3). The combined aerial photographs show that the area is located on the south part of Blitar Regency *JLS (Jalur Lintas Selatan)* Southern Cross Road. The aerial photographs were then overlaid with spatial data on Gurah KHDTK area, revealing two main polygons. The first polygon represents the protected forest area located in the lower part near the sea shore, while the second polygon represents the production forest area.

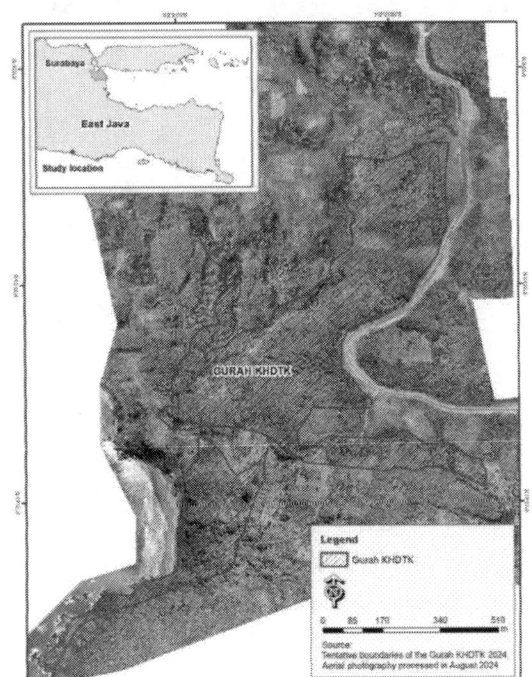

Fig. 3. Aerial Photograph of the Indicative Map of Gurah KHDTK in Blitar Regency.

3.2 Groundcheck tutupan lahan di KHDTK Gurah

The next step is conducting ground check of varied characteristics of land cover in the areas that have been mapped. Ground checks are conducted through direct field surveys and by sing UAVs to obtain aerial photography data for additional identification. The ground check locations are shown in Figure 4. For areas that are difficult to reach directly, verification is conducted using drone photography to facilitate the survey process and expand the identification scope. Based on the results of the ground check and aerial photo analysis, land cover in Gurah KHDTK area consists of several main classes: open land, forest, shrubland, and fields. The distribution of each land cover class can be visually observed in Figure 4, which illustrates the actual condition of land use in the area.

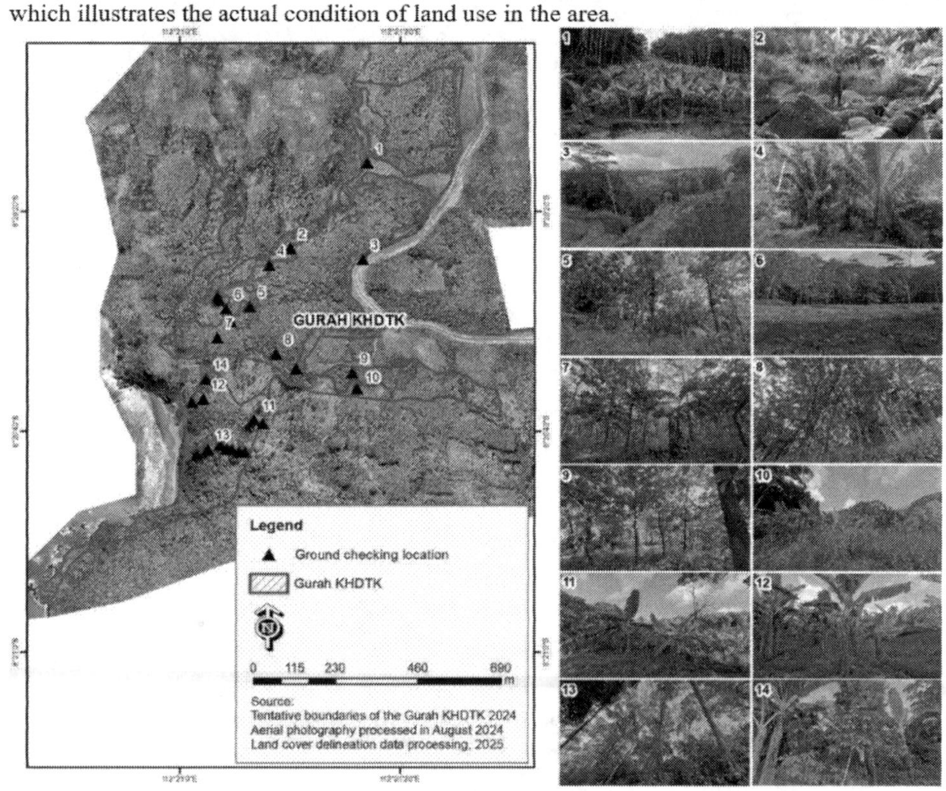

Fig. 4. Ground checking Spots and Land Cover Condition Display in Gurah KHDTK.

3.3 Land cover area delineation

The result of land cover delineation were obtained from a combination of several data sources. Those are aerial photographs, ground check results, and supporting data in the form of administrative data, KHDTK area boundaries, and DEMNAS data. Figure 5 presents the results of land cover delineation. Based on the analysis, several land cover classes are identified in Gurah KHDTK area, including primary forest, secondary forest, open land, shrubland, rice fields, and dry fields. Human activity shows the presence of open land, dry fields, and rice fields. The presence of community activity can be identified from the emergence of land cover classes such as open land, dry fields, and rice fields. Furthermore, indications of community use of protected forest areas by changing land functions were also found, resulting in some areas that should have functioned as protected areas change in designation. For example, the protected forest area around Gurah Beach is used as a banana and cassava plantation, and in the production forest area that has experienced landslides, it is planted with banana plants. These changes are not only related to land cover but also impact the provision of ecosystem services, such as water availability, soil fertility, and the protective function of forest areas. This situation indicates pressure on the ecological functions of the area, which requires further attention in sustainable management efforts.

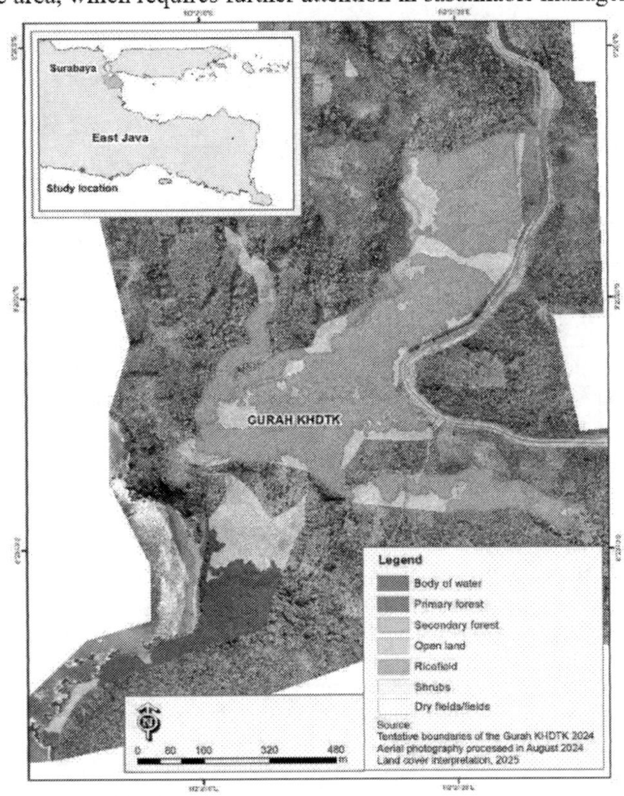

Fig. 5. Land Cover in KHDTK Gurah.

The Gurah KHDTK area is functionally designated as a protected forest and a permanent production forest. However, field identification results indicate different conditions. Significant land cover changes have occurred in both areas. Table 1 presents the results of the changes in Gurah KHDTK. The greatest change occurred in the protected forest area, which was converted to dry fields, up to 38.77%, while the production forest area also experienced changes to dry fields in 19.44%. This calculation is obtained from the land cover area (table 1) by digitizing the results of aerial photographs.

Table 1. Regional Functions and Regional Area Percentage

Function of Gurah KHDTK Area	Area (Ha)	Area Function Changing	Area (Ha)	Percentage (%)
Protected Forest Area	8.15	Water Bodies	0.05	0.61
		Primary Forest	4.34	53.25
		Secondary Forest	0.04	0.49
		Open Land	0.02	0.25
		Shrubland	0.54	6.63

Function of Gurah KHDTK Area	**Area (Ha)**	**Area Function Changing**	**Area (Ha)**	**Percentage (%)**
		Fields	3.16	38.77
		Total	**8.15**	**100.00**
Permanent Production Forest	33.45	Secondary Forest	25.5	76.23
		Open Land	0.69	2.06
		Rice Fields	0.01	0.03
		Shrubland	0.75	2.24
		Fields	6.5	19.44
		Total	33.45	**100.00**
Total	**41,60**		**41,60**	**100.00**

Source: Data processing results, 2025

3.4 Gurah Ecosystem Services in the Gurah KHDTK Area

Ecosystem services can be divided into four groups, namely provisioning services, regulating services, cultural services, and supporting services [9], [10]. Gurah KHDTK is an area that provides various important ecosystem services for the environment and surrounding communities. The protected forest serves as a habitat for diverse flora and fauna and has high scientific value, making it a suitable resource for education and research. Furthermore, the existing land cover contributes to the provision of food, both from agricultural land and from the utilization of non-timber forest products. The dominant forest cover in this area also plays a significant role in maintaining the sustainability of the water resources ecosystem, as indicated by the presence of springs that serve as a source of water for the surrounding community. In the coastal area of Gurah KHDTK, available ecosystem services including the provision of seafood sources, protection from natural disasters such as high waves, potential for recreation and tourism, and cultural services that thrive in coastal areas. Visually, Figure 6a shows ecosystem services in the form of a headland that protects the land from high waves, while Figure 6b shows a gently sloping plain that has been utilized by the community for banana and cassava farming. This demonstrates that Gurah KHDTK not only functioned as a conservation area but also provides tangible benefits in supporting food security, environmental protection, and improving the welfare of the surrounding community.

Fig. 6. Gurah KHDTK Ecosystem Services

3.5 Discussion

KHDTK is a forest area specifically designated for forestry research and development, forestry education and training, as well as religion and culture [11]. Gurah KHDTK serves important functions as a forestry education and training area [1]. Most of the forest area in

Gurah KHDTK area remains intact, although some locations have changed. This area has diverse landscape characteristics, including coastal areas, hilly morphology, and the Gurah sub-watershed. This diverse landscape has the potential to become environmental education media that can be utilized in various learning activities. For example, coastal areas can be used as educational platforms, as been applied in South Malang, where they have been utilized as outdoor learning media [12], [13], [14]. Similarly, the KDHTK of Gurah area has the potential to be designed as a landscape-based educational area with various educational objectives, such as at Modangan Beach, Jolosutro Beach, and the surrounding cultural areas. The design of this landscape-based environmental education model is expected to strengthen the function of Gurah KDHTK area as a forestry education center.

However, the main problem faced by Gurah KHDTK is changes in land cover that are not in accordance with the area's function. For example, protected forests that should be maintained are converted into fields or dry fields. This change not only reduces the area's ecological function but also increases the potential for natural disasters, such as landslides. This condition has been identified in several locations in Gurah KHDTK, where changes in land cover contribute to landslides. This is in line with a study in the Blukar Watershed, Kendal Regency, which showed that upstream areas with minimal human activity have high value in providing ecosystem services for disaster prevention and mitigation [15]. In other areas, changes in land cover have even led to the conversion of natural vegetation into built-up areas, which has an impact on the decline in ecosystem quality and the loss of a large portion of the value of ecosystem services [8].

In the context of Gurah KHDTK, the available ecosystem services can be seen from the current land cover conditions. The resulting ecosystem services include the food provision, recreation/ecotourism, education, disaster protection, cultural services, and supporting services for flora and fauna habitats. Based on previous studies [9], [10], ecosystem services in the Gurah KHDTK can be classified into four categories provisioning services, regulating services, cultural services, and supporting services. In terms of food provision, the presence of fields or dry fields, shrubs, rice fields, and water bodies contribute to the availability of community food sources. This condition is in line with findings in the Danau Sentarum National Park, which has high potential in providing food ecosystem services [6]. In addition, the coastal area of Gurah KHDTK also plays a role in disaster protection ecosystem services, particularly related to mitigating the risks of abrasion and tidal flooding. Meanwhile, the upstream area covered with dense vegetation functions to reduce the risk of landslides and flooding. This condition is in line with studies in other areas, where maintained land cover is able to provide ecosystem services in the form of flood prevention [15].

Besides the protection and food supply aspects, the potential for recreational and ecotourism ecosystem services in Gurah KHDTK is also enormous. Coastal areas can be developed into nature-based tourism destinations while still adhering to sustainability principles, while degraded lands can be rehabilitated through the implementation of agroforestry or reforestation systems. These efforts not only improving environmental conditions but also opening new opportunities for utilizing the area for recreational, educational, and ecotourism activities. Similar practices have been successfully implemented in various locations through a combination of land rehabilitation and agroforestry development, which has been proven to improve the quality of ecosystem services while supporting the economic sustainability of local communities [15].

Conclusion

Based on the research results, the slope conditions at KHDTK Gurah are dominated by slopes with very steep and steep conditions. Gurah KHDTK's main function are primarily as a forestry education and training area with diverse ecosystem service potential, including food provision, disaster protection, recreation/ecotourism, education, and culture. However, conditions on the ground indicate changes in land cover that are inconsistent with the area's function. The greatest changes occur in protected forest areas and permanent production forests, which are being converted to fields or dry fields. The greatest change occurred in the protected forest area, which was converted to dry fields, up to 38.77%, while the production forest area also experienced changes to dry fields in 19.44%. This condition has implications for the decline in the area's ecological function and an increased risk of disasters, such as landslides. Nevertheless, Gurah KHDTK still has significant potential for development as an environmental education and research area, as well as an ecotourism destination based on ecosystem services. Efforts to optimize the area's function can be carried out through designing a landscape-based education model, utilizing ecosystem services for educational activities, and strengthening the community's role in protecting and managing the area.

Acknowledgement

Gratitude to the Institute for Research and Community Service, State University of Malang, which has fully funded this research and to the parties who have assisted in this research.

References

1. M. LHK, "Keputusan Menteri Lingkungan Hidup Dan Kehutanan Republik Indonesia Nomor 885 Tahun 2024 Tentang Kawasan Hutan Dengan Tujuan Khusus Pendidikan dan Pelatihan Kehutanan Universitas Negeri Malang di Kabupaten Blitar Provinsi Jawa Timur Seluas 41.6 Ha," 2024, *Jakarta*.
2. C. M. Ariesta, D. E. Herwindiati, and M. D. Lauro, "Sistem Informasi Geografis Untuk Tutupan Lahan Hijau Pada Wilayah Bogor Berdasarkan Data Remote Sensing," *Comput. J. Comput. Sci. Inf. Syst.*, vol. 7, no. 1, pp. 1–9, 2023, doi: 10.24912/computatio.v7i1.16170.
3. R. M. Rumbiak and T. Gunawan, "Pemanfaatan Citra Penginderaan Jauh Untuk Identifikasi Perubahan Penutup Lahan Guna Mengetahui Arah Perkembangan Kota Manokwari Papua Barat 2010-2016," *J. Bumi Indones.*, vol. 7, no. 2, 2018.
4. M. M. Purnama, F. Pramatana, Y. Aini, and M. Soimin, "Analisis Tutupan Lahan Menggunakan Penginderaan Jauh di Kecamatan Kupang Tengah, Kabupaten Kupang, Provinsi Nusa Tenggara Timur," *J. Kehutan. Papuasia*, vol. 10, no. 1, pp. 96–106, 2024.
5. M. Rauzan and F. Yulianti, "Pemanfaatan Drone Untuk Identifikasi Penggunaan Lahan Di Dayah Raudhatul Quran Tungkop Kecamatan Darussalam Kabupaten Aceh Besar," *J. Pendidik. Geos.*, vol. VII, no. 1, pp. 105–113, 2022, doi: DOI:10.24815/jpg.v%vi%i.24400.
6. M. A. Hakim *et al.*, "Pemanfaatan Pesawat Tanpa Awak Untuk Pemetaan Dan Identifikasi Penutupan Lahan Pada Kawasan Hutan Pendidikan UNMUL," *J. AGRIFOR*, vol. XX, no. 1, pp. 1–14, 2021.
7. R. Septima S, H. Syahputra, I. Zulfa, I. Putra, and I. Irhamna, "Pemanfaatan Drone untuk Identifikasi Penggunaan Lahan Perkantoran di Kampung Kung Kecamatan Pegasing," *J. Nas. Komputasi dan Teknol. Inf.*, vol. 7, no. 5, pp. 1277–1285, 2024, doi: 10.32672/jnkti.v7i5.8079.
8. P. Widayani, H. N. Salsabila, and A. Andriantari, "Dampak perubahan penutup lahan terhadap nilai jasa ekosistem di Kabupaten Sleman Daerah Istimewa Yogyakarta," *Maj. Geogr. Indones.*, vol. 37, no. 2, pp. 104–113, 2023, doi: 10.22146/mgi.70636.
9. R. Costanza *et al.*, "The value of the world's ecosystem services and natural capital," *Nature*, vol. 387, no. 6630, pp. 253–260, 1997, doi: 10.1038/387253a0.

10. Department of the Environment Water Heritage and the Arts, *Ecosystem services : key concepts and applications*, vol. 1, no. 1. Department of the Environment, Water, Heritage and the Arts, Canberra, 2009.
11. KLHK, "Peraturan Menteri Lingkungan Hidup dan Kehutanan Nomor P.15/MENLHK/SETJEN/KUM.1/5/2018 tentang Kawasan Hutan dengan Tujuan Khusus," 2018. [Online]. Available: https://jdih.menlhk.go.id/kiosk/files/P.15-2018 KHDTK.pdf
12. A. Sahrina, F. Rosyida, and H. Masruroh, *Bentangalam Malang Selatan Dalam Pembelajaran Geografi*. Malang: Media Nusa Creative, 2024.
13. A. Sahrina *et al.*, "A Study of Tourism Objects in Supporting Fieldwork-Based Geography Learning in Sumbermanjing Wetan, Malang Regency, Indonesia," *IOP Conf. Ser. Earth Environ. Sci.*, vol. 1066, no. 1, 2022, doi: 10.1088/1755-1315/1066/1/012003.
14. A. Sahrina, F. Rosyida, Y. A. Wirahayu, P. Mahanani, and H. N. Ngoc, "Designing Fieldwork Material in Geography Learning through Potential Tourism and Landscape Feature (Case Study in Srigonco Village)," *Geosfera Indones.*, vol. 10, no. 1, pp. 98–112, 2025, [Online]. Available: https://doi.org/10.19184/geosi.v10i1.47186 Research
15. I. K. Fatmala *et al.*, "Analisis Jasa Ekosistem Pencegahan dan Penanggulangan Bencana Banjir di DAS Blukar Kabupaten Kendal," *J. Ilmu Lingkung.*, vol. 21, no. 4, pp. 965–973, 2023, doi: 10.14710/jil.21.4.965-973.

Feature Optimization for Rice Field Discrimination Using Sentinel-2A Imagery: A Case Study in the Adi Soemarmo Airport Area and its Surroundings

Indra Adi Saputra[1], *Hamim Zaky* Hadibasyir[2*], *Agus Anggoro* Sigit[1], and *Aditya* Saputra[3]

[1,2,3]Faculty of Geography, Muhammadiyah University of Surakarta, Surakarta 57162, Central Java, Indonesia

Abstract. Indonesia is known as an agrarian country with a very large area of rice fields. However, the high dependence on rice imports indicates an imbalance between land potential and domestic production. One of the main causes is the conversion of rice fields into built-up areas, especially in urban fringe areas such as Colomadu (Karanganyar) and Ngemplak (Boyolali), which are clear examples of urban expansion. This study aims to map rice fields in the area around Adi Soemarmo Airport, which has experienced significant land conversion, using Sentinel-2A imagery and the Random Forest algorithm. This algorithm was chosen due to its ability to handle complex data, automatically evaluate feature importance, and resist overfitting. A total of 16 features were used, consisting of 4 spectral features, 4 vegetation indices, and 8 texture features (variance, entropy, homogeneity, etc.). Feature contribution evaluation was conducted to form four feature configurations tested using the Out-of-Bag (OOB) method, with accuracy evaluated through Overall Accuracy (OA) and Kappa statistics. The results showed a gradual increase in accuracy, with the fourth configuration (features with the highest contribution) achieving the highest OA of 71.69% and a Kappa of 0.6562. These findings confirm that optimal feature selection greatly influences classification accuracy and supports more effective remote sensing-based agricultural mapping.

1 Introduction

Rice fields in Indonesia are considered very extensive, with a total reaching 7,463,948 hectares in 2019 based on data from the Ministry of Agrarian Affairs and Spatial Planning/National Land Agency (ATR/BPN) and the Ministry of Agriculture (Kementan). However, in reality, Indonesia still imports rice in large quantities. Data from the Food and Agriculture Organization (FAO) shows that rice is the main food commodity most imported by Indonesia. The Central Bureau of Statistics (BPS) recorded that the volume of rice imports continued to increase during the period from 2020 to 2023, reaching 3,062,857.6 tons in 2023. This high import figure reflects that domestic rice productivity has not been able to

* Corresponding author: hamim.zaky.h@ums.ac.id

meet national consumption needs. This indicates an imbalance between the potential of available land resources and actual production results. One of the factors that worsens this condition is the conversion of rice fields into built-up areas, which can threaten rice productivity and national food security [1]. This phenomenon mainly occurs in urban fringe areas that are experiencing massive urban expansion. For example, rice fields in Ngasem and Klodran Villages, Colomadu District, Karanganyar Regency, have been converted for toll road development [2]. Meanwhile, in Ngemplak District, Boyolali Regency, rice field conversion is estimated to reach 199.2 hectares, triggered by population growth and the strategic geographic location of the area for development [3].

This condition emphasizes the importance of systematic supervision and monitoring of land use changes. Information regarding land use plays a crucial role in supporting regional spatial planning and formulating food security strategies, including the estimation of production and distribution of rice commodities. Therefore, a method is needed that can consistently, broadly, and sustainably provide spatiotemporal data. Remote sensing is one of the most effective and efficient approaches in this context. This technology enables periodic observation of land changes without having to conduct time-consuming and costly field surveys.

In various studies, Sentinel-2A/B satellite imagery from the European Space Agency (ESA) is widely used for mapping crop types in small-scale agricultural areas in Asia [4]. This is because Sentinel-2A/B provides free sun reflectance data with a wide spectral range, high spatial resolution (10–20 meters), and a revisit time of every 5 days. The combination of visible, near-infrared (NIR), and short-wave infrared (SWIR) bands makes Sentinel-2A/B data highly potential for land cover analysis, especially in identifying vegetation and rice fields.

The use of this imagery still faces several challenges, particularly in distinguishing land objects with similar spectral characteristics, such as newly planted rice fields, wetlands, and shallow water bodies [5]. This issue can impact classification accuracy and spatial interpretation if not addressed with appropriate analytical methods.

Thus, the development of accurate and verified remote sensing-based land use mapping methods becomes an urgent need in order to provide high-quality geospatial information. This becomes a crucial foundation for sustainable regional planning, controlling rice field conversion, and strengthening national food security amidst the dynamics of development and population growth.

To overcome these challenges, the use of machine learning-based classification algorithms such as Random Forest (RF) is becoming increasingly relevant. Random Forest is known to be effective in building classification models and evaluating the importance of each feature used in the analysis by constructing several decision trees using various randomly selected data subsets and features [6]. This algorithm has the ability to handle high-dimensional data, address multicollinearity issues between variables, and is relatively resistant to overfitting risks [7]. In addition, this method is also equipped with an internal validation technique through the Out-of-Bag (OOB) approach to estimate the prediction error of the Random Forest (RF), which allows performance evaluation of the model without explicitly separating test data.

Most previous studies focusing on rice field mapping using remote sensing tend to use limited features without systematic evaluation of the contribution of each feature to the classification results. In fact, selecting relevant and informative features greatly affects overall model performance, which is a fundamental part of satellite imagery categorization. Therefore, the development of a classification approach that not only relies on the advantages of remote sensing data but also optimizes the potential of algorithms like Random Forest for analysis and feature selection is important in producing accurate and reliable spatial information to support effective decision-making and natural resource and spatial planning management.

A study in Kulon Progo successfully utilized a combination of Sentinel-2 MSI, Sentinel-1 SAR, and SRTM (elevation and slope data) to monitor the conversion of rice fields. Vegetation index transformations such as NDVI, NDBI, and NDWI from annual composite images of Sentinel-2 MSI were used to identify land types. Meanwhile, monthly composite images from Sentinel-1 were used to detect rice planting patterns during the flooded phase. Random Forest classification with 1,000 trees and a 70:30 training-testing data ratio showed high accuracy in rice field conversion mapping, reaching an accuracy of 96.16% in 2021 and 95.95% in 2017 [8].

This study aims to map rice fields using Sentinel-2A imagery with the Random Forest algorithm in the northwestern fringe area of Surakarta City, particularly in the Adi Soemarmo Airport area and its surroundings. This area has complex land cover and a relatively high rate of land conversion into built-up areas.

2 Methods

2.1 Study area

The research location is situated around Adi Soemarmo Airport and its surroundings, with a total area of approximately 115 km². Astronomically, this area is located at 110°41'21.93" – 110°48'28.54" E and 7°27'33.38" – 7°32'17.77" S, with an average elevation of 107 meters above sea level. The soil type in this area originates from volcanic materials of Mount Merapi and Merbabu, making it very fertile and ideal for rice fields.

The selection of this location is based on the rapid urbanization and infrastructure development around the airport, which has caused land use changes, particularly the conversion of rice fields. It is estimated that around 199.2 hectares of rice fields have been converted, driven by population growth and the strategic location of the area for development [3]. Its proximity to the city center makes this area important to monitor in order to control land conversion and implement sustainable land management practices.

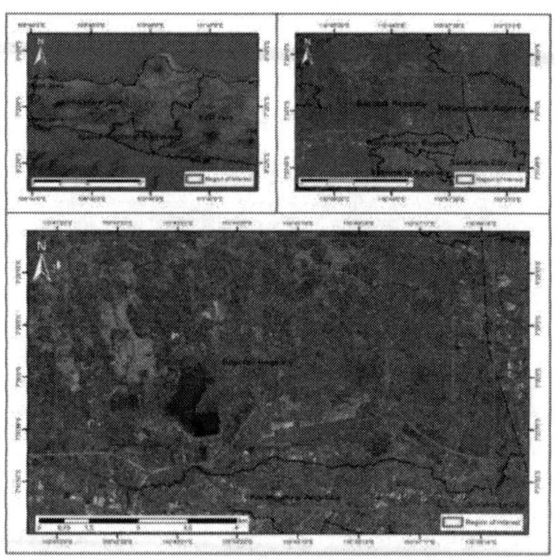

Fig. 1. Research Location

2.2 Data Collection

The data collection method in this study involves gathering satellite imagery and field data to support the analysis of rice field mapping. The satellite imagery will be obtained from the Copernicus Open Access Hub platform. The images used in this study were taken on October 4, 2024. A field survey will be conducted to verify land cover based on the classification results of the imagery, which was taken on October 25, 2024. The details are shown in the following table

Table 1. Materials Research

Data Type	Source	Acquisition Date.
Satellite Imagery: Sentinel-2A	Copernicus Open Access Hub	04 October 2024
Field Data	Direct Observation	25 October 2024

In addition to the materials, there are tools used to support the data collection, processing, and analysis in the satellite-based rice field mapping study. Each instrument plays a crucial role in generating accurate and reliable data and supporting various stages of the research, from field data collection to analysis using software. The following table presents the instruments used in this study

Table 2. Research Instruments

Name	Purpose/Usage
Jupyter Notebook	For advanced data processing and analysis using Python, supporting processing and visualization.
ArcGIS	For visualizing analysis results in the form of maps and performing sample data validation.
Timestamp Application	For field data collection, recording the time and location of image captures for verification.
GPS (Global Positioning System)	For ensuring accurate coordinates of field data collection locations.

2.3 Data Processing and Analysis

2.3.1 Image masking

Image masking is a method used to separate the area to be analyzed from regions outside the study area using shapefile data or Region of Interest (ROI). In this study, the masking technique is applied to the Sentinel-2A Level 2 data to define the specific study area, ensuring that the analysis is conducted only on relevant regions.

2.3.2 Surface Reflectance Correction

The solar radiation reflected by the Earth's surface to satellite sensors is modified by its interaction with the atmosphere, The Earth surface information from remote sensing sensors is presented as surface reflectance, which is influenced by factors such as the angle of incidence, energy, surface roughness, and materials. Surface Reflectance Correction is performed to correct for atmospheric effects on Sentinel-2A data. This correction method uses the L2A surface reflectance correction formula, which aims to obtain more accurate surface reflectance values. These reflectance values are crucial to ensure data quality before further analysis.

2.3.3 Extraction of Spectral Features and Vegetation Indices

a) Spectral Bands

In this study, the spectral bands used are Band 2 (Blue), Band 3 (Green), Band 4 (Red), and Band 8 (Near Infrared) with a spatial resolutioniof 10 meters [9], High spatial resolution imagery is crucial for detailed vegetation mapping, and these bands were selected because of their high resolution and ability to provide sufficient detail for specific vegetation mapping on rice fields. An overview of the Sentinel-2A spectral bands employed, including their spectral range and spatial resolution, is provided in Table 3.

Table 3. Sentinel-2A Spectral Bands Used in This Study

Band	Channel	Wavelength (nm)	Spatial Resolution (m)
Band 2	Blue	490-555	10
Band 3	Green	560-595	10
Band 4	Red	665-695	10
Band 8	NIR	842-957	10

b) Vegetation and Water Indices

Several vegetation indices used in this study include the Normalized Difference Vegetation Index (NDVI), Soil Adjusted Vegetation Index (SAVI), Enhanced Vegetation Index (EVI), and Normalized Difference Water Index (NDWI). These indices are applied to the imagery to highlight vegetation health and provide additional relevant information about rice field areas [10]. The specific formulas used to calculate these indices are shown in Table 4.

Table 4. Vegetation and Water Indices Formula Used in This Study

Indices	Formula
NDVI	$\frac{NIR - RED}{NIR + RED}$ (1)
SAVI	$\frac{(NIR - R)}{(NIR + R + L)} x\ (1 + L)$ (2)
EVI	$2{,}5\ x\ \frac{(NIR - R)}{(NIR + 6\ x\ RED - 7{,}5\ x\ BLUE) + 1}$ (3)
NDWI	$\frac{GREEN - NIR}{GREEN + NIR}$ (4)

2.3.4 Extraction of Texture Features using Gray Level Co-occurrence Matrix (GLCM)

The Gray Level Co-occurrence Matrix (GLCM) technique analyzes statistical features based on the grayscale values of an image, and this method is commonly used in applications such as face recognition. In this study, the GLCM matrix was created with a size of 5x5 to analyze the relationship between pixels based on grayscale values. Eight GLCM indicators used are Angular Second Moment (ASM), Contrast, Correlation, Homogeneity, Entropy, Dissimilarity, Variance, and Mean [11]. This technique helps in identifying different textures, such as smooth, rough, or complex texture variations, which is useful for distinguishing rice field areas, Eight GLCM texture indicators used in the feature extraction process are presented in Table 5 [11]

Table 5. GLCM Texture Features and Formulas Used in This Study

GLCM Indicator	Formula	
Angular Second Moment	$\sum_{i,j} P(i,j)^2$	(5)
Contrast	$\sum_{i,j} P(i-j)^2 P(i,j)$	(6)
Correlation	$\frac{\sum_{i,j}(i.j)P(i,j) - \mu_x\mu_y}{\sigma_x\sigma_y}$	(7)
Homogeneity	$\sum_{i,j} \frac{P(i,j)}{1+\lvert i-j\rvert}$	(8)
Entropy	$\sum_{i,j} P(i,j) log\,(P(i,j))$	(9)
Dissimilarity	$\sum_{i,j} \lvert i-j\rvert P(i,j)$	(10)
Variance	$\sum_{i,j}(i-\mu)^2 P(i,j)$	(11)
Mean	$\sum_i i \sum_j P(i,j)$	(12)

2.3.5 Dimension Reduction with Principal Component Analysis (PCA)

The Principal Component Analysis (PCA) method is applied to spectral features to reduce data dimensions and simplify the analysis by creating new uncorrelated components that can effectively represent the data. PCA generates principal components that retain the most important information from the spectral data (PC1, PC2, and so on). By performing dimensionality reduction, the processed data becomes more compact while still preserving significant information, this process is described by the following equation

$$\begin{bmatrix} PC1 \\ PC2 \\ \dots \\ PCn \end{bmatrix} = \begin{bmatrix} v11 & v21 & \dots & vn1 \\ v12 & v22 & \dots & vn2 \\ \dots & \dots & \dots & \dots \\ v1m & v2m & \dots & vnm \end{bmatrix} = \begin{bmatrix} DN^{l}_{MS1} \\ DN^{l}_{MS2} \\ \dots \\ DN^{l}_{MSn} \end{bmatrix} \quad (13)$$

Information

PCi (i = 1, 2, ..., n)	:	Principal components obtained through PCA
v_{nm}	:	Elements of the PCA transformation matrix, where each column is an eigenvector sorted in order
DN^{l}_{MSi} (i = 1, 2, ..., n)	:	Pixel values from spectral bands used to calculate the principal components

2.3.6 Unsupervised Classification with ISODATA

In this study, the unsupervised classification method using the ISODATA (Iterative Self-Organizing Data Analysis Technique) algorithm is applied to group data based on features derived from PCA. ISODATA is a modification of the k-means method that uses iteration to generate optimal clusters. The process continues until there is minimal change in the number of pixels in each cluster or until the maximum iteration is reached.

2.3.7 Sampling with Stratified Random Sampling

Stratified Random Sampling is applied to select samples from the ISODATA classification results. In this study, 50 samples are taken to represent each class of the clusters, ensuring an accurate representation of the entire study area without considering factors influencing the sample data collection [12].

2.3.8 Classification Analysis with Random Forest (RF)

The final classification is performed using the Random Forest (RF) algorithm, which is a decision tree-based machine learning method [13]. Random Forest is chosen for its ability to handle high-dimensional data and evaluate feature importance. RF builds multiple decision trees and combines their results to improve classification accuracy. This algorithm is also selected because it is not sensitive to overfitting.

2.3.9 Model Accuracy Evaluation

The classification model is then evaluated to determine its accuracy. This evaluation is conducted to assess the effectiveness of the feature combination in producing optimal classification results for paddy field mapping.

https://doi.org/10.1051/e3sconf/202565210003

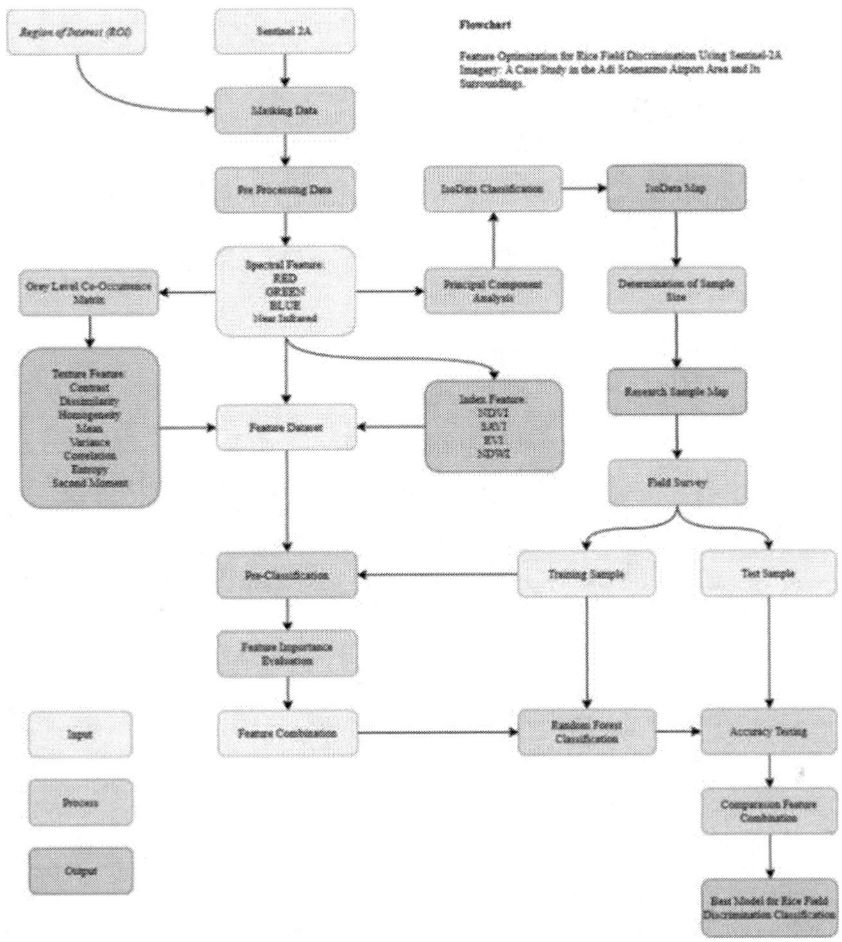

Fig. 2. Research Location

3 Result and Discussion

3.1 Features Used in the Study

In this study, various features from Sentinel-2A satellite imagery were utilized to enhance the accuracy of paddy field mapping. These features include spectral features, vegetation indices, and texture. Each feature was combined to find the best combination for creating an optimal model for paddy field mapping, using the Random Forest algorithm as the classification method. The Random Forest algorithm was selected due to its ability to handle various types of data and its capability to identify the most relevant features in the classification process. A list of the features used is presented in Table 6.

Table 6. Categories and Features Used

Category	Feature
Spectral	GREEN
Spectral	BLUE
Spectral	RED
Spectral	NIR
Indices	EVI
Indices	NDVI
Indices	NDWI
Indices	SAVI
Texture	Contrast
Texture	Correlation
Texture	Disimilarity
Texture	Entropy
Texture	Homogen
Texture	Mean
Texture	Second Moment
Texture	Variance

3.2 ISODATA Classification and Data Sampling Process

The ISODATA classification produced 10 land cover classes that reflect spectral variations from the Red, Green, Blue, and Near-Infrared (NIR) bands. Through the Principal Component Analysis (PCA) process, the spectral data dimensions were successfully reduced without losing essential information, resulting in a more segmented class distribution. This classification, as shown in **Fig.** 3, provides an initial overview of land cover variations such as vegetation, non-vegetation, water bodies, and built-up areas. The results of this unsupervised classification were utilized as a strategic preliminary step to guide the sampling process, thereby facilitating the identification of various land cover types in a representative manner. This integrated approach has been proven to improve classification accuracy by optimizing the strengths of each method [14].

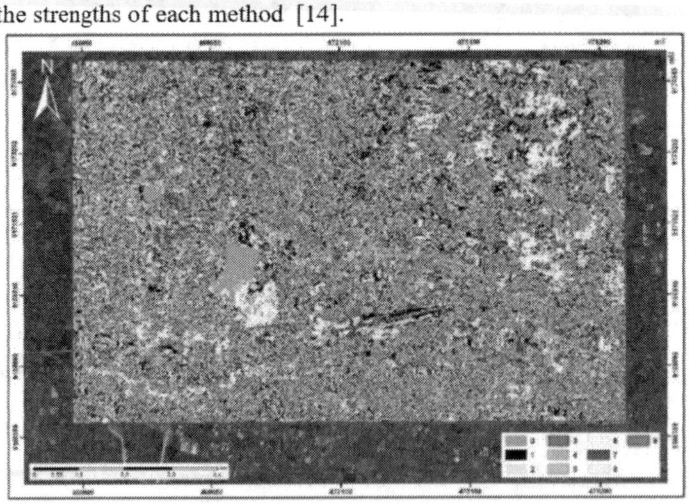

Fig. 3. ISODATA Classification

Training samples were selected using the Stratified Random Sampling method to ensure balanced representation for each class. After eliminating non-significant or redundant classes, a total of 8 classes with 550 samples were obtained and proportionally distributed. Initially, 500 samples were collected from 10 classes, and an additional 50 samples were added to achieve a more even distribution. The samples were then divided into 80% for training (440 samples) and 20% for testing (110 samples) [15]. This distribution was designed to ensure optimal representation of each land cover class and to enable objective evaluation of the model's performance, particularly in identifying paddy fields.

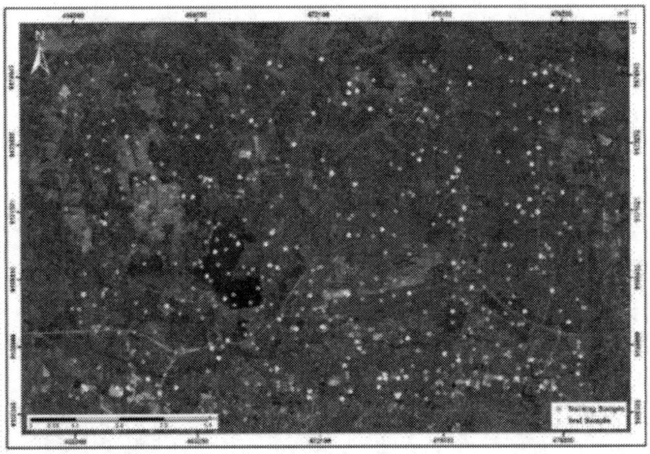

Fig. 4. Distribution Map of Samples.

3.3 Classification Model and Accuracy Test

In this study, four feature configurations were tested to obtain the best classification model using the Random Forest algorithm. Each feature configuration combines various combinations of spectral features, vegetation indices, and texture features. Random Forest was chosen for its ability to handle high data variability and its capacity to determine which features are most relevant in the classification process. Accuracy testing was performed by calculating the Overall Accuracy and Kappa value, with the results of the accuracy test presented in the table below.

Table 7. Overall Accuracy and Nilai Kappa

Feature Type	Overall Accuracy	KAPPA
Feature 1 (Spectral Features)	0.66544117647058	0.59728125800796
Feature 2 (Spectral Features + Indices)	0.68198529411764	0.61590074279650
Feature 3 (Spectral Features + Indices + Texture)	0.69852941176470	0.63709139426284
Feature 4 (The 8 Most Contributing Features from Feature 3)	0.71691176470588	0.65624717902425

The accuracy test results for rice field mapping with four feature configurations showed an improvement in model performance as more feature types were used, directly impacting the Overall Accuracy, which represents the percentage of data correctly mapped compared to the total reference data used in classification evaluation, Configuration 1 (spectral) yielded 66.54% accuracy and a Kappa of 0.5973, but struggled to distinguish rice fields in complex landscapes. Configuration 2 (spectral + vegetation indices) improved accuracy to 68.20% and Kappa to 0.6159, indicating that vegetation information enhances classification. Configuration 3 (spectral + vegetation indices + texture) further increased accuracy to 69.85% (Kappa 0.6371) by incorporating spatial pattern recognition. Configuration 4, consisting of the 8 best features from configuration 3, achieved the highest accuracy of 71.69% with a Kappa of 0.6562, showing that optimized feature selection reduces redundancy, increases efficiency, and enhances performance. These results highlight the importance of combining spectral, vegetation, and texture features for accurate rice field classification.Classification Model and Accuracy Test

3.3.1 Feature Contribution in Configuration 3

The evaluation of feature contribution in Configuration 3, which consists of the combination of Spectral Features + Vegetation Index Features + Texture Features using the Random Forest algorithm, is presented in the table below.

Table 8. Overall Accuracy and Nilai Kappa

Category	Feature	Contribution
Spectral	GREEN	0.1558421853560509
Spectral	BLUE	0.13050867931309768
Spectral	RED	0.10050052169664317
Texture	Mean	0.07926591043721785
Texture	Entropy	0.07436721911563171
Spectral	NIR	0.07104964285734316
Indices	NDWI	0.06448442864944783
Indices	NDVI	0.05306100194029199
Texture	Homogen	0.04794404762040372
Texture	Second Moment	0.04699764481122444
Indices	SAVI	0.0394959591859507
Indices	EVI	0.033854515207716776
Texture	Variance	0.027327323876889145
Texture	Correlation	0.026617178453695465
Texture	Contrast	0.02446322812860989
Texture	Disimilarity	0.02422051334978564

The evaluation of feature contribution in Configuration 3 using the Random Forest algorithm shows a variation in the feature contribution to rice paddy land classification. Spectral features provide a significant contribution, with the GREEN band having the highest contribution (0.1558) due to the distinct and characteristic green color of rice fields, which reflects healthy vegetation with high reflectance. This is followed by BLUE (0.1305) and RED (0.1005) as these bands also provide important information about soil and vegetation conditions, but are more sensitive to light changes and atmospheric conditions. Texture features also play an important role, with the Mean (0.0793) and Entropy (0.0744) features capturing the structural patterns of rice fields. Mean represents the average intensity of the texture, while Entropy measures the complexity of the texture patterns, helping distinguish rice fields from other land types. Vegetation index features such as NIR (0.0710) and NDWI (0.0645) also provide significant contributions, with NDWI helping differentiate rice fields based on water content. Other features like NDVI and SAVI, while contributing less, still play a role in improving classification accuracy. Overall, spectral and texture features provide the primary information for distinguishing rice fields, while features with smaller contributions help refine classification precision.

3.3.2 Feature Contribution in Configuration 4

Evaluation of feature contribution in Configuration 4, which consists of a combination of the 8 most significant features selected from Configuration 3, is presented in the table below.

Table 9. Feature Contribution in Configuration 4

Category	Feature	Contribution
Texture	Entropy	0.13524105599256556
Indices	NDWI	0.13381457702916255
Texture	Mean	0.12668182678083595
Spectral	GREEN	0.1259527778082647
Spectral	RED	0.12426712878567886
Spectral	BLUE	0.12146834218412941
Indices	NDVI	0.11930854918185356
Spectral	NIR	0.11326574223750964

The evaluation of feature contributions in Table 9, using the Random Forest algorithm, shows that texture and vegetation index features play a significant role in paddy field classification. The features Entropy (0.1352) and NDWI (0.1338) recorded the highest contributions. Entropy, which measures irregularity in texture patterns, is highly influential because it captures structural variations that help differentiate paddy fields. Meanwhile, NDWI reflects the water content in vegetation, which is crucial for identifying paddy fields, especially during the flooding phase.

The Mean feature (0.1267) also contributes significantly, representing the average intensity of texture, which aids in distinguishing land cover types. While spectral features such as GREEN (0.1260), RED (0.1243), and BLUE (0.1215) still provide important contributions, their influence is slightly reduced compared to texture and vegetation index features.

The change in the order of contributions between Feature Configuration 3 and Feature Configuration 4 can be explained by the more stringent feature selection in Feature Configuration 4. In this configuration, only the top 8 features, selected based on their contributions, were retained, leading to a dominance of texture and vegetation index features like Entropy and NDWI. In contrast, in Feature Configuration 3, spectral features were more dominant. This shift indicates that the model relies more on features that provide deeper insights into the land characteristics, such as texture and water content, rather than just spectral differences. This emphasizes that a feature combination focused on textural and hydrological aspects can improve classification accuracy, especially for land with similar spectral characteristics.

3.4 Confusion Matrix and Classification Analysis

3.4.1 Confusion Matrix for Feature Configuration 3

To evaluate the performance of the classification model based on the feature combination in Feature Configuration 3, the following confusion matrix is presented, illustrating the number of classification errors between different land cover categories.

Table 10. Confusion Matrix for Feature Configuration 3

Confusion Matrix								
Configuration 3	Water Bodies	Industrial Areas	Bare Land	Built-up Land	Residential Areas	Wet Rice Fields	Dry Rice Fields	Vegetation
Water Bodies	12	0	0	0	0	5	0	3
Industrial Areas	0	124	0	4	7	0	0	0
Bare Land	0	0	10	0	1	0	16	7
Built-up Land	0	13	11	5	8	0	0	0
Residential Areas	0	8	1	0	92	0	7	4
Wet Rice Fields	0	0	0	0	0	64	15	8
Dry Rice Fields	4	0	15	0	9	0	44	3
Vegetation	0	0	6	0	1	7	7	29

The results of the confusion matrix for feature configuration 3 provide a clear picture of the classification model's performance in mapping various land cover categories, including Wet Rice Fields and Dry Rice Fields. Overall, the model demonstrates good capability in distinguishing several land categories, although there are still challenges in differentiating certain land covers.

For the Wet Rice Fields category, 64 samples were correctly classified; however, 15 samples were misclassified as Dry Rice Fields and 8 samples as Vegetation. This indicates that the features used, particularly those related to texture and vegetation indices, are effective in identifying the characteristics of wet rice fields, although these misclassifications suggest some overlap between wet rice fields and other categories, such as dry rice fields or vegetation.

For the Dry Rice Fields category, the model also showed fairly good results, with 44 samples correctly classified, although there were some misclassifications into categories such as Bare Land, Water Bodies, and Vegetation.

Some land cover categories, such as Bare Land and Industrial Areas, face challenges in being distinguished due to having similar physical and spectral characteristics. The largest misclassification occurred in the Built-up Land category, which is often confused with Residential Areas and Industrial Areas.

Overall, this confusion matrix indicates that while the model with feature configuration 3 successfully classified most land cover categories well, there are still areas for improvement, particularly in differentiating between wet rice fields and dry rice fields, as well as built-up land from other categories

 https://doi.org/10.1051/e3sconf/202565210003

3.4.2 Confusion Matrix for Feature Configuration 4

To evaluate the performance of the classification model based on the feature combination in Feature Configuration 4, the following confusion matrix is presented, illustrating the number of classification errors between different land cover categories.

Table 11. Confusion Matrix for Feature Configuration 4

Confusion Matrix								
Configuration 4	Water Bodies	Industrial Areas	Bare Land	Built-up Land	Residential Areas	Wet Rice Fields	Dry Rice Fields	Vegetation
Water Bodies	12	0	0	0	0	5	0	3
Industrial Areas	0	128	0	1	7	0	0	0
Bare Land	0	0	7	0	3	0	18	4
Built-up Land	0	22	1	5	8	0	0	0
Residential Areas	0	9	0	0	98	0	3	2
Wet Rice Fields	0	0	4	0	0	59	14	12
Dry Rice Fields	4	0	9	0	8	0	51	2
Vegetation	0	0	3	0	2	8	8	30

The results of the confusion matrix from feature configuration 4 show an improvement in distinguishing several land cover categories compared to feature configuration 3, although challenges in differentiating Wet Rice Fields and Dry Rice Fields still persist.

For the Wet Rice Fields category, the model successfully classified 59 samples correctly; however, there were 14 misclassifications as Dry Rice Fields and 12 misclassifications as Vegetation. These errors indicate difficulties in distinguishing wet rice fields from other categories, particularly dry rice fields and vegetation.

For the Dry Rice Fields category, 51 samples were classified correctly, although there were 9 misclassifications as Bare Land, 2 as Vegetation, and 4 as Water Bodies. Nevertheless, these errors are fewer compared to feature configuration 3, although challenges remain in differentiating dry rice fields from categories with similar physical and spectral characteristics, such as Bare Land and Vegetation.

Misclassifications in the Built-up Land and Industrial Areas categories still occur, although not as frequently as in feature configuration 3. While the addition of features in configuration 4 has led to improvements, there is still room for enhancing the separation between similar categories, particularly Wet Rice Fields and Dry Rice Fields.

Overall, the model with feature configuration 4 demonstrates better performance compared to configuration 3, although challenges remain in distinguishing categories with very similar physical and spectral characteristics.

https://doi.org/10.1051/e3sconf/202565210003

3.5 Visualization of Rice Field Mapping from the Best Model for Each Feature Configuration

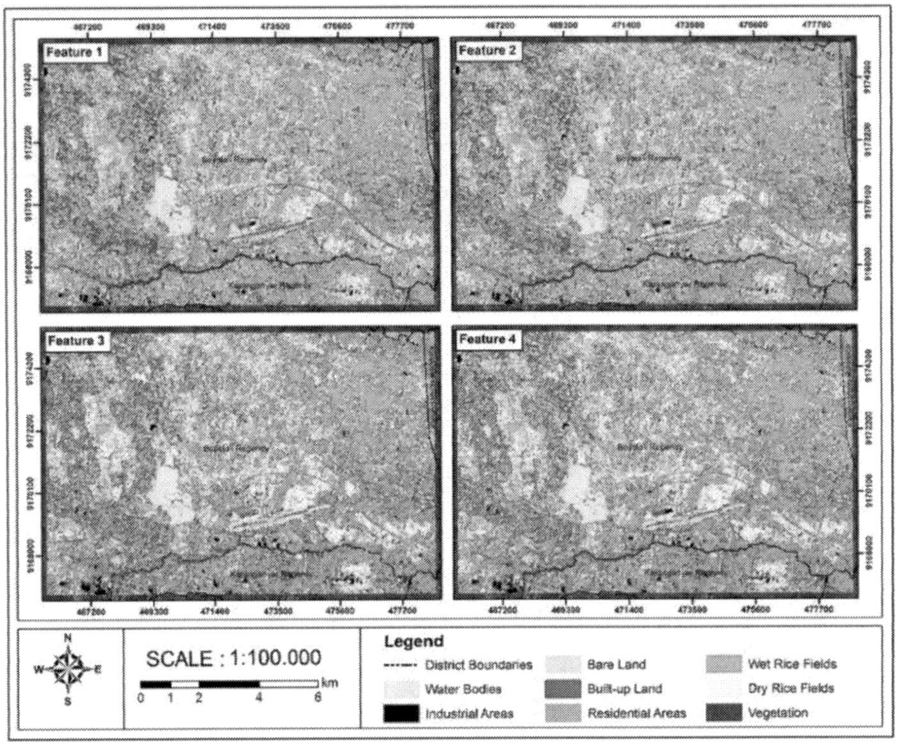

Fig. 5. Best Model for Each Feature Configuration

The visualization of land cover distribution globally is based on four feature configurations (Feature 1, Feature 2, Feature 3, and Feature 4). Overall, all features demonstrate the ability to identify various land cover categories, such as wet rice fields, dry rice fields, vacant land, vegetation, residential areas, industrial areas, and water bodies. However, there are significant variations in how each feature classifies regions, particularly in transitional or complex areas.

In Feature 1, the dry rice field areas (yellow) are still misclassified as residential areas (orange). The distribution of vegetation (dark green) is also broader compared to other features, indicating that this configuration may be more sensitive to dry vegetation or areas with low moisture. Conversely, Feature 2 presents results that are more focused on the presence of water bodies, with aquatic areas more clearly identified. This suggests that Feature 2 may be more sensitive to the spectral reflectance of water or high moisture levels.

Next, Feature 3 shows a broader distribution in the category of dry rice fields (yellow), which may indicate that this feature is more responsive to open or non-vegetated areas. This is evident in the central and northern parts of the map, where vacant land is more prominent compared to other features. On the other hand, Feature 4 displays sharper and more segmented classification results, particularly for residential areas (orange) and built-up areas (red). However, this pattern also indicates a higher variation in the classification of transitional areas, such as between wet rice fields, dry rice fields, and vacant land.

From a spatial perspective, this map shows that each feature configuration has specific advantages. However, the accuracy of each feature needs to be further validated using field

 https://doi.org/10.1051/e3sconf/202565210003

data to assess the significant differences produced by each feature configuration, as displayed in **Fig.** 6.

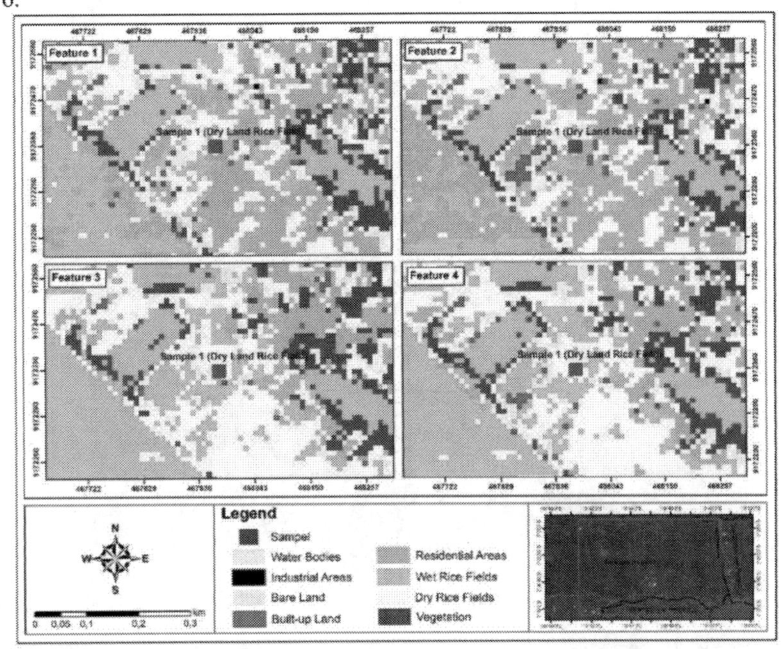

Fig. 6. Visualization of Classification Results for Four Feature Configurations in the Focus Area

In the focus area, classification using four feature configurations shows varying performance. Feature configuration 1, which relies solely on spectral bands (Red, Green, Blue, and NIR), often misclassifies dry rice fields (yellow) and wet rice fields (light green) as residential areas (orange). This occurs because the spectral reflectance of dry rice fields is similar to built-up surfaces, such as metallic or concrete roofs, particularly in the red and green bands. Feature configuration 2, which incorporates vegetation indices (NDVI, SAVI, EVI, NDWI), does not fully resolve this issue, as these indices are more sensitive to dense vegetation. As a result, dry rice fields with sparse vegetation and inundated rice fields still resemble residential areas.

Feature configuration 3, with the addition of GLCM texture features, significantly reduces misclassification. Texture features can distinguish the surface patterns of rice fields from the more structured surfaces of residential areas, thereby improving accuracy. Feature configuration 4, which combines the eight most contributive features (Entropy, NDWI, Mean, Green, Red, Blue, NDVI, and NIR), achieves the highest overall accuracy. However, at the local scale (**Fig 6**), feature configuration 3 produces sharper and more detailed visualizations.

Misclassification in the form of a salt-and-pepper effect remains across all configurations, with the highest noise occurring in configurations 1 and 2. In contrast, configurations 3 and 4 exhibit lower noise due to the integration of texture features, although some errors persist. Field validation conducted on October 25, 2024, confirms that the combination of texture features and vegetation indices provides the most accurate classification results

Fig. 7. Field Conditions of Survey Points

Based on **Fig 7**, the sample point indicates a dry rice field characterized by vegetation that is not inundated. The validation photo is consistent with the classification results, showing that the area surrounding the sample point is identified as a dry rice field, and its location matches the coordinates on the map, thereby strengthening classification accuracy.

In terms of visualization, feature configuration 3 provides clearer details, particularly in transition zones between dry rice fields, vegetation, and other land uses. Although feature configuration 4 achieves higher accuracy (OA: 71.69%; Kappa: 65.62%) than feature configuration 3 (OA: 69.85%; Kappa: 63.70%), the representation of feature configuration 3 corresponds more closely to field conditions. Therefore, feature configuration 4 is more suitable for large-scale analyses with an emphasis on global accuracy, whereas feature configuration 3 is better suited for local analyses requiring detail. Field validation and survey photos further confirm that feature configuration 3 offers a more representative classification of dry rice fields.

4 Conclusion

This study demonstrates that the combination of spectral features, vegetation indices, and texture derived from Sentinel-2A imagery can improve the accuracy of rice field mapping in the Adi Soemarmo Airport area. The overall classification accuracy increased progressively, from 66.54% in the initial configuration using only spectral features, to 68.20% with the addition of vegetation indices, then 69.85% with the inclusion of texture features, and finally reaching 71.69% in the fourth configuration with eight selected features. These results indicate that selecting the most relevant features can enhance model efficiency while maintaining high accuracy. Furthermore, the Random Forest algorithm proved effective in handling complex datasets, reducing overfitting, and performing optimal feature selection.

Although the classification results were generally satisfactory, challenges remain in differentiating wet rice fields, dry rice fields, and other land cover classes with similar spectral characteristics. For future research, accuracy improvements may be achieved by utilizing multitemporal data, which capture the phenological dynamics of rice growth, or multisensor data that provide complementary information beyond the optical spectrum, such as radar or thermal imagery. Developing an accurate mapping model is crucial for monitoring land-use change, supporting spatial planning, and promoting sustainable land management, particularly in areas affected by strategic infrastructure development

.

References

1. Sarono, R. Kurniawan, N.Y.A. Karim, H.Z. Hadibasyir, M.N. Fadhilla, M.A. Trisnanto, et al. Remote sensing and GIS integration for paddy production estimation in Bali Province, Indonesia. *ACRS 2015 - 36th Asian Conf Remote Sens Foster Resilient Growth Asia, Proc.* (2015)
2. I. Adinata, A.A. Sigit. Analisis Perubahan Penggunaan Lahan Di Kecamatan Colomadu Kabupaten Karanganyar Tahun 2009 dan 2019. *Univ Muhammadiyah Surakarta.* (2020)
3. U.N. Rohmah, Jumadi. Analisis Perubahan Penggunaan Lahan di Kecamatan Ngemplak Kabupaten Boyolali Tahun 2014 dan 2020. ,*Universitas Muhammadiyah Surakarta.* (2021)
4. N. You, J. Dong, J. Huang, G. Du, G. Zhang, Y. He, et al. The 10-m crop type maps in Northeast China during 2017–2019. *Sci Data.* **8**. (1). , 41. (2021). https://doi.org/10.1038/s41597-021-00827-9
5. M. Ozdogan. The spatial distribution of crop types from MODIS data: Temporal unmixing using Independent Component Analysis. *Remote Sens Environ.* **114**. (6). , 1190–1204. (2010). https://doi.org/10.1016/j.rse.2010.01.006
6. B. Alikhanov, B. Pulatov, L. Samiev. The Detection of Past and Future Land Use and Land Cover Change in Ugam Chatkal National Park, Uzbekistan, Using CA-Markov and Random Forest Machine Learning Algorithms. *Forum Geogr.* **38**. (2). , 121–137. (2024). https://doi.org/10.23917/forgeo.v38i2.4221
7. M. Belgiu, L. Drăguţ. Random forest in remote sensing: A review of applications and future directions. *ISPRS J Photogramm Remote Sens.* **114**. , 24–31. (2016). https://doi.org/10.1016/j.isprsjprs.2016.01.011
8. Y. Khoirurrizqi, R. Sasongko, N.L.E. Utami, A. Irbah, S. Arjasakusuma. Machine Learning-Based Rice Field Mapping in Kulon Progo using a Fusion of Multispectral and SAR Imageries. *Forum Geogr.* **37**. (2). , 134–148. (2023). https://doi.org/10.23917/forgeo.v37i2.20304
9. S. Delwart. ESA Standard Document. (1). (2015)
10. E. Diasmara. Analisa Indeks Vegetasi pada Data Sensor Satelit NOAA-AVHRR dan MODIS. (2007)
11. R.M. Haralick, K. Shanmugam, I. Dinstein. Textural Features for Image Classification. *IEEE Trans Syst Man Cybern.* **SMC-3**. (6). , 610–621. (1973). https://doi.org/10.1109/TSMC.1973.4309314
12. Y. Abidin. Strategi membaca teori dan pembelajarannya. *Bandung Rizqi Press Ahuja, Pramila.* (2010)
13. C.G. Indrayanto, D.E. Ratnawati, B. Rahayudi. Analisis Sentimen Data Ulasan Pengguna Aplikasi MyPertamina di Indonesia pada Google Play Store menggunakan Metode Random Forest. *J Pengemb Teknol Inf dan Ilmu Komput.* **7**. (3). , 1131–1139.

(2023)

14. D.I. Enderle, R.C. Weih Jr, R.C. Jr, D. IMEnderle, R.C. WeihJr. Integrating Supervised and Unsupervised Classification Methods to Develop a More Accurate Land Cover Classification. *J Ark Acad Sci*. **59**. , 10. (2005)
15. A. Nugroho. Analisa Splitting Criteria Pada Decision Tree dan Random Forest untuk Klasifikasi Evaluasi Kendaraan. *JSITIK J Sist Inf dan Teknol Inf Komput*. **1**. (1). , 41–49. (2022). https://doi.org/10.53624/jsitik.v1i1.154

Forecasting Rice Field Conversion in Boyolali Regency Using CA-ANN and Multitemporal Remote Sensing Data

Surya Fathurrahman[1], *Dewi Novita* Sari[2*]

[1]Faculty of Geography, Muhammadiyah University Surakarta, Surakarta 57162, Central Java, Indonesia
[2]Faculty of Geographic Information Science, Muhammadiyah University Surakarta, Surakarta 57162, Central Java, Indonesia

Abstract. Land conversion or land use change is a change in the function of land to another use caused by a change in the value of the land. Changes in land use occur due to various factors, such as population growth, urbanization, and infrastructure expansion, which cause rice fields to be converted into non-rice fields. In recent years, Boyolali Regency has experienced a decline in the area of rice fields due to land use change. This study aims to analyze the change in the area of rice field land use to non-rice field land and predict changes over the next 10 years. Cellular Automata - Artificial Neural Network (CA-ANN) is the method used in this analysis. Additionally, to analyze and predict changes in land use, driving factors such as distance from roads, distance from rivers, and distance from settlements are required. The results of land use changes in 2014 and 2024 show an increase in non-rice field land use of 9,846.29 ha (16.15%), while rice field land use decreased by 9,846.9 ha (20.26%), indicating substantial land changes. The results of the prediction modeling showed a significant increase in non-rice field land use of 72,774.03 ha, while rice field land use decreased by 36,818.45 ha. Modeling with CA-ANN showed strong accuracy results, with a Kappa index of 0.73350 and a correction of 87.94264%.

1 Introduction

Land conversion or land use change is a change in land use function to other uses caused by changes in the use value of land. The changing use value can be in the form of price levels or types of benefits, such as social benefits, public services, culture, and history [1]. Land use change occurs due to the pressure of various factors, such as population growth, urbanization, and infrastructure expansion that cause the conversion of agricultural land into non- agricultural land [2]. Land use change from agricultural land to non-agricultural land has various impacts. From an economic perspective, land conversion can increase people's income through livelihood diversification and public facility development. However, on the other hand, negative impacts include decreased food security, loss of agricultural ecosystems, and increased risk of environmental disasters such as flooding and soil erosion. Therefore, studies are needed that can predict land use change patterns in order to serve as the basis for more sustainable spatial planning.

* Corresponding author: dns104@ums.ac.id

Boyolali Regency, as one of the agricultural regions in Central Java, has experienced significant land use dynamics. Based on data from the Central Bureau of Statistics (BPS) of Boyolali Regency, the area of paddy fields has decreased from 22,778 hectares in 2014 to 22,703 hectares in 2023. This decline is caused by the increasing need for land for settlements, industry, and infrastructure development that has grown rapidly in the last decade. Data from the Central Bureau of Statistics (BPS) of Boyolali Regency shows that in the last 10 years, there has been a decrease in the area of paddy fields by 1.07% due to land conversion for settlements, industry, and infrastructure [3]. In Boyolali Regency, this phenomenon is driven by the increasing rate of urbanization that reaches 1.07% by 2023 as well as the development of strategic infrastructure such as the Solo-Kertosono toll road and new industrial areas.

Analysis of land use change by utilizing temporal spatial data is very useful, especially for monitoring locations that experience land use change by paying attention to visual appearance and estimating the increase and decrease in land area [4]. Spatial analysis with GIS can be used in studying land use change modeling in a certain period of time. In this study using CA-ANN or Cellular Automata-Artificial Neural Network modeling to project and predict land use change in Boyolali Regency in 2034, this research has also been conducted by Nabila (2023) in Sleman Regency using CA and ANN modeling which resulted in a Kappa accuracy value of 0.95621 and a percentage of prediction conformity with existing data of 93.52%. Previous research on land use change was also conducted by Tiara et al (2023) using CA-ANN (Artificial Neural Network - Cellular Automata) with spatial analysis using SPOT images and the MOLUSCE plugin in QGIS. Predictions for 2022-2042 show a trend of reducing rice fields and increasing settlements with an accuracy of 86.66% and using the maximum likelihood algorithm with a Kappa Index result of 0.83. Unlike previous studies, this study contributes to the assessment of land use change in Boyolali Regency, which plays a strategic role as a food buffer for Surakarta City and its surrounding areas. The dynamics of land change in this region not only have a local impact but also affect urban food security. The novelty of this research lies in the integration of more contextual driving factors, such as road network development, settlement patterns, and industrial expansion, which are closely related to population growth. The objectives of this study are (1) Analyzing the Change in Land Use Area of Rice Fields to Non-Rice Fields in 2014-2024. (2) Predicting Land Use Change of Sawan Area to Non-Sawah Area in the Next 10 Years (2034).

2 Methods

2.1 Research Location

This research is located in Boyolali Regency, Central Java Province, Indonesia with an area of 1,096.61 km2. Astronomically, the area is located at 110°41'21.93" - 110°48'28.54" East and 7°27'33.38" - 7°32'17.77" LS, with an average altitude of 107 meters above sea level. Based on its geographical location, Boyolali Regency is located in Java Island and does not directly border the sea. Boyolali Regency is directly adjacent to Semarang Regency, Salatiga City, Magelang Regency, Sleman, Klaten, Sukoharjo, Karanganyar, and Sragen. The region consists of 22 sub-districts. The largest sub-district is Juwangi sub-district with an area of 96.13 km^2, while Sawit sub-district has the smallest area of 18.71 km^2 [5].

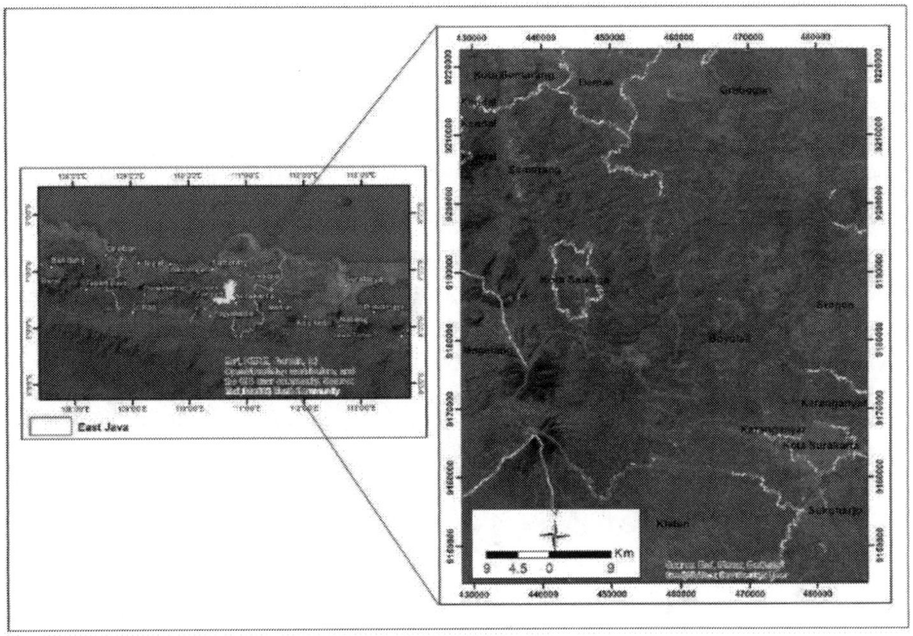

Fig. 1. Research Location

2.2 Tools and Materials

In this research, various types of spatial data were used to support the analysis of changes in the use of paddy fields to non-rice fields in Boyolali Regency. The data utilized includes 2014 and 2024 Landsat 8 satellite images to identify land use change over time. In addition, administrative boundary data of Boyolali Regency was used to geographically delimit the study area. Factors driving land use change were analyzed using shapefiles of road networks, river networks and settlements, all of which were obtained from the Geospatial Information Agency. The data sources used are guaranteed for reliability, including from USGS EarthExplorer for satellite image data. Predictions of change will be made until 2034 using CA-ANN modeling. The data collection method was carried out through several stages as follows:

Table 1. Research Materials

Data Name	Usefulness	Source
Landsat 8 Satellite Image 2014	Identifying land use change from paddy field to non paddy field	*https://earthexplorer.usgs.gov/*
Landsat 8 satellite image in 2024	Identifying land use change from paddy field to non paddy field	*https://earthexplorer.usgs.gov/*
Administrative Boundary of Boyolali Regency	Delimit the study area	Geospatial Information Agency
Shapefile Road network	A driving factor for land use change	Geospatial Information Agency
River Network Shapefile	A driving factor in land use change	Geospatial Information Agency
Settlement Shapefile	A driving factor in land use change	Geospatial Information Agency

In a satellite-based rice field mapping study, in addition to materials, there are tools used to assist with data collection, processing, and analysis. Each tool has an important role to play in producing accurate and reliable data, and they assist at various stages of the research, from field data collection to software analysis. The following table shows the instruments used in this study.

Table 2. Research Instruments

Name	Purpose/Usage
ENVI 5.3	To perform radiometric correction and guided classification (MLC)
ArcGIS	To display analysis results in map form and validate sample data.
Google Earth Pro	To test the accuracy of land use suitability processing
QGIS	To process the results of future prediction modeling

2.3 Data Processing and Analysis

2.3.1. Image Correction and Maximum Likehood Classification

Landsat 8 multitemporal satellite imagery data (2014 and 2024) was used to start processing data on land use change. Next, radiometric correction was performed to improve spatial accuracy and remove atmospheric distortions. After that, land use was classified using pixel- or object-based classification methods, such as Maximum Likelihood Classification, which is superior to other classifiers. This method is used to distinguish between rice fields and non-rice fields.

2.3.2. Plotting sample points for accuracy test and field validation

To test the accuracy of the image using the results of the maximum likehood land use classification. Sampling was done using the create random point tool in ArcGis 10.8. After that, field validation was conducted to ensure how the processing results correlated with actual field conditions. The 2014 land use change accuracy test used Google Earth Pro to compare the 2014 processing results with the actual situation. The 2024 land use change accuracy test also compared the processing results with the actual conditions on the ground.

The accuracy test uses Calculation Kappa where the accuracy test is conducted to determine whether the classification results made are in accordance with the reality in the field. The accuracy test involves comparing the unguided classification as the class and the guided classification as the reference data that is considered correct [6]. A good kappa test value is ≥ 0.85, so its accuracy can be considered very good, while a value of $<0.40-<0.70$ can be considered good. Furthermore, land use change was analyzed by spatial overlay between the 2014 and 2024 maps using ArcMap. The final result of this stage is a land use change map, which shows the shift of paddy fields to non-farmland within the time span.

2.3.3. Determining factors that influence change

The driving factors or driving variables in this study are important to determine the influence of land change and directly affect the accuracy and success of predictions. Factors that influence land change are distance to roads, settlements and rivers. Because accessibility is very important for residential locations, there is a tendency for people to settle near road networks to make it easier to carry out activities, as well as rivers that affect activities in rice field areas.

2.3.4. Processing of Land Use Change Prediction Modeling

In the prediction model, the data used is historical land change data from 2014 to 2024. In this model, driving factors such as distance to roads and settlement centers were used to influence the conversion of paddy fields to non-fields. To run the CA-ANN model, the MOLUSCE (Modules for Land Use Change Simulations) plugin was used in QGIS. MOLUSCE ran simulations based on previous land change patterns. The model produced a land use prediction map in 2034, which was then validated in 2024 with the land use prediction model, accounting for accuracy with an overall accuracy index, also known as the Kappa Index. This stage generates examples of how land use may change in the future.

2.3.5. Land Use Change Analysis

Land use change analysis was conducted by comparing multitemporal satellite imagery (2014 and 2024) to identify the conversion of paddy fields to non-farmland in Boyolali Regency. Satellite image data that has gone through geometric and radiometric correction is classified using the Maximum Likelihood Classification method to distinguish between paddy fields and non-rice fields [7]. The classification results were then validated to compare with conditions in the field to improve mapping accuracy. After the classification was completed, a spatial overlay was conducted between the 2014 and 2024 land use maps to determine the pattern of land change. This overlay allows analysis of how extensive land conversion is occurring and where the dominant locations of change occur.

2.3.6. Analysis of Prediction Modeling with CA-ANN

Through Cellular Automata-Artificial Neural Network (CA-ANN) modeling, this research uses spatial units in the form of grids to represent physical space and simulate changes over time. Historical land use change data (2014 and 2024) was analyzed along with geographic variables such as distance to major roads, rivers and settlements, reflecting location-based drivers of change. The Spatial approach examines how spatial characteristics influence the dynamics of land change. The utilization of the MOLUSCE plugin in QGIS is the technical means to implement this approach operationally [8]. The simulation in MOLUSCE considers the spatial relationship between grids as well as the pattern of affordability between regions, so that the 2034 prediction results are not only based on temporal statistics, but also spatial linkages, a characteristic in geographic analysis. Validation of the predictions using methods such as the Kappa Index was also performed on the spatial data, confirming that the entire analysis process is closely tied to a geographic spatial approach.

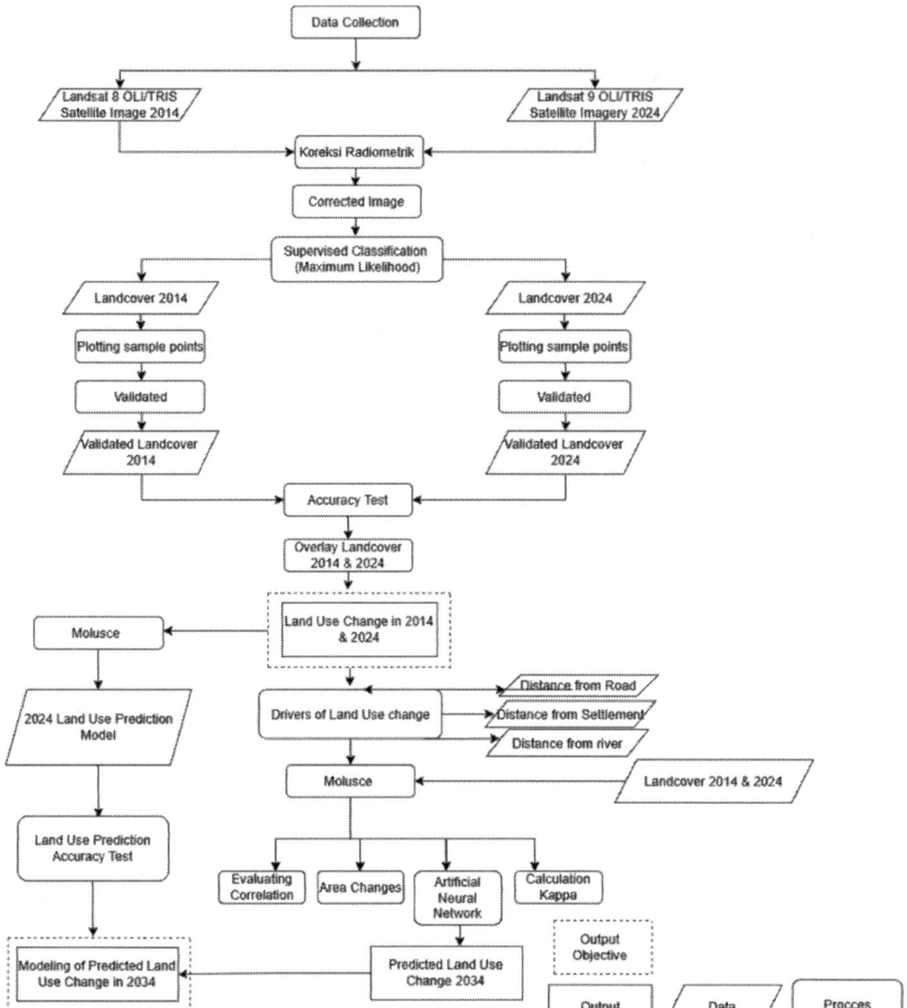

Fig. 2. Research Flow Chart

3 Results and Discussion

3.1 Land Change in 2014 and 2024

In order to obtain more accurate and integrated results, dynamic land use change requires continuous evaluation, analysis, and monitoring with social, economic, and geospatial data. Land use changes that occurred in Boyolali Regency in 2014 and 2024 have increased as well as decreased, the classification of land use is divided into two, namely: rice fields and non-rice fields which can be seen in the figure below.

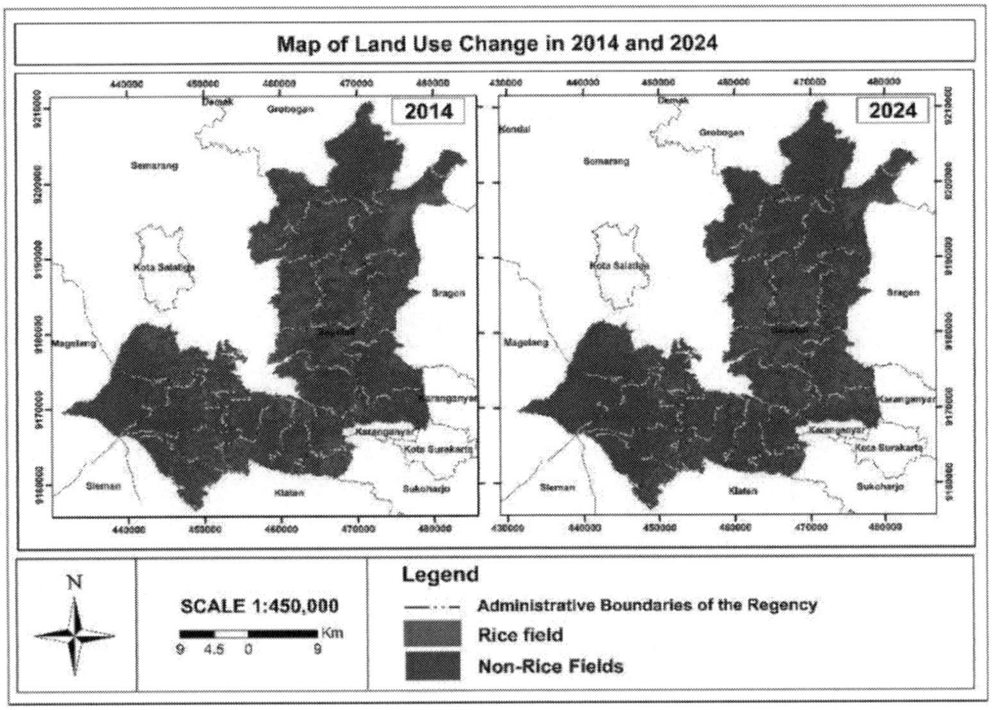

Fig. 3. Map of Land Use Change of Boyolali Regency in 2014 and 2024

Based on the map of land use change in 2014 and 2024 in Boyolali Regency, there is a significant shift from paddy fields to non-fields. In 2014, most areas in Boyolali Regency were still dominated by paddy fields, which were evenly distributed in almost all sub-districts, especially in the central and southern regions. However, in 2024, the area of paddy fields decreased dramatically, especially in sub-districts such as Ngemplak, Banyudono, Nogosari, Mojosongo, and Teras. This decline correlates with the increase in land conversion to non-agricultural uses, such as settlements, industry, and transportation infrastructure.

Ngemplak and Nogosari sub-districts, which are located close to the main access to Adi Soemarmo Airport and Surakarta City, have experienced the most conversion of paddy fields. Similarly, in Mojosongo and Banyudono, land use change is very striking because these areas are on the path of urban and suburban economic growth. On the other hand, sub-districts such as Tamansari, Selo, and Musuk that are located in the highlands or slopes of Mount Merapi show more minimal changes, due to topographic conditions that do not allow for large-scale development and the dominance of traditional agricultural functions.

This pattern of land change shows the tendency of increasing development pressure on productive agricultural land, especially in areas with high accessibility and proximity to centers of economic activity [9]. Urbanization and urban expansion, both from Surakarta and Boyolali city center, are the main drivers of land use change.

Table 3. Extent of LULC Changes in 2014 and 2024

LU	2014		2024		Changed Area	
	ha	**%**	**ha**	**%**	**ha**	**%**
Rice Fields	48601.30	**44.35%**	38754.40	**35.36%**	(9.846,9)	20.26%
Non-Rice Field	60991.72	**55.65%**	70838.01	**64.64%**	9.846,29	16,15%

Based on table 3. Quantitative changes in land use are presented detailing the area and percentage of each land use category. In 2014, the area of paddy fields was recorded at 48,601.30 hectares, while non-rice fields covered 60,991.72 hectares. In 2024, the area of paddy fields decreased to 38,754.40 hectares, which means a decrease of 9,846.9 hectares. In contrast, non-field land increased to 70,838.01 hectares, with an increase in area of 9,846.29 hectares. These results indicate that there has been a significant change in land use, especially land conversion from rice fields to non-fields within a decade in Boyolali Regency. This change reflects the conversion of agricultural land to non- agricultural uses, most likely for settlement, infrastructure, and economic development in the Boyolali region [10]. From a socio-economic perspective, the reduction of agricultural land has a direct impact on farmers' income and welfare, as limited production space reduces crop yields and decreases employment opportunities in agriculture. In addition, the reduction in agricultural area also has an impact on the decline in local food supply, which can threaten food security at the local and regional levels, given that Boyolali plays a crucial role as one of the food warehouses supporting the city of Surakarta and its surroundings. If this change in function continues without restriction, it will increase dependence on food supplies from outside the region and worsen the socioeconomic vulnerability of rural communities.

3.2 Prediction of Land Use Change in 2034

3.2.1 Evaluating correlation

Evaluating correlation is done by using *Pearson's Correlation* method on all existing driving factors such as roads, rivers, and settlements. *Pearson's Correlation* is a statistical matrix that measures the dependency and linear relationship between two random variables defined through variable X and variable Y. The results of the correlation test between roads, rivers, and settlements are shown in Table 3 below.

Table 3. Pearson's Correlation Test

Pearson's Correlation Test			
	Settlement	Road	River
Settlement	--	0.762264188	-0.09752
Road		--	-0.08716
River			--

The table above shows the results of the Pearson correlation test between settlements, roads and rivers. There is a strong positive relationship between settlements and roads with a correlation value of 0.762, indicating that the presence of roads is strongly associated with the presence of settlements. This is logical as accessibility is very important for settlement location, as people tend to settle near road networks for easier movement. In contrast, the negative correlations between settlements and rivers (-0.097) and between roads and rivers (-0.087) show very weak correlations, and may even be insignificant. This suggests that the presence of rivers does not greatly affect the distribution pattern of roads or settlements.

3.2.2 Area change

At this stage, a transition matrix is created that shows how likely land use change is. The transition matrix has a range of values from 0 to 1, with values between 0.01 and 0.99 indicating the possibility of land use change, while values between 0 and 1 indicate the possibility of no land use change or tend to remain. The following Table 4. presents the transition matrix of land use change in Boyolali Regency generated from Molusce processing.

Table 4. Transition Matrix

Transition Matrix		
	Rice Field	Non-Urban
Rice Field	0.54982653	0.45017347
Non-Urban	0.1988375	0.8011625

Table 4. shows the pattern of land use transition between the rice field and non-field categories, with diagonal values of 0.5498 for the rice field category and 0.8012 for the non-field category, each indicating the proportion of land that remains in the same category. In other words, around 54.98 percent of paddy fields remain as paddy fields, while 80.12 percent of non-field fields do not change their function. In contrast, 45.02 percent of paddy fields changed into non-field fields, and only 19.88 percent of non- field fields changed into paddy fields. This shows that there is a tendency to use paddy fields for other purposes, such as settlements, industry, and infrastructure. Due to the strategic function of paddy fields as a source of food production, this phenomenon is an important concern in the context of food security and environmental sustainability [11].

3.2.3 Transition potential modelling

The model training results show the parameters and performance of the classification model used in the study. The model uses (neighborhood) 1 pixel with a learning rate of 0.10 and 1000 maximum iterations. It also uses ten neurons in the hidden layer with a momentum of 0.05. According to the very small overall accuracy Δ value (0.00039), the model has reached convergence and shows no more significant improvement in the final iterations. The current validation Kappa value of 0.57308 indicates a moderate level of agreement between the model's predictions and the actual class, while the lowest validation value of 45.05% indicates variations in performance during training. Kappa values above 0.5 indicate that the model has reasonably good predictive ability, although not optimal. With the parameter configuration used, the model shows decent performance but still has room for improvement, for example with further parameter tuning or the addition of more representative training data. The following Table 5. presents the results of the training model generated from Mollusce processing.

Table 5. Model Training Results

Model Training Results	
Neighborhood	1 px
Learning rate	0.10
Maximum iterations	1000
Hidden layer	10
Momentum	0.05
Δ Overall accuracy	0.00039

Model Training Results	
Min validation overall accuracy	0.45054
Current validation Kappa	0.57308

3.2.4 *Cellular automata*

This Cellular automata processing stage produces a prediction map of land use change in 2034 using 2014 and 2024 land use data and several driving factors such as roads, rivers, and settlements. The results of the land use prediction map of Boyolali Regency in 2034 can be seen in Figure 3.

The land use prediction map of Boyolali District in 2034 shows that the majority of areas in the southwestern, central, and eastern parts of Boyolali District are non-farmland. This phenomenon is related to the Pearson correlation data, which shows a strong positive relationship between settlements and roads (correlation value (0.762) indicating that, as shown on the map, the growth of settlements is strongly influenced by the presence of the road network, which in turn encourages the conversion of paddy fields to non-farmland. In contrast, there is a weak negative correlation between settlements and rivers (-0.0975) and between roads and rivers (-0.0871) indicating that rivers are not the dominant factor. In the context of predicting future land use change, these results corroborate the notion that infrastructure (especially roads) plays an important role in driving the expansion of built-up areas. Theoretically, this is in line with the concept of land use change, which states that accessibility and economic development factors are the main drivers of land conversion.

3.2.5 *Kappa accuracy test results*

The prediction model above was further analyzed with a statistical calculation in the form of the Kappa accuracy test, the results of the classification model evaluation showed that the level of accuracy (% accuracy) was 87.94%. This shows that most of the classifications made by the model match the reference data. After subtracting the effect of conformity, the overall (total) Kappa value of 0.73350 is in the "substantial agreement" category according to the Kappa value interpretation standard. This value indicates high agreement between the classification results and the reference data. That the model was able to capture the class proportions (such as paddy field and non-hay field) very well is shown by the very high histogram capacity (0.96258). This means that the distribution of the number of pixels per class in the classification result is almost the same as the reference, but the Kappa location of 0.76201 indicates a high level of spatial fit, suggesting that the model is correct in the number of classes and spatial placement [12].

Changes that occur in developing countries are caused by a lack of awareness about the over-utilization of environmental resources and without considering future risks caused by the lack of a strong legal framework . Boyolali Regency is experiencing land use change so it requires prediction modeling for the next 10 years, this refers to Government Regulation Number 26 of 2008 concerning the National Spatial Plan, which serves as a standard for national development plans and spatial planning in provinces, districts / cities, and other countries. The results of processing land change prediction modeling using the MOLUSCE plugin in QGIS produce maps and land change prediction areas as shown below.

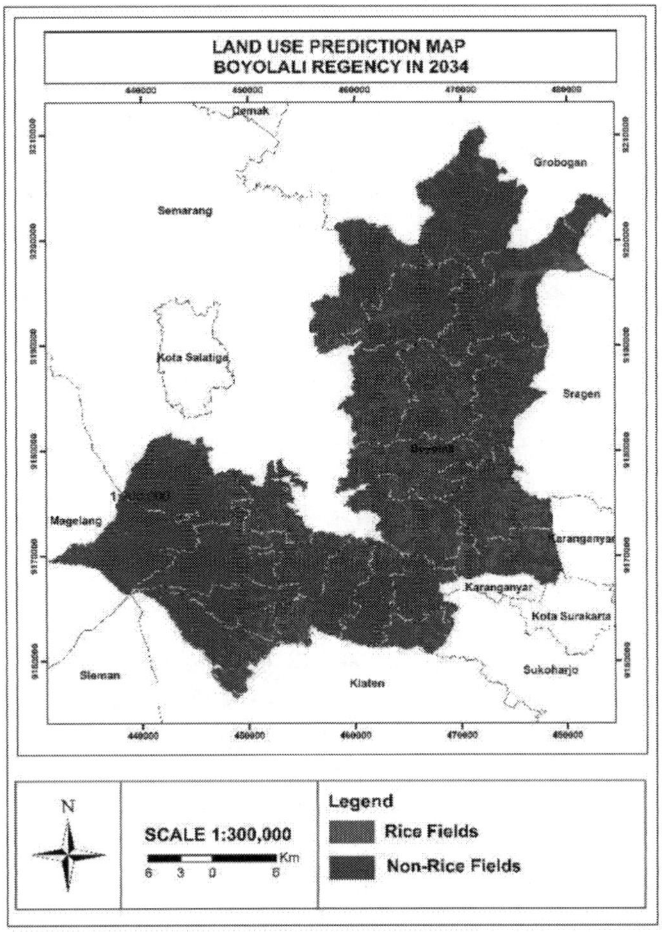

Fig. 4. Map of Predicted Land Use Change of Boyolali Regency in 2034

Based on the land use prediction map for 2034, it can be seen that the trend of conversion of paddy fields to non paddy fields is expected to continue significantly. The map shows that brown areas representing non-field land are increasingly dominating, especially in the western, southern, and eastern parts of Boyolali Regency. In contrast, the presence of paddy fields (green color) is increasingly fragmented and limited to a few sub-districts, especially in the northern and central regions.

Sub-districts such as Ngemplak, Nogosari, Mojosongo, Banyudono, Teras, and Sawit, which had previously shown a conversion trend in 2024, are predicted to lose most of their land function as paddy fields by 2034. This condition is caused by the expansion of residential areas, economic growth, and the pressure of intensive infrastructure development. The southwestern region that borders Magelang and Sleman regencies also shows a similar pattern of land change, along with the development of inter-regional connectivity.

Meanwhile, sub-districts such as Selo, Musuk, and Tamansari still retain most of their paddy fields, although the area is also shrinking. Geophysical factors such as topography and slope, as well as limited accessibility, are thought to be responsible for the low pressure of conversion in these areas. On the other hand, Wonosegoro and Wonosamodro, which are located in the north, still show the dominance of paddy fields, although it is possible to experience pressure in the future if there is no control of spatial utilization.

From a policy perspective, the results of this study emphasize the urgent need to implement Sustainable Food Agricultural Land (LP2B) in accordance with Law No. 41 of 2009. Given the significant projected changes in land use, the Boyolali Regional

Government needs to strengthen regulations protecting productive agricultural land by establishing LP2B zoning in the Regional Spatial Plan (RTRW). The implementation of LP2B will not only protect farmers from losing their livelihoods, but also ensure the sustainability of food production and support the achievement of sustainable development goals (SDG 2: Zero Hunger).

The phenomenon of agricultural land use change in Boyolali is in line with research findings in Buriram Province, Thailand. Research by Sangpradid (2023) shows that the use of Cellular Automata–Artificial Neural Network (CA-ANN) is effective in recognizing patterns of land use shift from agricultural to non-agricultural land due to residential expansion and infrastructure development [13]. These findings indicate that changes in agricultural land use are not only a local issue but also a regional challenge in Southeast Asia, with direct impacts on food security and sustainable spatial management.

Table 6. Predicted Area of Land Use Change in 2034

Land Use	**Area (ha)**	**Area (%)**
Rice Field	36818.45	33.60%
Non-Sawah	72774.03	66.40%

Table 6. shows that the rice field area is predicted to be 36,818.45 hectares, while non-field land reaches 72,774.03 hectares. Spatially, the map shows the expansion of non-field areas especially in the southern and central parts of the region, which indicates the land conversion towards non-agricultural uses such as settlements or infrastructure [14]. This data indicates an increasing trend of land conversion from paddy field to non- farm use in the future, potentially reducing the availability of agricultural land in the future, which could have an impact on food security and the welfare of rural communities.

4 Conclusion

Based on the results of the land change analysis and predictive modeling conducted, there are several important things related to land use change in Boyolali Regency, especially in the context of land conversion from paddy fields to non-fields. Land use change in 2014-2024 showed a significant decrease in the area of paddy fields by 20.26%, while non-field land increased by 16.15%. This reflects the pressure of regional development, urbanization, and land use change influenced by the strategic location of Boyolali Regency as a buffer zone for major cities such as Surakarta, Semarang, and Yogyakarta. Predictions of Land Use Change in 2034 through the Cellular Automata-based model and the influence of driving factors (roads, rivers, and settlements) show that non-rice field land will continue to grow, estimated to reach 72,774.03 hectares, while rice fields will only remain around 36,818.45 hectares. The strong positive correlation between settlements and roads (Pearson value 0.762) indicates that the presence of infrastructure is the main driver of built-up land expansion.

Acknowledgement

The author would like to express sincere gratitude to all parties who have helped and supported this research. Thanks are also extended to the Department of Geography at Muhammadiyah University Surakarta for the financial support provided, which enabled the author to publish this research. The author would also like to thank the supervisor who provided guidance and valuable information throughout the research process. The author would also like to thank colleagues who provided technical and moral support, both of which were crucial to the success of this research. Gratitude is also extended to the author's parents and family for their encouragement, hopes, and prayers, which have always been a source of strength. Finally, the author extends gratitude to all parties who have assisted and supported this research, both directly and indirectly.

References

1. T. Sarastika, Yusuf Susena, and Dwi Kurniawan, "Prediksi Konversi Lahan Pertanian Berbasis Artificial Neural Network-Cellular Automata (Ann-Ca) Di Kawasan Sleman Barat," *J. Tanah dan Sumberd. Lahan*, vol. **10**, no. 2, pp. 471–482, 2023, doi: 10.21776/ub.jtsl.2023.010.2.30.
2. E. F. Lambin *et al.*, "The causes of land-use and land-cover change: Moving beyond the myths," *Glob. Environ. Chang.*, vol. **11**, no. 4, pp. 261–269, 2001, doi: 10.1016/S0959-3780(01)00007-3.
3. *Badan Pusat Statistik Kabupaten Boyolali, 2023*. Kabupaten Boyolali.
4. M. Hartanto and A. A. Wibowo, "Analisis Perubahan Penutupan Lahan Periode Tahun 2014-2022 Serta Pengelolaannya Di Sub DAS Wiroko," *Dr. Diss. Univ. Muhammadiyah Surakarta*, 2024.
5. *Badan Pusat Statistik Kabupaten Boyolali, 2025*. Kabupaten Boyolali.
6. F. Muwardi and A. Fadlil, "Sistem Pengenalan Bunga Berbasis Pengolahan Citra dan Pengklasifikasi Jarak," *J. Ilm. Tek. Elektro Komput. dan Inform.*, vol. **3**, no. 2, p. 124, 2018, doi: 10.26555/jiteki.v3i2.7470.
7. D. Setiady and P. Danoedoro, "Prediksi perubahan lahan pertanian sawah sebagian Kabupaten Klaten dan sekitarnya menggunakan cellular automata dan data penginderaan jauh," *J. Bumi Indones.*, vol. **5**, no. 1, p. 79437, 2016.
8. E. Hariyati, "Prediksi Perubahan Tutupan/Penggunaan Lahan Tahun 2036 Di Kecamatan Kotabumi Utara Menggunakan Model Cellular Automata," *Univ. Lampung*, no. February, pp. 4–6, 2024.
9. V. N. Fikriyah, D. Danardono, M. I. T. Sunariya, M. Cholil, T. A. Hafid, and M. I. Ismail, "Spatio-temporal analysis of built-up area and land surface temperature in Surakarta using Landsat imageries," *Sustinere J. Environ. Sustain.*, vol. **6**, no. 2, pp. 92–101, 2022, doi: 10.22515/sustinerejes.v6i2.187.
10. Y. S. Dianti, "Analisis Konversi Lahan Pertanian Menjadi Perumahan Di Kecamatan Pohjentrek Kabupaten Pasuruan," *Swara Bhumi*, vol. **5**, no. 8, pp. 1–6, 2020.
11. S. Alhamda, H. Z. Hadibasyir, and M. GIS, "Analisis Perubahan Penggunaan Lahan Sawah dengan Daya Dukung Pangan Menggunakan Citra Sentinel 2A di Kabupaten Boyolali Tahun 2015 dan 2023," *Dr. Diss. Univ. Muhammadiyah Surakarta*, 2023.
12. M. Belihu, S. Tekleab, B. Abate, and W. Bewket, "Hydrologic response to land use land cover change in the Upper Gidabo Watershed, Rift Valley Lakes Basin, Ethiopia," *HydroResearch*, vol. **3**, pp. 85–94, 2020, doi: 10.1016/j.hydres.2020.07.001.
13. S. Sangpradid, "APPLICATION OF A MULTI-LAYER PERCEPTRON NEURAL NETWORK TO SIMULATE SPATIAL-TEMPORAL LAND USE AND LAND COVER CHANGE ANALYSIS BASED ON CELLULAR AUTOMATA IN BURIRAM PROVINCE, THAILAND," *Environ. Eng. Manag. J.*, vol. **22**, no. 5, pp. 917–931, 2023, doi: 10.30638/eemj.2023.074.
14. D. N. S. Putri Goeratanenq Sukmasari, "Analisis Kesesuaian Penggunaan Lahan

Terhadap Rencana Pola Ruang Tahun 2023-2043 Di Kawasan Perkotaa," *Dr. Diss. (Universitas Muhammadiyah Surakarta)*, pp. 1–10, 2024.

Evaluation of Drainage System at State University of Malang Campus 1

Vita Ayu Kusuma Dewi[1], *Mega Septia Sarda* Dewi[2], *Muhammad Abdul* Rahman[3], *Anie* Yulistyorini[4], *Syakira Salamah Kania* Mazuida[5]

[1,2,3,4,5]Department of Civil Engineering and Planning, Universitas Negeri Malang, Indonesia

Abstract. Rapid development at the State University of Malang (UM) has led to a reduction in water absorption areas, significantly increasing the risk of flooding. This issue frequently arises during heavy rainfall, exacerbated by suboptimal drainage channels and limited holding capacity, consequently disrupting campus mobility and academic activities. This research evaluates the holding capacity of existing drainage channels at UM using a quantitative descriptive method. Hydrological and hydraulic analyses indicate that, theoretically, the channels should be able to accommodate a design flood discharge of 11.9 m^3/second. However, field observations reveal that flooding persists due to an ineffective drainage system, characterized by a lack of inlets, damaged sections, and sedimentation. This study serves as a crucial foundation for UM to develop comprehensive and sustainable drainage solutions, ensuring uninterrupted campus operations amidst ongoing development.

1 Introduction

Malang is the second largest city in East Java Province. This status makes Malang a center for various fields, one of which is education. Malang is also known as the city of education because it is a popular choice for continuing studies and has many higher education institutions [1]. This will certainly have an impact on the population growth in Malang. According to data from the Central Statistics Agency in 2022, the population of Malang City was 846,126 with an annual growth rate of 0.14% [2]. Data on migrant residents also indicates that there are 330,000 students pursuing their education in Malang City. This will undoubtedly lead to changes in land use within the education sector, particularly in the campus area of Malang State University.

* Corresponding author: vita.ayu.kusuma.ft@um.ac.id •

https://doi.org/10.1051/e3sconf/202565211001

Malang State University (UM) is a higher education institution that has undergone continuous infrastructure development over the years, resulting in a reduction in water absorption areas. The reduction in water absorption areas will increase the amount of water runoff in the university area. UM is located in an area with high rainfall intensity and a high risk of flooding. Andrianto et al [3] argue that flooding can be caused by heavy rainfall, excessive water overflow, excessive water runoff, and poor drainage systems. A common problem during rainfall on this campus is excessive water runoff, leading to flooding everywhere. This issue is further exacerbated by other causes of flooding at UM, including inadequate drainage channels that cannot accommodate water and an insufficient drainage system on campus.

A drainage system is a water structure that serves to drain or divert water by removing excess water runoff from an area so that it can function optimally [4]. A university drainage system is a series of water structures that serve to reduce and remove excess water caused by high rainfall to water disposal areas such as rivers, lakes, and main channels in the campus area [5]. A university drainage system is a basic infrastructure requirement for the academic community at Malang State University.

Recognizing the importance of drainage, it is essential that its planning be well-structured and organized. The existing drainage system is crucial for accessibility and comfort within the campus area, as many areas are prone to flooding during heavy rains at this university [6]. The locations at Malang State University that frequently flood during rain are the Joint Lecture Building (GKB) A19-A20 and the parking lot. The impact of these floods is the disruption of mobility during academic activities. Drainage at Malang State University is a top priority because many buildings and other facilities influence water runoff. The performance of drainage systems from buildings to the connecting drainage network must be a key focus to ensure a safe and comfortable environment for the academic community at UM.

In the Malang State University area, the drainage system used is an open drainage system consisting of secondary channels that flow into primary/main channels, namely rivers [7]. In the UM area, secondary channels surround faculty buildings, mosques, lakes, and other structures. The condition of the drainage channels is partially no longer up to standard due to insufficient maintenance, repairs, and evaluations. Therefore, based on the above description, it is necessary to conduct a study on the evaluation of the drainage network's storage capacity in the University of Malang area. The objective of this study is to evaluate the storage capacity of the drainage network's "   " in the University of Malang using hydrological and hydraulic data analysis.

2 Method

The method used in this study is quantitative descriptive research. This study was conducted in the Campus I area of Malang State University on Jalan Semarang No. 5, Lowokwaru District, Malang City. Drainage storage capacity evaluation is a method used to determine the storage discharge condition based on existing field conditions [8]. Hydrological analysis is used to determine the design rainfall and flood discharge analysis. Hydraulic analysis is required to determine the condition, measurement, and analysis of the drainage channel's storage capacity [9]. The first step is surveying and data collection. Surveys are conducted to determine the actual conditions at the research site [10]. Surveys are carried out by obtaining rainfall data, existing topographic data of the channels, channel cross-section data, and existing dimensions of the drainage channels at Malang State University.

The second stage is hydrological and hydraulic analysis. Hydrological analysis includes rainfall data and determination of design rainfall. From these two aspects, data consistency tests will be conducted, rainfall calculations using Pearson's log and Gumbel methods, and data distribution using chi-square, Smirnov, and Kormakurof tests. Hydraulic analysis involves determining the drainage system's storage capacity.

The third stage involves calculating design flow data and generating output with an evaluation of the drainage channel's storage capacity. The evaluation is based on Q existing > Q planned. The final stage is drawing conclusions and providing recommendations regarding the evaluation of the drainage channel's storage capacity.

3 Results and Discussion

3.1 Design Flood Flow and Runoff Volume

Calculation of rainfall data analysis to obtain planned rainfall intensity. The planned rainfall used is the maximum rainfall data. The rainfall data used is 10 years of rainfall data (2012–2021) obtained from the nearest rainfall stations, namely Dau, Sukun, and Ciliwung. The maximum rainfall data is then analyzed into regional rainfall. The regional rainfall results are shown in the table 1.

Table 1. Regional rainfall

Year	Station Ciliwung	Sukun Station	Dau Station	Average Maximum Rainfall in the Region
	(mm)	(mm)	(mm)	(mm)
2012	138	125	97	138
2013	93	101	85	93
2014	125	134	100	125
2015	96	170	65	96
2016	64	122	75	64
2017	104	132	105	104
2018	97	94	95	97
2019	82	135	93	82
2020	97	144	135	97
2021	123	133	95	123

The data was then statistically analyzed to obtain the planned rainfall for the return period. Several factors that influence flood discharge values include planned rainfall, land cover type coefficient, rainfall intensity, and catchment area.

Based on the analysis of planned rainfall calculations, the planned rainfall heights for return periods of 2, 5, 10, 25, 50, 100, and 1000 years are presented in the table 2.

https://doi.org/10.1051/e3sconf/202565211001

Table 2. Design Rainfall

Return Period	Design Rainfall (mm)
	98.9
5	125
10	142.2
25	164.0
50	180.1
100	196.2

Based on the rainfall analysis of the design, the rainfall intensity values were obtained with the Figure 1.

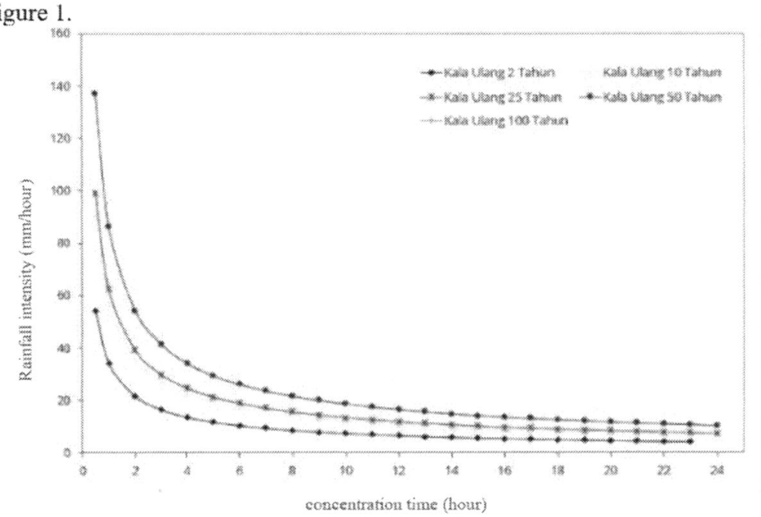

Fig. 1. Rainfall Intensity Graph

If the runoff coefficient C of 0.95 is taken from the 2-year design rainfall of 98.9 mm, the flood discharge for the entire UM area (463,992 m²) is 11.9 m³/s. The concentration time from the calculation is 0.209, resulting in a total runoff volume for the entire UM area of 8,953.6 m³.

3.2 Drainage Channel Network in UM

Drainage is a construction designed to reduce and remove excess water in an area so that the land can function optimally. It also serves to extend the economic life of physical facilities such as roads and conserve water resources. Currently, there are several drainage system infrastructures in the UM area, including open drainage channels, closed drainage channels, and water retention ponds (trapezoidal ponds).

At present, rainwater is not managed optimally and facilities that can be used to reduce the rate or volume of rainwater runoff, including facilities for treating rainwater on the UM campus, are still very limited. Rainwater falling on impermeable surfaces (parking lots, pathways, and campus roads) flows into rainwater runoff channels located around buildings that are directly connected to the main drainage system [11]. The drainage system at the UM campus is generally an open drainage channel system, with some channels being closed. Existing examples of open drainage channels for rainwater runoff on the UM campus are shown in Figure 2.

https://doi.org/10.1051/e3sconf/202565211001

Fig. 2. Open drainage channel at UM

In the initial stage, a survey was conducted that included measurements of elevation differences, cross-sectional (cross section) and longitudinal (long section) profiles of drainage channels and rivers on the UM campus. The overall mapping of existing drainage channels on the UM campus is presented in Figure 3 below. Based on field observations during the rainy season, the direction of rainwater flow generally flows toward the southern part of the campus.

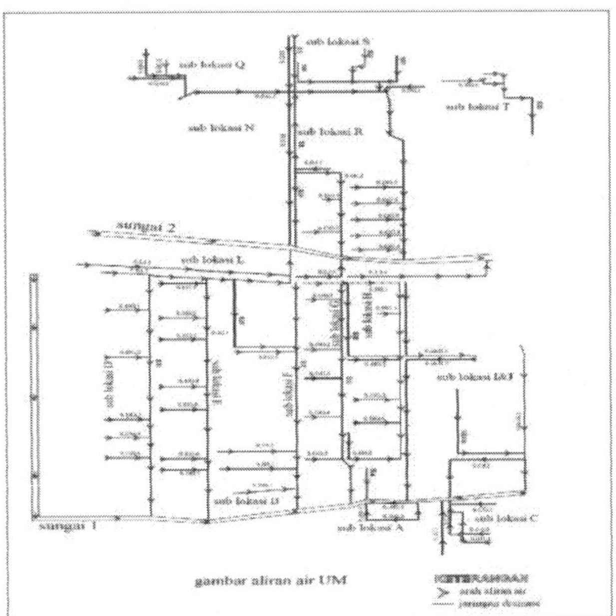

Fig. 3. Drainage Channel Network Map at UM

The results of the survey and channel capacity measurements are presented in the following table.

Table 3. Drainage channel capacity at UM.

Drainage Channel Section Name	Storage Capacity (m^3)
GKB Main River	5665.1
Behind FIP Canteen - D19 Parking Lot	4
FPsi - Mosque	330.75
FIK- Ambarawa	112.75
KPRI - FIS Canteen	5665.19
KPRI - FIS Canteen	72.5
Graduate School Front	240.12
Front of A19	371.40
FIP	140.39
Side FIS - FT	82.11
Rectorate	110.80
FIP Parking Lot	87.80
Next to the pool toward the river	11.19
In front of FIP - D19	118.20
Front channel of FPsi - FT	41.62
Rear bus parking lot toward SMP Lab	19.37
Trapezoidal Stage - FK	406.31
Behind the security post - monument	51.14
Tulip Dormitory Channel (south)	57.74
Cakrawala Street, west side	90
Behind Pasca - Sakri	778.92
Semarang Street Wall	234.55
Right side of the Rectorate, north side	94.91
Left side of Lili Dormitory - Library	108.07
Surabaya Gate, east side	17.87
Right Kopma	7.45
UM Lab Elementary School - UM Lab Middle School	36.30
Cakrawala Street, east side	80.27
East side of the abandoned building	4.57
FMIPA Branch - Cakrawala Street	99.40
Front of UPT PTIK	90.2
In front of KPRI UM	34.0
Lili Dormitory (south)	46.33
Lili Dormitory (North)	24.00
Tulip Dormitory (North)	8.24
Surabaya Gate West Side	5.89
Total storage capacity	10,167.53

The total storage capacity at UM is 10,167.53 m3. The flood discharge results were converted into discharge with a concentration time (Tc) of 0.204, resulting in a total overflow volume of 8,953.56. Based on the flood discharge calculations and storage capacity in each drainage section, the drainage channels are still capable of accommodating the flood discharge. However, based on field surveys, flooding occurred because water could not enter the drainage channels, one of the causes being the absence of an inlet pathway. There are several sections of the drainage channels that are cracked and damaged.

Addressing campus flooding requires a multi-faceted approach. Effective recommendations include implementing permeable paving and green infrastructure like rain gardens to increase water absorption and reduce surface runoff, upgrading and regularly maintaining existing drainage systems (clearing debris from storm drains and culverts) to ensure efficient water flow, and establishing a comprehensive early warning system and emergency response plan to prepare the campus community for flood events [12].

4 Conclusion and Recommendations

Based on the evaluation and analysis conducted, several conclusions were drawn. First, in terms of calculation, the existing drainage channel capacity at Malang State University is still capable of accommodating the planned flood discharge. However, in practice, flooding still occurs at several points due to the suboptimal drainage system, such as damaged channels, lack of water inlets, and sedimentation that obstructs the flow.

Based on these findings, the recommendations are to improve the drainage channel inlets, add infiltration wells or retention ponds, and repair the road pavement.

References

1. S. A. Dinarusta, Studi Deskriptif Tentang Pemilihan Tempat Nongkrong Mahasiswa Kota Malang," Perpustakaan Universtitas Airlangga, pp. 1–8. (2019).
2. Badan Pusat Statistik Kota Malang, Kota Malang dalam Angka 2023. (2023).
3. Ardianto, A. Junaidi, and B. Budi Irawan, Evaluasi Jaringan Drainase Kampus Universitas Dharma Andalas Menggunakan Software Storm Water Management Model," vol. 01, no. 01, pp. 13–25, Jun. (2021) doi: https://doi.org/10.47233/rivet.v1i01.230.
4. F. Ekananda, N. H. Pandja, and M. I. Rau, Evaluasi Saluran Drainase di Perumahan Alam Sinar Sari Kabupaten Bogor Jawa Barat," Jurnal Teknik Sipil dan Lingkungan, (2019) doi: https://doi.org/10.29244/jsil.4.3.219-232.
5. J.I., Sianipar, Evaluasi Sistem Drainase di Universitas Medan Area (Doctoral dissertation, Universitas Medan). (2022)
6. L. Lindawati, P. Irawan, and R. Nursani, Evaluasi Sistem Drainase Dalam Upaya Penggulangan Banjir Di Jalan A.H Nasution Kota Tasikmalaya Menggunakan Program EPA SWMM 5.1. Jurnal Siliwangi Seri Sains dan Teknologi, vol. 7, no. 2, pp. 41–51, (2021)
7. M. R. Mahendra, Aplikasi EPA SWMM 5.1 pada Perencanaan Sistem Drainase Kawasan Perumahan Aura Park Kecamatan Dau Kabupaten Malang, Jurnal Teknologi Sumber Daya Air, vol. 04, no. 01, p. 460, (2024)
8. H. Suryaman, Evaluasi Sistem Drainase Kecamatan Ponorogo Kabupaten Ponorogo. Jurnal Kajian Pendidikan Teknik Bangunan, 2(1/JKPTB/13). (2013)
9. R. Mantasa, S. Prastica, C. Maitri, C. Nugroho, and A. Hermawan, Analisis Banjir dan Perencanaan Desain Transportasi Sungai di Kota Bojonegoro, vol. 23, no. 2, pp. 91–101. (2017)

10. S. N. I, Sari. Perencanaan Drainase Perkotaan Wilayah Rawa Dengan Pendekatan Drainase Berkelanjutan Di Merauke. Journal of Scientech Research and Development, 6(1), 1108-1115. (2024).
11. M. A. Rahman, A. Sulistyorini, T. Rahayuningsih, E. Setyawan, V.A.K. Dewi, M.S.S. Dewi, Utilization of Rainwater Puddles at Campus I State University of Malang. In E3S Web of Conferences (Vol. 445, p. 01021). EDP Sciences. (2023).
12. Choi, E., & Kim. Sustainable campus flood management: A review of green infrastructure and water sensitive urban design strategies. Journal of Environmental Management, 265, 110467. (2020). doi: 10.1016/j.jenvman110467.

A Bibliometric Analysis of Reclaimed Asphalt Pavement for Green Infrastructure

Nuriyana Muthia Sani[1*], *Heri Sutanto*[2], *Bagus Hario Setiadji*[3]

[1]Doctoral Program of Environmental Science, School of Postgraduate Studies, Universitas Diponegoro, Indonesia
[2]Department of Physics, Faculty of Science and Mathematics, Universitas Diponegoro, Indonesia
[3]Department of Civil Engineering, Faculty of Engineering, Universitas Diponegoro, Indonesia

Abstract. Reclaimed Asphalt Pavement (RAP) has emerged as a strategic component in the development of sustainable and environmentally responsible infrastructure systems. As the global construction industry seeks innovative approaches to reduce environmental impacts and resource consumption, the reuse and recycling of asphalt materials have attracted growing academic and practical interest. This study aims to map the literature on RAP from 2015 to 2024 and to identify potential future research directions, particularly regarding its integration within green infrastructure initiatives. The bibliometric analysis was conducted using metadata obtained from the Scopus database and examined with the aid of the VOSviewer tool and RStudio through its Bibliometric package via the Biblioshiny web-based interface. The findings highlight key contributors in the field, including the most productive authors, influential journals, prominent institutions, and leading countries. The study also explores the network of co-authorship and keyword co-occurrence, revealing dominant research themes. While strong linkages were found between RAP and topics such as asphalt pavements, asphalt mixtures, and recycling, emerging areas—such as nature-based solutions, bio-based rejuvenators, and biomaterial by-products—remain underexplored and present valuable opportunities for advancing sustainable infrastructure research.

1 Introduction

For centuries, asphalt has been crucial for road construction and waterproofing due to its recyclability, strength, and adaptability. Modern asphalt technology now focuses on enhancing its sustainability and minimizing environmental impact, particularly in the context of green infrastructure development [1]. However, asphalt pavements face numerous challenges throughout their service life, including heat, fatigue, and cracking. Heavy traffic causes rutting and accelerated wear, while aging leads to raveling, and water/freeze-thaw cycles result in potholes [2]. Prompt intervention is essential to maintain pavement strength and effectiveness [3].

https://doi.org/10.1051/e3sconf/202565211002

* Corresponding author: nuriyanamuthia@gmail.com

As a petroleum derivative, asphalt's availability is diminishing, and its production is energy-intensive, contributing significantly to greenhouse gas emissions [4]. Consequently, identifying renewable alternatives and maximizing the recycling of old asphalt materials are key priorities for green infrastructure. Recycling Reclaimed Asphalt Pavement (RAP) has become a global standard for improving energy efficiency, economic performance, and environmental impact [5]. There's growing interest in using RAP and other recycled materials to develop more sustainable and cost-effective asphalt mixes for green infrastructure [6]. RAP's end-of-life material holds significant recycling potential due to its valuable asphalt binder and aggregates [7]. Rising aggregate costs and strict sustainable development policies underscore the importance of effective technology for regenerating high-content RAP for green infrastructure [8].

Significant experience has enabled recycling ratios approaching 100%. RAP offers substantial benefits for green infrastructure: 30% content optimizes blending, while 60% improves resistance to permanent deformation at lower stress levels [9], reducing annual lifecycle costs by 10% and carbon emissions by 30% [10]. Using 100% RAP can even save 50% in material costs compared to virgin mixes [11]. These advantages include financial gains, enhanced deformation stability, minimized non-renewable resource use, and reduced emissions with increased RAP content [12]. Life Cycle Assessment (LCA) and Life Cycle Cost Analysis (LCCA) quantify these environmental and economic impacts in green infrastructure applications [13]. For example, 40% RAP can reduce road pavement net present value costs by 18.3% (15-25% overall savings) [14] and cut life cycle costs by up to 26.2% and GHG emissions by up to 29% [13].

Effective reuse of RAP necessitates rejuvenating its asphalt component, which can yield mixtures comparable to or superior to fresh asphalt for green infrastructure. Various effective rejuvenators, including corn, waste engine, cashew nutshell, and soybean oils, restore aged asphalt properties and enhance performance [15]. While high dosages, often exceeding 10%, are typically required, these rejuvenators significantly improve mixture performance, such as enhancing fatigue cracking resistance in mixes with up to 30% RAP [6]. It can also reduce the viscosity of 100% RAP binder, restoring original aged binder characteristics for adequate mixture performance. However, blending hard, brittle aged RAP binder with virgin bitumen results in a stiffer final binder, negatively impacting workability, and increasing cracking susceptibility [7]. This stiffness stems from the deterioration of aged asphalt's viscoelastic properties, as volatile maltenes are lost or converted to less volatile asphaltenes, making the pavement stiffer, more brittle, and prone to cracking [8].

Amidst natural resource limitations and environmental impacts from the construction industry, sustainable development and improving road infrastructure quality are paramount for green infrastructure. This involves selecting high-quality materials, implementing appropriate construction technology, and planning for sustainable maintenance, while also integrating environmental considerations into construction and repair processes to enhance sustainability and cost efficiency. Strategies like asphalt modification with additives and RAP reuse with rejuvenators aim to produce high-quality, load-resistant, and age-inhibited asphalt suitable for green infrastructure applications.

Given the advancements and challenges in sustainable asphalt, this research employs bibliometric analysis to examine the current state and offer recommendations for Reclaimed Asphalt Pavement (RAP) in green infrastructure. The basis for the study needs strengthening by directly connecting the environmental issues of conventional asphalt to the necessity of a strategic bibliometric analysis. This analysis is crucial not just as a review, but as a mechanism to identify key research trends and uncover critical knowledge gaps that must be addressed to overcome technical obstacles and ensure the broad adoption of RAP as a viable green infrastructure solution. Utilizing Scopus data from 2015 to 2024, this study aims to map the landscape of sustainable RAP for green infrastructure. Specifically, current rejuvenation trends, along with publication patterns, prominent contributors (countries, sources, authors), and key research themes, are aimed to be illuminated by this study.

2 Research Methodology

This research employs scientific mapping (bibliographic analysis), a quantitative method assessing publication value through abstract terms and author-assigned keywords based on co-occurrence and co-citation [1]. The data for this research were collected from Scopus, which is considered a representative source due to its extensive database and robust metadata, covering a wide range of literature, including journals, articles, books, and conference papers. The selection criteria (e.g., author, title, publication year) significantly influence the final data and study outcomes, necessitating strong justification for search parameters [2].

The methodology aligns with the Preferred Reporting Items for Systematic Reviews and Meta-Analyses (PRISMA) 2020 flow diagram (Figure 1), ensuring transparency and reliability [2]. The three-step PRISMA process involved (i) identification, a total of 3,720 records were identified using "reclaimed" AND "asphalt" AND "pavement" keywords from Scopus. In other method, an additional 632 records were found using "reclaimed" AND "asphalt" AND "pavement" AND "rejuvenator" keywords from Scopus. After removing 21 irrelevant records, 3,699 records remained. (ii) Screening, the combined 3,699 records, 39 were excluded from the database results. Of the 632 records from other methods, 16 were not retrieved due to irrelevance. Eligibility, from 3,660 database reports, 920 were excluded for being outside the 2015-2024 timeframe, 230 for being non-article/non-conference types, and 83 for being non-English. Similarly, from 616 other-method reports, 81 were excluded by publication year, 59 by document type, and 10 by language. (iii) Included, a total of 2,427 studies from the database and 463 from other methods. The datasets, each exported as a CSV file, were analyzed using VOSviewer and RStudio (Biblioshiny interface). VOSviewer was used to visualize keyword co-occurrence and distribution with its clustering and mapping algorithms, while Biblioshiny analyzed publication patterns, prominent countries, and research themes using standard statistical methods. To ensure data consistency and robustness, a sensitivity analysis was performed by adjusting keyword occurrence thresholds. Additionally, a manual check was conducted in both software platforms to standardize synonymous and duplicate terms using their respective thesaurus and synonym functions.

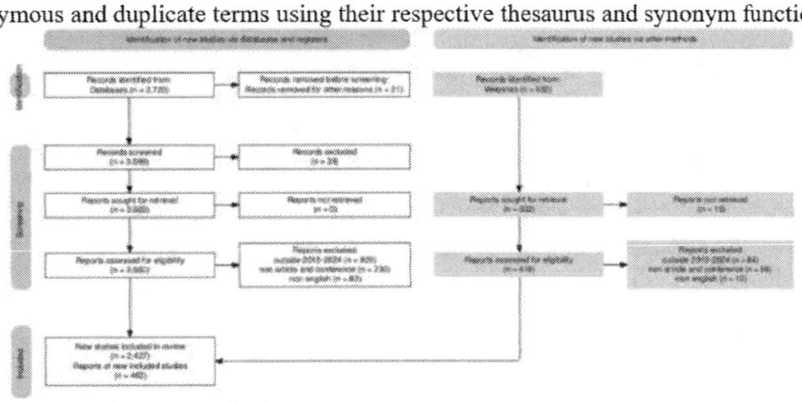

Fig. 1. PRISMA flow diagram [22].

3 Results and Discussion

https://doi.org/10.1051/e3sconf/202565211002

This research systematically gathered 2,427 articles from Scopus and 463 additional reports, all eligible for review, using the PRISMA 2020 method. A quantitative bibliometric analysis was then performed on this refined dataset, examining factors like publication year, document type, country of origin, journals, authorship, and keyword occurrences. This analysis was facilitated by graphical outputs from VOSviewer, Biblioshiny, and Microsoft Excel, aligning with common practices in quantitative data articulation [3]. Detailed findings are presented in the following sub-sections.

3.1 Analysis based on publication year

Research on Reclaimed Asphalt Pavement (RAP) saw a significant and consistent increase in publications from 2015 to 2024, totaling 2,427 articles. As shown in Figure 2, annual publications generally rose from 99 in 2015 to a peak of 370 in 2024, reflecting growing academic interest and the escalating global commitment to the United Nations' Sustainable Development Goals (SDGs), particularly those related to responsible consumption and production, has placed pressure on the construction industry to adopt more sustainable practices. Rising costs and limited availability of virgin materials (e.g., aggregates and bitumen) make the reuse of RAP an economically attractive solution.

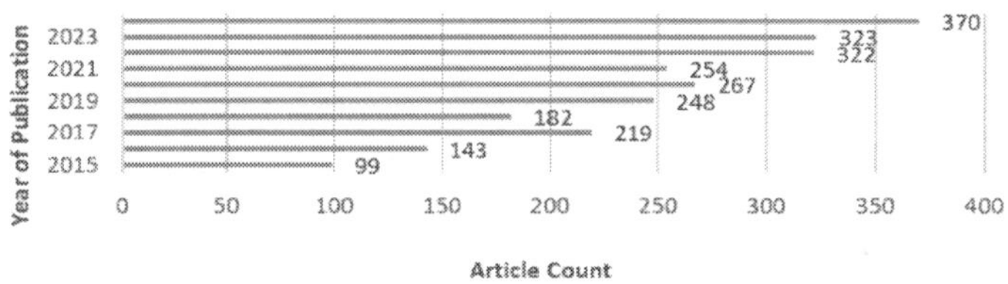

Fig. 2. Article distribution of research in RAP.

3.2 Document type analysis

Figures 3 illustrate the publication breakdown in RAP research from 2015 to 2024. Journal articles dominate, comprising 79% of publications, while conference proceedings account for 21%. Both types show a general upward trend. Journal articles grew from about 60 in 2015 to an estimated 280-300 in 2024, with conference proceedings also increasing from approximately 35 in 2015 to a consistent 60-80 papers between 2021 and 2024. This highlights journal articles as the primary dissemination channel for disseminating established research and in-depth findings, with conference contributions steadily adding to the field's knowledge and preliminary results. While the trend is generally upward, there is a slight dip in both article and conference publications in some years (e.g., around 2020). This anomaly, caused by the global pandemic in 2020-2021, likely impacted research activities, including laboratory work, fieldwork, and conference schedules, which would explain a temporary slowdown in publication output.

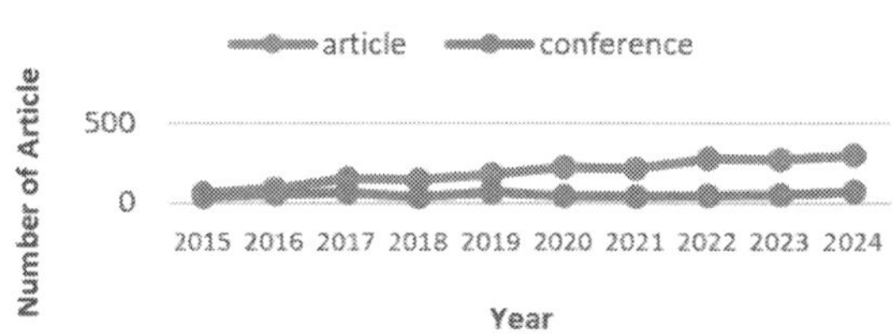

Fig. 3. Annual distribution of RAP journal articles and conference proceedings.

 https://doi.org/10.1051/e3sconf/202565211002

3.3 Country contribution analysis

Authors from 90 countries contributed to RAP research, with 20 countries publishing over 25 articles. As Figure 4 illustrates, the United States and China lead with over 400 publications each, predominantly through multi-country collaborations (MCPs), though both also have a notable number of single-country publications (SCPs). India follows with over 200 documents, also showing a strong inclination towards MCPs.

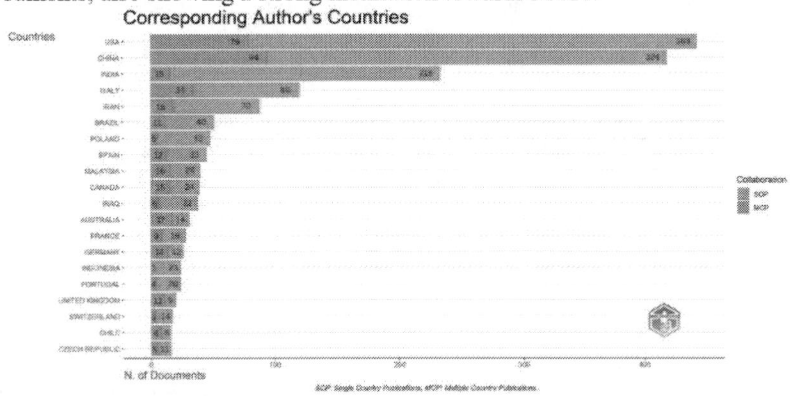

Fig. 4. Corresponding author's countries in RAP.

Beyond the leading contributors, countries like Italy, Iran, Brazil, Poland, and Spain each provided approximately 50 to 150 RAP documents, predominantly through multi-country collaborations (MCPs). Further down the list, nations including Malaysia, Canada, Australia, France, Germany, and the UK had fewer than 50 publications, yet MCPs still constituted a notable portion of their output. This consistent pattern underscores the global distribution of RAP research and the widespread engagement in international partnerships.

3.4 Analysis based on source

Figures 5a and 5b highlight the most productive and impactful sources in RAP research. "Construction and Building Materials" leads RAP research in both productivity (431 articles) and impact (H-index 61). Other key journals include "Journal of Materials in Civil Engineering" (152 articles, H-index 30), "Transportation Research Record" (131 articles), "Road Materials and Pavement Design" (121 articles, H-index 31), and "Journal of Cleaner Production" (96 articles, H-index 45). The data reveal that a high article volume does not guarantee a proportionally high H-index, demonstrating that impact (citations) is not always linear with productivity. This comparison illustrates that while a journal may produce a higher volume of articles, its H-index may not be proportionally high if those articles are not

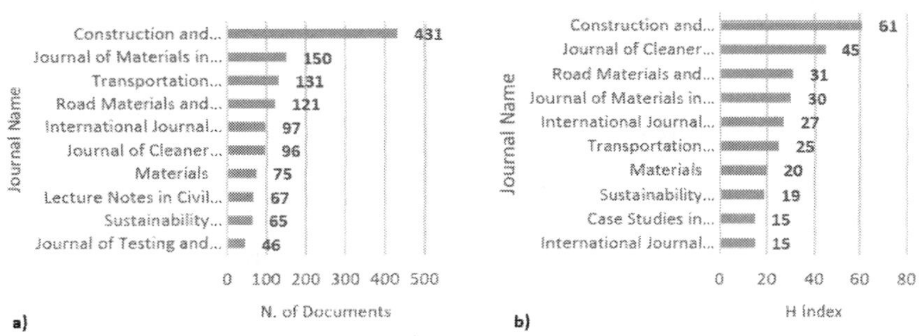

consistently cited at a high rate. Conversely, a journal with fewer, but more impactful, highly-cited articles can achieve a relatively higher H-index.

Fig. 5.a) Most relevant sources in RAP; **b)** Sources' local impact by H index in RAP.

3.5 Analysis based on authorship

Figures 6a and 6b offer contrasting views of author contributions in RAP research. Figure 6a identifies Zhang Y. as the most prolific author (45 documents), followed by Arulrajah A., Li X., and Xiao F. (28 each), and Canestrari F. (27). Conversely, Figure 6b, which uses the H-index to measure impact, ranks Singh S. highest (H-index 18), with Arulrajah A. (17) close behind. Canestrari F., Ma T., Maumanis M., and Zhang F. all share an H-index of 16, while Xiao F., Lo Presti D., Wu S., and Yang J. are at 15 and 14, respectively. This comparison demonstrates that high publication volume does not always correlate with the highest impact in the field.

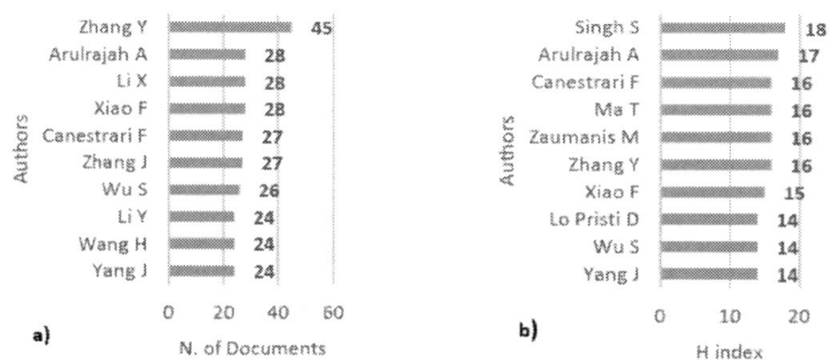

Fig. 6.a) Most relevant authors in RAP; **b)** Authors' local impact by H index in RAP.

3.6 Analysis based on keyword

Key insights into Reclaimed Asphalt Pavement (RAP) research come from keyword analysis. Table 1 shows "Reclaimed Asphalt Pavement," "asphalt pavements," and "asphalt mixtures" as the most frequent terms, underscoring RAP's core application and composition. Figure 7 visually and categorize keywords into four clusters based on each top three keywords: (i) sustainable pavement engineering and construction, with a focus on circular economy principles in road infrastructure ("reclaimed asphalt pavement," asphalt pavement," and "asphalt"), (ii) recycled asphalt mixtures and their properties/performance ("asphalt mixture," "recycling," and "mixture"), (iii) binder modification and performance in pavement applications, especially concerning material compatibility and mix design ("binders," "asphalt binders," and "blending"), and (iv) optimizing hot mix asphalt performance through additive technologies ("performance," "additives," and "hot mix asphalt").

Table 1. The highest-occurring ten keywords within the RAP research domain

Keyword	Occurrence	Keyword	Occurrence
Reclaimed Asphalt Pavement	1610	Mixtures	798
Asphalt pavements	1559	Binders	591
Asphalt mixtures	1133	Aggregates	362
Recycling	915	Reclamation	292
Asphalt	894	Temperature	280

Figure 8 clearly shows a consistent and growing interest in RAP research topics from 2015 to 2024. Keywords like "reclaimed asphalt pavement," "asphalt," and "recycling" remain highly prevalent, underscoring the sustained importance of this field. The increasing prominence of "sustainability," "rejuvenator," and "warm mix asphalt" over time indicates a shift towards more environmentally conscious and technologically advanced approaches to RAP utilization. The advancements in RAP recycling technology, such as the development of rejuvenators and specialized asphalt mixtures, have made it possible to produce high-quality pavements with high percentages of recycled material, further encouraging research and implementation.

In conclusion, the presented data strongly support the notion that RAP research is a critical component of developing green infrastructure. By focusing on recycling, material optimization, overcoming technical hurdles, and extending material lifecycles, RAP research directly contributes to more sustainable, resource-efficient, and environmentally friendly pavement solutions. The trends suggest a promising future for RAP as a cornerstone of sustainable urban and road development.

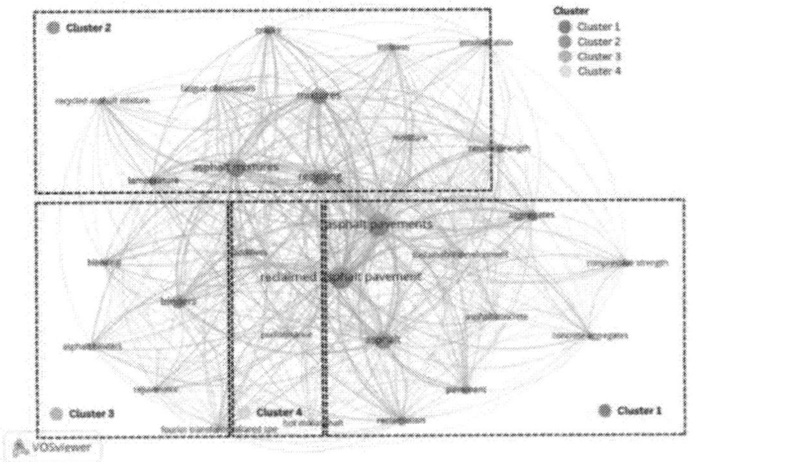

Fig. 7. Keyword co-occurrence by research clusters in RAP.

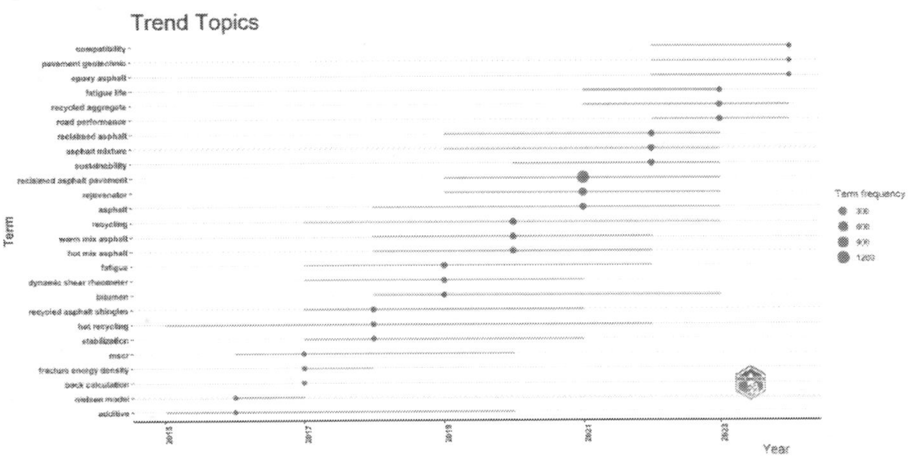

Fig. 8. Trends of RAP research topic from 2015 to 2024.

Focusing on RAP rejuvenator, Table 2 shows the most frequent research themes in the specific area of Reclaimed Asphalt Pavement (RAP) rejuvenators. Keywords like "Reclaimed Asphalt Pavement" (337 occurrences) and "Recycling" (179 occurrences) directly underpin the circular economy aspect of green infrastructure, emphasizing the reuse of materials. "Rejuvenator" (140 occurrences) itself is central, indicating the focus on restoring aged asphalt for continued use. The strong presence of "Asphalt pavements," "Asphalt mixtures," "Asphalt," and "Binders" highlights the technical focus on incorporating these rejuvenated materials into functional road surfaces, ensuring performance while maximizing recycled content. The occurrence of "Temperature" and "Blending" further points to the practical engineering considerations for effectively integrating rejuvenated RAP into new mixes.

Table 2. The highest-occurring ten keywords concerning RAP rejuvenators

Keyword	Occurrence	Keyword	Occurrence
Reclaimed Asphalt Pavement	337	Recycling	179
Asphalt pavements	316	Mixtures	160
Asphalt mixtures	243	Rejuvenator	140
Asphalt	185	Temperature	70
Binders	179	Blending	60

Figure 9, "Keyword co-occurrence by research clusters in RAP rejuvenator," reveals interconnected research areas vital for green infrastructure. Cluster 1 focuses on sustainable asphalt pavement construction and recycling, emphasizing RAP use in new mixtures for a circular economy. Cluster 2 delves into rejuvenating aged asphalt binders for sustainable pavement applications, crucial for reusing materials and reducing virgin resource demand. Cluster 3 highlights temperature-controlled technologies for producing and applying recycled asphalt mixtures, balancing environmental benefits with performance. The strong links between clusters, especially via "reclaimed asphalt pavements" and "rejuvenator," demonstrate a holistic commitment to maximizing recycled materials in sustainable road construction.

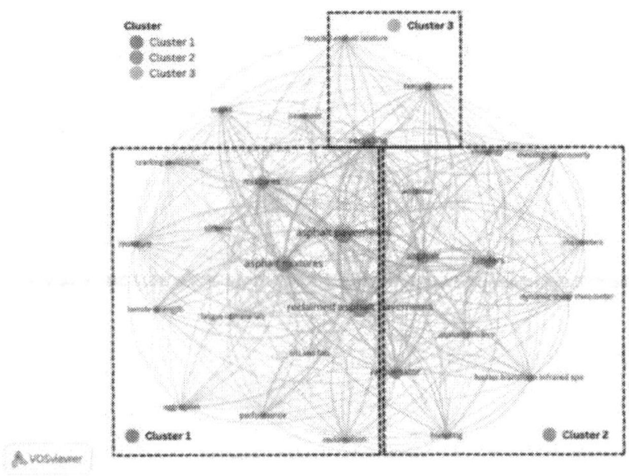

Fig. 9. Keyword co-occurrence by research clusters in RAP rejuvenator.

Figure 10, showing "Trends of RAP rejuvenator research topic from 2015 to 2024," highlights the evolving focus on sustainable pavement. "Reclaimed asphalt pavement," "rejuvenator," and "recycling" remain consistently prominent, while "circular economy" and "sustainability" (especially around 2021-2023) show increasing frequency, emphasizing the growing focus on environmental benefits. Furthermore, the interest in "waste cooking oil" and "waste vegetable oil" as potential rejuvenator sources demonstrates a proactive move towards even more environmentally friendly and bio-based solutions, further strengthening the green infrastructure relevance by exploring renewable and waste-derived materials. The strong trend of waste cooking oil and waste vegetable oil as rejuvenators will be directly tied to policy recommendations. This includes advocating for government and industry policies

 https://doi.org/10.1051/e3sconf/202565211002

that incentivize the circular economy by creating stable supply chains for waste-derived bio-binders and updating pavement specifications to accommodate these sustainable materials. The prominence of rejuvenator research directly implies a shift in industry practices from mere "cold-mix" recycling to high-performance hot-mix recycling. This requires industry investment in specialized rejuvenator dosing equipment and advanced testing protocols to ensure the quality of high-RAP mixtures, thereby setting a new, higher benchmark for sustainable pavement practices.

Collectively, these figures and the table paint a picture of a research domain deeply committed to developing advanced, high-performance, and environmentally responsible pavement solutions, making significant contributions to the principles of green infrastructure.

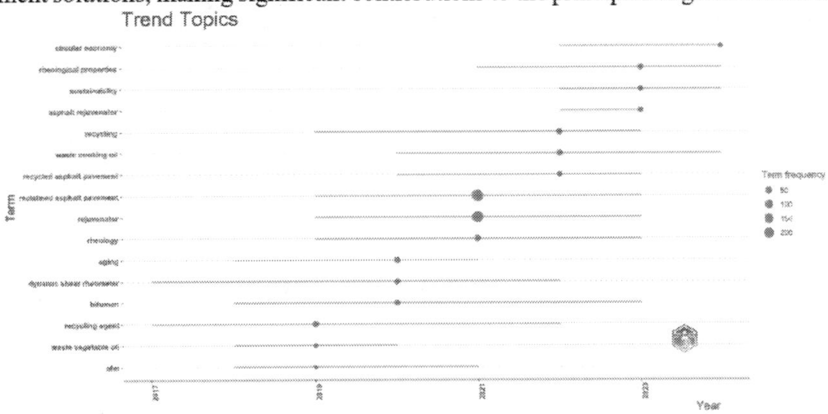

Fig. 10. Trends of RAP rejuvenator research topic from 2015 to 2024.

Table 3 details rejuvenator types studied from 2015 to 2024. While "others" leads (208 articles), indicating diverse emerging types, "oil base" (88), "bio-base" (65), and "bio composite" (61) are significantly established categories. "Chemical" rejuvenators account for 41 articles. This data highlights a clear interest in bio-derived materials within rejuvenator research.

Table 3. Rejuvenator type of RAP from 2015 to 2024

Rejuvenator Type	Oil base	Bio base	Bio composite	Chemical	Others
Total Article	**88**	**65**	**61**	**41**	**208**

The increasing focus on bio-based rejuvenators (evidenced by 65 "Bio base" and 61 "Bio composite" articles and the trend of "waste cooking oil" and "waste vegetable oil") offers crucial insights for sustainable green infrastructure through renewable resource utilization. The identified underrepresentation of nature-based solutions and certain biomaterial by-products is now explicitly framed as a strategic research agenda, fulfilling the study's objective by providing the roadmap for the next decade of RAP research. Future research should prioritize these nature-based solutions, aligning with SDG 9 (sustainable infrastructure) and SDG 12 (eco-friendly resource use), by investigating their fundamental chemical interactions with aged asphalt for optimized long-term performance. Future work should focus on pavement design and the development of low-carbon, waste-derived bio-binders. Exploring biomaterial by-products as raw materials further supports SDG 12 (waste reduction) and SDG 11 (sustainable urban environments), fostering a circular economy. Beyond material development, a comprehensive approach is necessary, encompassing Life Cycle Assessments (LCA), long-term field performance studies across various climates, and research into economic feasibility and policy frameworks to incentivize adoption.

4 Conclusion

Our bibliometric analysis confirms that global Reclaimed Asphalt Pavement (RAP) research strongly supports sustainable infrastructure. There is a growing emphasis on rejuvenators for performance enhancement, with a significant shift toward bio-based rejuvenators and biomaterial by-products, moving away from traditional oil-based options. This transition is crucial for developing greener infrastructure that aligns with global sustainability goals by conserving resources, reducing waste, and minimizing environmental impact. Future research should prioritize these nature-based solutions to ensure RAP's long-term performance and widespread adoption. This robust research directly advances SDG 9 (resilient infrastructure), SDG 12 (resource efficiency and waste reduction), and SDG 11 (sustainable urban environments), highlighting RAP's pivotal role in achieving a circular economy and enhancing the "green" credentials of future road infrastructure.

References

1. M. S. Ahmad and S. A. Ahmad, J. Eng. Appl. Sci. **69**, (2022)
2. Q. Zhang, D. Hou, Z. Li, H. Wang, and S. Dong, Polymers (Basel). **16**, (2024)
3. A. M. Alnadish, N. S. S. Singh, and A. M. Alawag, Polymers (Basel). **15**, (2023)
4. L. P. Thives and E. Ghisi, Renew. Sustain. Energy Rev. **72**, 473 (2017)
5. M. S. Dehghani, F. Giustozzi, G. W. Flintsch, and M. Crispino, Transp. Res. Rec. **2361**, 16 (2013)
6. X. Ye, Y. Chen, H. Yang, Y. Xiang, Z. Ye, W. Li, and C. Hu, Constr. Build. Mater. **439**, 137361 (2024)
7. M. Mohammadafzali, H. Ali, J. A. Musselman, G. A. Sholar, and W. A. Rilko, Adv. Mater. Sci. Eng. **2017**, (2017)
8. G. Xu, Y. Fan, Z. Shen, H. Wang, and J. Gao, Constr. Build. Mater. **412**, 134868 (2024)
9. I. N. Usanga, E. O. Inyang, and C. C. Ikeagwuani, Hybrid Adv. **10**, (2025)
10. M. Liu, S. Luo, W. Huang, Y. Yang, J. Hu, Z. Zhang, B. Zhang, and K. Du, Sustain. Mater. Technol. **45**, (2025)
11. M. Zaumanis, R. B. Mallick, and R. Frank, Resour. Conserv. Recycl. **92**, 230 (2014)
12. Qazi Aurangzeb and Imad L Al-Qadi, Transp. Res. Rec. **2456**, 161 (2014)
13. L. Yao, Z. Leng, J. Lan, R. Chen, and J. Jiang, J. Clean. Prod. **336**, (2022)
14. Y. Qiao, E. Dave, T. Parry, O. Valle, L. Mi, G. Ni, Z. Yuan, and Y. Zhu, Sustain. **11**, (2019)
15. Y. Chen, X. Li, and X. Tian, Buildings **14**, (2024)

Durability Performance of Fly Ash Geopolymer Mortar under Long-Term Sulfate Exposure

Syahidah Fani Nastiti[1*], *Farhan Ahmad* Givari[1], *Agil Fitri* Handayani[1], *Dzul Fikri* Muhammad[1], *Nindyawati* Nindyawati[1], *Andri* Kusbiantoro[2], and *Hidayatul* Khoiriyah[1]

[1]Universitas Negeri Malang, Department of Civil Engineering and Planning, Jalan Semarang no 5, Indonesia
[2]Universiti Tun Hussein Onn, Production Department, 91944 Les Ulis Cedex A, Malaysia

Abstract. The long-term durability of geopolymer mortar against sulfate attack is crucial for sustainable infrastructure, especially in aggressive environments. This study presents experimental findings on the performance of fly ash-based geopolymer mortars immersed in 5% sulfuric acid for 84 days. Two mix designs with different activator ratios (1:2 and 1:2.5) were evaluated through periodic testing of compressive strength, mass loss analysis, and sulfate content. The results showed that GMAk1 :2 mortar exhibited lower mass loss (maximum 6.53 g) and smaller reduction in compressive strength (from 44.80 Mpa to 39.63 Mpa) compared to GMAk1:2.5, which experienced a mass loss of up to 11.80 g and a strength decrease from 35.96 MPa to 26.08 MPa. The sulfate content in GMAk1:2 mortar remained relatively stable, indicating better resistance to sulfate ingress. These findings highlight that optimizing the activator ratio significantly enhances the sulfate resistance of geopolymer mortar, making it suitable for structural applications in chemically aggressive environments.

1 INTRODUCTION

The widespread degradation of concrete structures due to aggressive chemical environments, especially sulfate attack, poses significant challenges to infrastructure durability and sustainability. Sulfate ions can react with cement hydration products, leading to expansive by-products such as ettringite and gypsum, which compromise the integrity and service life of Portland cement-based materials [1]. Conventional cementitious systems are prone to sulfate-induced cracking, spalling, and strength loss, making them unsuitable for long-term service in sulfate-rich environments. This has directed attention towards alternative binders with enhanced chemical resistance, particularly those that offer environmental benefits and better durability performance [2,3].

Fly ash-based geopolymer mortars (FAGMs) have emerged as promising low-carbon alternatives to Ordinary Portland Cement (OPC) owing to their superior resistance to acid and sulfate attack low calcium content[4–6], dense microstructure, lower permeability, and reduced CO_2 emissions. Geopolymers utilize aluminosilicate-rich industrial by-products

* Corresponding author: syahidah.faninastiti.ft@um.ac.id

 https://doi.org/10.1051/e3sconf/202565211003

activated with alkaline solutions such as sodium silicate (Na_2SiO_3) and sodium hydroxide (NaOH) to form durable, three-dimensional amorphous networks [7]. The fly ash geopolymerization process results in sodium aluminosilicate hydrate (N-A-S-H) gel formation, which exhibits superior resistance to sulfate penetration compared to the calcium-rich phases in OPC.

Despite the established chemical resilience of geopolymers, their performance is strongly influenced by mix design parameters, particularly the ratio of Na_2SiO_3 to NaOH in the activator solution. This ratio governs the polymerization rate, gel formation, and pore structure development, all of which are critical to the long-term durability of geopolymer mortars in sulfate environments [7,8]. Recent findings suggest that while higher silicate content enhances mechanical strength, it may also alter porosity and susceptibility to chemical ingress over time [9]. Recent studies [5,7,10,11] have consistently demonstrated that the ratio of alkali activators, particularly the Na_2SiO_3/NaOH proportion, plays a decisive role in governing the microstructural densification, pore refinement, and overall chemical stability of geopolymers. These findings highlight the strong influence of activator composition on both the mechanical and durability properties of geopolymer systems.

However, systematic evaluation of sulfate durability under varying activator ratios, especially in fly ash-based geopolymer mortars cured under ambient conditions, remains limited and requires further investigation. Despite the recognized chemical resistance of FAGMs s, there is still limited understanding of how different Na_2SiO_3/NaOH activator ratios affect their long-term sulfate durability under ambient curing conditions. This knowledge gap is critical, as ambient curing better reflects practical field applications in infrastructure [6,12]. Therefore, this study aims to evaluate the long-term sulfate durability of fly ash-based geopolymer mortars with different activator compositions. Specifically, it investigates the influence of two distinct Na_2SiO_3/NaOH ratios on mass stability, strength retention, and microstructural behavior after continuous sulfate exposure.

2 MATERIALS AND METHODS

Two geopolymers mortar types with different ratios of activator solution were designed and evaluated. The following section contains information on the materials used (including chemical content), the prepared specimens and the conducted test. The treatment of the specimens immersed in sulfate solution is described in detail icluding the determination of the change value of compressive strength and sulfate content.

Table 1. XRF Test Result of fly ash composition.

Parameter	% wt
SiO_2 (silicon dioxide)	44.49%
Al_2O_3 (aluminum oxide)	24.86%
Fe_2O_3 (iron trioxide)	17.06%
CaO (calcium oxide)	5.70%
MgO (magnesium oxide)	2.41%

https://doi.org/10.1051/e3sconf/202565211003

Class F Fly Ash was used as the primary precursor material in this study. Based on the chemical analysis **(Table 1),** fly ash is predominantly composed of silica, alumina, and iron oxide, meeting the minimum 70% combined content required for Class F classification according to ASTM C168. The relaively low calcium oxide and magnesium oxide content further supports this classification. These chemical properties make the fly ash suitable for geopolymers synthesis, providing the necessary aluminosilicate source for geopolymerization.

The alkaline activator solution consisted of a mixture of sodium silicate (Na_2SiO_3) and sodium hydorxide (NaOH) at a concentration of 12 M. The ratios of Na_2SiO_3 to NaOH were employed (1:2 and 1:2.5), to investigate the influence of activator composition on geopolymer properties. Standard silica sand was used as the fina aggregate to provide consistency in the mix design. The selection of these materials aimed to optimize the geopolymer matrix's reactivity and mechanical performance, particularly under varied alkaline ratio. The material composition of geopolymer mortar were used fly ash, fine aggregate and alkali activator was shown at **Figure 1.**

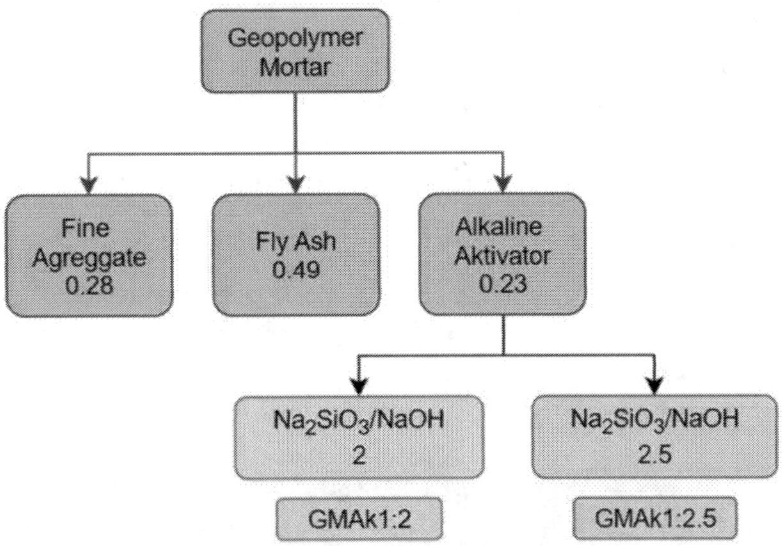

Fig. 1. Mix Design of geopolymer mortars (GMAk1:2 and GMAk1:2.5)

For each mix type, twenty one cube specimens (5 × 5 × 5 cm) were cast, with three specimens tested at each exposure age to ensure repeatability and statistical reliability, designated as GMAk1:2 and GMAk1:2.5, corresponding to the ratios of alkaline activator solution. The geopolymer mortar was poured into molds and allowed to cure before being subjected to durability testing. Each mix was designed to evaluate the effect of alkaline activator composition on the performance of fly ash-based geopolymers, particularly under aggresive chemical exposure.

To assess the acid resistance of the geopolymers composites, the specimens were immersed in a 5% sulfuric acid (H_2SO_4) solution for a total duration of 84 days. The experimental setup for the immersion of geopolymer mortar specimens in a sulfate solution is illustrated in the **Figure 2**, where cubic mortar samples (5×5×5 cm) are placed on a 3 cm thick horizontal wooden support inside an immersion tank. The specimens are partially submerged in a 5% sulfate solution with a controlled immersion depth of 5 cm, and the tank is sealed with a cover to minimize evaporation and contamination. To ensure solution stability, weekly pH monitoring is conducted; if significant pH fluctuations are observed, the existing sulfate solution is replaced with a freshly prepared 5% sulfate solution to maintain consistent exposure conditions throughout the test period. Additionally, temperature and humidity are maintained within normal room limits.

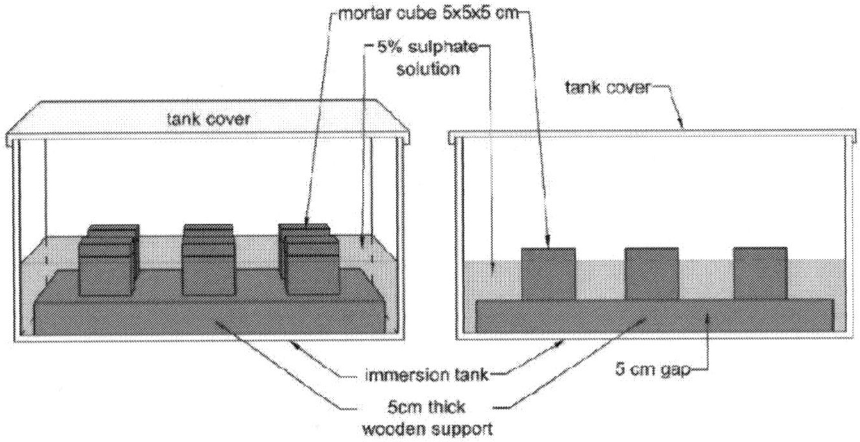

Fig. 2. Set-Up of mortar cube immersion in a 5% sulfuric acid (H_2SO_4) solution

The compressive strength evaluation of geopolymer mortar specimens was carried out at 14-day intervals using a digital compression testing machine, as illustrated in **Figure 3**. This test aimed to monitor the degradation of mechanical performance over time under sulfate exposure, where the mortar cubes were carefully placed between the loading plates and subjected to increasing axial force until failure. The recorded values provided insight into the structural stability and residual strength of the specimens after immersion. Such periodic testing ensured a consistent understanding of strength loss progression throughout the exposure period.

Fig. 3. Compressive strength testing

https://doi.org/10.1051/e3sconf/202565211003

Chemical and physical assessments were conducted to evaluate the durability of the mortar. As shown in **Figure 4**, periodic pH checks and gravimetric measurements were performed to determine mass loss, reflecting the extent of surface deterioration and leaching processes. Furthermore, the sulfate ion content in the solution was analyzed using spectrophotometry to quantify the level of chemical interaction between the aggressive environment and the binder matrix. These combined methods provided a comprehensive assessment of both the mechanical stability and chemical resistance of geopolymer mortar, enabling the correlation between microstructural changes and long-term performance under sulfate attack.

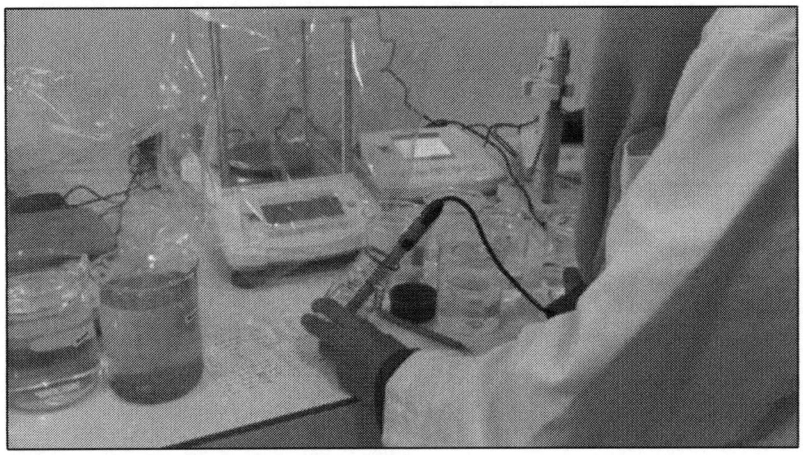

Fig. 4. pH checks and gravimetric measurements

3 RESULT AND DISCUSSION

3.1 Compressive strength

The compressive strength test results (**Table 2**) indicate that GMAk1:2 mortar maintained better mechanical integrity under prolonged sulfuric acid exposure compared to GMAk1:2.5. At 84 days, GMAk1:2 retained 81.47% of its original strength (39.63 MPa from an initial 48.64 MPa), whereas GMAk1:2.5 dropped to 58.95% (26.08 MPa from 44.24 MPa). The lower $Na_2SiO_3/NaOH$ ratio in GMAk1:2 may have promoted a more stable geopolymer network, possibly due to more balanced gel formation and reduced porosity. These findings are aligned with the literature stating that lower activator ratios can enhance durability by limiting the unreacted alkali content, which is prone to leaching under acid attack [13].

Table 2. Compressive strength of mortar geopolymer immersed in 5% sulfuric acid solution.

Mortar Type	Immersion in a 5% sulfuric acid solution (day)	Compressive strength (MPa)	Percentage of residual compressive strength (%)	Percentage of compressive strength loss (%)
GMAk1:2	0	48.64	100.00	-
	14	45.21	92.96	7.04
	28	44.84	92.19	7.81
	42	43.01	88.43	11.57

Mortar Type	Immersion in a 5% sulfuric acid solution (day)	Compressive strength (MPa)	Percentage of residual compressive strength (%)	Percentage of compressive strength loss (%)
	56	42.12	86.60	13.40
	70	40.13	82.51	17.49
	84	39.63	81.47	18.53
GMAk1:2.5	0	44.24	100.00	-
	14	43.20	97.65	2.35
	28	42.03	95.00	5.00
	42	37.49	84.75	15.25
	56	34.81	78.69	21.31
	70	33.36	75.41	24.59
	84	26.08	58.95	41.05

The strength decline is gradual for both samples but more pronounced in GMAk1:2.5 after 42 days (**Figure 5**), suggesting that prolonged exposure exacerbates the damage to less compact matrices. According to previous study, geopolymer mortars with lower solution-to-binder ratios and higher curing temperatures exhibit improved strength retention under sulfate attack[13]. The stronger performance of GMAk1:2 supports this, indicating that matrix density and polymerization efficiency are key factors for durability.

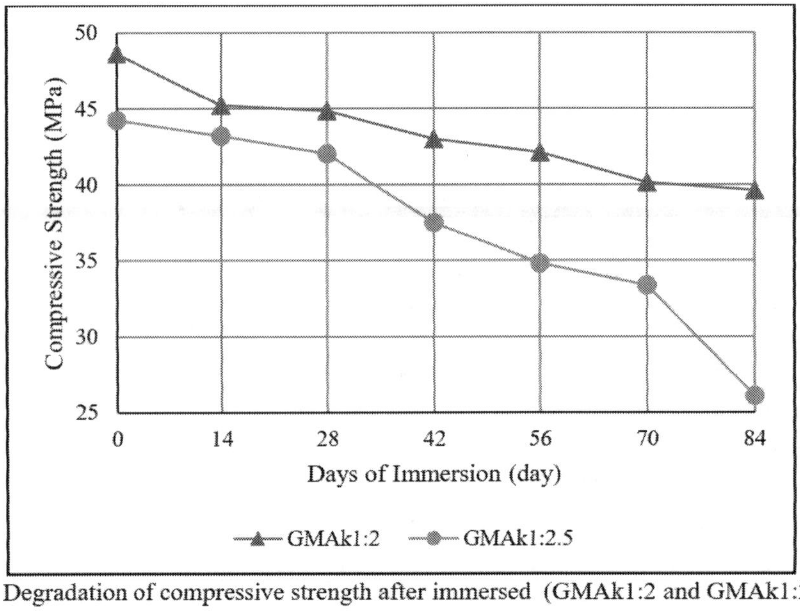

Fig. 5. Degradation of compressive strength after immersed (GMAk1:2 and GMAk1:2.5)

Additionally, the fly ash used in this study, rich in SiO_2 and Al_2O_3 and low in CaO, contributes to the formation of N-A-S-H gels that are more stable under acidic conditions. This is corroborated by multiple studies indicating that calcium-poor geopolymers

https://doi.org/10.1051/e3sconf/202565211003

demonstrate better acid and sulfate resistance due to the absence of reactive Ca phases that form gypsum or ettringite upon exposure [14].

3.2 Mass Loss

Mass loss trends further corroborate the durability superiority of GMAk1:2 over GMAk1:2.5, shown at **Table 3**. After 84 days of immersion, GMAk1:2 experienced only 2.46% mass loss (6.53 g), while GMAk1:2.5 lost 4.04% of its initial mass (10.40 g). The slower rate of deterioration in GMAk1:2 suggests a denser and less permeable matrix, which inhibits ion ingress and leaching. Previous findings support that lower permeability significantly restricts the penetration of aggressive agents like SO_4^{2-}, reducing the extent of degradation[14].

Table 3. Mass loss mortar geopolymer immersed in 5% sulfuric acid solution.

Mortar Type	immersion in a 5% sulfuric acid solution (day)	Mortar Mass Loss (gr)	Percentage Change in Initial Weight and Final Weight (%)
GMAk1:2	0	0. 00	0
	14	2. 93	1. 09%
	28	4. 27	1. 60%
	42	4. 70	1. 77%
	56	5. 03	1. 96%
	70	7. 80	2. 95%
	84	6. 53	2. 46%
GMAk1:2.5	0	0. 00	0
	14	8. 43	3. 16%
	28	8. 97	3. 40%
	42	9. 33	3. 49%
	56	10. 80	4. 21%
	70	11. 80	4. 37%
	84	10. 40	4. 04%

Interestingly, GMAk1:2.5 recorded an early mass loss spike by day 14 (3.16%) and continued to rise, indicating rapid initial degradation and microstructural disintegration (**Figure 6**). This early-stage deterioration may be attributed to excessive Na_2SiO_3 content, which, while increasing initial reactivity, potentially introduces unreacted alkalis or poorly bonded gel structures vulnerable to acidic erosion[13].

The correlation between higher silicate ratios and increased porosity has been highlighted, that mortars with higher silicate content exhibited increased microcracking and gypsum formation when immersed in sulfuric acid. This behavior aligns with the results observed in GMAk1:2.5, reinforcing the importance of optimal activator balance.

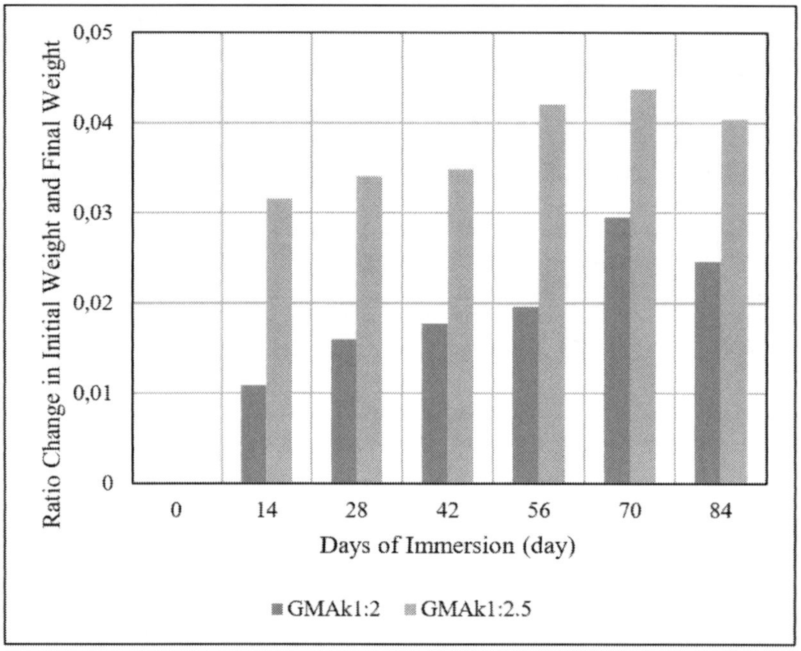

Fig. 6. Change in initial and final weight of mortar geopolymer (GMAk1:2 and GMAk1:2.5)

3.3 Sulfate content

Sulfate ion uptake analysis revealed that GMAk1:2 showed more consistent and lower variation in sulfate content throughout the 84-day period, ranging between 41.56–47.77 mg/g (shown at **Table 4**). Meanwhile, GMAk1:2.5 exhibited more fluctuation, peaking at 55.51 mg/g. The relative sulfate stability in GMAk1:2 suggests better resistance to sulfate ingress and chemical degradation, likely due to denser microstructure and more effective geopolymerization.

Table 4. sulfate content of mortar geopolymer immersed in 5% sulfuric acid solution.

Mortar Type	immersion in a 5% sulfuric acid solution (day)	Sulfate content (mg/g)
GMAk1:2	14	47.11
	28	41.56
	42	44.97
	56	47.77
	70	44.95
	84	47.56
GMAk1:2.5	14	8.99
	28	54.66
	42	34.09
	56	55.51

Mortar Type	immersion in a 5% sulfuric acid solution (day)	Sulfate content (mg/g)
	70	45.38
	84	46.03

Geopolymer matrices with well-formed N-A-S-H gels offer significantly reduced permeability to sulfate ions compared to OPC systems, as they lack CH and C-A-H phases prone to sulfate reactions [6,13,15]. The reduced sulfate uptake in GMAk1:2 is thus indicative of superior matrix integrity and minimized reactive sites.

High sulfate absorption in GMAk1:2.5 also corresponds with its elevated mass loss and reduced strength, suggesting gypsum or ettringite formation within pores that contributes to structural breakdown. Studies by Kwasny et al. emphasize that even small amounts of reactive calcium in the mix can catalyze such destructive reactions under sulfate exposure[13].

3.4 Correlation Analysis

There is a clear inverse relationship between sulfate content and compressive strength, as well as a direct relationship between sulfate content and mass loss. Mortars with higher sulfate content exhibited greater mass loss and lower residual strength, particularly in GMAk1:2.5. This indicates that sulfate ingress facilitates chemical reactions that break down the geopolymer network, leading to material deterioration (**Figure** 7).

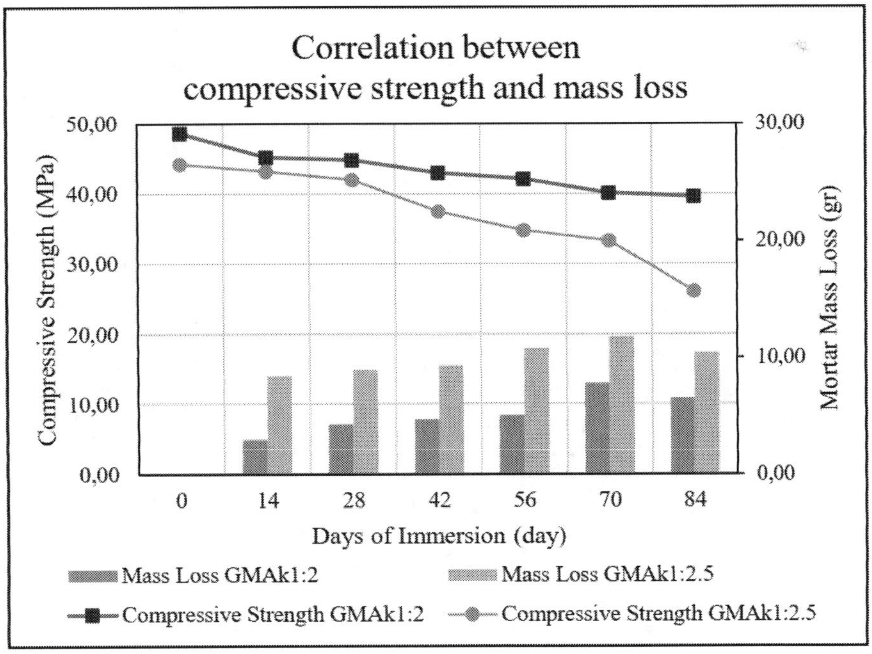

Fig. 7. Correlation between compressive strength and mass loss (GMAk1:2 and GMAk1:2.5)

The superior strength retention of GMAk1:2 can be attributed to more homogeneous N-A-S-H gel formation, which reduces pore connectivity and enhances resistance against sulfate ingress. Similar observations were made by Oluwafemi et al. [7], who reported that optimized $Na_2SiO_3/NaOH$ ratios improve polymerization efficiency, resulting in denser matrices and higher residual strength under aggressive environments. Li et al. [11] further demonstrated that activator ratio adjustments significantly influence microstructural evolution, as confirmed through SEM and XRD analyses, aligning with the improved durability observed in GMAk1:2. The data validates the use of sulfate content as an early indicator of degradation potential in aggressive environments.

Mass loss and sulfate content analysis also highlighted the durability advantage of GMAk1:2, with reduced degradation compared to GMAk1:2.5. The higher silicate ratio in GMAk1:2.5 likely increased microcracking and introduced loosely bound alkalis, accelerating leaching and facilitating gypsum/ettringite formation. This degradation mechanism is consistent with the findings of Aygun and Uysal [10], who emphasized that excessive silicate content can destabilize the matrix and promote chemical ingress despite higher initial reactivity. Interestingly, the anomalous fluctuations in sulfate uptake observed in GMAk1:2.5 further support the presence of unstable gel phases, a trend also noted in seawater-exposed geopolymer systems [11].

The comparative analysis with existing studies strengthens the interpretation of the present findings. Siva Sai Hoshithaa et al. [5] reported that geopolymer concretes with optimized activator compositions exhibited superior sulfate and chloride resistance relative to OPC, underscoring the advantages of aluminosilicate-based binders in aggressive media. This comparison reinforces the role of activator ratio optimization not only in enhancing chemical stability but also in reducing long-term maintenance costs and environmental impacts.

Present study contributes to the growing body of evidence that fly ash-based geopolymers, when designed with optimal activator ratios, offer a sustainable alternative to OPC for infrastructures exposed to harsh chemical environments. The findings suggest that optimizing activator ratios not only enhances durability but also lowers lifecycle costs and reduces environmental burdens, in line with recent sustainability-focused studies [10]. Future research should evaluate durability performance beyond 84 days, extend testing to other aggressive environments such as chloride and carbonation exposure, and incorporate microstructural characterization techniques to validate the degradation mechanisms.

4 CONCLUSION

This study demonstrates that fly ash-based geopolymer mortar performance under acidic and sulfate-rich environments is highly dependent on the alkaline activator ratio. GMAk1:2 mortar exhibited superior durability, as evidenced by higher compressive strength retention (81.47%), lower mass loss (2.46%), and more stable sulfate content (~47 mg/g). In contrast, GMAk1:2.5 experienced accelerated degradation, with higher sulfate ion absorption and pronounced structural decay. The correlation among compressive strength, mass loss, and sulfate content underlines their interdependence in evaluating geopolymer resistance. These findings confirm that optimizing the activator composition is critical in enhancing long-term chemical durability, making GMAk1:2 a promising candidate for structural applications in aggressive environments. The superior durability of GMAk1:2 highlights its potential use in sewer systems, marine infrastructure, and foundations exposed to sulfate-rich soils.

Future research should aim to extend the exposure periods beyond 84 days in order to capture the long-term durability performance of fly ash-based geopolymer mortars under sustained sulfate attack. In addition, exploring the behavior of these materials in other aggressive environments, such as chloride ingress, carbonation, or combined chemical exposures, would provide a more comprehensive understanding of their durability profile. Complementary microstructural analyses, including SEM, XRD, and FTIR, are also essential to validate the degradation mechanisms inferred from mechanical and chemical performance data. Overall, this study contributes to the development of low-carbon and durable construction materials by valorizing industrial waste such as fly ash, thereby supporting the advancement of resilient and sustainable infrastructure.

References

1. P. W. Ariyadasa, A. C. Manalo, W. Lokuge, V. Aravienthan, A. Gerdes, and J. Kaltenbach, Cem Concr Res **194**, (2025)
2. M. Albitar, M. S. Mohamed Ali, P. Visintin, and M. Drechsler, Constr Build Mater **136**, 374 (2017)
3. Y. Aygörmez, O. Canpolat, M. M. Al-Mashhadani, M. Uysal, and F. Şahin, in *Sustainable Construction Materials and Technologies* (2019)
4. Md. H. Islam, D. Law, Y. Patrisia, and C. Gunasekara, Case Studies in Construction Materials **23**, e05036 (2025)
5. T. S. S. Hoshitha, T. C. S. Rao, and S. Ricci, in *Transportation Research Procedia* (Elsevier B.V., 2025), pp. 115 - 122
6. F. El Inaty, H. Nasreddine, A. Djerbi, L. Gautron, M. Marchetti, M. Quiertant, and O. O. Metalssi, Results in Engineering **27**, (2025)
7. J. Oluwafemi, O. Ofuyatan, A. Adedeji, D. Bankole, and V. Bitrus, Case Studies in Construction Materials **22**, (2025)
8. D. de Klerk, A. Naghizadeh, S. O. Ekolu, and M. Welman-Purchase, Constr Build Mater **468**, (2025)
9. H. Konduru, S. Karthiyaini, and M. Shanmugasundaram, Case Studies in Construction Materials **22**, (2025)
10. B. F. Aygun and M. Uysal, Materials Today Sustainability **29**, (2025)
11. K. Li, D. Zhang, Q. Li, and Z. Fan, Cem Concr Res **115**, 545 (2019)
12. K. K. D. A. Wijesekara, M. Sadique, I. Carnacina, A. Fielding, and G. C. Bojczuk, Cleaner Waste Systems **11**, (2025)
13. H. E. Elyamany, A. E. M. Abd Elmoaty, and A. M. Elshaboury, Constr Build Mater **184**, 111 (2018)
14. J. Kwasny, T. A. Aiken, M. N. Soutsos, J. A. McIntosh, and D. J. Cleland, Constr Build Mater **166**, 537 (2018)
15. R. A. Mahmood, E. Delik, N. U. Kockal, and B. E. Tefon-Öztürk, Next Materials **8**, (2025)

https://doi.org/10.1051/e3sconf/202565211004

Analysis of Groundwater Availability Potential Using Euclidean Distance in the Suburbs of Yogyakarta City (study: Banguntapan District)

Valencio Evanio[1*], *Linda* Prasetyorini[1], *Ery* Suhartanto[1], *Dian* Chandrasasi[2], and *Rachmat Fajar* Lubis[3]

[1,2]Water Resources Engineering Department, Universitas Brawijaya, 65142, Malang, Indonesia
[3]Research Centre for Limnology and Water Resources, Badan Riset dan Inovasi Nasional (BRIN), 15314, Serpong, Indonesia

Abstract. As one of the areas directly adjacent to the outskirts of Yogyakarta City, the Banguntapan District is increasingly affected by various rapid urban activities. So that period 2006 - 2024, more and more vegetation areas have experienced changes in function into building areas. This phenomenon is a warning for the sustainability of shallow groundwater in the study area in the future. Through a descriptive approach, the novelty of this research is the first study to integrate Euclidean Distance analysis from multi-temporal satellite imagery for water potential assessment so that the study aims to determine the potential availability of shallow groundwater based on changes in distance between vegetation areas, both herbaceous, shrubby, and tree-type vegetation, to the existence of water sources (wells). The research variables consist of land cover, vegetation type, rainfall intensity, soil and rock conditions, land slope, and fluctuations in groundwater depth (well depth). The results of the study indicate that in the period 2006-2024, there was a conversion of green land and open areas (rice fields/gardens/fields) into residential and business areas, reaching more than 45%, which resulted in a reduction in the area of the very high potential water catchment zone, covering only 12.8% of the total area. As a result, the reduced groundwater recharge has reached 316,752.52 m^3/year, and the reduced groundwater infiltration volume rate has reached 26,036,055.022 m^3/year. This situation aligns with the results of an analysis of 14 observation wells, which showed variations in well water level decline ranging from 1.72 m to 8.47 m. The results of this study enrich the analysis of shallow groundwater potential using Euclidean Distance analysis, which correlates with the distance between vegetation areas, providing valuable information for spatial planning policies, especially in suburban areas, to maintain the sustainability of groundwater resources.

1 Introduction

Water is one of the most essential human needs for survival. With the dynamics of urban development and climate change, efforts to protect water resources, particularly maintaining shallow groundwater potential in suburban areas, will face increasing challenges. Banguntapan District is one of the regions directly adjacent to the outskirts of Yogyakarta City. Over time, all life activities in the study area are affected by rapid urban development,

* Corresponding author: evansika08@student.ub.ac.id

which will create water demand issues in the future. Therefore, an analysis of static groundwater potential and water demand is necessary, calculated based on the assumption of static groundwater flow that can be pumped from aquifers [1].

Increasingly, people are aware that the suburban phenomenon will impact groundwater resources. The reduction in vegetated land will increase the potential for flooding, necessitating calculations of critical infiltration levels based on land use, slope gradient, and infiltration rate [2]. In this study, the measurement of potential shallow groundwater availability is based on the presence of vegetation between regions. The distance between vegetation areas, whether herbs, shrubs, or trees, will affect the infiltration area, which is assumed to be a source of groundwater recharge. Changes in land use and climate change will affect the quality and quantity of potential groundwater availability. Peri-urban areas, located between rural and urban areas, present a more significant challenge. Rapid growth in these peri-urban areas often leads to environmental degradation. Rapid suburban growth, characterized by the transformation of open spaces into residential and commercial areas, significantly reduces the amount of natural vegetation that can absorb rainfall.

The inherent capacity of the soil to recharge groundwater reserves has been compromised by the presence of buildings [3], preventing the influx of well water for daily needs and access to sufficient groundwater. Climate anomalies, such as shorter rainy seasons and prolonged droughts, exacerbate this problem by placing further pressure on groundwater recharge systems [4].

The rapid expansion of settlements in Banguntapan District, which borders the center of Yogyakarta City, is a clear example of this issue. Temporal analysis of satellite imagery shows substantial land conversion from rice fields and other open spaces to built-up areas between 2006 and 2024. Furthermore, areas covered by vegetation, such as herbs, shrubs, and trees, have a direct impact on water infiltration rates. Areas with higher vegetation density typically absorb more water [5,6]. However, soil structure, permeability, and rock type remain crucial in determining how well water infiltrates.

A mechanistic understanding of the impact of specific vegetation types on infiltration is limited by the use of remote sensing proxies or coarse land-use classifications in many studies to describe the impacts of vegetation [7]. Furthermore, the use of geographic information systems (GIS)-based multicriteria decision analysis (MCDA) methods is becoming increasingly common in groundwater potential mapping. However, these methods are not consistently used in conjunction with field measurements and validation, which can result in scale discrepancies between remote sensing data and point observations [8]. The information gaps that hinder efficient water resource management in rapidly changing peri-urban areas are exemplified by the lack of comprehensive sub-district-scale quantification of groundwater recharge losses and a limited understanding of vegetation-specific infiltration dynamics [9].

This study fills the research gap by utilizing Geographic Information Systems (GIS) and multitemporal satellite imagery to conduct a comprehensive spatial analysis of groundwater potential distribution in Banguntapan District. This study aims to: (1) quantify the loss of open space and vegetation on groundwater recharge at the sub-district scale, (2) determine the specific effects of different vegetation types (herbs, shrubs, and trees) on groundwater infiltration potential, and (3) establish the spatial distribution of groundwater potential zones using integrated GIS analysis and field validation. This study statistically assesses groundwater infiltration and recharge capacity by synthesizing time series of satellite imagery using Euclidean distance analysis between 2006 and 2024, field measurements from 14 strategically located wells, meteorological data, and spatial modeling. Additionally, this study employs the Analytical Hierarchy Process (AHP) and Weighted Linear Combination (WLC) analysis.

The findings of this study will facilitate the integration of Euclidean distance analysis into the spatial analysis of water absorption and groundwater recharge in order to protect critical groundwater recharge areas while enabling adequate urban growth.

2 Methods

This study employs a descriptive approach to analyze the potential availability of unconfined groundwater in the peri-urban areas of Yogyakarta, specifically the Banguntapan District, a transition zone from rural to urban land use. The methodology used includes field observations, satellite imagery analysis, and spatial data modeling to examine two main aspects: groundwater recharge and water absorption. These aspects are influenced by open space cover, vegetation type, rainfall intensity, soil and rock conditions, land slope, and groundwater depth.

This research began with a document review to collect secondary data from official institutions, including satellite imagery (2006 and 2024), rainfall data, soil type data, and groundwater level data. The ArcGIS application utilizes Geographic Information System (GIS) technology to view high-resolution satellite imagery annually. Vivid Standard from Maxar Technologies served high-resolution maps in the 2006 product that had a Ground Sample Distance (GSD) resolution of 30cm and a spatial accuracy of CE90 5m. The Pléiades Neo HD15, from Airbus, was used to map in 2024. It uses AI to improve the 30cm native satellite data to a resolution of 15cm. In interpreting these two satellite images, they were projected in the Universal Transverse Mercator (UTM) system, zone 49S, on the World Geodetic System 1984 (WGS 84) vertical datum.

Image interpretation serves to depict open spaces and vegetation zones, including herbs, shrubs, and trees. This visual interpretation was conducted through fieldwork involving direct observation of well conditions and environmental characteristics. A total of 14 observation wells were selected through purposive sampling to represent various vegetation conditions across the district.

Spatial analysis was used to assess the proximity of groundwater infiltration zones using GIS-based Euclidean distance. This method calculates the shortest straight-line distance from a specific point or feature (open space and vegetated areas) to potential infiltration zones. In identifying groundwater infiltration zones, the Euclidean distance measurement technique provides a direct visualization of geographic elements such as open spaces, vegetated areas, and geological formations that generate potential infiltration capacity. This analysis allows for the measurement of groundwater recharge potential in the suburban areas of Banguntapan District, Yogyakarta.

Spatial analysis of Thiessen polygons was used to model rainfall distribution. This technique enables an accurate spatial representation of rainfall patterns, which is essential for understanding how precipitation contributes to groundwater recharge in the study area. Meanwhile, slope analysis and surface soil classification further refine the assessment of groundwater recharge potential. Steeper slopes typically result in higher surface runoff, reducing infiltration opportunities, while flatter areas enhance groundwater recharge. Soil classification helps determine permeability, with sandy soils facilitating greater infiltration than clay-rich soils. These factors are integrated to provide a comprehensive understanding of recharge dynamics.

Vegetation cover (herbs, shrubs, and trees), soil conditions, and rock types are analyzed to determine water infiltration. Vegetation plays a significant role in moderating water infiltration rates. Herbs have minimal root depth, thus allowing water infiltration beneath them. In contrast, shrubs and trees contribute significantly to infiltration due to their deeper root systems and editorial effect, which slows surface runoff. These characteristics were evaluated in conjunction with soil and rock types to delineate zones with high infiltration potential.

These characteristics are evaluated along with soil and rock types to delineate high-potential infiltration zones. Groundwater table depth is measured using the Spline Interpolation Method to identify spatial variations and potential zones of groundwater availability. This method will produce a map depicting variations in groundwater depth, providing valuable insights into the spatial distribution of shallow groundwater.

Groundwater table depth was measured using the Spline Interpolation Method to identify spatial variations and potential zones of groundwater availability. This geostatistical method produces a smoothed surface map that depicts variations in groundwater depth, providing valuable information about the spatial distribution of shallow aquifers.

A Geographic Information System (GIS)-based Euclidean Distance technique was used to calculate straight-line distances from selected points to recharge zones. This calculation provides a quantitative measure of proximity, facilitating the evaluation of potential groundwater recharge areas. A Weighted Linear Combination (WLC) assessment model was applied, incorporating the Analytical Hierarchy Process (AHP) to assign weights based on the relative importance of each variable criterion. Variables with a greater influence on groundwater recharge, such as vegetation cover and open space, received higher scores.

This multi-criteria decision-making approach ensured a holistic evaluation of factors influencing groundwater availability. Based on this study's objectives, several hypotheses were formulated to guide the research. First, changes in land cover, particularly the reduction of open space in the outskirts of Yogyakarta City, are hypothesized to have significantly reduced the capacity of groundwater recharge zones.

When open areas are converted to built-up areas, the natural infiltration process is disrupted, resulting in reduced water availability for recharging groundwater reserves. Second, vegetation type is crucial in determining infiltration rates, with herbs, shrubs, and trees contributing differently to water absorption. Among these, shrub vegetation is hypothesized to have the most significant impact due to its unique root structure and water retention capacity, which effectively increases infiltration rates compared to other vegetation types.

The analysis of water recharge and reduction in open land from 2006 to 2024 was conducted by calculating groundwater reserves equivalent to groundwater recharge from rainfall (RC).Recharge to the aquifer can be calculated using the recharge approach as a percentage of average annual rainfall [10], with the following formula:

$$RC = RF \times A \times RC\ (\%) \tag{1}$$

Where RF is the average annual rainfall in the catchment area calculated using the Thiessen Polygon method (m), A is the catchment area (m^2), and RC is the recharge percentage, by %. In this study, the recharge percentage for the Yogyakarta Special Region—a mixed volcanic and sedimentary terrain—was set at 27%.

Analysis of the amount of water infiltration due to the decline in herbaceous, shrubby, and tree vegetation from 2006 to 2024 was conducted by calculating the groundwater infiltration coefficient, which is equated with the groundwater infiltration rate/volume. The infiltration coefficient is the amount of rainfall that flows as infiltrated water out of the total rainfall. To obtain the infiltration coefficient (C) for a watershed, use the following formula:

$$C = \frac{(I \times 365 \times A)}{(P \times A)} \tag{2}$$

The C value, which is the infiltration coefficient (%), is calculated by dividing the product of the baseflow infiltration rate (I, m/day), days per year, and catchment area (A, m^2) by the product of the annual rainfall (P, m) and the catchment area (A) [11].

The annual groundwater infiltration rate can be calculated by plugging the infiltration coefficient value from the above equation into the following formula:

$$R = \sum(A \times P \times C) \tag{3}$$

The formula used is the annual groundwater infiltration rate (R) in units of 10^6 m^3 by calculating the total infiltration surface area (A, in 10^6 m^2), the average annual rainfall in the area (P, in meters), and the specific infiltration coefficient in the study area (C, in percentage) [11].

Finally, it is hypothesized that urban expansion and the corresponding reduction in infiltration areas have significantly altered the spatial distribution of unconfined groundwater potential. The zone of high potential for groundwater availability is expected to shrink, moving further away from the city limits as suburban areas rapidly develop. This hypothesis addresses the interrelated dynamics between land use change, vegetation cover, and groundwater potential in a rapidly urbanizing suburban landscape. The analysis method is presented in Table 1 below.

Table 1. Analysis Method

Qualitative Analysis	**Quantitative Analysis**
A visual interpretation of Satellite imagery, namely, delineation of open space and vegetation zones, herbaceous types, shrubs, and trees	Tiered and weighted rating (weighting) Based on guidelines or empirical previous research, the weighing factor with a value of 1 (one) shows a small effect, and a value of 5 (five) has a large impact on the availability of groundwater reserves (well water level).
Thiessen Polygon Namely, the measurement of rainfall distribution parameters	Overlay Weighted Sum (Analyze water recharge class and water absorption class) Overlap mathematical functions on maps • Groundwater Recharge Class = (open space × 5) + (rainfall × 4) + (slope × 2) [12] • Water absorption Class = (shrub vegetation area × 5) + (tree vegetation area × 4) + (herbaceous vegetation area × 2) + (surface soil state × 3) + (composing rock state × 5) + (depth of well water level × 1) [9,12]
Euclidean Distance Namely, measurement of distance parameters of open space covered and range of vegetation, herbaceous types, shrubs, and tree types	Analyze the amount of groundwater recharge based on rainfall, wide open space, and the percentage of groundwater recharge in 2006 and 2024, so that the changes can be obtained $RC = RF \times A \times RC\ (\%)$

 https://doi.org/10.1051/e3sconf/202565211004

Slope based on satellite imagery from DEMNAS, namely, the measurement of slope parameters	Analyze the reduced water absorption rate based on the infiltration coefficient, range of area vegetation, and rainfall in 2006 and 2024, so that the changes can be obtained $R = \Sigma (A \times P \times C)$
Interpolating Spline Methods Namely, measurement of groundwater level (well water level) depth parameters	Analyze the potential distribution of unconfined groundwater (unconfined aquifer) with the intersection overlay technique

The comprehensive methodology for evaluating groundwater availability potential in Banguntapan District employs a multitemporal systematic framework that integrates spatial analysis with environmental data to elucidate the complex dynamics of shallow groundwater resources. This methodological framework utilizes Geographic Information System (GIS) techniques, specifically the Euclidean Distance principle integrated with Weighted Linear Combination (WLC) and Analytical Hierarchy Process (AHP), to assess variations in groundwater potential from 2006 to 2024, as illustrated in the following diagram. Figure 1.

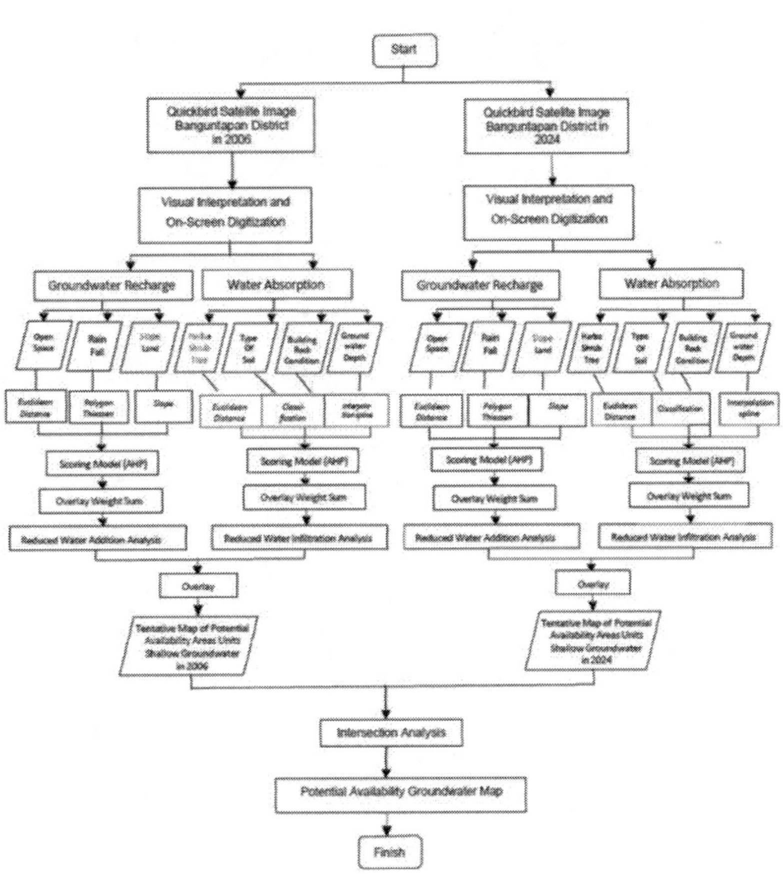

Fig. 1. Research Flow

3 Results and Discussion

3.1 Satellite Image Interpretation

Banguntapan District is one of the districts bordering Yogyakarta City, with an area of 2,786.53 hectares divided into eight villages, with a high population density (5,238 people/km^2) that continues to increase, resulting in a reduction in open space and the presence of vegetation. This can be seen visually in multitemporal satellite imagery, recorded on June 8, 2006, and March 13, 2024, as shown in Figure 2.

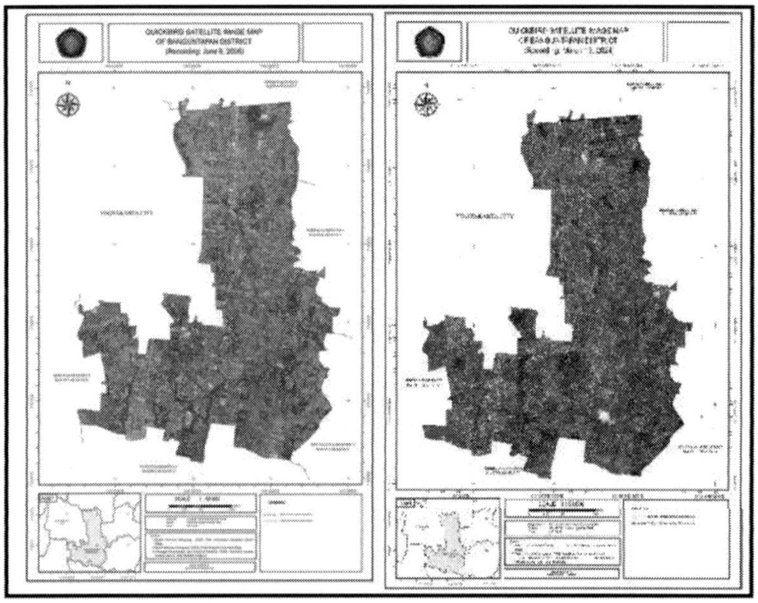

Fig. 2. Comparison of the Satellite Imagery in the Suburbs of Yogyakarta City (Banguntapan District)

Figure 2 provides a detailed overview of land use change in Banguntapan District between 2006 and 2024. This map shows the increasing distribution of built-up areas, the reduction of green open spaces, and areas of vegetation. As a densely populated region, Banguntapan is experiencing rapid development, as evidenced by the dominance of built-up areas in several villages, such as Banguntapan and Baturetno.

Research indicates that the rate of land use change in the Yogyakarta region reaches 3.8% per year, with Banguntapan District being one of the fastest-changing areas. This is reinforced by research showing that urbanization pressures have converted nearly 45% of agricultural land into built-up areas in the past six years. Spatial analysis reveals an uneven pattern of change, with a higher concentration of development in the northern and western regions. The remaining green areas are increasingly fragmented, forming vegetation zones between built-up areas. This transformation not only impacts the physical aspects of the landscape but also has significant environmental implications, such as the reduction of water catchment areas.

The characteristics of land use change in Banguntapan District reflect the comprehensive challenges in managing urban development. The balance between infrastructure growth and environmental preservation is a crucial issue that requires a sustainable planning approach.

Therefore, it is crucial to maintain natural water catchment areas amidst development pressures to ensure the sustainability of water resources in the region.

Analysis shows that water recharge decreased along with the reduction in open areas from 2006 to 2024, from 1,684.15 hectares to 1,056.77 hectares. Therefore, based on formula (1), groundwater recharge decreased by 316,752.52 m^3/year. Meanwhile, water infiltration decreased along with the reduction in vegetation area from 2006 to 2024, from 607,143 hectares to 481,612 hectares. Therefore, based on formulas (2) and (3), water infiltration decreased by 26,036,055.022 m^3/year. Thus, groundwater availability in Banguntapan District is decreasing, as shown in Figure 3.

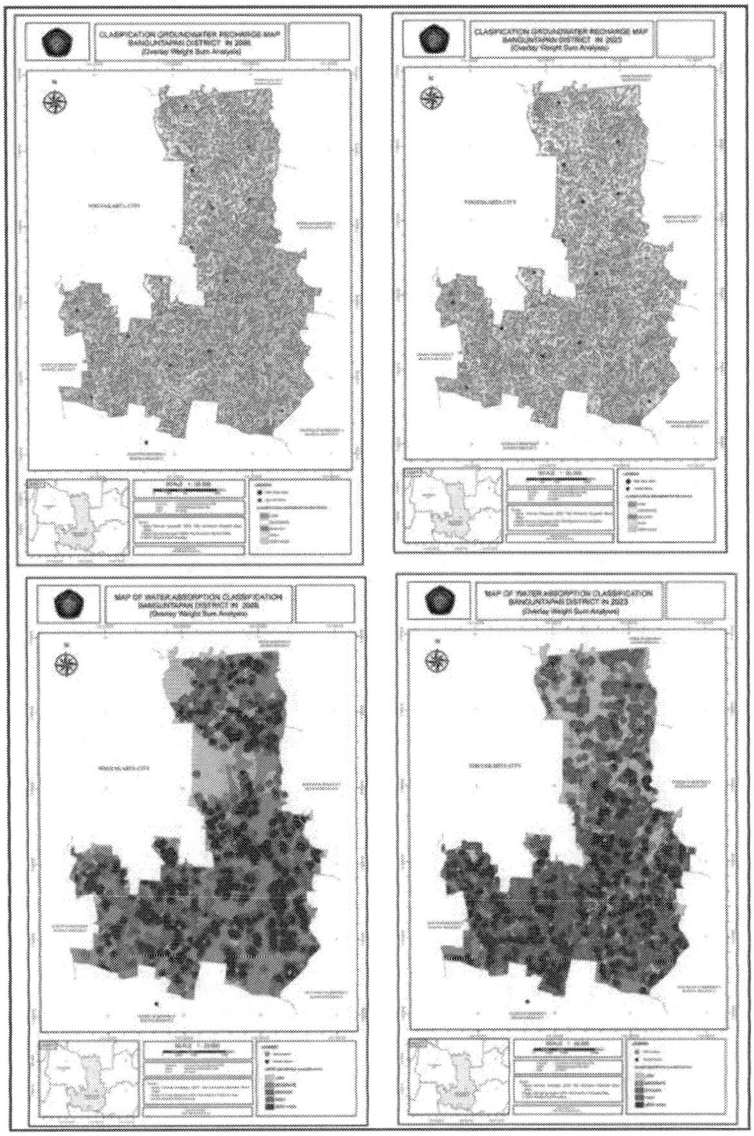

Fig. 3. Classification Map of Groundwater Recharge and Water Infiltration Using Euclidean Distance Analysis in the Suburbs of Yogyakarta City (Banguntapan District)

https://doi.org/10.1051/e3sconf/202565211004

The results of the Euclidean Distance analysis, visualized in Figure 3, show the distribution of groundwater recharge rates based on proximity to open land and water infiltration rates based on proximity to herbaceous vegetation, shrubs, and trees.

Zones with very high water infiltration are found in areas with abundant vegetation, while low water infiltration dominates built-up areas. This distribution demonstrates the importance of open land in supporting groundwater recharge in Banguntapan District. Land with low water recharge is mostly located near built-up areas with minimal vegetation.

The study area is a suburban area encompassing seven villages: Tamanan, Wirokerten, Jambidan, Potorono, Banguntapan, Baturetno, and Singosaren. The analysis indicates that the area has very high groundwater potential, approximately 8.33%. This suggests that the massive growth of buildings has reduced the effectiveness of open land in absorbing water. Table 2 below presents the distribution of potential groundwater availability using GIS intersection analysis techniques until 2024, with classifications of very available, sufficiently available, and less available.

Table 2. Areas of Potential Availability of Shallow Groundwater in the Suburbs of Yogyakarta City

Village	Area (ha)	Area of Potential Availability of Shallow Groundwater (ha)					
		Very Available (ha)	Area Percentage (%)	Sufficiently Available (ha)	Area Percentage (%)	Less Available (ha)	Area Percentage (%)
Tamanan	381.01	40.17	10.54	340.84	89.46	0	0
Wirokerten	367.51	43.88	11.94	323.63	88.06	0	0
Jambidan	358.51	45.04	12.56	313.47	87.44	0	0
Potorono	403.64	43.42	10.76	360.22	89.24	0	0
Banguntapan	596.67	14.86	2.49	406.98	68.21	174.83	29.30
Baturetno	550.02	32.27	5.87	513.62	93.38	4.13	0.75
Singosaren	129.17	12.54	9.71	108.00	83.61	8.63	6.68
Total Area	2786.53	232.18	8.33	2366.76	84.94	187.59	6.73

Although the study area has undergone significant changes with the addition of numerous buildings, most residents still rely on dug well water for their daily needs. Therefore, observations were made on changes in shallow groundwater availability in 14 wells selected by purposive sampling, representing seven village areas, as shown in Table 3.

Table 3. Distribution of Observation Wells in Banguntapan District

No	Well	UTM Coordinate (Zone 49 S)		Depth of Well Water Advance (m)						
		X (m East)	Y (m South)	Jul 2023	Agt 2023	Sept 2023	Okt 2023	Nov 2023	Des 2023	Mar 2024
1	A	435105.12	9134598.34	2.70	3.20	3.64	3.70	3.51	3.15	2.25
2	B	434672.37	9132883.29	1.99	2.45	3.27	3.28	3.12	2.68	1.72
3	C	432766.62	9133239.25	3.10	3.85	4.24	4.32	4.11	3.83	2.77
4	D	434287.22	9137281.57	4.90	6.55	7.68	7.81	7.70	7.52	7.08
5	E	434267.47	9135415.35	5.93	6.40	6.88	7.13	7.01	6.82	6.41

https://doi.org/10.1051/e3sconf/202565211004

No Well		UTM Coordinate (Zone 49 S)		Depth of Well Water Advance (m)						
		X (m East)	Y (m South)	Jul 2023	Agt 2023	Sept 2023	Okt 2023	Nov 2023	Des 2023	Mar 2024
6	F	434165.82	9138830.45	6.65	7.05	8.28	8.47	8.28	7.97	6.82
7	G	435634.18	9137864.34	3.25	3.85	4.32	4.44	4.31	3.99	3.11
8	H	435641.70	9136573.32	3.50	4.30	4.81	4.72	4.52	4.27	3.29
9	I	434759.36	9136362.80	4.25	5.70	6.22	6.25	6.08	5.89	5.05
10	J	435112.13	9133685.13	3.50	4.20	4.40	4.49	4.17	3.71	2.32
11	K	433547.95	9134622.65	4.25	5.80	6.27	6.38	6.26	6.04	5.18
12	L	431903.43	9131774.21	3.40	4.40	4.94	4.88	4.69	4.48	3.52
13	M	436393.32	9131435.62	3.20	4.10	4.44	4.37	4.25	4.02	3.25
14	N	431569.12	9133890.92	4.30	5.65	6.82	6.98	6.81	6.57	5.62

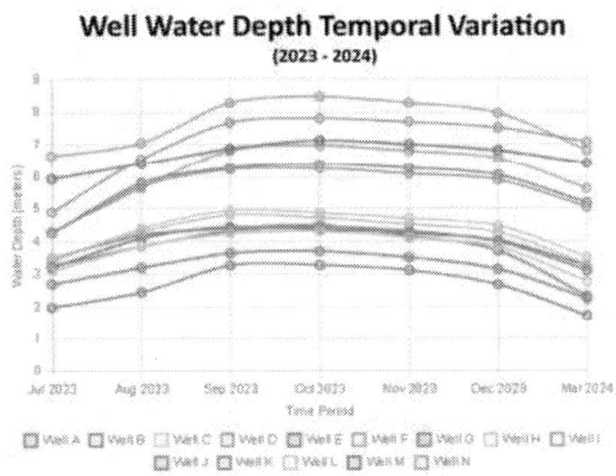

Fig. 4. Well Water Depth Fluctuation

Analysis of groundwater level data in Banguntapan District revealed significant variations in 14 observation wells from July 2023 to March 2024. These wells were spread across several areas with UTM zone coordinates of 49, covering variations in groundwater level depths ranging from 1.72 m to 8.47 m.

The general pattern observed was an increase in groundwater level depths from September to November 2023, followed by a decrease from December 2023 to March 2024.

Environmental conditions around the wells, seasonal rainfall factors, and groundwater use can contribute to these variations. Some wells, such as wells D and F, exhibited greater fluctuations in groundwater level depths, with peak depths reaching 7.81 m in October 2023 and then decreasing to 7.08 m in March 2024. Meanwhile, other wells, such as wells B and J, exhibited a more gradual decline.

The analysis revealed significant temporal variations in groundwater levels, with water depths fluctuating throughout the months. Wells located in more vegetated or exposed areas (e.g., wells F and G) show relatively higher water levels, particularly during the rainy season, such as in December 2023, when groundwater levels generally increase. Conversely, wells located near built-up areas (e.g., wells A and B) show lower water levels and smaller increases, indicating the impact of reduced exposure on groundwater recharge. A seasonal pattern is evident, with higher water levels during the rainy season (e.g., October and December 2023) and lower water levels during drier months (e.g., March 2024).

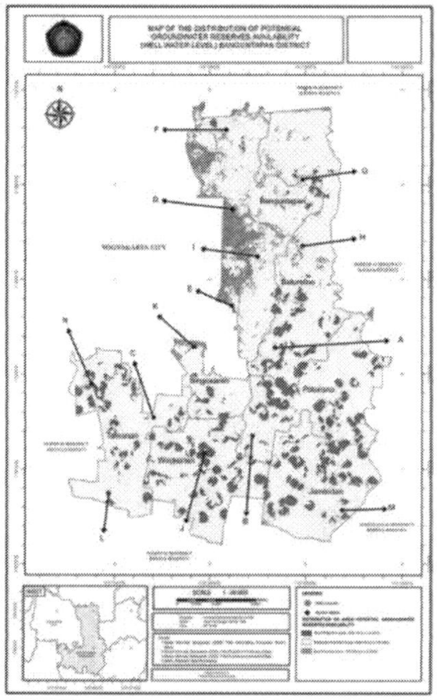

Fig. 5. Distribution Map of Well Water Levels and Potential Availability of Shallow Groundwater

Table 4. Distribution of Observation Wells in Banguntapan District

No	Well	Depth of Well Water Advance (m)	Range Covered (m)	Percentage of the Potential Availability of Groundwater Reserves			Percentage Area of Around the Well		Measurement of Average Ground water Level Increase (Nov 2023 – Mar 2024) (m)
				High (%)	Medium (%)	Low (%)	Open Space & Vegetation (%)	Other (Building) (%)	
1	A	3.21	250	28.2	71.8	-	78.3	21.7	1.45
2	B	2.7	250	3.1	96.9	-	81.5	18.5	1.56
3	C	3.79	250	6.3	93.7	-	80.2	19.8	1.55
Low									
6	F	7.71	250	-	94.7	5.3	63.7	36.3	1.65
7	G	3.91	250	8.8	91.2	-	77.4	22.6	1.33

No	Well	Depth of Well Water Advance (m)	Range Covered (m)	Percentage of the Potential Availability of Groundwater Reserves			Percentage Area of Around the Well		Measurement of Average Ground water Level Increase (Nov 2023 – Mar 2024) (m)
				High (%)	Medium (%)	Low (%)	Open Space & Vegetation (%)	Other (Building) (%)	
8	H	4.25	250	-	87.5	12.5	61.5	38.5	1.43
9	I	5.69	250	-	76.2	23.8	55.2	44.8	1.2
High									
11	K	5.79	250	-	63.2	36.8	72.7	27.3	1.2
12	L	4.39	250	5.2	94.8	-	71.7	28.3	1.36
13	M	3.98	250	28.8	87.2	-	79.4	20.6	1.12
14	N	6.19	250	41.6	58.4	-	67.2	32.8	1.36

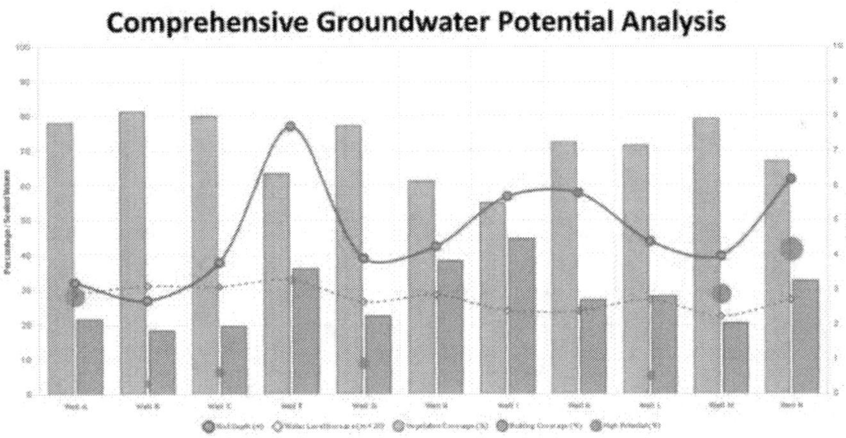

Fig. 6. Groundwater Availability Analysis

Table 4 presents well distribution data at 14 different locations, with a consistent 250-meter spacing between wells in Banguntapan District, focusing on variations in groundwater table depth between November 2023 and March 2024. Analysis of the observation wells reveals significant complexity in groundwater dynamics, including striking variations in hydrogeological conditions. Wells categorized as "high" and "low" exhibit fundamental differences in environmental characteristics and groundwater reserve potential, as follows: (1) Well F has a depth of 7.71 meters, with a land composition dominated by open space (63.7%) and buildings (36.3%). (2) Well K exhibits different dynamics, with a depth of 5.79 meters and a more complex land composition. (3)

The average groundwater level rise between November 2023 and March 2024 ranged from 1.12 to 1.65 meters, indicating significant fluctuations in groundwater reserves. (4) Well N showed the highest potential, with 41.6% in the high category, while several other wells showed a predominance of moderate conditions, reaching 96.9%. (5) The land composition around the wells also influences groundwater dynamics. Most locations have a high percentage of open space and vegetation (55.2%-81.5%), which is crucial for infiltration and groundwater availability.

The results also indicated that wells with low groundwater potential, such as wells A and B, experienced a significant reduction in surrounding vegetation and open space. These wells showed increases in groundwater levels of approximately 1.45 m and 1.56 m, respectively, indicating lower recharge rates, likely due to inadequate natural infiltration areas such as open space and vegetation. In contrast, wells in areas with taller vegetation and open space, such as wells K and L, exhibited significantly higher increases in groundwater levels, reaching 1.7 m and 2.12 m, respectively.

The measurement results also indicate that areas around wells with high groundwater availability potential have relatively fewer buildings, as seen in wells K and N, which have relatively high percentages of open space and vegetation (72.7% and 67.2%, respectively). Therefore, this study demonstrates that the spatial variability of groundwater potential requires a comprehensive water resources management approach.

Wells with low groundwater availability potential, such as wells A and B, experienced a significant reduction in surrounding vegetation and open space. These wells showed increases in groundwater levels of approximately 1.45 m and 1.56 m, respectively, indicating lower water recharge rates, likely due to insufficient natural infiltration areas such as open space and vegetation. In contrast, wells in areas with higher vegetation and open space, such as wells K and L, showed significantly higher increases in groundwater levels, amounting to 1.7 m and 2.12 m, respectively.

3.2 Climate Change Impacts on Groundwater Dynamics in Banguntapan District: Rainfall Patterns and Drought Implications

Vegetation significantly influences groundwater recharge by increasing infiltration and preventing surface runoff [12]. The data also shows that areas around wells with high groundwater potential are less urbanized, as seen in wells K and N, which have relatively high percentages of open space and vegetation (72.7% and 67.2%, respectively).

According to the Yogyakarta Special Region Environment and Forestry Service (2022), Yogyakarta Province only has 39.54% of its 318,600 hectares of land covered by vegetation. Conversely, the vegetation levels in Banguntapan District around wells K and N are 67.2% and 72.7%, respectively. This indicates that Banguntapan is better off, even though it is measured at the sub-district level as a buffer zone for Bantul Regency. The Land Cover Quality Index (IKLH) for Yogyakarta Province is only 58.82%, with an unattainable target of 66.20% by 2022, which is 8–14 percentage points below the Banguntapan vegetation level [13]. This also indicates the high level of vegetation in the Banguntapan District compared to the provincial scale.

The areas in wells K and N show significant increases in groundwater levels, supporting the theory that open space and natural vegetation facilitate groundwater recharge. In contrast, areas closer to the urban boundary with higher building coverage, such as wells F and G, show minimal increases in groundwater levels, further highlighting the negative impact of urbanization on groundwater recharge.

The interaction between climate change and potential groundwater availability in Banguntapan District is a crucial environmental challenge with far-reaching consequences for future water resource management. Recent climate trends indicate complex transformations in precipitation patterns and the hydrological cycle, significantly impacting groundwater recharge and sustainability.

3.3 Climate Trend Analysis (2006-2024)

Based on the current situation and potential future impacts of groundwater availability in Banguntapan District due to weather conditions, as analyzed with NASA Power data, the table climate trend analysis is presented in Table 5.

Table 5. Climate Trend Analysis in Banguntapan District

Year Range	Average Annual Rainfall (mm)	Rainfall Days	Dry Spell Duration (days)	Temperature Variation (°C)
2006-2010	2,1	120-135	30-45	18,05 – 34,77
2011-2015	1,95	110-125	45-60	19,06 – 34,22
2016-2020	1,8	95-110	60-75	19,41 – 34,46
2021-2024	1,65	85-100	75-90	19,35 – 34,46

Table 5 shows a progressive decline in annual rainfall, a reduction in rainy days, and a concerning lengthening of the dry season. These trends indicate significant hydroclimate transformations driven by regional climate change dynamics [14]. This study highlights the critical implications of a shorter rainy season for groundwater recharge. Analysis in Banguntapan District shows a substantial decrease in the effective rainfall period, which has the potential to impact water resource availability.

There has been a decrease in the duration of the rainy season [15]. This phenomenon also occurred in the study area between 2010 and 2024, with a duration of approximately 5 to 6 months. During this period, the phenomenon was exacerbated by the occurrence of extreme dry seasons lasting more than 90 consecutive days, resulting in a significant reduction in groundwater recharge rates. The progressive decline in rainfall and the longer dry period were directly correlated with variations in groundwater levels observed in 14 observation wells. In particular, wells in areas with less vegetation and closer to urban areas exhibit more pronounced water level fluctuations, indicating increased vulnerability to the impacts of climate change. Therefore, the Euclidean distance method for measuring the effective distance of wells to vegetation using GIS software provides clear correlations with groundwater level fluctuations.

4 Conclusion

A comprehensive study of groundwater availability potential in the outskirts of Yogyakarta (Banguntapan District) reveals the complex dynamics of shallow groundwater availability influenced by 3 main factors: massive " buildings are growing ", reduction in herbaceous, shrubby, and tree vegetation areas, and climate change. Spatial and temporal analysis shows significant regional landscape transformation, with green and agricultural land conversion reaching 45% between 2006 - 2024. This directly impacts groundwater recharge capacity, where zones with very high water catchment potential only cover 12.8% of the total area. So, the reduced groundwater recharge has reached 316,752.52 m^3/year, and the reduced groundwater infiltration volume rate has reached 26,036,055.022 m^3/year. Variations in groundwater depth in 14 observation wells (ranging from 1.72 m to 8.47 m) indicate a close relationship between land cover, vegetation, and groundwater availability. Climate projections show a significant decrease in rainfall (3-5% per decade) and a potential reduction in groundwater recharge capacity of up to 40% in the urban zone by 2035, threatening the sustainability and resilience of groundwater availability potential.

References

1. Salamah SD, Purnama ILS, Primacintya VA. Groundwater potency for domestic demand in Banguntapan District, Bantul Regency. E3S Web Conf. 2020;200:02011. doi:10.1051/e3sconf/202020002011
2. Budianta W, Firmansyah W, Setijadji LD. The influence of urbanization on the water infiltration zone in Banyumanik Sub-District, Semarang City, Central Java, Indonesia. KURVATEK. 2024 Apr 5;9(1):7–16. doi:10.33579/krvtk.v9i1.4698
3. Owuor SO, Butterbach-Bahl K, Guzha AC, Rufino MC, Pelster DE, Díaz-Pinés E, et al. Groundwater recharge rates and surface runoff response to land use and land cover changes in semi-arid environments. Ecol Process. 2016;5(1). doi:10.1186/s13717-016-0060-6
4. Maliva R. Climate and groundwater primer. In: Climate change and groundwater: Planning and adaptations for a changing and uncertain future. Cham: Springer; 2021. p. 15-34. doi:10.1007/978-3-030-66813-6_2
5. Boldrin D, Knappett J, Leung A, Brown J, Loades K, Bengough A. Modifying soil properties with herbaceous plants for natural flood risk-reduction. Ecol Eng. 2022;180:106668. doi:10.1016/j.ecoleng.2022.106668
6. Seifollahi-Aghmiuni S, Kalantari Z, Egidi G, Gaburova L, Salvati L. Urbanisation-driven land degradation and socioeconomic challenges in peri-urban areas: Insights from Southern Europe. AMBIO. 2022;51(6):1446-1458. doi:10.1007/s13280-022-01701-7
7. Yan X, Sun J, Li S, Ning T, Li Z. Indirect effects of revegetation dominate groundwater recharge change at the basin scale. J Clean Prod. 2024;456:142396. doi: 10.1016/j.jclepro.2024.142396
8. Kaewdum N, Chotpantarat S. Mapping potential zones for groundwater recharge using a GIS technique in the lower Khwae Hanuman sub-basin area, Prachin Buri Province, Thailand. Frontiers in Earth Science. 2021;9:717313. doi:10.3389/feart.2021.717313
9. Yunagardasari C, Paloloang AK, Monde A. Model infiltrasi pada berbagai penggunaan lahan di Desa Tulo Kecamatan Dolo Kabupaten Sigi. AGROTEKBIS: Jurnal Ilmu Pertanian. 2017;5(3):315-323.
10. Bakker, Pulawski. Metode Perhitungan Imbuhan Air Tanah sebagai Prosentase Curah Hujan Tahunan Rata-Rata. 1983.
11. Asdak C. Hidrologi dan Pengelolaan Daerah Aliran Sungai. Yogyakarta: Universitas Gadjah Mada (UGM) Press; 2023.
12. Wibowo M. Model penentuan kawasan resapan air untuk perencanaan tata ruang berwawasan lingkungan. J Hidrosfir. 2006;1(1):1-7.

13. Dinas Lingkungan Hidup dan Kehutanan Daerah Istimewa Yogyakarta. Rancangan akhir RENSTRA 2022-2027: rencana strategis Dinas Lingkungan Hidup dan Kehutanan Daerah Istimewa Yogyakarta. Yogyakarta: Dinas Lingkungan Hidup dan Kehutanan Daerah Istimewa Yogyakarta; 2023. 177 p.
14. Mensah FO, Alo CA, Ophori D. Hydroclimatic trends and streamflow response to recent climate change: an application of discrete wavelet transform and hydrological modeling in the Passaic River Basin, New Jersey, USA. Hydrology. 2024 Mar 25;11(4):43. doi:10.3390/hydrology11040043.
15. Zhang Q, Zhou B, Dong F, Liu Z, Ostovari Y. Spatial variability of groundwater quality for freshwater production in a semi-arid area. Water. 2021;13(21):3024. doi:10.3390/w13213024

Integrated Flood Control Strategy in New Capital City of Nusantara (IKN) Indonesia

Vita Ayu Kusuma Dewi[1], *Titi* Rahayuningsih[2], *Wartini* Ng[3], *Eko* Suwarno[4], *Mega Septia Sarda* Dewi[5], *Muhammad Abdul* Rahman[6], *Syafiqa Raudatur* Nadya

[1,2,4,5,6,7]Department of Civil Engineering and Planning, Universitas Negeri Malang, Indonesia
[3]School of Life and Environmental Science & Sydney Institute of Agriculture, The University of Sydney, Australia

Abstract. Indonesia's New Capital City, Nusantara (IKN) is situated in a region prone to hydrological complexities and potential climate change impacts, effective flood management is paramount for its sustainable development and the safety of its future inhabitants. This research synthesizes existing research on various approaches, including structural measures like dams, levees, and retention ponds, alongside non-structural measures such as early warning systems, land-use planning, and community-based disaster preparedness. The principles of Integrated Water Resources Management (IWRM) and its applicability in designing a resilient urban environment are explored, emphasizing the interconnectedness of water systems with urban planning, infrastructure, and ecological considerations. Special attention is given to incorporating nature-based solutions (NBS), such as restoring wetlands and creating green infrastructure, which offer co-benefits like biodiversity enhancement and improved water quality. The importance of governance frameworks and stakeholder participation in ensuring the successful implementation and long-term sustainability of flood control initiatives is also considered. Ultimately, this research aims to provide a comprehensive overview of integrated flood control strategies, offering valuable insights for, urban planners, and engineers involved in the development of IKN.

1 Introduction

Indonesia's ambitious undertaking to establish its new capital, Nusantara (IKN), in East Kalimantan represents a monumental step towards equitable national development and a sustainable urban future. However, the chosen location, while strategically advantageous in many respects, lies within a region inherently susceptible to hydrological complexities and the escalating threats of climate change [1]. This inherent vulnerability makes effective and integrated flood management not merely a logistical consideration but a paramount prerequisite for IKN's enduring viability and the safety of its future populace. The historical and scientific record unequivocally demonstrates that unchecked urban expansion in flood-prone areas, without robust and adaptive strategies, leads to significant economic losses, ecological degradation, and profound human suffering [2]. Therefore, a comprehensive understanding and proactive implementation of advanced flood control measures are indispensable for IKN to truly embody its vision as a smart and sustainable forest city [3].

The imperative for a sophisticated approach to flood control in IKN stems from several critical factors. Firstly, the dynamic nature of hydrological systems in equatorial regions, characterized by intense rainfall variability and complex riverine networks, necessitates solutions that go beyond traditional, singular interventions. Scientific literature increasingly advocates for Integrated Water Resources Management (IWRM), recognizing the intrinsic interconnectedness of water systems with urban planning, infrastructure development, and ecological preservation. This holistic perspective is crucial for IKN, where rapid urbanization must not compromise the delicate balance of the surrounding natural environment. Secondly, the accelerating impacts of global climate change, including more frequent and intense extreme weather events, present an unprecedented challenge to conventional flood defense mechanisms [2]. Futureproofing IKN against these unpredictable climatic shifts demands adaptable strategies that incorporate resilience and foresight.

Moreover, while structural measures like dams and levees have historically played a role in flood mitigation, their limitations, particularly concerning environmental impact and long-term sustainability, are well-documented. Contemporary scientific discourse and practical experience increasingly highlight the efficacy of non-structural measures, such as robust early warning systems, intelligent land-use planning, and empowered community-based disaster preparedness initiatives, in complementing physical infrastructure [4]. Crucially, the integration of nature-based solutions (NBS), including wetland restoration and the creation of green infrastructure, offers a paradigm shift in flood management. These solutions not only provide effective flood attenuation but also deliver significant co-benefits, such as enhancing biodiversity, improving water quality, and fostering urban liveability, aligning perfectly with IKN's ecological aspirations [5].

Finally, the success of any flood control strategy hinges on robust governance frameworks and active stakeholder participation. Academic discourse consistently emphasizes that top-down approaches often falter without the genuine involvement of local communities, experts, and various governmental and non-governmental entities [6]. For IKN, a project of national significance, ensuring broad participation and clear institutional responsibilities is essential for effective implementation and long-term maintenance of flood control initiatives. This research, therefore, aims to synthesize existing knowledge on these diverse yet interconnected approaches, providing a comprehensive overview of integrated flood control strategies. It seeks to offer valuable insights for policymakers, urban planners, and engineers tasked with the critical mission of developing a resilient and sustainable New Capital City of Nusantara for generations to come.

2 Geographical and Climatic Overview of IKN

Nusantara's chosen location on the eastern coast of Borneo, specifically within the Penajam Paser Utara and Kutai Kartanegara regencies, presents a distinct topographical profile that significantly influences its hydrological behavior. The area is characterized by a blend of rolling hills, flat lowlands, and an intricate network of rivers and smaller tributaries that drain into the Makassar Strait. Key river systems, such as the Karang Mumus, Mahakam, and Sepaku rivers, are vital to the region's ecology and water supply, but also serve as potential conduits for floodwaters [7]. The presence of extensive peatlands and mangrove forests in coastal areas further complicates hydrological dynamics, acting as natural sponges but also posing challenges for drainage if not managed carefully.

Climatically, IKN falls within the tropical rainforest zone, experiencing high annual rainfall with distinct wet and dry seasons, though often subject to inter-annual variability influenced by phenomena like the El Niño-Southern Oscillation (ENSO). Mean annual rainfall can exceed 2,500 mm, with intense, short-duration downpours common, particularly during the monsoon peaks. Historical flood data for the broader East Kalimantan region, though sometimes localized and fragmented, indicates a recurrent pattern of inundation, especially in low-lying areas and along riverbanks, often triggered by prolonged heavy rainfall or a combination of rainfall and tidal influences in coastal zones [8]. Understanding these specific patterns of rainfall intensity and distribution, alongside the capacity and characteristics of the existing river systems and their floodplains, is fundamental to designing effective flood control strategies for IKN.

The vulnerability of IKN to flooding is poised to intensify significantly under projected climate change scenarios. Global and regional climate models for Southeast Asia consistently predict alterations in precipitation patterns, leading to more extreme rainfall events and, paradoxically, longer dry spells [2]. For IKN, this translates into an increased likelihood of flash floods and riverine flooding due to higher volumes of water in shorter periods, overwhelming existing drainage infrastructure and natural absorption capacities. Concurrently, sea-level rise (SLR) presents a substantial long-term threat to IKN's coastal areas. As a coastal city, even modest increases in sea level, combined with potential land subsidence, can exacerbate tidal flooding, impede riverine drainage, and lead to saltwater intrusion into freshwater sources.

The implications for urban development are profound. Traditional engineering solutions may prove insufficient against a backdrop of escalating extreme weather. Climate change-induced shifts necessitate a dynamic and adaptive approach to urban planning, including revising flood hazard maps, updating building codes to withstand more severe inundations, and investing in resilient infrastructure that can endure future climatic stresses. Ignoring these projections would risk locking IKN into a development pathway highly susceptible to increasingly frequent and severe flood disasters, undermining its long-term sustainability and the safety of its inhabitants.

3 Evolution of Flood Control Approaches in Indonesia

3.1 Traditional Structural Method

Historically, Indonesia, like many rapidly developing nations, has predominantly relied on structural measures to mitigate flood risks. These conventional engineering solutions aim to control water flow physically and protect high-value areas. Dams, such as those in various river basins across Java and Sumatra, have been constructed for multi-purpose functions, including flood control, irrigation, and hydropower generation. While effective in regulating river discharge and providing storage for large flood events, their limitations include significant upfront costs, potential environmental impacts like habitat alteration and sediment accumulation, and the creation of a false sense of security in downstream communities.

Dikes and levees have been extensively employed along major rivers and coastal areas, particularly in densely populated urban centers like Jakarta, to prevent overbank flow. These structures offer direct protection but can also disconnect rivers from their natural floodplains, intensifying flood peaks downstream and increasing the severity of breaches if they fail. Retention ponds, or polders, have been implemented in urban areas to temporarily store stormwater runoff, reducing immediate inundation. While beneficial for localized flood reduction, their capacity is often limited, and their effectiveness depends heavily on proper maintenance and the rate of urban expansion [9]. The environmental footprint of these traditional measures, including land acquisition, ecological disruption, and the potential for increased vulnerability in unforeseen extreme events, has increasingly prompted a re-evaluation of their sole reliance.

3.2 Non-Structural Method

Recognizing the inherent limitations and often unsustainable nature of purely structural interventions, Indonesia has gradually incorporated non-structural measures into its flood management repertoire. These strategies focus on reducing vulnerability and exposure to floods without altering natural water flow channels. Early warning systems (EWS), for instance, have been developed and deployed in several river basins, leveraging hydrological modeling and real-time monitoring to provide timely alerts to communities at risk. The efficacy of EWS is directly tied to the robustness of the data collection, accuracy of forecasts, and, critically, the effectiveness of communication channels and community response protocols.

Land-use zoning regulations and building codes are fundamental non-structural tools aimed at guiding urban development away from high-risk floodplains and ensuring that new constructions are resilient to anticipated flood levels. While regulatory frameworks exist, their enforcement can be challenging, particularly in rapidly urbanizing areas with informal settlements. Public awareness campaigns and community-based disaster preparedness programs are crucial for fostering a culture of resilience. Initiatives like flood drills, education on safe evacuation routes, and the establishment of local disaster management teams empower communities to take proactive steps, significantly reducing casualties and damages during flood events [10]. However, the long-term efficacy of these measures hinges on sustained funding, continuous public engagement, and strong institutional support.

3.3 Integrated Water Resources Management (IWRM) in Urban Contexts

The deficiencies of fragmented, single-sector approaches to water management have driven a global paradigm shift towards Integrated Water Resources Management (IWRM). IWRM promotes the coordinated development and management of water, land, and related resources in order to maximize economic and social welfare in a truly equitable manner without compromising the sustainability of vital ecosystems [11]. For IKN, a city being built from scratch, the principles of IWRM offer an unprecedented opportunity to embed flood resilience at the very core of its urban fabric, rather than as an afterthought [4].

In an urban context, IWRM for flood resilience emphasizes the interconnectedness of water systems with urban planning, infrastructure, and ecological considerations. This means moving beyond merely channelling floodwaters away, towards strategies that (a) Work with natural processes: Leveraging natural floodplains, restoring wetlands, and promoting permeable surfaces to absorb and slow down water flow. (b) Integrate multi-sectoral planning: Ensuring that decisions related to housing, transportation, green spaces, and waste management all consider their implications for the water cycle and flood risk. (c) Foster collaborative governance: Involving all relevant stakeholders – from government agencies and private developers to local communities and environmental groups – in the planning and decision-making processes. (d) Emphasize adaptive management: Recognizing that climate conditions are changing and planning for flexibility and continuous adjustment of strategies.

For IKN, applying IWRM means designing a resilient urban environment where flood control is not a standalone engineering discipline, but an inherent aspect of sustainable urban design. This involves smart land-use allocation that respects natural hydrology, investments in green infrastructure that doubles as flood mitigation, and a governance structure that facilitates cross-sectoral coordination and community empowerment. This approach aims not just to protect the city from floods, but to enable the city to live with water in a sustainable and resilient manner.

4 Nature-Based Solutions (NBS) for Flood Resilience in Urban Environments

As Indonesia embarks on the development of Nusantara (IKN), there is a compelling opportunity to move beyond conventional flood control methods and embrace Nature-Based Solutions (NBS). These innovative approaches leverage the power of ecosystems and natural processes to manage water, offering not only flood resilience but also a myriad of co-benefits essential for a sustainable urban future [5].

In the context of flood management, NBS refers to actions that protect, sustainably manage, and restore natural or modified ecosystems to address societal challenges effectively and adaptively, while simultaneously providing human well-being and biodiversity benefits [5]. Fundamentally, NBS for flood control work by harnessing natural hydrological functions. Instead of constructing hard infrastructure to resist water, NBS aims to mimic or enhance natural processes that absorb, store, and slow down water flow. This can involve restoring natural sponges like wetlands, increasing infiltration through permeable surfaces, or creating natural barriers that dissipate flood energy.

NBS can be broadly classified into several types based on the ecosystem they utilize and their primary function [12]. These include:

- Green Infrastructure: Incorporating natural elements into urban areas, such as green roofs, permeable pavements, urban parks, and street trees, to manage stormwater runoff at its source.
- Ecosystem Restoration: Re-establishing degraded ecosystems like wetlands, floodplains, or forests to restore their natural flood retention and water purification capacities.
- Sustainable Land Management: Practices that reduce soil erosion and improve water infiltration in agricultural or forested areas upstream, thereby mitigating downstream flood risks.
- Coastal Protection: Utilizing natural coastal ecosystems like mangrove forests, coral reefs, or sand dunes to buffer against storm surges and coastal erosion.

The core principle is to view nature not as something to be controlled, but as an ally in creating a resilient and liveable urban environment.

For IKN, integrating NBS into its urban fabric is not merely an option but a strategic imperative that aligns with its vision of a "smart and sustainable forest city." The blank slate provided by a new capital offers unparalleled opportunities to embed NBS principles from the very initial stages of planning and design.

Firstly, strategic land-use planning must prioritize the preservation and restoration of existing natural floodplains, wetlands, and forests within and surrounding the IKN area. These natural assets should be designated as core components of the city's blue-green infrastructure network, not as developable land. Design principles should embrace "working with water", meaning layouts that allow for natural water flow, stormwater harvesting, and the creation of multi-functional green spaces that serve both recreational and flood retention purposes. For instance, parks and public spaces can be designed as temporary retention basins during heavy rainfall events [13].

Secondly, the development of IKN's infrastructure can seamlessly integrate NBS. This includes mandating permeable pavements for roads, sidewalks, and parking areas;

promoting green roofs and vertical gardens on buildings to reduce runoff; and implementing rain gardens and bioswales along streets and within residential areas. The extensive tree cover envisioned for IKN can be strategically designed to enhance urban forestry's role in stormwater interception and infiltration.

Finally, community acceptance and participation are crucial for the long-term success of NBS in IKN. Educating future inhabitants about the benefits of living with nature-based solutions and involving them in their maintenance and enhancement can foster a sense of ownership and responsibility. By embracing NBS, IKN can become a global exemplar of urban resilience, demonstrating how a thriving modern city can harmoniously coexist with its natural environment, effectively managing flood risks while enhancing ecological health and human well-being.

5 Challenges and Opportunities in Implementing an Integrated Flood Control Strategy for IKN

Implementing a comprehensive integrated flood control strategy in IKN will inevitably face several hurdles. Firstly, financial constraints loom large. While the Indonesian government is committed to the IKN project, the sheer scale of infrastructure development, especially for sophisticated integrated systems and nature-based solutions, will demand substantial and sustained investment. Reports from the National Development Planning Agency (Bappenas) have consistently highlighted the need for diverse funding sources, including significant private sector participation and international climate finance, to meet the IKN development targets [14].
Secondly, land acquisition issues, though potentially less complex than in densely populated existing cities, could still pose challenges. The vast area designated for IKN includes existing communities, indigenous lands, and diverse ecosystems. Ensuring equitable and transparent land acquisition processes that respect local rights, and ecological integrity is paramount to avoid social conflicts and project delays. Media reports and academic analyses often point to land tenure complexities as a recurring challenge in large-scale development projects across Indonesia. Mismanagement of land can disrupt natural flood pathways and increase vulnerability. [15]
Thirdly, technological limitations and the availability of skilled human resources may present obstacles. While Indonesia is progressing digitally, the implementation of advanced smart city solutions for real-time flood monitoring, predictive modeling, and sophisticated data analytics requires specialized expertise and cutting-edge infrastructure that might still need further development or significant investment in capacity building. This gap in local technical capacity for complex integrated systems necessitates robust training and knowledge transfer programs.

Finally, coordination difficulties among agencies and issues of public acceptance are common challenges in large-scale infrastructure projects. Integrated flood control demands seamless collaboration across various ministries (e.g., Public Works and Housing, Environment and Forestry, National Development Planning), local governments, and even private sector entities. Overlapping mandates or a lack of clear communication channels could hinder effective implementation [15]. Furthermore, ensuring public understanding and acceptance of new, sometimes less conventional, flood control measures like nature-based solutions is vital. Communities must be actively involved in the planning and maintenance, overcoming potential skepticism or resistance to change, as emphasized by local government and community engagement efforts in other Indonesian disaster risk reduction programs.

Despite these challenges, IKN's status as a greenfield development offers unprecedented opportunities to implement cutting-edge, sustainable, and resilient flood control measures. This is a chance to avoid the retrofitting challenges faced by older cities and to design flood resilience in from day one.

One significant opportunity lies in leveraging new technologies and smart city solutions. IKN can integrate sophisticated IoT (Internet of Things) sensors for real-time rainfall monitoring, river levels, and urban drainage performance. This data, fed into advanced predictive models and AI-driven analytics, can enable highly accurate flood forecasting and dynamic management of water infrastructure. The IKN Authority's commitment to a "smart city" concept provides a fertile ground for these innovations, allowing for capabilities like automated pump stations or intelligent floodgate management based on real-time data. The digital backbone of IKN can also facilitate seamless communication for early warning systems, disseminating information rapidly and precisely to affected areas [3].

Furthermore, IKN has the unique position to fully adopt international best practices in integrated urban water management and nature-based solutions. This includes learning from pioneering "sponge cities" like Wuhan, China, or Copenhagen, Denmark, which have successfully integrated green infrastructure for stormwater management, demonstrating reduced flood risk and improved urban liveability. The strong emphasis on establishing a "forest city" provides a natural synergy for extensive reforestation and wetland restoration efforts within and around IKN, not just for ecological preservation but specifically as vital components of the flood attenuation system [3]. This commitment to green infrastructure offers significant co-benefits like enhanced biodiversity, improved air and water quality, and creating a more liveable urban environment, truly embodying the "sustainable" aspect of IKN.

6 Conclusion and Recommendations

To ensure the long-term effectiveness and sustainability of IKN's integrated flood control strategy, several key recommendations for future research and policy directions emerge (a) develop a comprehensive digital hydrological twin for IKN, (b) pilot and Monitor Nature-Based Solutions Effectiveness, (c)establish a robust cross-sectoral governance framework. (d) incentivize private sector participation in green infrastructur. (e) Develop Comprehensive Capacity Building Programs. By proactively addressing these challenges and capitalizing on the unique opportunities, Indonesia can position IKN as a global exemplar of urban flood resilience, demonstrating how forward-thinking planning and an integrated approach can safeguard a new capital city against the growing threats of a changing climate.

References

1. R. Muharomah, B.I. Setiawan, V.A.K., Dewi, & N.I Hasanah. Identification of hydroclimatic patterns and trends in the new capital of Indonesia. In BIO Web of Conferences (Vol. 123, p. 04007). EDP Sciences. (2024)
2. Intergovernmental Panel on Climate Change (IPCC). Climate Change 2023: Synthesis Report. Contribution of Working Groups I, II and III to the Sixth Assessment Report of the Intergovernmental Panel on Climate Change. IPCC. (2023).
3. National Capital Authority. Vision and Principles of Nusantara. https://www.ikn.go.id/en . (2025)
4. United Nations Office for Disaster Risk Reduction (UNISDR). (2015). Sendai Framework for Disaster Risk Reduction 2015-2030. UNISDR.

5. International Union for Conservation of Nature (IUCN). Nature-based Solutions for Climate Change Adaptation: A Guide for Local Action. IUCN. (2020).
6. E. Feitelson, J. Chenoweth. Water poverty: towards a meaningful indicator. Water policy, 4(3), 263-281. (2002).
7. R. Plummer., D. Armitage. A resilience-based framework for evaluating adaptive co-management: empirical insights from a coastal management initauthoiative. Ecology and Society, 15(4), 23. (2010)
8. N. A. D. M. Hilal, K. Komariah, A. H. Ramelan. The multifaceted implications and challenges of relocating Indonesia's capital city: A comprehensive review of socio-economic, environmental, urban planning, and policy considerations. Sustinere: Journal of Environment and Sustainability, 8(3), 375-396. (2024).
9. S. Fawzy, A. I. Osman, J. Doran, D.W. Rooney. Strategies for mitigation of climate change: a review. Environmental Chemistry Letters, 18(6), 2069-2094. (2020)
10. E. Permanasari., M. Wijaya, V. Feliren, F. Khikmah, A. Lechner, A., Virani, R. Saputra. Enhancing urban resilience through integrated flood policy and planning: a mixed-methods evaluation of retention ponds for flood mitigation in South Bandung. AQUA—Water Infrastructure, Ecosystems and Society, 74(2), 267-282. (2025).
11. United Nations Office for Disaster Risk Reduction (UNDRR). (2017). Strengthening Community Resilience through Disaster Risk Reduction Education. UNDRR.
12. A. Fulazzaky. Challenges of integrated water resources management in Indonesia. Water, 6(7). (2014)
13. J.A. Castellar, L. A. Popartan, J. Pueyo-Ros, N. Atanasova, G. Langergraber, I. Säumel, V. Acuna. Nature-based solutions in the urban context: Terminology, classification and scoring for urban challenges and ecosystem services. Science of the Total Environment. (2021).
14. L. Xiaodan, K. Yamaguchi. Issues and strategies for designing flood resilient public space to achieve a balance between public amenity and stormwater management infrastructure. Urban and Regional Planning Review, 10, 197-223. (2023).
15. Kompas. Tantangan dan Peluang Pembangunan IKN tahun 2025. [Online news article, https://ikn.kompas.com/read/2024/12/28/190000587/tantangan-dan-peluang-pembangunan-ikn-tahun-2025?page=all . (2023).

Land Surface Temperature Dynamics due to the Development of Batang Integrated Industrial Estate (2014 & 2024)

Firdausha Amalia[1], *Dewi Novita* Sari[2*]

[1]Faculty of Geography, Muhammadiyah University Surakarta, Surakarta 57162, Central Java, Indonesia
[2]Faculty of Geographic Information Science, Muhammadiyah University Surakarta, Surakarta 57162, Central Java, Indonesia

Abstract. This study examines the relationship between land use change and Land Surface Temperature (LST) in the Batang Integrated Industrial Estate (BIIE), Central Java, using Landsat 8 imagery from 2014 and 2024. Land cover classification was performed to identify spatial changes, the Normalized Difference Vegetation Index (NDVI) was calculated to assess vegetation cover, and thermal infrared bands were processed to estimate LST. Random sampling of 50 points per year was applied for validation, and simple linear regression was used to evaluate the correlation between vegetation cover and LST. The results indicate a substantial conversion of agricultural and vegetated areas into built-up and industrial land. The mean LST increased from 28.3 °C in 2014 **to** 33.9 °C in 2024, reflecting a rise of more than 5°C within a decade. Correlation analysis shows a significant negative relationship between NDVI and LST, where the correlation coefficient value indicates that a decrease in vegetation cover is accompanied by an increase in surface temperature. This finding shows that industrial development contributes directly to regional temperature increases, and emphasizes the importance of integrating environmentally friendly infrastructure into spatial planning.

1 Introduction

Batang Regency is one of the regions in Central Java Province with an area of 788.65 km² divided into 15 sub-districts, and is currently undergoing accelerated infrastructure development to increase regional economic growth. One of the main projects is the development of the 4,300-hectare Batang Integrated Industrial Estate (BIIE), which is part of the National Strategic Project (PSN) and was initiated by the central government to attract foreign investment [8]. This area is an important point of industrial development on the north coast of Java and is planned to be fully operational by 2023. Currently, the Batang Integrated Industrial Estate (BIIE), which is part of the National Strategic Project (PSN), is being built in Batang Regency, Central Java, with an area of 4,300 hectares. The purpose of developing this area is to attract foreign investment and improve the country's economic competitiveness. However, converting vegetative land into industrial areas causes an increase in land surface temperature, one of the ecological impacts.

The growth of industrial areas in Indonesia contributes significantly to national economic development, but also has complex environmental impacts, particularly in relation to land use change and rising surface temperatures. The conversion of agricultural land and vegetation to built-up areas often reduces ecosystem function, increases habitat fragmentation, and triggers the heat island effect in the area. Previous studies have shown

[*] Corresponding author: dns104@ums.ac.id

that rapidly growing industrial areas, such as the Kendal and Cikarang Industrial Zones, contribute to increases in Land Surface Temperature (LST) due to reduced vegetation cover and the expansion of impervious surfaces.

However, comparable studies on the Batang Integrated Industrial Zone are still lacking. BIIE is one of the largest national strategic projects in Central Java, which has been developed since 2014 and is expected to become a new center of industrial growth. Research with a long-term temporal approach (2014 & 2024) is almost non-existent, even though this area is experiencing accelerated development with possible environmental impacts. This research gap needs to be filled in order to gain a comprehensive understanding of land use dynamics and surface temperature changes in new industrial areas.

In line with this gap, this study aims to: (1) Analyze changes in land use in the Batang Integrated Industrial Zone Batang in 2014 and 2024, (2) Analyze Land Surface Temperature (LST) in the Batang Integrated Industrial Zone in 2014 and 2024, and (3) Analyze the correlation between changes in land use area and temperature changes in the Batang Integrated Industrial Zone in 2014 and 2024.

The literature review reinforces this research framework through various aspects. Land use change has been extensively studied as a result of industrial development, which usually results in a reduction in productive vegetation cover [1]. Second, NDVI as a vegetation indicator shows a high response to changes in vegetation cover and is consistently used to evaluate land degradation [10]. Third, LST is a thermal indicator that can be obtained from the thermal infrared channel of Landsat imagery, and is strongly influenced by land use type: vegetation tends to be cooler, while built surfaces tend to be hotter [11; 12]. Previous studies in Indonesia have confirmed that industrial zones accelerate LST increases. For example, [2] reported a significant increase in surface temperature in the Kendal Industrial Area, while [8] found a similar pattern in coastal areas. However, research remains more focused on older industrial areas and large cities, such as Semarang, Jakarta, and Surabaya [Hayati]. Studies at the district level, especially in new industrial areas such as KITB, are still scarce.

Thus, this study contributes to two aspects. Theoretical contribution (novelty): it offers a temporal analysis of land use and LST in KITB, which was previously unavailable, and strengthens empirical evidence regarding the relationship between NDVI and LST. Practical contribution: it provides relevant spatial data to support sustainable land use policies and reduce the environmental impact of industrial development in Central Java.

2 Methods

2.1 Research Location

This study is located in Batang Regency, Central Java Province, Indonesia with an area of 788.65 km². Astromonically, the area is located at 6° 51' 46“ to 7° 11' 47” South latitude and 109° 40' 19“ to 110° 03' 06” East longitude. Based on its geographical location, Batang Regency is located on the island of Java and directly borders the Java Sea, besides that Batang Regency is directly adjacent to Kendal Regency, Wonosobo Regency, Banjarnegara Regency, Pekalongan Regency, and Pekalongan City. The region consists of 15 sub-districts, with the largest sub-district being Subah with an area of 83.52 km², while the smallest sub-district is Batang with an area of 28.11 km².[3].

https://doi.org/10.1051/e3sconf/202565212001

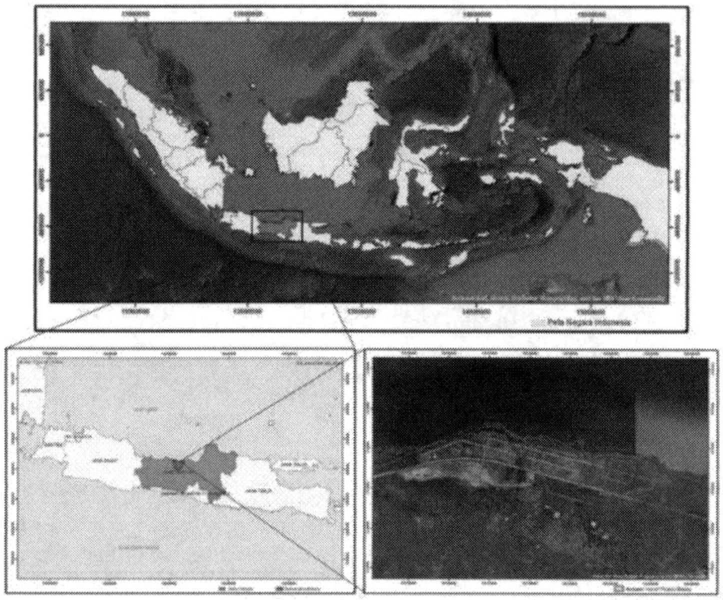

Fig. 1. Research Location

2.2 Tools and Materials

In this research, various types of spatial data were used to support the analysis of industrial estate development on land surface temperature changes in Batang Regency. The data used included Landsat 8 satellite images from 2014 and 2024 to identify land use change and land surface temperature change over time. In addition, data on the administrative boundaries of Batang Regency were used to provide geographical boundaries of the studied area. Factors driving land use change were analyzed using shapefiles of road networks, river networks obtained from the Geospatial Information Agency. The data sources used were assured, including from USGS EarthExplorer for satellite image data. The data collection method was carried out through the following steps:

Table 1. Research Materials

Data Name	Usefulness	Source
Landsat 8 Satellite Image 2014	Identify land use change and land surface temperature	*https://earthexplorer.usgs.gov/*
Landsat 8 satellite image in 2024	Identify land use change and land surface temperature	*https://earthexplorer.usgs.gov/*
Administrative Boundary of Batang Regency	Delimit the study area	Geospatial Information Agency
Shapefile Road network	A driving factor for land use change	Geospatial Information Agency
River Network Shapefile	A driving factor in land use change	Geospatial Information Agency

In satellite-based land mapping studies, there are tools used to help collect, process and analyze data. Each tool plays an important role in producing accurate and reliable data. And these tools help in various stages of research, from field data collection to software analysis. The following is a table of instruments used in this research:

Table 2. Research Instruments

Name	Purpose/Usage
ENVI 5.3	To processing image data
ArcGIS	To processing spatial data
Google Earth Pro	To test the accuracy of suitability of land use
Field observation	To record field observations

2.3 Data Processing and Analysis

This research applies a quantitative approach based on remote sensing and spatial analysis. the primary data consist of Landsat 8 OLI/TIRS imagery from 2014 and 2024, downloaded from the United Geological Survey (USGS). The analysis involved several stages:

2.3.1 Image Pre-processing

Radiometric and atmospheric corrections were performed, followed by image clipping to the Batang Regency boundary and coversion of Digital Number (DN) tp reflectance and brightness temperature.

2.3.2 Land Use Classification

A supervised classification was carried out using the *Maximum Likelihood* algorithm. Validation employed ground truth points and visual interpretation via Google Earth Pro.

2.3.2.1 Overall Accuracy: 87% (2014) and 85% (2024)

2.3.2.2 Kappa Coefficient: 0.86 (2014) and 0.84 (2024).

These values indicate a high level of classification reability (substantial agreement).

2.3.3 NDVI (Normalized Difference Vegetation Index) Calculation

NDVI was derived from the red and near-infrared (NIR) bands to assess vegetation density.

2.3.4 Land Surface Temperature (LST)

LST was computed from the thermal bands 10 and 11 by incorporating surface emisivity.

The emisivity (ε) is the radiation surface compared to a dark object at an identical temperature [6].

Surface temperature values are obtained by utilizing the thermal channel of landsat and extracted using the Mono-window Brightness Temperature algorithm. The LST value is influenced by the wavelength that is very sensitive to surface temperature, namely thermal infared [9].

2.3.5 Correlation Analysis

The relationship between NDVI and LST was tested using linear regression with 50 randomly selected sample point per year, following the *Central Limit Theorem.*

2.3.6 Spatial Analysis

The analysis technique in this research uses spatial data analysis, namely by using Landsat 8 satelite image to analyze land surface temperature. This satellite data processing uses software sch as ArcGIS, ENVI. In addition, geometric and radiometric correlations were made to ensure the data had spatial accuarcy and representative pixel values. A change analysis was also conducted to compare LST temperature values between 2014 and 2024 in the study area. Land cover change analysis was conducted using image classification methods to identify land cover change, as well as analyzing the relationship between land cover change and land surface temperature change.

2.3.7 Simple Linear Regression Analysis

Simple linear regression analysis is a method used to measure the relationship between one dependent variable and an independent variable. This simple linear regressions analysis is used to determine the directions of the relationship between the independent cariable and the dependent variable, whether it has a positive or negative relationship and to predict the values of the dependent variable if the value of the independent variable increases or decreases. The independent variable in this study is the area of change if Batang integrated industrial area, while the dependent variable in this study is LST.

2.3.8 Quantitative Descriptive Analysis

Quantitative descriptive analysis is a research method used to draw convlusion from the result of simple linear analysis that has been done before. Quantitative decriptive analysis in this case explain the relationship between the influence of land use change, in this case the development of industrial estates, on land surface temperature in Batang Regency.

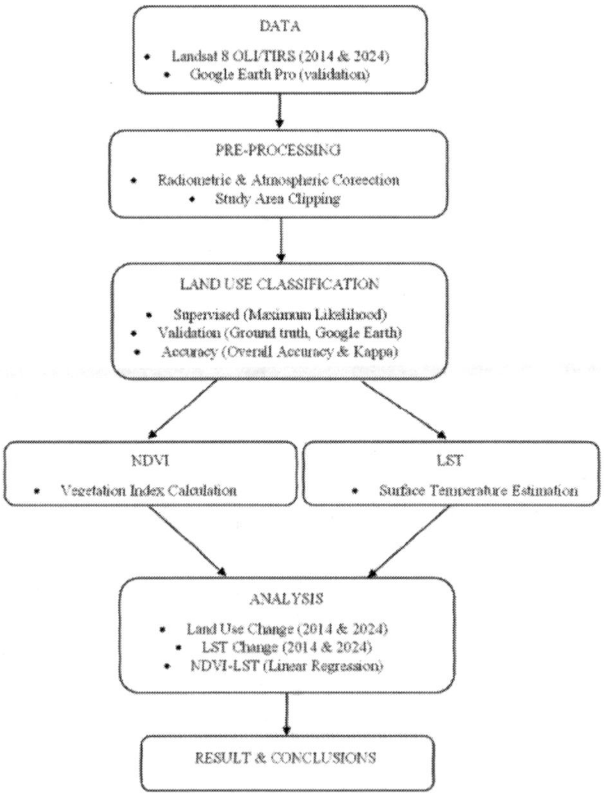

Fig. 2. Research Flow Chart

 https://doi.org/10.1051/e3sconf/202565212001

3 Results and Discussion

3.1 Land Change in 2014 and 2024

Land use is a form of land utilization carried out by humans, both permanent and non-permanent. Land use changes occur because of human needs to utilize existing land. Land use in Batang Regency, especially in the Batang Integrated Industrial Estate in 2014 and 2024 has changed, in this case land use is divided into 6 types of land use, including built-up land, forests, water bodies, plantations, rice fields, and vacant land. Which can be seen in the picture below:

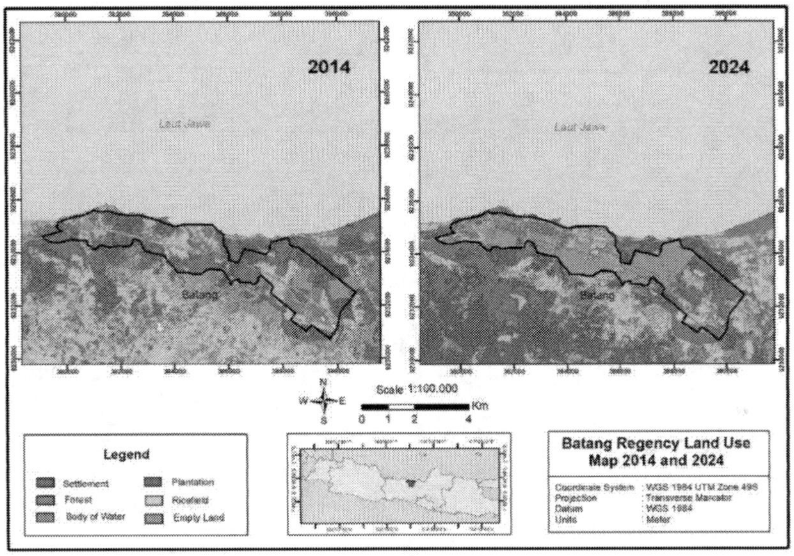

Fig. 3. Map of Land Use Change of Batang Regency in 2014 and 2024

The 2014 and 2024 Batang District land use maps show that over the past ten years, the northern area of Batang District, particulary the area now home to the Batang Integrated Industrial (BIIE), has been dominated by agricultural land, plantations and some forest. Light green, purple and dark green colors dominate this area, indicating that the vegetation there is still extensive and productive, both in agriculture and forestry.

However, in the 2024 map it is clear that areas that used to consist of rice fields and plantations have significantly changed into settlements and built-up areas show in red. Following the industrial development plan, residential areas are more widely spread and concentrated in the northern region. In addition, the yellow vacant land is increasing. These may be areas that are in preparation or transition for industrial development and supporting infrastructure.

The direct impact of industrial development is the transformation of land from natural vegetation to settlements and vacant land. in addition to reducing the area of agriculture and forests, it also disrupts ecological balances such as the water cycle and local microclimate. The map sgows that these changes cover the core of the industrial estate as well as buffer areas such as Gringsing and Banyuputih sub-districts. This pattern is in line with findings in other coastal areas, such as in Puring Sub-district, Kebumen Regency, where development led to a decrease in agricultural land and an increase in residential areas and infrastructure [8]

Therefore, it can be concluded that the development of BIIE has caused a major shift in land use in the northern area of Batang Regency. This shift is characterized by an increase in the area of residential land and vacant land, as well as a drastic decrease in the area of plantations and rice fields. This transformation is also one of the factors causing the increase in Land Surface Temperature (LST), as shown in the 2024 land surface temperature map.

Table 3. Extent of LULC Changes in 2014 and 2024

LULC	2014	2024
	Area (ha)	**Area (ha)**
Settlement	1.092.160	2.422.593
Forest	30.771.940	36.881.383
Water Body	6.876.937	2.566.851
Plantation	9.639.394	14.954.151
Paddy Field	22.312.420	19.080.265
Vacant land	8.171.302	2.958.920

Data showing significant changes in land use in Batang Regency from 2014 to 2024 shows significant dynamics caused by development, especially with the emergence of the Batang Integrated Industrial Estate (BIIE). Illustrating the rapid expansion of settlements, residental land increased rapidly by 1.330.433 hectares, or approximately 121,82%. This is very likely directly related to infrastructure needs, labor changes, and the development of the industrial estate around BIIE. On the other hand, forest land increased by 6.109.443 hectares, or 19,85%, as a result of forestation or land reclassification.

In contrast, there was a significant decrease in the land area of water bodies by 4.310.086 hectares (-63,79%). The decrease in the area of paddy fields by 3.232.155 hectares (-14,49%) also shows that the function of agricultural land has also changed due to the development of industrial estates. Plantation land actually increased by 5.314757 hectares, or 55,14%, indicating that the agro-industry, industry has moved or developed outside the area directly affected by the BIIE development.

Overall, these land use change trends indicate a strong shift towards industrialization and urbanization, with significant ecological and social impacts fot the people and environment of Batang District. The significant increase in settlements and decrease in the amount of water and paddy fields show that the BIIE development has had an impact on local and regional land use.

This pattern is consistent with the case in the Kendal Industrial Area, where since 2016, changes in agricultural land use have reached thousands of hectares, thereby impacting local food security [2]. In Pekalongan, the pattern of change is more focused on coastal areas, which include the conversion of fish ponds and mangrove forests into residential areas and trade zones [5]. These contextual differences indicate that geographical typology influences the direction of land change. BIIE is more focused on

https://doi.org/10.1051/e3sconf/202565212001

the agricultural sector, as is Kendal but on a smaller scale, while Pekalongan is more prone to coastal ecosystem degradation.

3.2 Land Surface Temperature (LST) in 2014 and 2024

Land surface temperature is the temperature of radiation on the land surface, which indicates how hot the earth's surface is at a particular location. Changes in land surface temperature occur due to human activities such as land use transformation, especially as a result of the conversion of vegetative land such as forests, rice fields, and plantations into built-up land such as settlements, industrial areas, and supporting infrastructure in an area. Land surface temperature in Batang Regency, especially in the Batang Integrated Industrial Estate (BIIE) in 2014 and 2024, has changed, in this case the land surface temperature is divided into 5 classes, including very low, low, medium, high, and very high. For more details can be seen in the figure below:

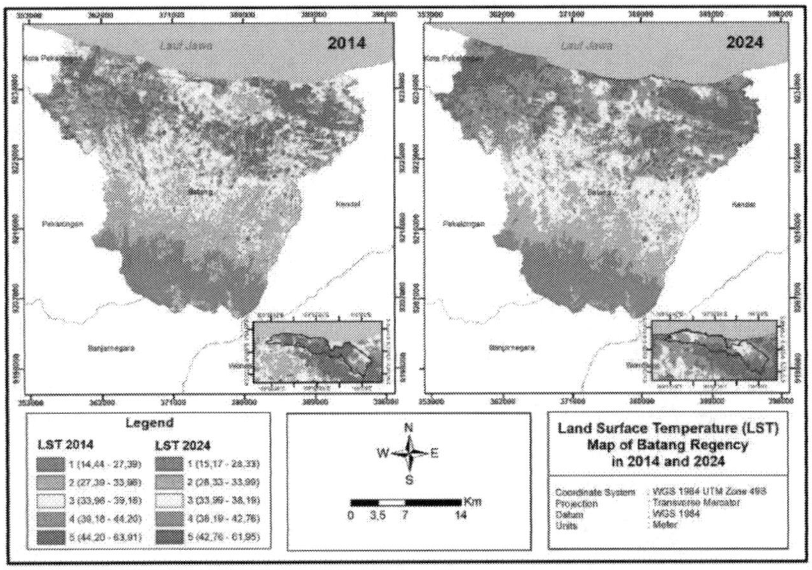

Fig. 4. Map of Land Surface Temperature (LST) of Batang Regency in 2014 and 2024

Based on the Land Surface Temperature (LST) maps of Batang Regency in 2014 and 2024, there has been a significant increase in land surface temperature in the span of ten years, especially in the Batang Integrated Industrial Estate (BIIE) development. This change is closely related to the dynamics of land use, intensification of infrastructure development, and the growth of built-up areas in the region.

Thermal conditions in Batang District were generally stable in 2014. Temperatures in the northern part of what is now BIIE were mostly low to moderate, ranging between 14,44°C and 39,18°C. The green and yellow colors that dominate this area indicate the presence of natural vegetation cover such as agricultural land, rice fields, secondary forests, and smallholder plantations. This vegetation cover serves to reduce surface temperature through the process of evapotranspiration. Meanwhile, the very low temperatures in the southern region of Batang Regency, especially in hilly areas and protected forests, indicate that areas with good vegetative cover and minimal

https://doi.org/10.1051/e3sconf/202565212001

development activity have a natural ability to regulate surface temperatures and keep the local microclimate stable.

The 2024 ESG map shows a significant increase in land surface temperature, especially in the northern area where BIIE is built. Orange to red colors, indicating high and very high temperature classes (38,19°C), begin to dominate the area. The 2024 land use map shows that the increase in the area is very consistent with significant land conversion. In the development of industrial parks such as BIIE, large-scale land clearing, surface paving, construction of factory buildings, roads and logistics facilities are common. These surface temperatures are very high during the day and do not decrease significantly in the evening due to their nature of absorbing heat intensively and releasing heat slowly. In addition, the loss of vegetation reduces evapotranspiration which is a natural process to cool down. As a result, the overall thermal balance is disturbed.

The increase in temperature caused by the constrction of the industrial estate to this area of BIIE and surrounding areas such as Gringsing and Banyuputih sub-districts, as well as parts of central Batang. In this case, the temperature of the built-up area is always higher than that of the surrounding vegetation. This shows that even in buffer areas that do not experience physical development directly, large-scale industrial development can have dignificant thermal impacts. In the LST than 2024 map, there is a marked difference between the northern and southern areas of Batang District. The north has high temperatures, but the south has low temperatures due to the remaining forest cover. This shows the ecological role of vegetation as a heat sink and temperature regulator for the region.

In addition to local environmental impacts, rising surface temperatures have broader consequences. These include increased energy consumption for cooling, reduces thermal comfort for aurrounding communities, and increased stress on micro-ecosystems and agriculture. Moreover, these thermal impacts will continue to expand if the development of industrial estate is not balanced with adequate green open spaces. Therefore, at every stage of future industrial estate development, it is imperative to apply sustainable development principles, such as the implementation of green infrastructure, vegetative buffer zones, and microclimate temperature management.

In comparison, Kendal shows a smaller increase in temperature because industrial development is more focused and has not yet reached the scale of BIIE. Pekalongan is affected by a number of factors, including mangrove degradation, sea surface temperature increase, and coastal settlement development. Thus, BIIE has a unique position because its thermal influence is directly driven by massive large-scale industrial development.

3.3 Correlation

The relationship between the effect of land use change on LST is carried out by calculating simple linear regression. The calculation using linear regression is land use as the independent variable (X) and LST as the dependent variable (Y), the data included in this calculation is the area of land use based on class classification and LST temperature data in Batang Regency.

https://doi.org/10.1051/e3sconf/202565212001

3.3.1 The relationship between the effect of land use area on LST in Batang Regency in 2014

Simple linear regression calculations on the relationship between the influence of land use area on LST resulted in a correlation value (R) of 0,999. The coefficient of determination (R Square) is 0.9990487, thus the influence of the independent cariable on the dependent cariable is 99% while the remaining 1% influenced by other factors outside the large variable which in this case is not explained in this study.

Table 4. Correlation between area (ha) of land use and LST in 2014

	Area(Ha)	*LST*
Area(Ha)	1	
LST	0,99979777	1

Based on the results of the correlation calculation above, it shows that there is a strong relationship between the independent cariable (land use) and the dependent variable (LST), this very strong relationship menas that there is a very strong positive relationship between the two variables. The correlation calue is very close ti 1, indicating that the wider the land use class, the higher the ESG. This relationship is unidirectoral so it can be concluded that changes in the area of a land cover greatly affect changes in surface temperature. This is especially true for development activities such as in industrializes areas. In addition, this finding enhances the understandig of the direct relationship between surface temperature increase and land transformation from vegetation to built-up land such as settlements and industries. Nonetheless, this correlation does not automatically indicate a cause-and-impacts the thermal environment in the study area.

3.3.2 The relationship between the effect of land use area on LST in Batang Regency in 2024

Simple linear regression calculations on the relationship between the influence of land use area on LST resulted in a correlation value (R) of 0,9997. The coefficient of determination (R Square) is 0.9997872, thus the influence of the independent cariable on the dependent cariable is 99% while the remaining 1% influenced by other factors outside the large variable which in this case is not explained in this study.

Table 5. Correlation between area (ha) of land use and LST in 2024

	Area(Ha)	*LST*
Area(Ha)	1	
LST	0,99893554	1

Based on the results of the correlation calculation above, it shows that there is a strong relationship between the independent variable (land use) and the dependent variable (LST), this very strong relationship means that there is a very strong positive relationship between the two variables. The correlations value menas that the larger the area of a land use class, the higher the land surface temperature value generated. This study is in line with several previous tudies that show that land use change, especially the transformation of vegetative land into built-up land, contributes significantly to the increase in LST. Due to their ability to evapotranspire and absorb heat, vegetation plays an important role in reducing remperature [12]. In contrast, built-up land such as store and reflect heat into the atmosphere, thus causing surface temperatures to increase [11].

Between 2014 and 2024, the development of the Batang Integrated Industrial Estate (BIIE) will have a noticeable impact on land use change and land surface temperature in Batang District. Cause-and-effect relationships were used to examine the correlation between the BIEE development and changes in land surface temperature. Industrial development changes the land function from vegetative land to built-up land, and as vegetation cover is lost, this change in land function results in an increase in surface temperature. In addition, the effects of large-scale infrastructure development increase heat emissions from construction activities, vehicles, and machinery.

Overall, it can be said that while the development of industrialized areas such as BIIE is helpful for local economic growth, it also brings about ecological changes, especially the increase in land surface temperature and the destruction of natural land cover. Therefore, to mitigate the negative impacts of development without hindering the economic progress of the region, sustainable spatial planning and environmental policies are needed, including the implementation of green infrastructure, preservation of vegetative buffer zones, and strict supervision of excess land conversion.

When compared to Kendal and Pekalongan, the correlation pattern is relatively similar, but differences in the intensity of land change make BIIE more noticeable. In Kendal, the loss of agricultural land has resulted in an increase in temperature, although it is still within controlled limits [2]. In Pekalongan, the loss of mangroves has further increased temperatures and exacerbated the threat of coastal disasters [5]. In BIIE, because development is concentrated in a large flat area, there is a more linear and stronger correlation between land change and temperature increase. The strength of this relationship underscores the importance of implementing mitigation policies, because without effective spatial planning interventions, the pattern of temperature increase in BIIE will continue exponentially [13].

The impact of this research is very significant in the context of spatial planning and green infrastructure policy development in the Batang Integrated Industrial Zone. Large-scale industrial development must be accompanied by spatial planning that not only focuses on economic growth but also considers environmental capacity and community welfare. In spatial planning, the existence of green open spaces must be maintained at a minimum of 30% in accordance with the provisions of Law No. 26 of 2007 [7], with even distribution to maintain temperature balance and the ecological function of the area. In addition, the implementation of buffer zones in the form of green belts around industrial areas is an important tool in reducing pollution, maintaining water absorption, and preventing heat spread from built-up areas.

In terms of green infrastructure, the development of BIIE should apply the principles of eco-industrial zones with environmentally friendly designs, such as the use of green roofs, green walls, and reflective materials to reduce heat absorption [4; 13). In addition, the creation of ecological networks through urban accommodation, industrial parks, and riparian vegetation restoration is crucial to maintaining ecosystem connectivity and reducing the risk of environmental damage [12]. The use of renewable energy and sustainable transportation should also be prioritized to create low-carbon industrial areas.

4 Conclusion

Based on the results of the analysis of land use change on land surface temperature that has been carried out, the following conclusions can be drawn:

1. The BIIE and surrounding areas experienced significant changes in their land use structure over ten years. Its land area increased dramatically from 1,092,160 ha in 2014 to 2,422,593 ha in 2024, indicating the development of industrial development. In contrast, paddy fields, water bodies and vacant land experienced a significant decrease in area. This indicates that open land and productive land have turned into built-up areas. This change is part of the development trend towards strong urbanization and industrialization.
2. Land surface temperature across the BIIE region is increasing, especially in the northern region which mainly consists of residential and industrial areas. In 2014, the LST was predominantly in the low to moderate temperature class (14.44°C - 39.18°C), but by 2024 more areas with high to very high temperature class (above 39.18°C) are expanding. This shows the direct effect of land use change on surface temperature, which increases as vegetation cover decreases.
3. The results of the analysis show that there is a positive correlation between changes in land use area and changes in LST. Land such as plantations and settlements show an increase in surface temperature. However, a decrease in land area such as paddy fields and vacant land tends to be associated with a decrease in the natural cooling capacity of the region. This correlation suggests that the development of industrial areas without careful ecological planning may worsen the thermal conditions of the area and potentially lead to urban heat island in the future.

References

1. Anderson, J. R., Hardy, E. E., Roach, J. T., & Witmer, R. E. (1976). *A land use and land cover classification system for use with remote sensor data* (Vol. 964). U.S. Government Printing Office. https://doi.org/10.3133/pp964
2. Baihaqi, H. F., Prasetyo, Y., & Bashit, N. (2019). Analysis of the development of the Kendal Industrial Area on land surface temperature changes (Case study: Kendal Industrial Area, Kendal Regency). *Jurnal Geodesi Undip, 9*(1), 176–186. https://doi.org/10.14710/jgundip.2020.26162
3. Coordinating Ministry for Economic Affairs. (2014). *Carrying the concept of the Smart & Sustainable Industrial Estate, the Batang Integrated Industrial Area becomes the front line of investment in the industrial sector* [Government report]. Coordinating Ministry for Economic Affairs. Retrieved from https://www.ekon.go.id/unduh/publikasi/5671/mengusung-konsep-the-smart-sustainable-industrial-estate-kawasan-industri-terpadu-batang-menjadi-garda-terdepan-investasi-sektor-industri
4. Grimmond, S. (2007). Urbanization and global environmental change: local effects of urban warming. The Geographical Journal, 173(1), 83–88. https://doi.org/10.1111/j.1475-4959.2007.232.x
5. Hayati, A. R. N. (2019). *Utilization of Landsat 8 imagery to identify land surface*

temperature changes in Ngawi Regency in 2015, 2017, and 2019 (Doctoral dissertation, ITN Malang).

6. Ibrahim, F., Atriani, F., Wulan, T. R., Putra, M. D., & Maulana, E. (2016). Comparison of Landsat 8 TIRS brightness temperature extraction without and with atmospheric correction for estimating surface temperature. In *National Geography Seminar UMS 2016: Efforts to Reduce Disaster Risk Related to Climate Change* (pp. 568–576). Solo, Indonesia. https://doi.org/10.13140/RG.2.2.15218.30402
7. Ministry of Agrarian Affairs and Spatial Planning/National Land Agency. (2007). Law of the Republic of Indonesia Number 26 of 2007 on Spatial Planning. Jakarta: Ministry of ATR/BPN.
8. Mungiza, U., & Sigit, A. A. (2021). *Analysis of land use change in the coastal area of Puring District, Kebumen Regency in 2009 and 2020* (Doctoral dissertation, Universitas Muhammadiyah Surakarta). Universitas Muhammadiyah Surakarta. http://eprints.ums.ac.id/id/eprint/97298
9. Nugroho, U. C., & Domiri, D. D. (2017). Identification of land surface temperature distribution of the geothermal area in Mount Ungaran using Landsat 8 imagery. *International Journal of Remote Sensing and Earth Sciences (IJReSES), 12*(2), 143. https://doi.org/10.30536/j.ijreses.2015.v12.a2708
10. Rouse, J. W., Jr., Haas, R. H., Schell, J. A., & Deering, D. W. (1974). Monitoring vegetation systems in the Great Plains with ERTS. In S. C. Freden, E. P. Mercanti, & M. A. Becker (Eds.), Proceedings of the Third Earth Resources Technology Satellite-1 Symposium, Vol. 1, Technical Presentations (pp. 309-317). NASA. NASA Publication Number SP-351. Document ID 19740022614.
11. Voogt, J. A., & Oke, T. R. (2003). Thermal remote sensing of urban climates. *Remote Sensing of Environment, 86*(3), 370–384. https://doi.org/10.1016/S0034-4257(03)00079-8
12. Weng, Q. (2004). Thermal infrared remote sensing for urban climate and environmental studies: Methods, applications, and trends. *ISPRS Journal of Photogrammetry and Remote Sensing, 64*(4), 335–344. https://doi.org/10.1016/j.isprsjprs.2009.03.007
13. Sun, Q., Wu, Z., & Tan, J. (2019). The relationship between land surface temperature and land use/land cover in Guangzhou, China. Environmental Earth Sciences, 59(6), 1347–1355. https://doi.org/10.1007/s12665-009-0112-4.

Correlation of Built-up Land Density, Transportation Network, and Activity-Travel Patterns with Activity Space: A Temporal Study

Sakinah Fathrunnadi Shalihati[1,2*], *Andri* Kurniawan[3], *Sri* Rum Giyarsih[3], and *Djaka* Marwasta[3]

[1]Doctoral Student, Faculty of Geography, Universitas Gadjah Mada, 55281 Yogyakarta, Indonesia
[2]Geography Education, Universitas Muhammadiyah Purwokerto, 53182 Banyumas, Indonesia
[3]Faculty of Geography, Universitas Gadjah Mada, 55281 Yogyakarta, Indonesia

Abstract. The issue of sustainability in urban land use is increasingly relevant in the context of preparedness for future crises, based on the experience of community spatial dynamics before and during the COVID-19 pandemic. Studies that temporally examine the relationship between the built environment, transportation systems, and individual activity-travel patterns are still limited. This study explores changes in activity-travel patterns and activity spaces on weekdays and weekends between before and during the pandemic, and investigates the correlation of built-up land density, transportation network characteristics, and activity-travel patterns with activity spaces using Spearman's rank analysis. Activity space is measured through two indicators: how far the distributions of out-of-home activity locations are from the home location (*Ih*), and the diversity level of out-of-home activity locations from their center of out-of-home activity locations (*Ic*). The results show that commercial area density (r = -0.004 to -0.241) and time allocation at home for mandatory activities (r = -0.014 to -0.473), maintenance (r = -0.021 to -0.291), and leisure (r = -0.008 to -0.487) are negatively related to activity space. In contrast, time allocated out-of-home for maintenance (r = 0.038 to 0.662) and leisure (r = 0.139 to 0.625) activities was positively correlated. Time allocated for mandatory activities out-of-home was positively correlated (r = 0.242 to 0.588) with *Ih*, while non-motorized mode use showed a positive correlation with *Ic* (r = 0.041 to 0.212). These relationships were generally consistent temporally, although not always significant. Meanwhile, road network characteristics made relatively little contribution to activity space, while trip duration, motorcycle use, number of trips, and trip chains actually strengthened during the pandemic period (r = 0.767 to 0.911, $p < 0.05$). These findings deepen our understanding of the relationship between the built environment and spatial mobility, and provide a basis for formulating land management policies that support efficient, adaptive, and sustainable mobility.

* Corresponding author: sakinah2019@mail.ugm.ac.id

1 Introduction

The rapid development of urban areas necessitates effective land use management and transportation systems that can efficiently and sustainably support community mobility. The COVID-19 pandemic has become a significant turning point, exposing the vulnerability of urban systems to sudden social change and providing long-term lessons for urban planning practices. The Yogyakarta Metropolitan Area (YMA) serves as a strategic representation in this context, given the complexity of its activities as a center for education and tourism, its intensive spatial growth, mobility, and accessibility issues, and the challenges of spatial management during mobility restriction policies during the pandemic [1–3].

Mobility restriction policies in the YMA during the pandemic, such as working and studying from home, limiting visits to public spaces, and implementing social distancing, have had a direct impact on community behavior. As many as 27.81% of residents decided not to travel, 55% of public facilities were closed, and 63.18% of social activities were suspended [3]. A shift in mobility from public transportation to private vehicles has also occurred, marked by an increase in motorcycle and car ownership of more than 3% each and a decrease in public transportation passengers of 71.48% [4]. These changes also impact activities and visits to public spaces, as reflected in the 84.08% of micro and small businesses reporting a decrease in visitors [5]. These dynamics not only reflect temporary adaptations but also the changing patterns of mobility and use of public spaces that have the potential to have long-term impacts on spatial planning and transportation, thus emphasizing the importance of strengthening the resilience of urban systems to face similar crises in the future.

The dynamic urban context emphasizes the urgency of understanding the role of physical environmental characteristics, such as built-up land density and transportation network structure, as well as individual activity-travel patterns in shaping daily space use and interaction patterns. This understanding provides a crucial foundation for adaptive spatial planning and transportation policies that support long-term urban resilience. The space utilized for daily interactions and reflecting individual adaptation to limitations of space, time, and transportation resources is called activity space, conceptually understood through a space-time prism [6,7]. Analysis of activity space is a crucial indicator for understanding spatial connectivity and community mobility behavior, as well as providing a basis for regional planning strategies [8]. However, studies examining the dynamics of the temporal relationship between physical environmental characteristics, activity-travel patterns, and activity space, particularly in the context of disruption caused by the pandemic, are still limited. This research aims to fill this gap.

This research aims to examine changes in activity-travel patterns and activity space on weekdays and weekends between before and during the pandemic, while also evaluating the relationship of built-up land density, transportation networks, and activity-travel patterns with activity space. This study was conducted to provide an empirical basis for urban planning that is responsive to changing mobility patterns while supporting long-term efficiency, sustainability, and resilience to potential future crises.

2 Method

2.1 Data Collection Sources and Techniques

This research was designed as a quantitative study with a comparative longitudinal survey approach, comparing individual activity-travel data before and during the pandemic. Data during the pandemic period were collected through a daily activity-travel log survey of 109

https://doi.org/10.1051/e3sconf/202565212002

respondents at YMA from July 8, 2021, to August 24, 2021, coinciding with the surge in COVID-19 Delta variant cases. All respondents were among the 400 respondents (± 0.03% of the YMA population) previously involved in a study by Dharmowijoyo et al. (2022) on the influence of space-time prisms on electric motorcycle adoption. The phase of gathering data for that study spanned from October 8, 2019, until January 21, 2020 (a time before the pandemic emerged) and included daily activity-travel logs, choice experiments, travel satisfaction, and socio-demographic information. Respondents were selected using the probability proportional to size method, which gives greater weight to areas with large populations, and random sampling for individual selection.

Convenience sampling was used for re-measurement during the pandemic, considering respondents' willingness to participate again and ease of access during mobility restrictions. Health protocols were strictly implemented throughout the survey process. Data collection procedures were consistent across both periods. The survey team assisted respondents in completing activity-travel diaries at 15-minute intervals for four consecutive days (Thursday to Sunday), along with addresses and location coordinates, which were verified using a Global Positioning System (GPS). During the pandemic, recording was also facilitated through Google Forms and remote communication via telephone and WhatsApp, without changing the core procedures. This approach streamlined recording while allowing for direct verification between respondents and surveyors, thus maintaining data validity.

To understand the characteristics of built-up land and transportation networks in the respondents' residential areas, secondary data from 2020-2021 was used. Land use spatial data was obtained from the YMA Land and Spatial Planning Office, while transportation network spatial data was sourced from the Humanitarian OpenStreetMap (OSM).

Differences in sampling techniques have the potential to introduce bias in comparing results between periods, particularly regarding the representativeness of data during the pandemic. To minimize this impact, the analysis focused on the respondent group with the dominant characteristics shown in Table 1. The composition of respondents was relatively balanced by gender, with the majority being actively employed, married, highly educated, without access to a private car, and coming from middle-income households with 3-4 family members. This profile indicates that most respondents are of productive age and have the potential to be actively involved in activities out-of-home.

Limited private car ownership among respondents indicates a reliance on other modes of transportation, which strengthens the relevance of this study to changes in activity-travel patterns due to the pandemic. Meanwhile, the characteristics of the respondents' residential environment were operationalized through the average density of commercial areas, residential areas, bus stops, intersections, and road length per square kilometer in the zone of origin (the sub-district where the respondents reside). In general, the environment is mostly residential rather than commercial and has facilities and infrastructure that support connectivity and mobility.

https://doi.org/10.1051/e3sconf/202565212002

Table 1. Respondent Characteristics (n= 109)

Characteristics	Percentage or Mean
Male (%)	49.54
Female (%)	50.46
Employed (%)	76.15
Students (%)	6.42
Non-Employed (%)	17.43
Married (%)	57.80
Higher Education (%)	69.72
No Access to a Private Car (%)	66.97
Part of Middle-Income Household (3-6 million rupiah/month) (%)	55.96
Part of Household with 3-4 Members (%)	64.22
Commercial Area Density in the Home Zone (km2/km2) (Mean)	0.07
Residential Area Density in the Home Zone (km2/km2) (Mean)	0.32
Bus Stop Density in the Home Zone (stops/km2) (Mean)	20.49
Road Intersection Density in the Home Zone (intersection/km^2) (Mean)	495.39
Road Section Density in the Home Zone (km/km^2) (Mean)	0.49

2.2 Activity Space

Activity space measurement has been developed through various approaches, one of which is the two-index method, which is designed to observe the dynamics of spatial expansion over time and is able to differentiate sizes even though the geometric shape is similar. The advantage of this method lies in determining the activity center based on the preference for activity locations out-of-home. The two-index method emphasizes two main aspects: how far the distributions of out-of-home activity locations are from the home location (Ih) and the diversity level of out-of-home activity locations from their center of out-of-home activity locations (Ic). The formulation is simple, with distance as a key element (Figure 1). Initially, calculations were performed using Euclidean distances between activity locations. However, to reflect realistic spatial conditions and reduce bias, a road network-based approach was used [10].

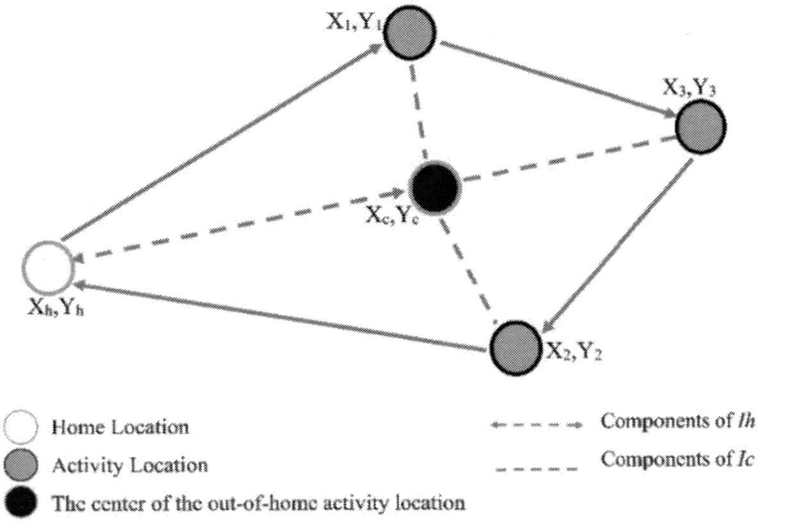

Fig. 1. Two-index method.

Formulation to estimate *Ih* and *Ic*:

$$I_h = L^2 = (X_h-X_c)^2+(Y_h-Y_c)^2, \text{ where } X_c=\frac{1}{3}\sum_{i=1}^{3} X_i, \text{ and } Y_c=\frac{1}{3}\sum_{i=1}^{3} Y_i \quad (1)$$

$$I_c=\sum_{i=1}^{3}\{(X_i-X_c)^2+(Y_i-Y_c)^2\} \quad (2)$$

Using ArcGIS software, the estimation of *Ih* and *Ic* based on the road network is executed through multiple distinct phases. The procedure starts by pinpointing the central coordinates for the out-of-home activity area, which is done by averaging the Universal Transverse Mercator (UTM) coordinates from all the locations where out-of-home activities occur. Subsequently, the distance of the most direct route is computed, both from the residence to this central point and from each individual out-of-home activity site to the center, following the sequence detailed in Figure 3. These individual shortest-route distances are then employed independently to determine the *Ih* and *Ic* values. Figure 4 provides a visual representation of how the shortest distance is derived, with steps (3) and (4) outlining the process for establishing the coordinates of the out-of-home activity center and steps (5) and (6) detailing the calculation of Ih and Ic.

 https://doi.org/10.1051/e3sconf/202565212002

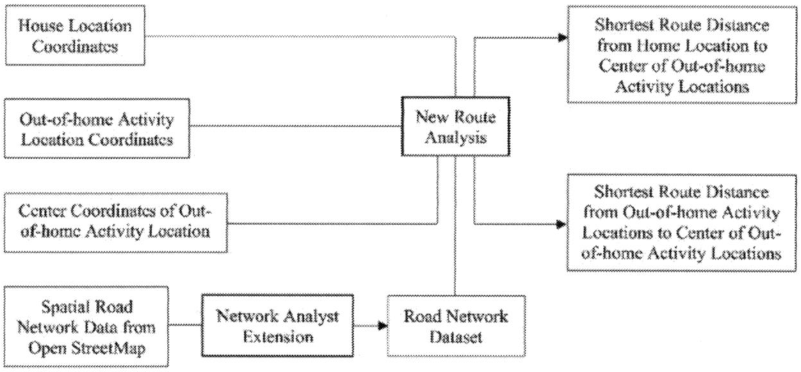

Fig. 2. Procedures in ArcGIS.

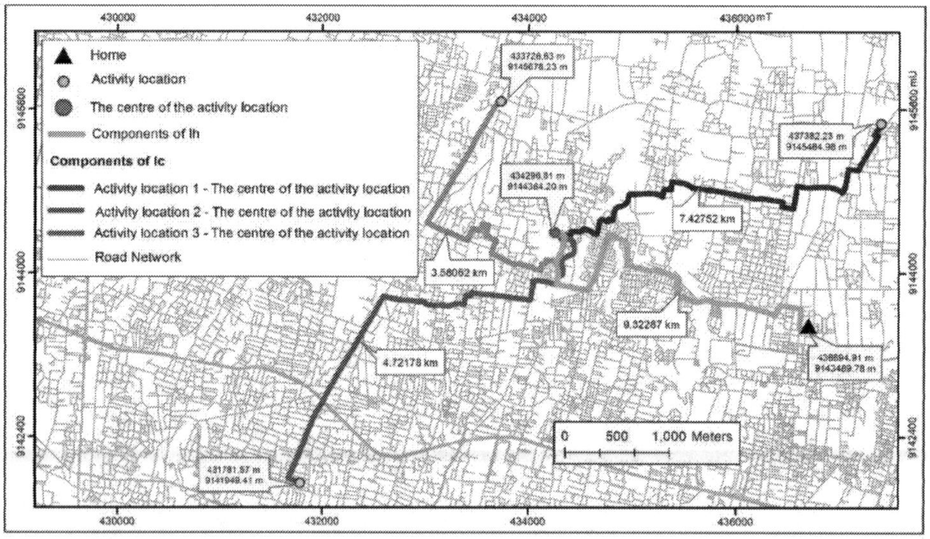

Fig. 3. Illustration of obtaining the shortest distance using ArcGIS.

Determining the coordinates of the center of the location of out-of-home activities in Figure 3:

$$X_C = \frac{437382.23+433726.63+431781.57}{3} = 434296.81 \quad (3)$$

$$Y_C = \frac{9145464.96+9141949.41+9145678.23}{3} = 9144364.20 \quad (4)$$

The estimates of *Ih* and *Ic* from Figure 3 are:

$$I_h = (9.32287)^2 \approx 86.92 \text{ km}^2 \quad (5)$$

$$I_c = (7.42752)^2 + (4.72178)^2 + (3.58062)^2 \approx 90.28 \text{ km}^2 \quad (6)$$

2.3 Data Management and Analysis

The use of time for various activities recorded in respondents' daily activity-travel logs presents challenges in data management and analysis. To simplify the diversity of the data, activities are classified based on temporal and spatial characteristics into three main categories: mandatory activities, maintenance activities, and recreational activities (Table 2). Mandatory activities have a high level of temporal and spatial certainty, encompassing both meeting basic needs within the home and activities out-of-home that involve interaction with other people or specific objects. Maintenance activities are carried out in between mandatory activities to meet daily physiological, biological, and household needs, adjusted to the individual's time. Recreational activities aim to meet cultural, psychological, and relaxation needs. Maintenance and free activities are classified as discretionary activities, which can be carried out in-home or out-of-home according to the individual's preferences and time availability [11].

Table 2. Classification of activities

Activity	Mandatory	Maintenance	Leisure
In-home	Sleep, personal activities, eating and drinking, studying, and working	Cooking, washing, cleaning the house, and taking care of the children	Entertainment activities such as watching TV, listening to the radio or similar, browsing social media, and socializing
Out-of-home	Working in-office, working out-of-office, studying at school, studying out-of-school, taking children to and from school	Shopping for daily needs and personal care at various service facilities (the hospital, bank, post office, and others)	Entertainment activities such as watching movies or similar, social and family activities, sports, organizational and volunteer activities, recreational activities, and shopping trips

Trip duration and transportation modes used during the trip were also identified from daily activity-travel records. Consistent modes of travel before and during the pandemic were private cars, private motorcycles, and non-motorized transportation (walking or cycling). Mode preference was calculated as the percentage of each mode of use relative to the total travel time.

Spatial data related to built-up land and road networks were managed using ArcGIS. Commercial and residential areas were calculated using the "calculate geometry" feature in the attribute table, road segment lengths were obtained using the geometry attribute in the "shape_length" field, the number of intersections was calculated using the "junction" feature in network analysis, and the number of bus stop locations was identified using the "count point" function in spatial join. All these elements were proportioned to the area of the respondent's residential sub-district to obtain the density values for each element.

A descriptive quantitative analysis was conducted to present demographic and socioeconomic characteristics, as well as average built-up land density, transportation network density, activity-travel patterns, and activity spaces. The Wilcoxson test was used to identify changes in activity-travel patterns and activity space, while the Spearman Rank Correlation was used to analyze the relationship of built-up land density, transportation network density, and activity-travel patterns with activity space. The analysis was conducted separately for weekdays, weekends, and all days before and during the pandemic. A guide to interpreting Spearman's rank correlation coefficient values is presented in Table 3.

Table 3. Interpretation of Spearman Rank Correlation Coefficient Values

Value (+ or -)	Interpretation
0.00 – 0.19	Very weak
0.20 – 0.39	Weak
0.40 – 0.59	Moderate
0.60 – 0.79	Strong
0.80 – 1.00	Very strong

Although data collection procedures were consistent across both periods, this study acknowledged several methodological limitations. Differences in sampling techniques could potentially introduce representativeness bias; the short survey period (four days) may limit temporal variation in activity-travel, and self-reported daily activity recording poses the risk of recall and reporting bias. However, these limitations were minimized through GPS verification, surveyor assistance, and a focus on respondents' dominant characteristics, supported by the use of secondary spatial data and road network-based analysis. Taking these measures into account, the study results still provide a relevant picture of changes in activity-travel patterns and activity spaces, as well as the dynamics of the relationship between environmental characteristics and activity-travel patterns with activity spaces on weekdays and weekends before and during the pandemic.

3 Result

3.1 Changes in Activity-Travel Patterns and Activity Spaces

The pandemic significantly changed how people spent their time, as found in the YMA (Table 4). Activities in-home increased across almost all aspects. Time spent on mandatory activities at home, including work or study, increased, particularly on weekdays, reflecting adaptation to mobility restrictions through work-from-home or remote learning. On weekends, which provide respite from work or study responsibilities, the increase in time spent on mandatory activities did not show a significant change. A more dramatic increase occurred in maintenance and leisure activities at home across all temporal conditions, particularly weekends, where time spent nearly doubled compared to before the pandemic. This indicates that individuals are not only allocating time for obligations but also taking advantage of opportunities to increase recreational activities at home as part of adapting to mobility restrictions.

Activities out-of-home experienced the opposite, with a significant decline. Time spent on mandatory, maintenance, and leisure activities out-of-home decreased simultaneously, as did travel time, which was approximately half the pre-pandemic level. This indicates that individuals are significantly affected by mobility restrictions, reducing travel out-of-home. The shift from out-of-home to in-home activities and the drastic reduction in travel indicate that the pandemic has driven changes in individual activity patterns, with the intensity of the changes being more pronounced on weekdays. A drastic decrease in leisure activities and eating out was also found in Kanto, Japan [12]. The findings of this study extend the literature by showing that changes in activity due to the pandemic not only affect activities outside the home but also affect activities inside the home, as well as travel time, with clear variations across weekdays and weekends.

Table 4. Time Use for Activity-Travel Before and During the Pandemic (Mean)

Time Use	Pandemic	Weekdays	Weekends	All Days
In-home Mandatory Time (minutes)	Before	671.63	685.73	678.68
	During	733.28*	703.28	718.28*
In-home Maintenance Time (minutes)	Before	75.96	102.87	89.42
	During	148.62*	194.24*	171.43*
In-home Leisure Time (minutes)	Before	189.84	252.39	221.11
	During	335.30*	391.44*	363.37*
Out-of-home Mandatory Time (minutes)	Before	259.47	124.75	192.11
	During	102.94*	26.35*	64.64*
Out-of-home Maintenance Time (minutes)	Before	60.96	51.81	56.39
	During	31.86*	25.94*	28.90*
Out-of-home Leisure Time (minutes)	Before	77.68	112.71	95.19
	During	35.78*	47.41*	41.59*
Travel Time (minutes)	Before	104.45	109.75	107.10
	During	52.22*	51.33*	51.78*

* p-value < 0.05 significant differences from before the pandemic

In addition to changes in time utilization, YMA also experienced significant changes in travel patterns and activity spaces. During the pandemic, individuals tended to reduce their use of motorcycles and cars, while non-motorized transportation such as walking and cycling also decreased overall, although not as sharply as motorized modes and consistently across time periods. This suggests that the pandemic influenced overall travel decisions, rather than just mode choice.

The number of trips across time periods also decreased dramatically, as did the number of destinations reached per trip (trip chains). Individual activities became more concentrated in areas closer to home, with locations for out-of-home activities spread more closely together, indicating that travel patterns became simpler and the frequency of out-of-home activities decreased. This aligns with findings in the Netherlands, where the number of trips and trip distances decreased, but the use of active transportation and private cars increased [13]. This study extends the literature by showing that the pandemic not only reduced the number of trips and altered the percentage of transportation modes used but also reduced trip chains and narrowed individuals' activity spaces.

Table 5. Travel Patterns and Activity Spaces Before and During the Pandemic

Aspects	Pandemic	Weekdays	Weekends	All Days
Percentage of Motorcycle Use	Before	67.10	50.75	58.92
	During	45.14*	34.77*	39.96*
Percentage of Car Use	Before	7.05	10.39	8.72
	During	2.11*	3.83*	2.97*
Percentage of Non-motorized Transportation Use	Before	8.16	14.48	11.32
	During	5.96	9.56	7.76*
Number of Trips	Before	4.12	3.70	3.91
	During	1.36*	1.09*	1.23*
Trip Chain	Before	1.97	1.74	1.85
	During	0.69*	0.57*	0.63*
Ih	Before	110.68	267.79	189.23
	During	33.25*	98.77*	66.01*
Ic	Before	83.08	243.04	163.06
	During	2.32*	20.65*	11.49*

* p-value < 0.05 significant differences from before the pandemic

3.2 Relationship of Built-up Land Density, Transportation Networks, and Activity-Travel Patterns with Activity Space

Analysis of the relationship of built-up area density, transportation networks, and activity-travel patterns with activity space revealed varying patterns on weekdays and weekends, both before and during the pandemic (Figures 4 and 5). In general, built-up area density and transportation networks had relatively weak effects, while activity-travel patterns showed a more significant relationship with activity space.

Before the pandemic, commercial area density across all temporal categories had a very weak negative correlation with *Ih* and *Ic*. This trend strengthened during the pandemic, although it remained relatively weak. The correlation on weekends was insignificant, so the tendency for commercial area density to limit activity space cannot yet be considered significant on weekends. Meanwhile, residential area density, which initially had a very weak but significant negative correlation with *Ih*, shifted to a positive and insignificant correlation during the pandemic. This indicates that as individuals live in increasingly dense residential areas, there is a tendency to travel farther to activity locations during the pandemic, possibly as an effort to avoid crowds, although this was not statistically significant.

The transportation network also showed a very weak correlation with activity space, with the direction of the correlation varying depending on the day. Before the pandemic, road section, road intersection, and bus stop density tended to have negative correlations with *Ih* and positive correlations with *Ic*. Of these three indicators, only weekday bus stop density significantly influenced *Ic*. During the pandemic, most correlations with *Ic* became negative, except for weekend bus stops, but all were insignificant. This indicates that individuals were less likely to align activity locations with transportation networks during the pandemic.

The relationship of activity-travel patterns with activity space is more pronounced. Activities at home, whether mandatory, maintenance, or leisure, consistently show a negative correlation with activity space, with very weak to moderate strength. Before the pandemic, activities at home tended to narrow activity space, with a significant preponderance on weekends. During the pandemic, this negative correlation strengthened, particularly for mandatory and leisure activities at home, which clearly limit *Ih* and remain a dominant factor in limiting activity space. The correlation results during the pandemic align with mobility restriction policies that enforce working and studying from home and limit access to public recreational spaces.

Mandatory, maintenance, and leisure activities out-of-home before the pandemic showed very weak to weak correlations. Mandatory activities out-of-home encourage an expansion of *Ih*, maintenance activities increase *Ic*, especially on weekends, while leisure activities encourage an expansion of both *Ih* and *Ic*. This indicates that discretionary activities play a role in expanding spatial variation, with a preference for locations farther from home on weekends than on weekdays.

Travel time before the pandemic showed a moderate correlation with *Ic* across all temporal conditions, as well as with *Ih* on weekdays. Transportation modes also exhibit varying effects: cars contribute weakly but significantly to *Ih*, particularly on weekends; motorcycles play a weak but not yet dominant role; while non-motorized modes consistently limit *Ih*, although they slightly increase *Ic* across various temporal conditions. The most dominant factor expanding *Ic* before the pandemic was the number of trips, with a very strong correlation. Trip chains also show a similar pattern in aggregate, confirming that trip intensity and interconnectedness were key determinants of activity space before the pandemic.

During the pandemic, time allocated to out-of-home activities, trip duration, transportation mode use, number of trips, and trip chains showed positive correlations ranging from very weak to very strong and were significant almost universally. The exceptions were non-motorized travel to *Ih* on weekdays and weekends, and cars and non-motorized travel to *Ic* on weekends. The most dominant factors expanding *Ih* were the number of trips, trip chains, travel time, and motorcycle use. Meanwhile, time allocated to mandatory activities, maintenance, and leisure out-of-home consistently played a supporting role to the dominant factor, both before and during the pandemic.

Thus, time allocated to out-of-home activities, trip duration, transportation mode use, number, and trip chains were predominantly positively and significantly correlated with *Ih* and *Ic*, particularly on weekends. This correlation strengthened and was almost always significant across various temporal situations during the pandemic. The strength of the relationships formed also varied by day but tended to be significantly stronger during the pandemic. This indicates that the need or desire for out-of-home activities, time availability, and access to transportation modes drive the formation of activity spaces under various temporal conditions. Despite mobility restrictions, these do not alter decision-making patterns but rather encourage adaptation through simpler and more limited preference adjustments.

The density of commercial and residential areas, the availability of road networks, time allocation for mandatory activities, maintenance, freedom at home and away from home, travel time, number of trips, trip chains, and motorized and non-motorized transportation modes have dynamic daily relationships with activity spaces [14]. The findings of this study not only corroborate these studies but also emphasize the variation in relationships that emerged, both in terms of day-to-day grouping and due to the pandemic situation. The inconsistency in the direction and strength of the relationships on weekdays and weekends before and during the pandemic indicates variations in spatial patterns influenced by daily needs, activity types, time allocation, and applicable policies.

https://doi.org/10.1051/e3sconf/202565212002

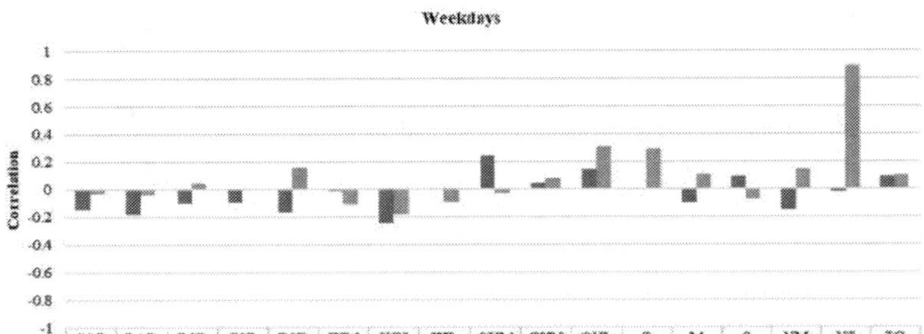

	CAD	RAD	RSD	RID	BSD	IHM	IHN	IHL	OHM	OHN	OHL	T	M	C	NM	NT	TC
Ih	-0.149*	-0.177*	-0.103	-0.096	-0.166*	-0.014	-0.248*	-0.008	0.242*	0.038	0.139*	-0.002	-0.103	0.094	-0.153*	-0.029	0.09
Ic	-0.035	-0.037	0.039	0.006	0.151*	-0.109	-0.177*	-0.096	-0.036	0.071	0.308*	0.285*	0.107	-0.074	0.139*	0.888*	0.102

* = significant (p-value < 0.05)

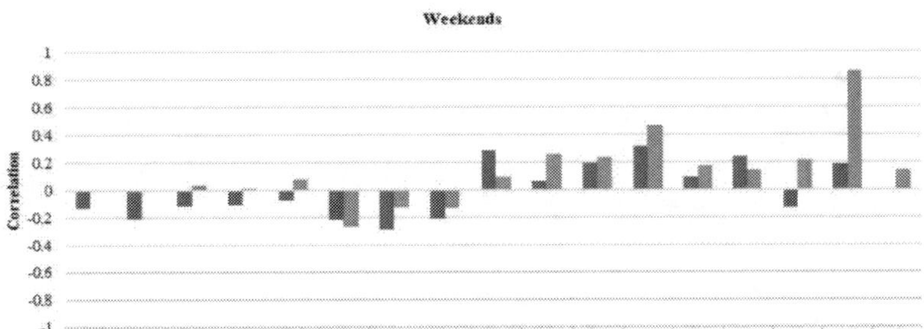

	CAD	RAD	RSD	RID	BSD	IHM	IHN	IHL	OHM	OHN	OHL	T	M	C	NM	NT	TC
Ih	-0.128	-0.205*	-0.112	-0.105	-0.072	-0.217*	-0.291*	-0.206*	0.286*	0.067	0.191*	0.313*	0.094	0.236*	-0.13	0.183*	-0.011
Ic	-0.004	0.000	0.031	0.011	0.082	-0.268*	-0.122	-0.128	0.095	0.258*	0.229*	0.466*	0.174*	0.139*	0.212*	0.861*	0.136*

* = significant (p-value < 0.05)

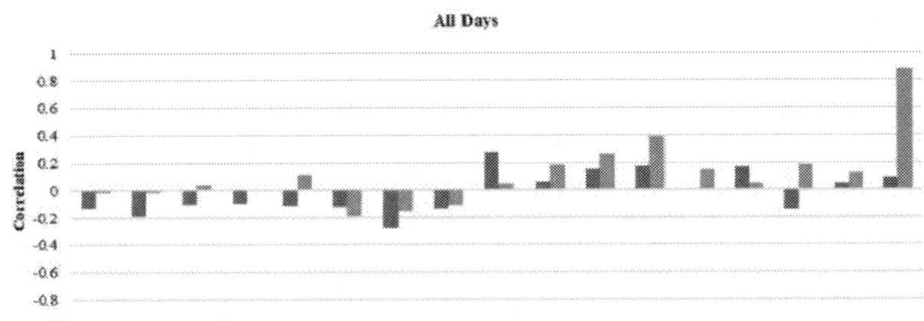

	CAD	RAD	RSD	RID	BSD	IHM	IHN	IHL	OHM	OHN	OHL	T	M	C	NM	NT	TC
Ih	-0.133*	-0.188*	-0.105*	-0.099*	-0.118*	-0.122*	-0.28*	-0.135*	0.27*	0.057	0.153*	0.168*	0.004	0.166*	-0.145*	0.046	0.084
Ic	-0.02	-0.017	0.034	0.007	0.113*	-0.187*	-0.154*	-0.11*	0.045	0.174*	0.259*	0.383*	0.144*	0.04	0.176*	0.123*	0.872*

* = significant (p-value < 0.05)

CAD: Commercial Area Density
RAD: Residential Area Density
RSD: Road Section Density
RID: Road Intersection Density
BSD: Bus Stop Density
IHM: In-Home Mandatory
IHN: In-Home Maintenance
IHL: In-Home Leisure
OHM: Out-of-Home Mandatory
OHN: Out-of-Home Maintenance
OHL: Out-of-Home Leisure
T: Travel time
M: Motorcycle
C: Car
NM: Non-motorized
NT: Number of Trips
TC: Trip chain

Fig. 4. Correlation Results for the Before Pandemic Period.

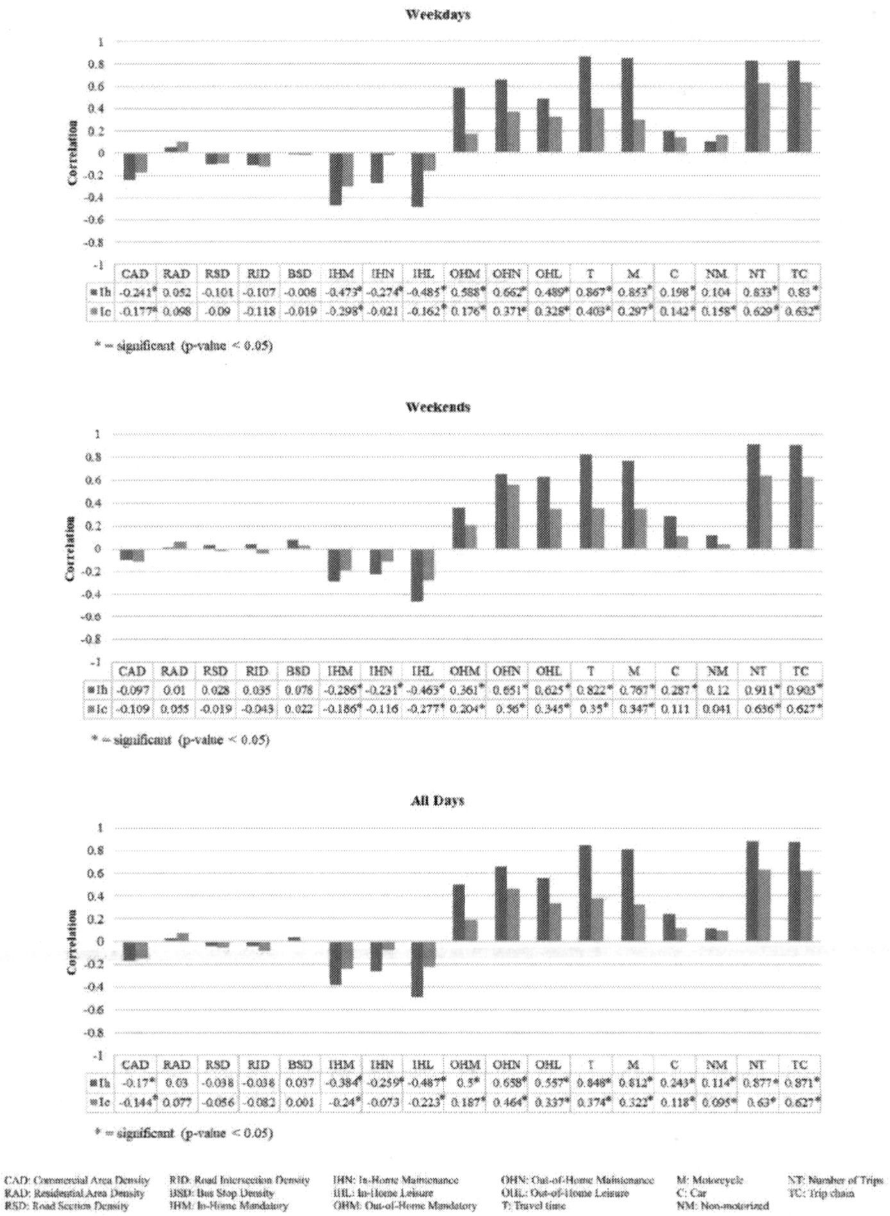

Fig. 5. Correlation Results for the During Pandemic Period.

4 Discussion

The COVID-19 pandemic forced significant changes in time allocation for various activities, travel patterns, and activity spaces on weekdays and weekends. Activity-travel pattern factors had a stronger influence than physical factors.

Before the pandemic, on weekdays, the number of trips and trip chains were the dominant factors expanding activity spaces, particularly *Ic*. A high correlation value ($r > 0.8$) indicates that the greater the number of trips and the more complex the trip chains, the wider *Ic*. Meanwhile, on weekends, this pattern remained, albeit with slightly lower intensity, as it tended to concentrate on recreational or social locations. Activities out-of-home, particularly mandatory activities out-of-home, strengthened the reach of pre-pandemic activities, while home-based activities actually limited activity spaces.

During the pandemic, this pattern shifted. The factors that became more prominent were trip duration and motorcycle use, both of which expanded *Ih* with a strong correlation ($r \sim 0.8$). This is understandable, given that motorcycles are more flexible and adaptable to mobility restrictions. Meanwhile, the number of trips and trip chains remains consistently important, despite reduced frequency due to activity restrictions. Activities for maintenance and leisure out-of-home continue to support the expansion of activity space, with a strengthening correlation (r rising to 0.662). Meanwhile, home-based activities further limit activity space due to the increased proportion of time spent at home during the restriction period.

In terms of transportation modes, motorcycles play the most significant role in shaping activity space. Cars show a lower and inconsistent correlation, while non-motorized modes tend to be weak or even negative. This indicates that in the YMA context, motorcycles remain the backbone of community mobility. Meanwhile, physical factors, such as built-up land density and transportation networks, show weak and inconsistent correlations. This confirms that the availability of physical infrastructure does not necessarily result in greater utilization of activity space if it is not balanced by community behavioral responses. In other words, physical development without the support of active use risks stagnation in regional functions.

So, what are the implications for urban and transportation planning? These findings emphasize that urban and transportation planning cannot focus solely on physical aspects. The effectiveness of spatial planning and mobility systems is highly dependent on people's spatial behavior. Therefore, future strategies need to integrate behavior-based approaches, for example, by adjusting public transportation schedules to the rhythm of community activities, strengthening low-emission modes, providing digital infrastructure to support long-distance activities, and developing mixed-use areas to encourage healthier, more efficient, and environmentally friendly mobility behavior.

The pandemic has provided important lessons for preparedness for similar situations in the future. Urban areas can be designed with flexible mobility systems, resilient digital infrastructure, and transportation policies that adapt to changing community mobility needs. Through real-time monitoring of spatial behavior, planners can anticipate shifts in land use and travel patterns and adjust interventions more quickly.

5 Conclusion

The activity space, represented by how far the distributions of out-of-home activity locations are from the home location (*Ih*) and the diversity level of out-of-home activity locations from their center of out-of-home activity locations (*Ic*) in this study, has a stronger relationship with activity-travel patterns than the physical characteristics of the environment. The number of trips had a very high correlation on weekdays before the pandemic ($r = 0.888$ for *Ic*), and this relationship remained dominant and even increased during the pandemic ($r = 0.877$ for *Ih*; $r = 0.630$ for *Ic*). The trip chain was also strongly correlated ($r = 0.872$ for *Ic* before the pandemic; $r = 0.871$ for *Ih* and $r = 0.627$ for *Ic* during the pandemic). Other factors supporting the expansion of activity space include time allocation for mandatory activities, maintenance, and leisure out-of-home ($r = 0.308$ for *Ic* before the pandemic; $r = 0.361$ to $r = 0.662$ for *Ih* during the pandemic), travel time ($r = 0.822$ to $r = 0.867$ for *Ih*), and motorcycle use (r =

0.767 to r = 0.853 for *Ih*). Conversely, time allocation for home-based activities consistently limits activity space, while cars and non-motorized transportation modes show weak effects. Physical factors, such as built-up land density, are not always significant; transportation networks more dominantly prove to be insignificant, with low correlations (r = -0.241 to r = 0.052 for *Ih*; r = -0.177 to r = 0.098 for *Ic*).

These findings confirm that the success of spatial planning and mobility systems is more determined by people's behavior in allocating time, choosing transportation modes, and determining activity locations than by physical conditions alone. Thus, future urban and transportation planning strategies must integrate a deep understanding of people's spatial behavior. To support these findings, further research is needed to explore the simultaneous interrelationships between socioeconomic characteristics, built-up land density, transportation networks, and activity-travel patterns in shaping activity spaces. Furthermore, studies on travel motivations and the reasons for choosing activity locations are also crucial to strengthening a comprehensive understanding of people's spatial behavior.

6 Acknowledgement

The writers extend their appreciation to the Universitas Gadjah Mada Research Directorate for offering financial assistance through the 2021 Final Project Recognition Grant, which made it feasible to acquire data during the COVID-19 pandemic and to develop this paper. Appreciation is also extended to Dimas Bayu Endrayana Dharmowijoyo, S.T., M.Eng., Ph.D., a lecturer and researcher in the field of activity-travel behavior, for providing access to data before the COVID-19 pandemic.

References

1. S. A. Bawana, R. Rachmawati, Evaluasi Lokasi Eksisting Halte Bus Rapid Transit Trans Jogja. J. Manaj. Transp. Logistik. **07**, 1–12 (2020). http://dx.doi.org/10.25292/j.mtl.v7i1.351
2. Y. Herwangi, P. Pradono, I. Syabri, I. Kustiwan, Making the Connection between Transport Disadvantage and Motorcycle Usage of Low-Income People in Yogyakarta Urbanized Area, Indonesia. J. East. Asia Soc. Transp. Stud. **11**, 90–109 (2015). https://doi.org/10.11175/easts.11.90
3. BPS Provinsi D.I. Yogyakarta, Perilaku Masyarakat Pada Masa PPKM Darurat D.I. Yogyakarta (BPS Provinsi D.I. Yogyakarta, Provinsi D.I. Yogyakarta, 2021)
4. Dinas Perhubungan Provinsi D.I Yogyakarta, Transportasi Dalam Angka 2022 (Dinas Perhubungan Propinsi D.I Yogyakarta, Provinsi D. I. Yogyakarta, 2022)
5. Badan Pusat Statistik Provinsi Daerah Istimewa Yogyakarta, Profil industri mikro dan kecil Daerah Istimewa Yogyakarta 2020 (Badan Pusat Statistik, Provinsi Daerah Istimewa Yogyakarta, DI Yogyakarta, 2022)
6. H. J. Miller, A Measurement Theory for Time Geography. Geogr. Anal. **37**, 17–45 (2005). https://doi.org/10.1111/j.1538-4632.2005.00575.x
7. H. J. Miller, Time Geography and space-time prism. edited by D. Richardson, N. Castree, M. F. Goodchild, A. L. Kobayashi, W. Liu, and R. A. Marston (John Wiley & Sons, New Jersey, 2017)
8. M. Dijst, Action Space as Planning Concept in Spatial Planning. Netherlands J. Hous. Built Environ. **14**, 163–182 (1999). https://doi.org/10.1007/BF02496820
9. D. B. E. Dharmowijoyo, M. Z. Irawan, N. C. Kresnanto, S. F. Shalihati, A. Illahi, Policy Incentive, Spatiotemporal, and Intrinsic Motivation Effects on Electric Motorcycle Adoption. Wait. Approv. from Transp. Res. Part A. 1–25 (2022).

http://dx.doi.org/10.2139/ssrn.4035095

10. S. F. Shalihati, A. Kurniawan, S. R. Giyarsih, D. Marwasta, D. B. E. Dharmowijoyo, Daily Activity Space for Various Generations in the Yogyakarta Metropolitan Area. Sustain. **14**, 1–24 (2022). https://doi.org/10.3390/su142013011
11. D. B. E. Dharmowijoyo, Y. O. Susilo, A. Karlström, L. S. Adireja, Collecting a Multi-Dimensional Three-Weeks Household Time-Use and Activity Diary in the Bandung Metropolitan Area, Indonesia. Elsevier Transp. Res. Part A Policy Pract. **80**, 231–246 (2015). https://doi.org/10.1016/j.tra.2015.08.001
12. G. Parady, A. Taniguchi, K. Takami, Travel Behavior Changes during the COVID-19 Pandemic in Japan: Analyzing the Effects of Risk Perception and Social Influence on Going-out Self-Restriction. Transp. Res. Interdiscip. Perspect. **7**, 100181 (2020). https://doi.org/10.1016/j.trip.2020.100181
13. M. de Haas, R. Faber, M. Hamersma, How COVID-19 and the Dutch 'Intelligent Lockdown' Change Activities, Work and Travel Behaviour: Evidence from Longitudinal Data in the Netherlands. Transp. Res. Interdiscip. Perspect. **6**, 100150 (2020). https://doi.org/10.1016/j.trip.2020.100150
14. D. B. E. Dharmowijoyo, Y. O. Susilo, A. Karlström, Relationships among Discretionary Activity Duration, Its Travel Time Spent and Activity Space Indices in the Jakarta Metropolitan Area, Indonesia. J. Transp. Geogr. **54**, 148–160 (2016). https://doi.org/10.1016/j.jtrangeo.2016.05.012

AUTHOR INDEX

EDP Sciences – Web of Conferences
17, Avenue du Hoggar
Parc d'Activité de Courtabouf
BP 112
F-91944 Les Ulis Cedex A
France

CURRAN ASSOCIATES INC. proceedings.com

ISBN 979-8-3313-2722-4
9 798331 327224

1st Conference on Traditional and Renewable Energy Sources: Perspectives and Paradigms for the 21st Century (TRESP 2021)

E3S Web of Conferences Volume 250 (2021)

Prague, Czech Republic
22 – 23 January 2021

edp sciences